Periodic Table of the Elements

	IIIA	IVA	VA	VIA	VIIA		2 4.003 **He** 0.126 Helium
	5 10.81 **B** 2.34 Boron	6 12.01 **C** 2.26 Carbon	7 14.01 **N** 0.81 Nitrogen	8 16.00 **O** 1.14 Oxygen	9 19.00 **F** 1.505 Fluorine		10 20.18 **Ne** 1.20 Neon

IB	IIB	13 26.98 **Al** 2.70 Aluminum	14 28.09 **Si** 2.33 Silicon	15 30.97 **P** 1.82 Phosphorus	16 32.06 **S** 2.07 Sulfur	17 35.45 **Cl** 1.56 Chlorine	18 39.95 **Ar** 1.40 Argon
28 58.71 **Ni** 8.9 Nickel	29 63.54 **Cu** 8.96 Copper	30 65.37 **Zn** 7.14 Zinc	31 69.72 **Ga** 5.91 Gallium	32 72.59 **Ge** 5.32 Germanium	33 74.92 **As** 5.72 Arsenic	34 78.96 **Se** 4.79 Selenium	35 79.91 **Br** 3.12 Bromine	36 83.80 **Kr** 2.6 Krypton
46 106.4 **Pd** 12.0 Palladium	47 107.9 **Ag** 10.5 Silver	48 112.4 **Cd** 8.65 Cadmium	49 114.8 **In** 7.31 Indium	50 118.7 **Sn** 7.30 Tin	51 121.8 **Sb** 6.62 Antimony	52 127.6 **Te** 6.24 Tellurium	53 126.9 **I** 4.94 Iodine	54 131.3 **Xe** 3.06 Xenon
78 195.1 **Pt** 21.4 Platinum	79 197.0 **Au** 19.3 Gold	80 200.6 **Hg** 13.6 Mercury	81 204.4 **Tl** 11.85 Thallium	82 207.2 **Pb** 11.4 Lead	83 209.0 **Bi** 9.8 Bismuth	84 (210) **Po** (9.2) Polonium	85 (210) **At** — Astatine	86 (222) **Rn** Radon
110	111	112	113	114	115	116	117	118

119
Supermetal

64 157.3 **Gd** 7.89 Gadolinium	65 158.9 **Tb** 8.27 Terbium	66 162.5 **Dy** 8.54 Dysprosium	67 164.9 **Ho** 8.80 Holmium	68 167.3 **Er** 9.05 Erbium	69 168.9 **Tm** 9.33 Thulium	70 173.0 **Yb** 6.98 Ytterbium	71 175.0 **Lu** 9.84 Lutetium
96 (247) **Cm** Curium	97 (247) **Bk** Berkelium	98 (249) **Cf** Californium	99 (254) **Es** Einsteinium	100 (253) **Fm** Fermium	101 (256) **Md** Mendelevium	102 (254) **No** Nobelium	103 (257) **Lr** Lawrencium

CHEMISTRY IN MODERN PERSPECTIVE

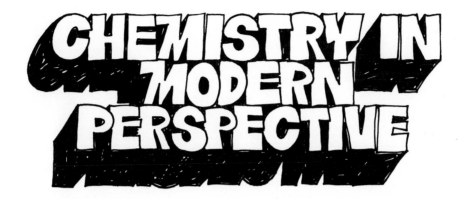

CHEMISTRY IN MODERN PERSPECTIVE

GLEN GORDON / WILLIAM ZOLLER

UNIVERSITY OF MARYLAND

▲ ADDISON-WESLEY PUBLISHING COMPANY

READING, MASSACHUSETTS • MENLO PARK, CALIFORNIA • LONDON • AMSTERDAM • DON MILLS, ONTARIO • SYDNEY

This book is in
THE ADDISON–WESLEY SERIES IN CHEMISTRY

Consulting Editor
Francis T. Bonner

Cover photo: EPA–DOCUMERICA: Alexander

ISBN 0-201-02561-2
CDEFGHIJ-HA-798

To Connie and Vivian
for their patience and support

PREFACE

In 1969 we were asked to develop a chemistry course for non-science majors. We found that most syllabuses and textbooks available at that time were simply for watered-down versions of General Chemistry instead of courses designed specifically for the intended audience. We felt that a new approach was needed—one which would be quite interesting to the students and, at the same time, prepare them as well as possible to evaluate the various scientific and technological issues with which all of us, as citizens, are confronted in our increasingly complex society. Since environmental quality was becoming a topic of considerable interest to students, we decided that a course that emphasized the chemical aspects of energy production and environmental quality would be highly appropriate. As there was no adequate text material for that type of course, we developed our own lecture notes, which have gradually evolved into the manuscript for this textbook, after being tested by use for seven offerings of a one-semester course at the University of Maryland.

Since we realize the importance of providing the appropriate scientific background for meaningful examination of environmental and technological issues, we have employed the technique of first introducing a scientific principle and then illustrating it with examples of application to "real world" problems. Using such an approach, we find that we maintain high student interest throughout the term, reinforce the student's understanding through the use of dramatic examples, and encourage student responsiveness.

One can extend the concept-and-application technique to those courses that have an associated laboratory, as ours does. Experiments which are designed specifically for the audience and which are complementary to the text material

are far more appealing than the standard general chemistry experiments. We have designed a set of 22 experiments for use in our course (see G. E. Gordon, W. H. Zoller, and J. C. Ingangi, *J. Chem. Educ.*, 1974). These experiments include several that deal with household products (antacids, bleach, food additives in meat, vinegar, carbonated beverages) and others on measurements of environmental quality parameters (O_2 content of water, chemical oxygen demand, chloride in water, phosphate in water and detergents, lead in air, paint, soil and ceramics). The laboratory experiments are published in a separate manual.

Since we initiated work on this text, a number of texts have appeared on the themes of "chemistry and man" or "chemistry and environment." However we feel that our approach is unique in that most of the others emphasize the negative aspects of chemistry as it affects society, i.e., air and water pollution and the disposal of wastes. By contrast, we view the field as one dealing with the question, "How can chemistry be used to enhance the quality of life?" In this broader sense, we deal not only with the clean-up of problems resulting from our technological societies, but also with the more "positive" question of the provision of resources, the food, energy and other products needed to develop or maintain a high quality of life.

Our text is particularly timely today in that we devote a considerable amount of attention to the important question of energy sources—both the present-day sources that are in short supply and the new types of sources being considered for development. In Chapter 4 we deal extensively with energy needs and sources, the problems of thermal pollution and hydroelectric plants, and the prospects for two future energy sources: solar and geothermal. The whole question of petroleum, which is of crucial importance today, is discussed in considerable detail in Chapter 14. The reserves and usage rates of fossil fuels are discussed along with those of other mineral resources in a "limits to growth" treatment in Chapter 12. Environmental degradation resulting from production and use of fossil fuels is covered in Sections of Chapter 9 (global effects), 10 (SO_2 and particulates) and 14 (auto emissions, oil spills). Nuclear energy and its hazards are treated in Chapters 5 and 6. Obviously, in a field that is changing so rapidly, we cannot hope to be completely up-to-date, but the fundamentals contained herein should make it possible for students to understand today's newspapers.

The scientific framework of the book begins with introductory chapters on the interactions of science, technology and society, some general physical and chemical properties, and concepts involving energy and mass. The chemical discussions follow a progression from the smallest submicroscopic units, nuclei, to atoms, molecules, properties of some chemical groups, and ending with the highly complex molecules of organic and biochemistry.

We have not limited ourselves to just those concepts needed to understand the applications of science. Rather, we believe that some feeling for the excitement and intellectual curiosity of science can be conveyed best by discussion of some concepts beyond the realm of everyday experience; e.g., quantum theory and probability, relativity, "black holes." In addition, we have included considerable material from scientific disciplines other than chemistry, such as geology, astronomy, physics, meteorology, etc. Since the problems discussed cut across the boundaries of many dis-

ciplines, such material is needed for an understanding of them. Furthermore, since this may be the only college-level science course that most nonscience majors take, it is entirely appropriate that they receive some exposure to sciences related to chemistry.

Some readers may question our early placement and extensive discussion of nuclear chemistry. We feel, however, that one of the most central issues facing our society over the next few years will be that of nuclear versus nonnuclear sources of energy. Furthermore, the constant threat of nuclear weapons is a topic that cannot be ignored. It is thus essential that educated persons know a good bit about these topics, which are so often misunderstood. Most nonscientists have some conception of the workings of a fossil-fuel plant, but most have little understanding of the operation of a nuclear plant. Therefore, we feel it is important to teach students the origin of nuclear energy, the production of fission products in reactors and their radioactive decay, the behavior of important fission products in the environment and in the body, and the effects of radiation upon biological species.

Chapter 6, on fusion and nuclear reactions in stars, is also rather unique for a text at this level. Too often, we chemists fail to stop and wonder why our Earth is made up in the particular composition in which we find it. We heartily recommend coverage of this chapter, although it can be avoided without detraction from the remainder of the book (see "Critical Path" below).

In teaching the course, we keep the mathematical requirements to a bare minimum. However, we have included enough material on stoichiometry, gas laws, equilibria, etc., that instructors desiring a more quantitative approach will have the necessary material (including quantitative problems and an appendix on mathematical operations).

We cover most of the material in the book in one semester at the rate of one chapter per week. However, there is so much material included that one could easily use the text for two semesters if the various topics are covered in more depth than we do in our lectures. In a two-semester course, one would have time to have students do extra reading on certain topics from a supplementary readings book.

We have given numerous references, both in footnotes and in reading lists at ends of chapters, to articles and books that give more detailed information on the topics discussed. Many of these would serve as useful outside reading for students or as a source of background information for instructors. The main purpose of the footnotes is that of documenting the many facts and figures presented. We feel that a major failing of many environmental discussions is a lack of quantitative detail. One of the special talents that chemists can bring to the study of the environment is a quantitative evaluation of the magnitudes of various effects. We have tried to present as much data for this approach as possible.

CRITICAL PATH APPROACH TO USING THE TEXT

For the convenience of instructors who prefer to leave out some of the topics in a one-semester course, we list below a "critical path" which contains those portions of the book that we feel are essential for a logical presentation of the basic scientific framework of the text:

Chapters 2 and 3

Sections 4–1 through 4–4

Sections 5–2 and 5–3

Chapter 7

Sections 8–1 through 8–4 (Section 8–5 in highly quantitative courses)

Sections 10–1 through 10–7

Sections 11–1, 11–2, and 11–5

Sections 12–1 through 12–5

Chapter 13, except Section 13–12

Sections 14–1 through 14–3

Chapter 15

One can add to this "bare bones" basic structure as many applications as time and interest dictate. It should be noted that one can vary the order of topics in the Critical Path. For example, many instructors may prefer to jump from Sections 4–1 through 4–4 to Chapters 7 and 8 on atomic and molecular structure, leaving the extensive nuclear discussions of Chapter 5 for a later time.

ACKNOWLEDGMENTS

We wish to thank the many people who have aided us in the preparation of this text: Francis Bonner, our Consulting Editor, whose reviews were always incisive and humorous; G. Ronald Husk, Robert West, Carl Moeller, Paul Haake, Howard Knachel, Shelton Bank, R. D. Gaines, John Gilbert, Stephen Monti, James Mitchell, T. W. Perry, Richard Goldsby, David Martin, Chester Holmlund, Carl Rollinson, and the late Theos J. Thompson, who reviewed all or sections of the manuscript.

Despite the considerable help received from others, however, the final decisions on the inclusion and treatment of topics were ours and we take full responsibility for them.

We also want to thank the Addison-Wesley editorial and production staffs, whose efforts made our manuscript into a textbook; the many University of Maryland students who carried around the heavy mimeographed versions of the text and whose comments and reactions helped us so much in editing the text; John C. Ingangi and the many teaching assistants who helped us develop the course and our Department Chairman, Professor Joseph T. Vanderslice, who gave us the opportunity and a completely free hand to develop it.

We would particularly acknowledge our gratitude to the late Professor Charles D. Coryell of MIT, our friend and mentor, who was a tremendous inspiration in teaching, in the ways and uses of science, and in life in general.

We thank the typists who did such an excellent job of typing the several versions of the manuscript: Elizabeth Shell, Ellen Klavon, Mary Wason and Karen Junghans. We are grateful to Mr. Daniel Hogg for his constructive criticism of our use of language.

Finally we thank our families, Connie, Karl and Kirstie, and Vivian, Eileen, Danny, Charlie, and Kay for their patience and support during the many hours spent in writing the book.

College Park, Maryland
November, 1974

G. E. G.
W. H. Z.

CONTENTS

1 SCIENCE, TECHNOLOGY, AND OUR ENVIRONMENT

Earth as photographed by Apollo 10 astronauts from a distance of 250,000 miles. (NASA photo.)

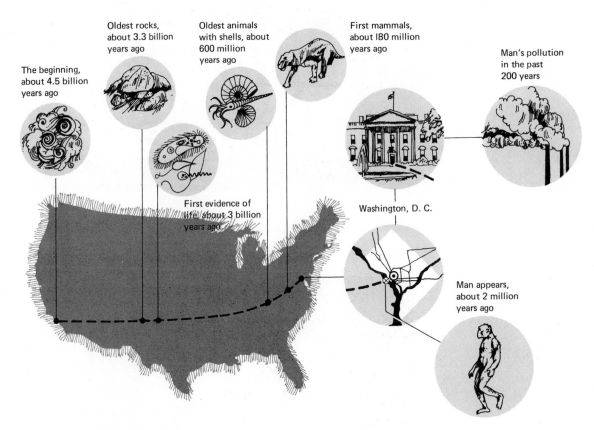

Fig. 1–1 A trip across the United States is analogous to the time scale of events on Earth. Man's pollution in the past 200 years corresponds to only the last 7 inches of the cross-country trip.

About 4.5 billion years ago, clouds of intergalactic dust, the debris of stars that existed prior to our sun, came together to form our solar system. Over the enormous time span since then, the dust has condensed in certain regions to form the sun, the planets, and their satellites. The earth has fractionated into a core, mantle, and crust, the hydrosphere (water bodies) and atmosphere have developed, and single-celled organisms have appeared and evolved into various animals, ultimately man. At first, man lived much as the animals do, existing on food gathered from the naturally growing plants and animals he was clever enough to hunt and kill. Gradually he became agricultural by learning that he could clear land of its natural growth and plant seeds of his own choice. He domesticated animals to do some of his manual labor and provide steady supplies of milk and meat. Slowly, with his campfire as a primitive kiln and blast furnace, he learned to make objects of pottery, glass, and a few metals. By the time of Christ, after one or two million years of existence, man was able to work about seven metals. Finally, within the last two hundred years, man has become "technological," greatly increasing both his understanding of his surroundings and his ability to alter it, always for what appeared to be desirable ends. Until recently, we have largely neglected the side effects of our technology, which have a great capacity for inadvertent modification of our surroundings, generally in undesirable ways. Now, with our great ability to change the thin, fragile layer of water, land, and air in which we live on Earth, we are faced with a grave question: how much longer can man and other forms of life survive on this planet?

The time between the formation of the solar system and the appearance of "intelligent" life on Earth was unimaginably long. By contrast, it has taken technological man a mere moment in history to "foul his nest." We can get some notion of the great life span of the solar system by comparing it to a trip from Los Angeles, California, to the White House in Washington, D.C., a distance of 2644 miles (see Fig. 1–1). The oldest rocks found on the present surface of the earth were formed about three billion years ago, corresponding to Tucumcari, New Mexico, on our cross-country trip. The earliest evidence of primitive life forms is in those oldest rocks, and it is dated by the age of the rocks. Perhaps life on Earth goes back even further than three billion years, but no older rocks are known. Rocks and life forms may well have existed earlier on the earth's surface, but they may have been buried and modified by the vast natural processes of erosion and sedimentation.

Fossils from animals with shells (and, later, bones) do not appear until about Bristol, on the Tennessee-Virginia border (about 600 million years ago, at the start of the Cambrian period). The earliest mammals appear at about Harrisonburg, Virginia, and overlap the age of dinosaurs, which extends to about 40 miles outside Washington. The earliest evidence for man does not appear until we reach the Potomac River, a mile or two from the White House. It took man about 1.5 million years of development before he became "technological." As soon as man started killing animals in large numbers and clearing land of its natural growth, he was altering his environment. But the magnitude of his effects has grown fantastically with the relatively recent increase in human population and the development of large-scale mining, manufacturing, and agricultural processes. Nearly all his serious damage to the planet has been done during the 200 or so years since the start of the industrial revolution, a period corresponding to *the last 7 inches of our 2644-mile cross-country trip!* When we compare the relative time scales of evolution and technological change, we see how dubious is the assumption, made by some, that evolution will equip mankind to cope with the changes made by technology.

During the past 200 years, human intelligence has produced many beneficial developments: machines that perform much of the hard labor and routine calculations previously done by man; devices and techniques to make use of electricity; rapid communication around the earth by telephone,

St. Louis under smog. (Photo from the St. Louis Post Dispatch, by permission of the Environmental Protection Agency.)

radio, and television; rapid travel around the earth and, recently, to our natural satellite, the moon; the release of the energy in atomic nuclei; and elimination of many diseases.

In 1948 and 1952, serious air pollution episodes in Donora, Pennsylvania, and London, England, suggested that man's technology could have some very damaging side effects; similar effects were indicated by the growing Los Angeles smog problem and the poorer quality of our nation's waters. A few people took note of these warnings, but the quality of our environment was not a major issue. Most scientists and engineers were not much concerned.

If there is any landmark of the beginning of widespread awareness of man's deleterious effects on the surroundings, it is surely the publication of Rachel Carson's *Silent Spring*, first as a series of articles in *The New Yorker* in 1962. During the years after World War II, DDT and similar persistent pesticides were viewed as modern miracles. Small quantities of these chemicals could kill the entire population of certain insect species over huge areas and could thus control the spread of insect-borne diseases and prevent insect damage to food crops. But Rachel Carson, demonstrating the power of a single individual armed with the facts, alerted us to the dangers connected with use of pesticides. Many insect species evolve enhanced resistance to the pesticides, requiring increased dosage or new types of pesticides. Other animals, especially the predators of the insects, may be killed as effectively as the target insects. The pesticides cause the eggs of some bird species to have very fragile shells that often break before the young can be hatched, thus threatening the existence of the species. The pesticide compounds are not easily broken down in nature and thus work their way up the food chain to man. We are now accumulating potentially hazardous amounts of pesticides in the fat of our bodies. As with many environmental problems, the solution is not easy. If we suddenly stopped all uses of DDT, millions of people around the world would die from starvation or insect-borne diseases. Instead, it is probably better to tolerate some of the risks of DDT while replacing it as quickly as possible with other pest-control strategies like the ones discussed in Chapter 13.

Man has many other undesirable effects upon his environment. He has created cities that, with all their crowding, noise, and air pollution, have become almost intolerable. Many of our rivers and lakes have become choked with algae and filled with toxic industrial chemicals and harmful pathogens from human and animal wastes.

Man's scale of activities has so increased that we have now polluted our environment on a global basis. Although nuclear energy can be used beneficially to supply needed power and fresh water to underdeveloped nations, at least two countries possess enough nuclear weaponry to make the planet uninhabitable. Widespread use of chemical and biological weapons could endanger all life on the planet. The concentration of carbon dioxide in the atmosphere has increased by about 10% during this century because of extensive burning of fossil fuels (coal, oil, and gas). We are not sure what the effects of this modification will be, but (as discussed in Chapter 9) global climatic changes could result. Oil slicks, pieces of plastic, and other mementos of civilization are now found in the middle of the ocean.

Vaccines, sulfa drugs, antibiotics, and other medicines have been developed to cure and prevent many previously fatal diseases, but many products, both medicines and food additives, have been released for public consumption without adequate testing. In an alarming example, several thousand babies were born badly deformed because their mothers took the sedative Thalidomide during pregnancy. To their credit, U.S. officials had delayed approval of Thalidomide in this country because they suspected that it might have harmful effects.

Trash and debris on the banks of the Brandywine River in Indiana. (U.S. Dept. of Agriculture–Soil Conservation Service photo by Erwin W. Cole.)

In ancient times, change came very slowly. For example, it may have taken a community several hundred years to learn that one could make a more effective stone blade by chipping on both sides of the stone rather than on just one side. Communications were so primitive that it took many years for the knowledge of such developments to pass from one community to the next. By today's standards, the ancient rate of "progress" seems painfully slow, but the slow adoption of new developments protected mankind in general from unsuccessful experiments in isolated communities.

By contrast, today a newly developed product can be tested for a few months or years and then distributed to tens of millions of consumers in only a few weeks. After being widely used for another few months or years, however, it may be found to contain some inherent but unsuspected danger. To our knowledge, we have so far not poisoned a major segment of our population, but we may have shortened some lives. It is usually quite difficult to determine the effects of many years of exposure to low levels of food additives and chemicals in air and water pollution. If we continue on our present course, a mass disaster seems statistically unavoidable. Only time will tell what disasters we have already initiated. In 1969, cyclamates (artificial sweeteners) were withdrawn from public consumption because large doses of cyclamates were found to cause cancer in rat bladders—this after an estimated three-fourths of the U.S. population had been consuming cyclamates throughout the 1960s!

Many environmental problems present us with difficult choices between alternatives, all with risks or known drawbacks. Consider, for example, the problems of population explosion and birth-control techniques. Millions of women around the world are taking birth-control pills, although many women suffer uncomfortable side effects and the long-range effects of "the Pill" are not fully understood. But what is the alternative? The explosion of world population, as estimated in Fig. 1–2, is a major factor looming in the background of all environmental problems. If developing nations are to raise their standard of living or even ensure the survival of their people, effective and acceptable methods *must* be found by which to reduce or halt population growth. Those of us in the developed nations, where doubling times of population are typically 60 years or more,* don't feel the direct effects of population growth so acutely. But in some of the underdeveloped countries, e.g., in Central and South America, population doubling times are as low as twenty years. Think of that—it means that those countries must double their food supplies, other resources, and facilities every twenty years just to maintain their present low standards of living! How can they find additional resources and capital to improve the lot of their people? When viewed in this way, the possible dangers of effective birth-control agents seem small in relation to the mass starvation and poverty of a majority of the world's people that will almost certainly result from uncontrolled population increases. Strong birth control measures are a necessity.

Improvements in medicine and nutrition have raised man's life expectancy in the United States from 47.3 years in 1900 to 68.2 years in 1950 (see Fig. 1–3); yet despite continued medical advances, it increased by less than two years between 1950 and 1966. It now appears to have leveled off at about

---

*   P. R. Ehrlich, *The Population Bomb* (Ballantine Books, New York, 1968).

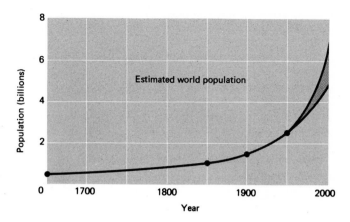

Fig. 1–2   Estimates of world population during recent history and projections into the future, based on pessimistic (upper curve) and optimistic (lower curve) assumptions.

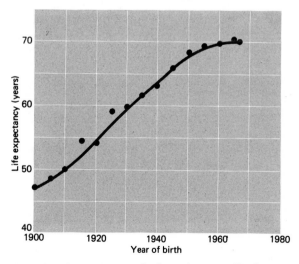

Fig. 1–3   Life expectancy at birth as a function of birth year.

70 years. Have we reached some limit determined by man's natural aging processes, or are subtle effects of our technological way of life counteracting medical advances? For example, we know that the average life expectancy is being held down to some extent by the now middle-aged generation of heavy smokers of cigarettes. If we assume that a third of the population become heavy smokers and that the average heavy smoker loses about eight years of life, we can calculate that the use of cigarettes depresses the average by about two years.

Despite our common assumption that we in the United States have the highest standard of living in the world, we rank thirteenth among the nations of the world in infant mortality (22 per 1000 live births versus 13 in Norway). We rank seventh in maternal mortality and about fourteenth in life expectancy!* In large part, our deficiencies in these areas stem from inadequate health care for poor people. However, certain aspects of our environment, such as the stresses of complex urban life, additives in our food, and pollution of our air and water, may contribute to our excessive mortality.

In view of the problems that man has created for himself by his applications of science and technology, it is not surprising that many—particularly among the young—have grown skeptical, rebellious, or depressed about modern society. They live under the threat of nuclear annihilation, overwhelmed by large organizations and the complexities of impersonal institutions and surrounded by an increasingly polluted environment. Many feel that by so-called "rational" thought we have led ourselves to "irrational" ends—our stockpiles of weapons, massive defoliation of a small country's vegetation by a superpower, our "organization man" style of life. There are strong currents of anti-establishment, anti-technology, and even anti-science feeling among many young people today.†

Although many of the criticisms of modern, technological society are surely valid, we (the authors) feel there are great dangers in blanket indictments of science and technology and moves to discontinue work in these endeavors. At this stage in the life of our planet, we can't all

simply go live in the woods and exist in harmony with nature—there are too many of us and we would starve in the attempt! Man must use his wits to alter the environment (e.g., by growing of crops) in ways that will assure his survival. Despite the rapid growth of scientific knowledge in recent decades, we still don't know how to produce the energy (with minimal pollution) and grow the food required for our burgeoning population. Viewing the situation more positively, we should be able to "have our cake *and* eat it"—by retaining the good features of our technological life and eliminating their detrimental side effects, now that we have become aware of them. A great deal more scientific knowledge will be needed before we can accomplish that objective, but moves in that direction have begun among scientists. Just as the 1950s emphasized nuclear research and the 1960s concentrated on space research, the 1970s may ultimately be known for environmental research.

In this connection, we should emphasize one point regarding the nature of science—science itself (at least in the authors' view) is amoral. It is a search for truth, and those of us engaged in it find intellectual stimulation in the quest. We would like to communicate to others its spirit of objective inquiry. The knowledge gained from scientific investigation can be used either beneficially or destructively. For example, a new fundamental result in biochemical research might be used either to develop a cure for cancer or to devise a biological weapon. Clearly all of us, scientist and citizen alike, must make greater efforts to insure that scientific results are applied in constructive rather than harmful ways. One encouraging sign that our society is paying more attention to the uses of science and technology is the debate over the merits of continued development of the supersonic transport (SST) that occurred in the U.S. Congress during 1971. We don't wish to debate the issue here (see Chapter 9), but we would like to point out that extensive citizen participation in the debate was a new phenomenon. A few years earlier, all decisions would have been left to the "experts." There has also been strong citizen involvement in such national issues as antiballistic missiles (ABM's), the banning of nuclear weapons tests, improvement of automobile safety, local issues such as freeways vs. rapid transit, and the fluoridation of water.

In writing this book, we have several objectives. First, we want to give you, the student, enough scientific background to understand the complex environmental problems with which we are faced. Few of you are likely to become

---

*   I. J. Lewis, "Government Investment in Health Care," *Scientific American* **224** (4), 17–25 (April 1971).
†   See, for example, Theodore Roszak, *The Making of a Counter Culture* (Doubleday, Garden City, N.Y., 1969).

Young people filling bags with litter in Fort Smith, Arkansas. (EPA–Documerica photo by Jim Olive.)

scientists and deal directly with environmental research. But all of us, as citizens, will be increasingly called upon to make decisions on complex scientific and technological issues—which gasoline we should buy, which household laundry detergent, whether to vote for the bond issue for a new municipal incinerator. As we shall see in the following chapters, many of these issues are so complex that there is disagreement even among experts. We will have succeeded in our purpose if you acquire a sound base for making judgments of these critical issues. It's very easy to point a finger at the detergent makers and demand that they take the phosphates out of detergents. But it's not so easy to determine the consequences of phosphate substitutes!

Our second objective in writing this book is to give you some idea of the ways of science and scientists—their often haphazard way of stumbling into important discoveries rather than always following the logical procedure described as "the scientific method." Further, we would like to share with you some of the human element of science—the personality conflicts and competitive spirit that exist among

The supersonic Concorde on a test flight from its home base in Gloucestershire, England. (Photo by permission of the British Aircraft Corporation.)

scientists (both *The Double Helix* and *Lawrence and Oppenheimer,* listed in the bibliography, are fascinating books in this respect). We want to convey some feeling for the interpretation of scientific data—measurements and their errors, pitfalls in the use of statistics, and so on.

We have arranged this book in a scientifically logical progression from the nucleus to the atom, from the atom to the molecule, and finally to complex organic and biochemical structures and processes. Whenever possible, we have used environmental problems or everyday applications of chemistry to illustrate the chemical principles. We have tried to discuss the principles themselves in ways that will make them interesting and enlightening to you. We hope you will find the principles just as exciting and relevant as the problems that involve them.

It is entirely appropriate that we discuss environmental problems in the context of chemistry. Although, in the end, we are most concerned about the effects of the problems upon biological species, chemical and biochemical reactions play a central role in nearly every problem.

With the increasing degradation of our environment, it is very easy to become depressed about the prospects for improving the quality of our surroundings. Thus we would like-to close this chapter by pointing out that the rate of change in our environment may be slowing down, and that if we act quickly and wisely, there may still be time to preserve the environment.

Many of our most pressing environmental (and social) problems are closely related to the extreme rapidity with which our knowledge has increased and been applied over the past several decades, especially since World War II. We are suffering from a condition that Toffler has termed "future shock."[*] These developments have come so rapidly that we have had great difficulty in assimilating the changes into our society and our personal lives. Consider just a few of the enormous transitions of science and technology that have occurred in the past century: the speed of communication has increased by a factor of ten million, the destructive power that can be released by a single man in war has gone up about one million-fold, and the speed at which a person can travel has gone from about 20 mph to 2000 mph if we limit ourselves to airplanes or about 25,000 mph if we consider space travel.

John Platt has made the very interesting point that, for many of the changes that have occurred, we have gone about as far as we can go—that is, we have almost reached the theoretical limit.[†] We now communicate with people all over the world and send signals into space at the speed of light—theoretically the highest velocity at which any signal can be transmitted. Once we have the power to annihilate everyone on Earth, further increases in weaponry are almost irrelevant. The speed at which ordinary people (nonastronauts) can travel will perhaps increase from about 600 mph to 1800 mph on supersonic airplanes.

Changes in our science and technology will probably be much smaller in the near future than they have been in the recent past. If we can survive until we have learned to live with what we have achieved, life on the planet may continue for many centuries. However, Platt warns that the time available for us to come to terms with our "progress" is severely limited. The greatest danger at present is the threat of all-out nuclear war, for which he feels the "half-life" is about 20 to 30 years; that is, there is a 50% probability that we will have a nuclear war within the next 20 to 30 years. That is one view. Clearly, we must prepare ourselves well to attack all the problems of disarmament, overpopulation, food shortages, environmental damage, and a large number of social ills. Informed and interested citizens are a vital key to their solution.

SUGGESTED READINGS

E. S. Barghoorn, "The Oldest Fossils," *Scientific American* **224** (5), 30–53 (May, 1971).

Lincoln Barnett and the Editors of Life, *The World We Live In* (Time, Inc., New York, 1963), Vols. I–III. A well-written and beautifully illustrated account of the solar system, the earth, and the development of life on earth.

J. D. Bresler, ed., *Environments of Man* (Addison-Wesley, Reading, Mass., 1968). Paperback.

Rachel Carson, *Silent Spring* (Houghton Mifflin, Boston, 1962). Available in paperback (Crest, New York, 1964).

Barry Commoner, *The Closing Circle* (Knopf, New York, 1971).

Barry Commoner, *Science and Survival* (Viking Press, New York, 1966).

N. P. Davis, *Lawrence and Oppenheimer* (Simon and Schuster, New York, 1968). Available in paperback (Fawcett Publications, Greenwich, Conn., 1969).

P. R. Ehrlich, *The Population Bomb* (Ballantine Books, New York, 1968). Paperback.

P. Farb and the Editors of Life, *Ecology* (Time, Inc., New York, 1963).

Wesley Marx, *The Frail Ocean* (Ballantine Books, New York, 1967). Paperback.

W. T. Pecora, "Science and the Quality of Our Environment," *Science and Public Affairs* (formerly the *Bulletin of the Atomic Scientists*) **26** (8), 20–23 (Oct. 1970).

J. D. Watson, *The Double Helix* (Atheneum, New York, 1968).

---

[*]   Alvin Toffler, *Future Shock* (Random House, New York, 1970).

[†]   J. R. Platt, "What We Must Do," *Science* **166**, 1115–1121 (1969).

2
ELEMENTS, ATOMS, AND MOLECULES

A medieval alchemist, after a painting by David Teniers. (Courtesy of The National Library of Medicine.)

2–1 CHEMISTRY AND ALCHEMY

Chemistry is classically defined as the study of the properties of various materials and the transformations that occur when different substances are brought together and allowed to interact. That definition, however, does not adequately describe the scope of modern chemistry, as the field has become so intertwined with related sciences that one can scarcely mark out the boundaries of the field. Biochemists join with molecular biologists to study the chemical nature of the subunits of cells of biological tissue. Geochemists touch upon geology and astronomy in studying the chemical composition of rocks, meteorites, and samples from the moon's surface. Physical chemists study the structure and properties of materials, as do their colleagues in physics, the chemical and solid-state physicists. These few examples indicate the breadth of chemistry and the difficulty of delineating its periphery.

Chemistry is an ancient science. Throughout history men have wondered to what extent material can be subdivided before it loses its identifying properties. Consider a substance that appears uniform to the naked eye, e.g., some crystals of salt. If they are broken into smaller and smaller pieces, do they continue to have the properties of salt down to the smallest units we can make? Or do we eventually reach some small unit of salt that, when broken down further, is no longer like salt? The earliest known suggestion of finite small units was made by Democritus (ca. 450 B.C.), who hypothesized that all matter consists of small, indivisible building blocks that he called *atoms* (from the Greek, "uncuttable").

Whether or not one believed in the existence of atoms, there was the further question: Are the infinite number of different substances around us made up of a small number of fundamental substances ("elements") mixed together in varying proportions? Or is there an infinite number of kinds of elements? For many centuries, the theory of Aristotle was in vogue. He thought all substances were made up of the four "elements" earth, air, fire and water, combined in various proportions.

From the Grecian era throughout the Middle Ages and up to about the eighteenth century, the study of chemistry was largely in the hands of alchemists. Some of these men would be considered scientists by today's standards, but others were quacks and frauds—the sort who today advertise cancer cures or sell "genuine simulated moon dust." Alchemists operated in the fuzzy areas between astrology and religion, metal working and medicine. They did not discover much information of a fundamental nature, although they produced some practical applications and developed equipment and techniques for studying chemical reactions.

Alchemical studies were dominated by two major projects: the search for the "philosopher's stone," a material that could be used to convert "base metal" (e.g., lead) into gold, and a search for the "elixir of life," a substance that would prolong life and cure diseases. When all matter was thought to be made up of four elements in different proportions, it was natural to suppose that all one needed to do was to mix some substances in the correct proportions in order to make gold. Today we understand why their search for the philosopher's stone was vain. One cannot make gold (which is itself an element) by mixing things together and heating them up. High-energy projectiles must be used to change the nucleus, the tiny center of the atom, in order to make gold from another substance (see Chapter 5).

Nuclear chemists and physicists have invented the "philosopher's stone" in the form of cyclotrons and other nuclear accelerators, whose projectiles can induce nuclear reactions in other elements and convert them to gold. This is, however, not a practical method. We estimate that the gold produced would cost about $600 billion per ounce! The price of natural gold is about $100 per ounce.

2–2 ATOMS AND ELEMENTS

Today we know that all substances are composed of atoms. There are 105 known types of atoms, each of which we refer to as an *element*. All atoms of a given element behave in the same way in chemical reactions. Each atom of the element sodium, for example, has the same chemical properties as every other sodium atom. Only 83 of the 105 elements exist in reasonable quantity on Earth; the balance of them have been made by man. Thus the infinite variety of substances that exist in our surroundings arises from various combinations of just these 83 types of atoms.

Atoms are much too small to be seen with the naked eye. With the best optical microscopes, one can only pick out objects containing at least thousands of atoms. However, in 1970, Professor Albert V. Crewe of the University of Chicago used a scanning electron microscope, a highly sophisticated apparatus, to observe individual atoms.* He was not able to

---

* A. V. Crewe, "A High Resolution Scanning Electron Microscope," *Scientific American* **224** (4), 26–35 (April 1971).

observe their structure in detail, but he did produce a photographic plate in which atoms of thorium, separated by as little as 10^{-7} cm on long organic molecules, stood out clearly as bright dots (Fig. 2–1).

The radii of atoms are normally measured in angstroms (1 Å = 10^{-8} cm) and range between about 0.5 and 2.0 Å. Atoms are so small that they are difficult for us to visualize. A one-inch row of atoms would contain about 250,000,000 of them—a little more than the population of the United States! A dime expanded to the size of the United States would have atoms about the size of billiard balls (see Fig. 2–2).

The atom itself has structure—it consists of a nucleus with electrons circulating around it. Most of the mass of the atom is concentrated in the nucleus, whose mass is about 2000 times as great as the electrons around it. The radius of the nucleus is about 10,000 times smaller than that of an atom. In the example of a dime the size of the United States, the nucleus would still be too small to be seen with the naked eye. If we further expand the billiard ball to the size of the Houston Astrodome, the nucleus would be the size of a grapefruit (see Fig. 2–3). The radius of an atom is determined by the maximum distances of the circulating electrons, each of which is smaller than the tiny nucleus. Thus all atoms, including those of such "solid" objects as bricks, metal rods, and wooden blocks, consist almost entirely of empty space!

If we can just barely "see" an atom with the best electron microscope, how can we measure the size of the nucleus, which is 10,000 times smaller? True, no one has "seen" a nucleus, but there are ways in which to infer its properties.

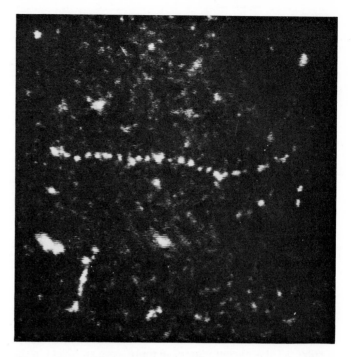

Fig. 2–1 Long-chain molecules in which thorium atoms alternate with benzene rings (see Section 13–4). The bright dots in the chain are produced by individual atoms of the heavy element thorium ($_{90}$Th), which are separated by several angstroms. The photo was made by A. V. Crewe with a scanning electron microscope. (Courtesy of the Enrico Fermi Institute.)

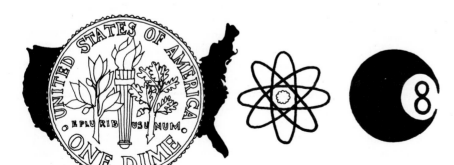

Fig. 2–2 A dime the size of the United States would have atoms the size of billiard balls.

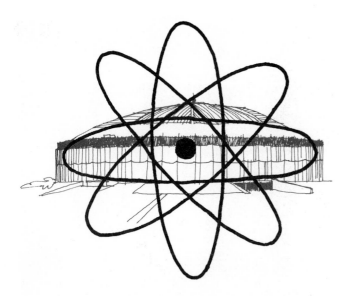

Fig. 2–3 An atom the size of the Astrodome would have a nucleus the size of a grapefruit.

One of the most important methods is that of bouncing projectiles off the nucleus. Suppose there were an object in the center of a pool table beneath a hanging canopy—so that we couldn't see the object, but billiard balls could go under the canopy and emerge at any angle (see Fig. 2–4). We could learn a lot about the unseen object by shooting billiard balls under the canopy and seeing what happened to them. If the object were a bowling ball, some of the billiard balls that struck it would be scattered at wide angles. Those that hit it "head on" would come straight back. Even if we did not hit the bowling ball hard enough to knock it out where it could be seen, we could determine its size and mass by observing the fraction of billiard balls scattered and their speeds going in and coming out. If the central object was smaller than the billiard balls, say a marble, none of the billiard balls would come straight back, but we could still determine the marble's properties from the angles and speeds of the billiard balls emerging on the other side of the canopy. Of course, in this case, the marble could be knocked out from under the canopy where it could be seen. Even if we did not pick it up, we could calculate its mass by noting its speed after being hit by a billiard ball of known speed. If the central object was a small lump of Jello, none of the balls would come back to us and there would be little wide-angle scattering. Some of the balls would be diverted slightly and slowed down.

In 1911, Ernest Rutherford used essentially this method to learn about the nature of atomic nuclei. Instead of billiard balls, he used *alpha (α) particles* (the nuclei of helium atoms), which are emitted by some of the naturally occurring radioactive elements such as uranium and radium (see Section 5–3). He caused a narrow beam of α particles to strike a target of very thin gold foil. He observed the scattered α particles with a zinc sulfide (ZnS) crystal that gave off a small flash of light observable with a microscope when it was struck by an α particle as shown in Fig. 2–5. He moved the zinc sulfide crystal to various angles and counted the flashes produced by the α particles. Most of the α particles went right through the foil with little or no deflection, demonstrating the "emptiness" of the gold atoms. Many of the α particles were scattered at wide angles, and some came straight back out of the foil. The gold nuclei are much larger than the α particles, and so the situation is similar to that of billiard balls bouncing off bowling balls. From these experiments he determined that the nucleus of the atom is very dense and small. So enormous is the density of nuclear material that a marble made out of solid nuclear material would weigh about 500 million tons! Today we still study nuclei by scattering, but now instead of using α particles from radioactive decay, we use high speed α particles and other projectiles from cyclotrons and similar electronuclear machines.

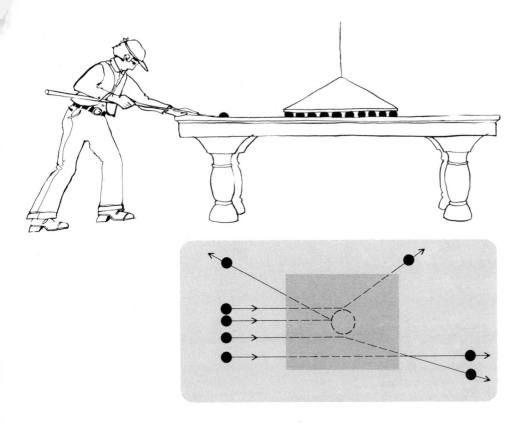

Fig. 2–4 We can learn about an object hidden under a canopy hanging over a pool table by shooting billiard balls under the canopy and observing how they scatter. A similar method is used to find out about the nuclei of atoms.

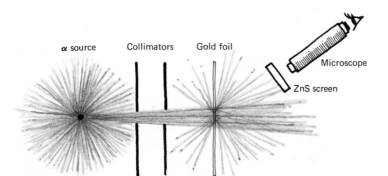

Fig. 2–5 Schematic diagram of the Rutherford scattering experiment which determined that the positive charge of an atom resides on a small dense nucleus at the center of the atom. The collimators are devices that permit passage of only those particles that are in a parallel beam.

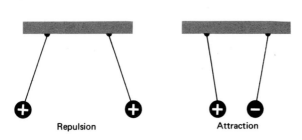

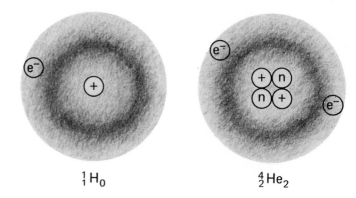

Fig. 2–6   The familiar pith ball experiment with electric charges demonstrates that like charges repel and unlike charges attract each other.

Fig. 2–7   Simplified structures of hydrogen and helium atoms.

Electrons are attracted to the positively charged nucleus because of the universal attraction of objects having opposite electric charges. This effect is often demonstrated with pith balls hanging on strings (see Fig. 2–6). First, a piece of amber rubbed vigorously with cat fur acquires excess electrons from the fur. Since the charge of the electron is defined as negative, the amber becomes negatively charged. If we touch a pith ball (suspended by an insulating thread) with the amber, some of the electrons flow onto the pith ball, giving it a negative charge. If we similarly charge a second pith ball and bring it near the first one, the pith balls repel each other. Just as two magnetic south poles are mutually repellent, so are two charges of the same sign. The force of repulsion, i.e., the strength of the "push" that one ball exerts on the other, may be expressed quantitatively by Coulomb's law:

$$\text{force} = \frac{q_1 q_2}{r^2}, \qquad (2\text{–}1)$$

where q_1 and q_2 are the charges on two objects and r is the distance between them. [The units of q and r determine the units of force. If charge is in electrostatic units (esu) and r in cm, force is given in dynes; see Appendix C.] We see that the force decreases rapidly with increasing distance.

If we touch the second pith ball with a glass rod that has been rubbed with silk, the second ball becomes charged positively; that is, it becomes deficient in electrons. Since unlike charges attract each other, the pith balls now pull toward each other, with a force as given by Eq. (2–1), but now the sign of the force is negative, since the force is attractive.

Atoms are similar to the pith balls in that the electrons are negatively charged and nuclei are positively charged; thus the electrons are attracted to the nucleus and repelled by each other. All electrons carry the same amount of charge, 1.6×10^{-19} coulomb (or 4.8×10^{-10} esu). The small size of this charge can be appreciated if we consider a common application of electricity. Electrical current is the motion of electrons through a wire: a current of 1 ampere is produced by the passage of 1 coulomb per second (C/s) past a point on the wire. An iron draws about 8 amps; if it takes about 15 minutes (900 seconds) to iron a shirt, this means that about

$$\frac{9 \times 10^2 \text{ s} \times 8 \text{ C/s}}{1.6 \times 10^{-19} \text{ C/electron}} = 4.5 \times 10^{22} \text{ electrons} \qquad (2\text{–}2)$$

pass through the iron.

Atomic nuclei consist of *neutrons* and *protons*. The neutrons are uncharged, and each proton has a charge equal and opposite to that of the electron. In a neutral atom, i.e., one that has no net charge, the number of electrons circulating around the nucleus is equal to the number of protons in the nucleus. Each element is defined by its *atomic number* (designated Z), which is the number of protons in the atom's nucleus. The simplest atoms are those of hydrogen, with atomic number $Z = 1$. Neutral atoms of ordinary hydrogen each have a nucleus consisting of a single proton, with a single electron circulating around it. Other "isotopes" of hydrogen are deuterium and tritium, which have one and two neutrons, respectively, in the nucleus in addition to the proton. The neutral hydrogen atom may be indicated symbolically as $_1$H. The subscript 1 preceding the symbol is sometimes included to remind us that the atom has atomic number $Z = 1$; it is not necessary, however, because as chemists we know that hydrogen is element number 1, as shown on the periodic table inside the front cover.

The symbol $_2$He stands for the element helium, whose atomic number is $Z = 2$. That is, helium has two protons in the nucleus and two electrons circulating around it when the atom is neutral (see Fig. 2–7). The highest atomic number among elements that occur in nature on Earth is 92, that of the element uranium, $_{92}$U, whose neutral atoms have 92 electrons around the nucleus.

The chemical behavior of an atom is determined almost exclusively by the number of electrons that it contains when neutral; the number of electrons, in turn, is determined by the amount Z of positive charge on the nucleus. Thus every atom of an element of atomic number Z contains Z protons in its nucleus and has the same chemical behavior as all other atoms of that element. Added neutrons make the nucleus heavier, but being uncharged, they do not change the charge Z of the nucleus. Every element has several different kinds of nuclei which differ in the number of neutrons N, but all have Z protons. For example, all chlorine nuclei have 17 protons, but naturally occurring chlorine has some nuclei with 18 neutrons and some with 20 neutrons. Atoms of the same element but with different numbers of neutrons are called *isotopes* (from the Greek "same place").

Neutrons and protons have about the same mass, and both are much heavier than electrons; in fact, a proton is about 1837 times as heavy as an electron. Since most of the mass of an atom is that of the neutrons and protons, we define the *mass number A* as the total number of neutrons and protons in the nucleus, $A = N + Z$. The first chlorine isotope mentioned above would have $A = 35$, and the other would have $A = 37$. They would be indicated symbolically as

$$^{35}_{17}Cl_{18} \quad \text{and} \quad ^{37}_{17}Cl_{20},$$

where the superscript indicates the mass number A and the subscript following the chemical symbol represents the neutron number N (see Fig. 2–8). The N value is not usually given, as it can be obtained by subtracting Z from A. Furthermore, it causes confusion in a chemical equation because it occupies

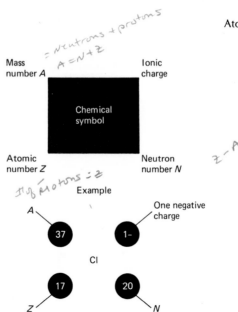

Fig. 2–8 Complete nuclear designation of an isotope, with atomic number Z, neutron number N, and mass number A surrounding the chemical symbol of the element. The concept of ionic charge will be discussed later in the book.

the position normally reserved for the number of atoms in a molecule. The mass number is important in nuclear chemistry, because isotopes with different mass numbers often have very different nuclear properties. For example, ^{38}Cl does not normally exist in nature as it is radioactive. Chlorine-38 atoms spontaneously transform themselves into atoms of the neighboring element, argon (the isotope $^{38}_{18}$Ar, in particular—see Chapter 5).

2–3 ATOMIC WEIGHTS

Atoms are so small that it is inconvenient to use their actual masses in calculations. Neutrons and protons each weigh about 1.67×10^{-24} g, and electrons are nearly two thousand times lighter, 9.1×10^{-28} g. As noted above, the weight or "mass" of an atom is nearly proportionate to the *mass number A* of the atom, i.e., the total number of neutrons and protons in the nucleus of the atom. For example, an atom of $^{12}_{6}C_{6}$ weighs approximately twelve times as much as an atom of $^{1}_{1}H_{0}$. In part because the neutron and hydrogen atom do not weigh the same and for other reasons discussed in Chapter 5, this relationship is not exact. Thus we need to have some quantity similar to the mass number by which to relate the mass of one atom to those of others. It might seem natural

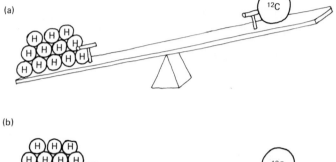

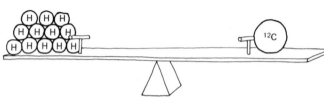

Fig. 2–9 (a) Twelve atoms of ^{1}H weigh slightly more than one ^{12}C atom. (b) 11.907 atoms of ^{1}H weigh exactly the same as one ^{12}C atom.

to "weigh" atoms in terms of ^{1}H atoms, that is, indicate how many times heavier the given atom is than an atom of ^{1}H. The *relative* mass scale for atoms, called the *atomic weight* or *mass scale*,* is similar to this. However, instead of setting the mass of ^{1}H as 1.000000, chemists and physicists define the atomic mass scale by calling the atomic mass of ^{12}C exactly 12.000000. On this scale, the atomic mass of ^{1}H is 1.007825 *atomic mass units* (amu). This means that the mass of a ^{1}H atom is 1.007825/12.000000 (or 0.083985) that of a ^{12}C atom, and that a ^{12}C atom has a mass of 12.000000/1.007825 (or 11.907) times that of a ^{1}H atom, instead of exactly 12 times a ^{1}H atom (see Fig. 2–9).

The atomic masses of the isotopes are very useful to nuclear chemists and physicists, who often study properties of individual isotopes of elements. But since the various isotopes of a particular element behave almost identically in chemical reactions, chemists find it more convenient to work with *chemical atomic weights*, which take into account the fact that many elements are made up of more than one isotopic form. For most such multi-isotopic elements, the fractions of the various isotopes, the isotopic abundances, present in a sample of the element are the same throughout

all areas and media of the earth. For example, regardless of where we obtain a sample of chlorine, be it from the salt of the ocean or from a bottle of hydrochloric acid (HCl) in the laboratory, 75.53% of the atoms will be the isotope ^{35}Cl and 24.47% will be ^{37}Cl. To obtain the chemical atomic weight of chlorine, we take an average of the atomic masses of the two isotopes, weighted by their isotopic abundances:

Isotope	abundance		Atomic mass		
^{35}Cl:	0.7553	×	34.96885	=	26.41
^{37}Cl:	0.2447	×	36.96590	=	9.046 (9.05)

Chemical atomic weight of chlorine: $= 35.46$

The periodic table of the elements inside the front cover lists the atomic number Z of each element and its chemical atomic weight, obtained as shown above for elements that have more than one naturally occurring isotope. For several elements, the chemical atomic weight is given as an integer within parentheses. These elements have no stable isotopes; that is, all their isotopes are radioactive and decay to isotopes of other elements, as in the example of ^{38}Cl decay to ^{38}Ar noted above. Many of these elements do not exist naturally on earth and have been synthesized by man, e.g., all the elements above $_{92}$U. The whole number listed for each of these elements represents the mass number of its most prominent isotope.

Since chemical reactions rarely favor one isotope of an element relative to others, the isotopic distribution is, for most elements and for all practical purposes, the same for all natural terrestrial samples of the element. There are some exceptions to this general rule, however. Since deuterium, ^{2}H, is twice as heavy as ^{1}H, the former enters into some chemical reactions more slowly than the latter. When water is broken down by electricity into its components hydrogen and oxygen, ^{1}H reacts more readily than ^{2}H, causing the water left behind to become slightly enriched with molecules containing one or two ^{2}H atoms (^{2}H$_2$O is called *heavy water*).

Since atoms are very small and the number of them in any volume we can measure in a chemistry laboratory is extremely large, it is inconvenient to work with actual numbers of atoms. However, it would be handy to work with quantities *proportional to* the number of atoms. Chemists in the nineteenth century devised a unit of measure called the *gram atomic weight*, or simply the *gram-atom*, for measuring quantities of atoms. *We define the gram-atom as the chemical atomic weight of an element expressed in grams.*

---

* We use atomic "weight" and "mass" interchangeably, although atomic mass is probably to be preferred (Appendix B).

Although it may not be immediately obvious, a gram-atom of a given element contains the same number of atoms as a gram-atom of any other element. Consider hydrogen and carbon as examples. Their chemical atomic weights are 1.008 and 12.01, respectively. Thus 1.008 g of H atoms constitutes a gram-atom of hydrogen and 12.01 g of C atoms makes up a gram-atom of that element. Since each carbon atom weighs about twelve times as much as a hydrogen atom, 12.01 g of carbon contains the same number of atoms as 1.008 g of hydrogen (see Fig. 2–10).

The same considerations are valid for all other elements. For example, the chemical atomic weights of sodium (Na), aluminum (Al), and gold (Au) are 22.99, 26.98, and 196.97, respectively; thus a 22.99 g sample of Na, a 26.98 g sample of Al, and a 196.97 g sample of Au each contains the same number of atoms, within limits of error of our measurements. The actual number of atoms contained in a gram-atom is called *Avogadro's number*; it is 6.023×10^{23} atoms/gram-atom. Although chemists rarely need to know the number of atoms in a sample, the value can be computed from Avogadro's number if desired. Suppose one had a 1 g sample of gold—that's 1/196.97, or 0.005077 of a gram-atom. It would contain

$$0.005077 \times 6.023 \times 10^{23} = 3.058 \times 10^{21} \text{ atoms of gold.}$$

The concept of gram-atoms is used to calculate the amounts of various elements that must be mixed together to make a chemical reaction occur with maximum efficiency. For example, consider the reaction of carbon with oxygen to make carbon monoxide:

$$C + O \longrightarrow CO. \qquad (2–3)$$

(The reaction as written is oversimplified; oxygen gas has pairs of atoms linked together as O_2 molecules. Only at high temperatures, as in an automobile engine, will single oxygen atoms exist to take part in a reaction of this type.) The written reaction states that one *atom* of carbon reacts with one *atom* of oxygen to form one *molecule* of carbon monoxide. But if we have a 6-g sample of carbon, how many grams of oxygen would be required to exactly use up the carbon so that neither carbon nor oxygen were left over? Since we now know that a gram-atom of any element contains the same number of atoms as a gram-atom of any other element, we can also interpret Eq. (2–3) to mean that one *gram-atom* of carbon reacts with exactly one *gram-atom* of oxygen. Since the atomic weight of carbon is 12, the 6-g sample represents

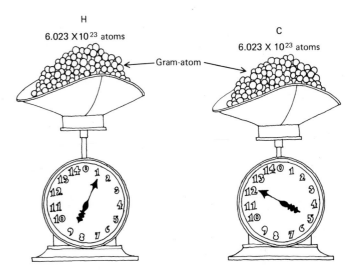

Fig. 2–10 A gram-atom of H weighs 1.008 g and a gram-atom of C weighs 12.01 g. Each contains 6.023×10^{23} atoms.

half a gram-atom ($\frac{6}{12} = \frac{1}{2}$ gram-atom of carbon) and thus will require half a gram-atom of oxygen; we can readily see that half of 16 g of oxygen is 8 g:

$$\frac{1}{2}\text{ gram-atom} \times \frac{16 \text{ grams oxygen}}{\text{gram-atom}} = 8 \text{ grams oxygen.}$$

2–4 MOLECULES, COMPOUNDS, AND CHEMICAL REACTIONS

For reasons that will be discussed in more detail in Chapter 8, very few elements occur in nature as individual atoms. Instead, the atoms of the element combine with each other or with atoms of other elements to form *molecules*. Molecules range in size from the simplest diatomic molecules like carbon monoxide, which have just two atoms bonded together (see Eq. 2–3), to extremely large molecules containing millions of atoms that we encounter in biochemical systems.

Many pure elements occur naturally in the form of diatomic molecules, for example, N_2, O_2, and Cl_2. Some pure elements can exist in more than one molecular form. For example, oxygen normally exists as O_2, which makes up about 20% of our atmosphere. Another form of pure oxygen is ozone, O_3, which is present in minute amounts in the atmosphere. Although both species of oxygen contain only atoms of oxygen, the difference in molecular forms gives them

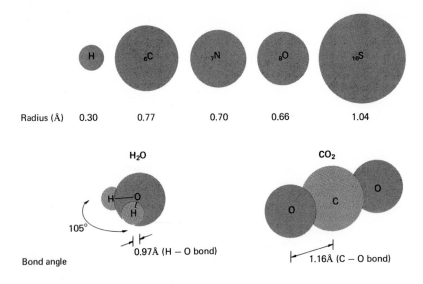

Radius (Å)   0.30   0.77   0.70   0.66   1.04

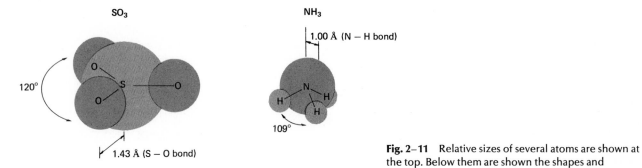

H_2O

105°

Bond angle

0.97Å (H − O bond)

CO_2

1.16Å (C − O bond)

SO_3

120°

1.43 Å (S − O bond)

NH_3

1.00 Å (N − H bond)

109°

Fig. 2–11 Relative sizes of several atoms are shown at the top. Below them are shown the shapes and dimensions of several molecules made up from those atoms.

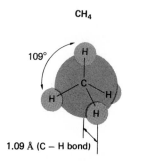

CH_4

109°

1.09 Å (C − H bond)

different chemical properties. Ozone, which is present in photochemical smog, is destructive to rubber and other materials.

In the solid form, pure metals occur as three-dimensional lattices (or "crystals") of atoms bonded together. The only elements that normally occur as individual atoms are the noble gases, which are listed in the right-hand column of the periodic table. Unlike other elements, the noble gas atoms have little tendency to exchange or share electrons with other atoms to form chemical bonds.

The sizes of molecules formed by combining two or more atoms are approximately what one would predict by bringing the atoms together until they touch each other. Molecules with three atoms may be linear (that is, the nuclei of the atoms are in a straight line) like CO_2 (carbon dioxide), or bent like H_2O (see Fig. 2–11). Molecules having four or more atoms may be planar (flat) like SO_3 (sulfur trioxide), or three-dimensional like NH_3 (ammonia), which has the shape of a pyramid. In larger molecules, especially those containing carbon, there are structures in which a central atom (e.g., C in CH_4, methane) is at the center of a tetrahedron formed by four surrounding atoms.

A *pure compound* is a substance in which essentially all the molecules are identical (see discussion of impurities in Section 3.7). For example, in very pure water all the molecules are H_2O; that is, each one is made up of two hydrogen atoms bonded to an oxygen atom. In very pure sugar (sucrose), nearly all the molecules have the chemical formula $C_{12}H_{22}O_{11}$. A pure compound can be depended on to always have the same chemical composition.

In many cases the atoms of these molecular units remain bonded together whether they are in the solid, liquid, or gaseous form, or even if they are dissolved in some other liquid, e.g., sugar dissolved in water. However, some compounds are *ionic*. *An ion is an atom or group of atoms that has gained or lost one or more of its electrons* and is thus electrically charged. Sodium has a nuclear charge $Z = 11$; therefore, neutral atoms of sodium have eleven electrons circulating about the nucleus to exactly cancel the nuclear charge. In reacting with atoms of other elements, a sodium atom gives up one of its electrons to the other species to become a sodium *ion*; i.e., it retains only ten electrons and acquires a net charge of 1+ unit:

$$Na \longrightarrow Na^+ + e^-. \qquad (2-4)$$

atom   positive   electron
          sodium ion

Conversely, a neutral atom of chlorine containing 17 electrons often acquires an additional electron in the course of a chemical reaction, becoming a negatively charged *chloride* ion:

$$Cl_2 + 2e^- \longrightarrow 2Cl^-. \qquad (2-5)$$

neutral   electrons   negative
molecule   chloride ions

(Here we have properly written the equation with Cl_2 molecules, as that is the normal form of pure chlorine.)

When sodium metal reacts with chlorine gas to form the compound NaCl (sodium chloride, common table salt), each sodium atom gives up one of its electrons to a chlorine atom so that the product consists of equal numbers of Na^+ and Cl^- ions:

$$2Na + Cl_2 \longrightarrow 2Na^+ + 2Cl^-. \qquad (2-6)$$

In solid form, sodium chloride exists as a crystal lattice of regularly spaced Na^+ and Cl^- ions (see Fig. 2–12). In such ionic crystals, one cannot identify individual molecular units; i.e., there is no particular Cl^- ion associated with each Na^+ ion. Instead, each Na^+ ion is surrounded by six Cl^- ions, equidistant from it, and each Cl^- ion is surrounded by six Na^+ ions. When sodium chloride is dissolved in water, the Na^+ ions are each surrounded by a sheath of water molecules and remain detached from the Cl^- ions. Although there are no distinct molecules in ionic compounds, we can handle them in the same way as "molecular" compounds in the calculations discussed below.

Now that we know what molecules are, we are able to interpret chemical reactions more fully and calculate yields of their products. Equation (2–3) states that one atom of carbon reacts with one atom of oxygen to form one molecule of carbon monoxide. If we start with one gram-atom, containing Avogadro's number of atoms, of each reactant, we obtain Avogadro's number of CO molecules. The mass of this quantity of molecules is the compound's *gram-molecular weight*. By correspondence with the definition of the gram-atom given above, *the gram-molecular weight of a compound is defined as the molecular weight expressed in grams*, where the *molecular weight is the sum of the chemical atomic weights of the atoms in its molecules*. The molecular weight of CO is the sum of 12.01 (the atomic weight of C) and 16.00 (for oxygen), or 28.01. Thus a gram-molecular weight, or *mole* as it is normally abbreviated, of carbon monoxide is 28.01 g of CO. *A mole of any compound contains the same number*

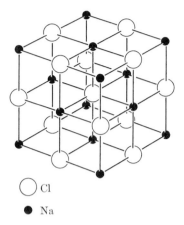

○ Cl

● Na

Fig. 2–12   Drawing of a sodium chloride crystal. [From Bruce H. Mahan, *University Chemistry*, 2nd. ed., (Reading, Mass.: Addison-Wesley Publishing Co., 1969), p. 91. Reprinted by permission.] The photo by Grant Heilman shows several NaCl crystals.

of molecules as a mole of any other compound, a number which is also equal to the number of atoms contained in a gram-atom of a pure element.

With this definition of the mole, we can also interpret Eq. (2–3) to mean that one gram-atom of carbon combines with one gram-atom of oxygen to form *one mole* of carbon monoxide. If we prefer to express the relationship in terms of weights of the various species, we can substitute their chemical atomic and molecular weights and state that 12.01 g of carbon reacts with 16.00 g of oxygen to form 28.01 g of carbon monoxide:

C	+	O	$\longrightarrow$	CO
1 atom		1 atom		1 molecule
1 gram-atom		1 gram-atom		1 mole (gram-molecular weight)
12.01 g		16.00 g		28.01 g

As stated above, Eq. (2–3) was written in over-simplified form for illustrative purposes. Since oxygen gas occurs in the form of O_2 molecules, the reaction is properly written as

$$C + \tfrac{1}{2}O_2 \longrightarrow CO. \qquad (2-7)$$

This change does not alter the number of oxygen atoms or the total weight of oxygen involved in the reaction; it merely recognizes the fact that pure oxygen occurs as diatomic molecules rather than individual atoms. If one prefers not to work with half-integer quantities one can double the quantities of Eq. (2–7):

$$2C + O_2 \longrightarrow 2\,CO. \qquad (2-8)$$

which may be interpreted as stating that two atoms of carbon react with one molecule of oxygen (containing two atoms) to form two molecules of carbon monoxide (see Fig. 2–13). Since the molecular weight of O_2 is $2 \times 16.00 = 32.00$, a mole of O_2 is 32.00 g; thus the half mole of O_2 in Eq. (2–7) is 16.00 g, just as the one gram-atom of oxygen atoms used above.

We can apply the same type of calculations to any other chemical reaction. One gram-atom of sodium weighs 22.99 g, and one mole of Cl_2 and NaCl weigh 70.90 g and 58.44 g, respectively. Therefore, Eq. (2–6) can be interpreted to mean that $2 \times 22.99 = 45.98$ g of Na reacts with 70.90 g of Cl_2 to form two moles, $2 \times 58.44 = 116.88$ g of NaCl. The fact that NaCl does not contain specific molecular units does not alter the number of Na^+ and Cl^- ions or the total weight of NaCl obtained.

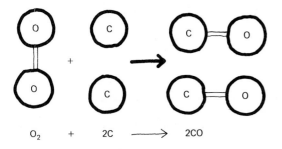

Fig. 2–13 Reaction between carbon and oxygen to form carbon monoxide.

Chemical reactions do not destroy or create atoms of the various elements. Thus, we must always work with *balanced* chemical equations; that is, for each element there must be the same number of atoms on both sides of the arrow in the equation. We balance the equation by adjusting the coefficients, the numbers in front of each chemical species that indicate how many gram-atoms or moles of each are involved. In Eq. (2-8), the balancing was quite straightforward, but in more complicated cases we must be careful to balance the equations properly. For example, if we know that O_2 reacts with aluminum, Al, to form aluminum oxide, Al_2O_3, we can write

$$Al + O_2 \longrightarrow Al_2O_3$$

and then apply coefficients to balance the equation:

$$2Al + \tfrac{3}{2}O_2 \longrightarrow Al_2O_3$$

Or if you prefer to avoid half-integers, we can write

$$4Al + 3O_2 \longrightarrow 2Al_2O_3. \qquad (2\text{–}9)$$

Also, if any charged species are involved, i.e., positive or negative ions or free electrons, one must make sure that the net charges of species on the two sides are the same.

One can multiply the amounts of all the species involved in a reaction by any factor. Just as a cook can double the amounts of all the ingredients in a cake recipe, so we can multiply the number of moles of each species in a chemical reaction by any number. A 1.2 g sample of carbon, for example, would be 0.1 gram-atom of carbon. According to Eq. (2-7), a 1.2 g sample of carbon would, therefore, react with

0.05 mole of O_2 (weighing 0.05 mole $\times$ 32.00 g/mole = 1.6 g of oxygen) to produce 0.1 mole of CO, weighing 0.1 mole $\times$ 28.01 g/mole = 2.8 g.

The most important distinction between pure compounds and mixtures of various substances is that a pure compound *always* has the same chemical composition; that is, the weights of the elements it contains always occur in the same ratios. For example, one mole of carbon monoxide weighs 28.01 g, of which 12.01 g is carbon and 16.00 g is oxygen, giving a weight ratio of carbon to oxygen of 12.01/16.00 = 0.75. In pure samples of carbon monoxide, the ratio C/O always has this value.

We could as easily calculate the fraction of the weight of the compound contributed by each element. Let us consider CO as an example:

$$\text{weight fraction C} = \frac{\text{atomic weight of C}}{\text{molecular weight of CO}}$$

$$= \frac{12.01}{28.01} = 0.429 \text{ or } 42.9\%. \qquad (2\text{–}10)$$

$$\text{weight fraction O} = \frac{\text{atomic weight of O}}{\text{molecular weight of CO}}$$

$$= \frac{16.00}{28.01} = 0.571 \text{ or } 57.1\%.$$

Likewise the weight fraction of aluminum in Al_2O_3 is given by

$$\text{weight fraction Al} = \frac{2 \times \text{atomic weight of Al}}{\text{molecular weight of } Al_2O_3}$$

$$= \frac{2 \times 26.98}{(2 \times 26.98) + (3 \times 16.00)}$$

$$= \frac{53.76}{53.98 + 48.00} \qquad (2\text{–}11)$$

$$= \frac{53.76}{101.98} = 0.5272 \text{ or } 52.72\%.$$

In making a cake or other mixture, one can often change the composition of the final product by altering the proportions of ingredients. However, a pure compound must always be made up from elements in specific proportions. For example, water can be made by combining hydrogen and oxygen gases:

$$2H_2 \quad + \quad O_2 \quad \longrightarrow \quad 2H_2O. \qquad (2-12)$$

| 2 moles | 1 mole | 2 moles |
| 4.03 g | 32.00 g | 36.03 g |

Thus a 32.00 g sample of O_2 would require exactly 4.03 g of H_2 for reaction. If more hydrogen were present, it would simply be left over as excess. If less hydrogen were present, all of it would be used up, but less than 36.03 g of H_2O would be formed and some O_2 would be left over. In both cases, the weight ratio of hydrogen to oxygen in the water would be the same, $4.03/32.00 = 0.126$.

Under different conditions it is possible to form different compounds with the same set of elements. In addition to forming water, hydrogen and oxygen can also, under different conditions, form the compound hydrogen peroxide, H_2O_2. The weight ratio H/O is exactly half the value in H_2O_2 that it has in H_2O, but pure binary (two-element) compounds of H and O with weight ratios in between these values are not possible. That is, there are no pure compounds having formulae such as $H_{1.4}O_2$ or $H_2O_{1.7}$. Similarly, in the presence of large amounts of oxygen, carbon reacts to produce carbon dioxide, CO_2, instead of carbon monoxide, CO.

Pure compounds are also distinct from mixtures in that their properties (e.g., melting point, density) are always the same, whereas those of mixtures are generally dependent on the amounts of the various components of the mixture. We will discuss this point in more detail after considering the states of matters and properties of materials.

SUGGESTED READING

I. D. Garard, *Invitation to Chemistry* (Doubleday, Garden City, N.Y., 1969). Available in paperback.

K. J. Laidler and M. H. Ford-Smith, *The Chemical Elements* (Bogden and Quigley, Tarrytown-on-Hudson, N.Y., 1970). Paperback.

B. H. Mahan, *University Chemistry*, 2nd ed. (Addison-Wesley, Reading, Mass., 1969).

J. R. Partington, *A Short History of Chemistry*, 3rd ed. (Macmillan, London, 1957). Paperback.

G. T. Seaborg and E. G. Valens, *Elements of the Universe* (E. P. Dutton, N.Y., 1958).

PROBLEMS AND QUESTIONS

1. Briefly define the following terms.

   a) atom
   b) nucleus
   c) molecule
   d) gram-atom
   e) mole (gram-molecular weight)
   f) compound
   g) Avogadro's number
   h) ion

2. a) How many neutrons and protons are contained in the nucleus of a ^{208}Pb atom?

   b) What specific modifications of the ^{208}Pb nucleus would be required to transform it into that of an atom of ordinary gold?

3. The nucleus of atom #1 contains 20 protons and 20 neutrons, whereas that of atom #2 contains 20 protons and 24 neutrons.

   a) How many electrons would be in a neutral atom of #1? of #2?

   b) How many electrons would be in a +2 charged ion of #1? of #2?

4. Balance the following chemical equations.

   a) $H_2 + Cl_2 \longrightarrow HCl$.
   b) $Br_2 + Zn \longrightarrow ZnBr_2$.
   c) $V + O_2 \longrightarrow V_2O_5$.
   d) $C + S \longrightarrow CS_2$.
   e) $Y + F_2 \longrightarrow YF_3$.

5. a) How many *gram-atoms* are contained in a 5 g sample of gold ($_{79}Au$)?

   b) How many *atoms* are contained in a 5 g sample of gold?

   c) How much does a single atom of gold weigh?

6. Calculate the molecular weight of each of the following compounds.

   a) C_3H_8 (propane)
   b) SO_2 (sulfur dioxide)
   c) KBr (potassium bromide)
   d) NH_3 (ammonia)
   e) Fe_2O_3 (ferric oxide)

7. a) What fraction of the weight of Fe_2O_3 is represented by iron?

b) What is the weight of iron in a 10 g sample of Fe_2O_3?

c) Geologists often report the metal concentrations of rocks not by the actual amount of the metal present, but in terms of the amount of oxide formed by the metal. If, for example, a rock were reported to contain 1% by weight of Fe_2O_3, what would be the weight of iron in the rock?

8. Suppose we wish to determine the atomic weight of a newly discovered element, symbol X, whose ions are known to combine with Cl^- ions via the reaction

$$X^+ (aq) + Cl^- (aq) \longrightarrow XCl (s).$$

If 52.6 g of X^+ is found to combine with 7.09 g of Cl^- to form 59.69 g of XCl precipitate, what is the chemical atomic weight of X? (This is the way in which chemical atomic weights of the elements were originally determined.)

9. What is the chemical atomic weight of gallium, which has two isotopes, $^{69}_{31}Ga$ and $^{71}_{31}Ga$.

Isotope	Mass	Abundance
$^{69}_{31}Ga$	68.926	60.16%
$^{71}_{31}Ga$	70.925	39.84%

10. Hydrogen and oxygen combine (burn) to form water.

a) Write a balanced equation for the reaction.

b) If 5.0 g of hydrogen gas is burned with 6.0 g of oxygen, which gas will not be entirely consumed?

c) How many grams and how many moles of water will be produced?

11. A chemical compound is analyzed and found to contain sodium, nitrogen, and oxygen in the following amounts: $_{11}Na$, 69 g; $_7N$, 42 g; $_8O$, 144 g. What is the simplest chemical formula of the compound?

3 FORMS AND PROPERTIES OF MATTER

Biochemists measure the flow rate of milk through a homogenizer.
(U.S. Dept. of Agriculture photo by Jim Strawser.)

3–1 CHARACTERIZATION OF MATERIAL

In Chapter 2 we learned about the submicroscopic nature of material—the nuclei and electrons that make up atoms, the combining of atoms to form molecules, and the relative weights of various types of atoms and molecules. Although knowledge of the submicroscopic character of atoms and molecules is essential to an understanding of chemical behavior, that knowledge is not sufficient for a person to study materials in a typical chemistry laboratory, especially one designed for undergraduate chemistry courses. Many highly complex instruments such as cyclotrons, electron microscopes, and spectrometers (see Chapter 7) are needed to observe many of the properties discussed in Chapter 2. Only the finest research laboratories are equipped with most or all of these facilities.

Suppose we are asked to characterize a sample of an unknown material with simple equipment available in most laboratories. Even without the sophisticated equipment mentioned above there are many simple macroscopic observations we can make on the bulk material to learn a lot about it. For example, by noting the physical state of the material (i.e., solid, liquid or gas) and measuring a few of its properties such as the melting point, boiling point, density, etc., we can usually determine whether the substance is a pure compound or a mixture. As noted in Chapter 2, a pure compound (or element) always exhibits the same properties when measured under the same conditions, whereas the properties of mixtures depend on the proportions of various compounds present in the mixture. Furthermore, if the substance is found to be a pure compound, we can often identify the compound by measuring a few of its properties and searching handbooks for a compound that fits our observations.

Below we discuss density, an easily measured property of material. Then we discuss the physical states of matter, the solid, liquid and gaseous states, as well as certain types of mixtures: true solutions, as well as colloids, aerosols, and suspensions, which are frequently encountered in the "real" world outside the laboratory, especially in the atmosphere and bodies of water. Some additional properties of material, especially electrical and thermal conductivity, are discussed in connection with metals in Chapter 12.

3–2 THE DENSITY OF MATTER

Density is *the weight of a substance per unit volume*. Densities of solids vary over a wide range—among the lightest are balsa wood, 0.08 g/cm³, and polyurethane foam, 0.014 g/cm³, a foamy material used for protective packaging. The densities of pure metals vary from a low of 1.74 g/cm³ for magnesium to such high values as 13.6 g/cm³ for mercury ($_{80}Hg$), 11.4 g/cm³ for lead ($_{82}Pb$), and 19.07 g/cm³ for uranium ($_{92}U$). In general, the metals with high atomic and mass numbers are more dense because the increasing weight of their atomic nuclei is accompanied by a less than proportional increase in the volume occupied by the atoms.

One of the conveniences of the metric or SI units (see Appendix B) for chemists is that weight and volume units have been defined so that the density of liquid water is 1 g/cm³ (also, 1 g/ml). This is strictly true only at 4.0 °C, since thermal expansion for most materials causes them to become less dense with increasing temperature. However, the variation of the density of water is small enough that we can ignore it in all but the most precise laboratory measurements.

Density determines which objects can float in a given liquid. When an object is placed in a liquid it displaces a volume of liquid equal to the volume of the object beneath the surface (see Fig. 3–1). If the object floats, it sinks into the liquid only until it displaces a weight of the liquid that is equal to its own total weight. This can occur only if the object has lower density than the liquid. If the object is denser than the liquid, an equal volume of liquid weighs less, and the object sinks.

Density is a property that can be used to identify substances and verify their purity. Centuries ago Hieron II, king of Syracuse, suspected that his aides had removed much of the gold from his crown and substituted silver, a cheaper metal. He asked the scientist Archimedes to investigate. Archimedes realized that he could make a simple check by measuring the crown's density. If an object has a simple shape, one can measure its edges to compute the volume, weigh it, and then calculate the density. Of course, one could not do this with an ornate crown. However, Archimedes cleverly realized that he could measure the volume of the crown by submerging it in a full container of water (a bathtub, so legend has it) and then measuring the volume of water that overflowed. He did so and discovered that the crown did contain impurities with densities less than that of gold. We still use Archimedes's method in the laboratory to measure the density of irregular objects.

Density measurements are used in everyday applications, particularly in connection with automobiles. The fraction of antifreeze in the radiator coolant determines its freezing point. Since the major component of antifreeze, ethylene

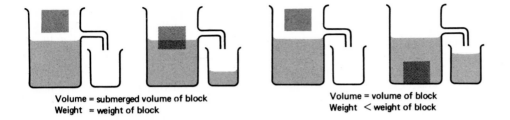

Volume = submerged volume of block
Weight = weight of block

Volume = volume of block
Weight < weight of block

Fig. 3–1 At left, the object floats because it weighs less than an amount of water having the same volume. Its weight is equal to that of the water it displaces. The object at right sinks, because it is heavier (denser) than the displaced water.

glycol, is denser than water, one can determine the antifreeze content by measuring the density of the fluid. The quality of an automobile storage battery is largely determined by the concentration of sulfuric acid in the battery fluid, which is also routinely measured via the density of the fluid.

The density of substances is also involved with environmental problems, particularly oil slicks. Certain fractions of petroleum are less dense than water. Thus, when an oil tanker is wrecked at sea or spills oil in transfer, much of the oil floats on top of the water, spreading out into a large, thin film, or "slick." The coating of oil on the water is very harmful to the biological life in the water, as we shall discuss in detail in Chapter 14.

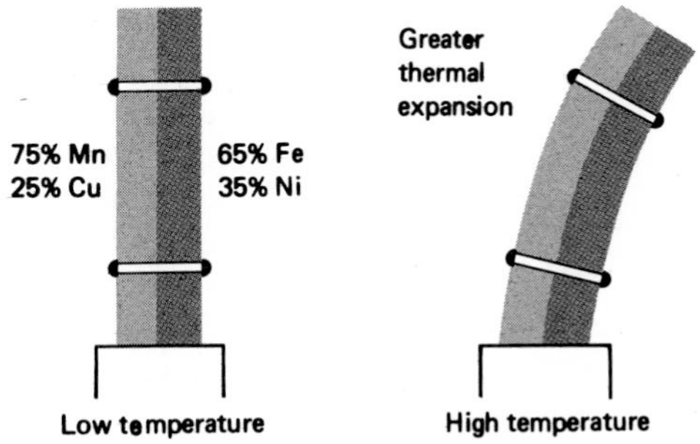

Greater thermal expansion

75% Mn
25% Cu

65% Fe
35% Ni

Low temperature   High temperature

Fig. 3–2 The bimetallic strip in a thermostat bends at high temperature because the metal on the left expands more than the one on the right.

3–3 SOLIDS AND LIQUIDS

Near absolute zero of temperature (0 K or −273°C), the lowest temperature that is theoretically realizable, nearly every pure compound or element forms a crystalline solid. The atoms or ions occur at regular intervals in a three-dimensional lattice, as described above for crystals of sodium chloride (Fig. 2–12).

If a crystal is warmed, the atoms, ions, or molecular units of the crystal begin to vibrate faster and faster about their average positions in the lattice. The vibration causes a *thermal expansion* of the crystal. A metal rod, for example, gets longer as its temperature is increased. The increase in the length of a metal rod with increasing temperature can, in fact, be used to measure temperature. Thermostats that control home heating systems are constructed by fastening together strips

of two different metals or alloys that expand at different rates when heated (see Fig. 3–2). The different changes of length cause the bimetallic strip to bend and touch a contact, completing an electrical circuit that controls a furnace or air conditioner.

If we raise the temperature of a solid to the *freezing point* or *melting point* (these are the same temperature), the vibrations of the atoms, ions, or molecules become so violent that the lattice cannot hold together, and the solid melts to form a liquid. The melting point is another characteristic property of a pure compound that is often used for identification and for an indication of purity. In general, compounds or elements of very low molecular or atomic weight in which the species are bound to each other with weak attractive forces have low melting (and boiling) points. Hydrogen (H_2), nitrogen (N_2), and oxygen (O_2) melt at -259, -209 and $-219°C$, respectively, not far above absolute zero. Helium (He) remains liquid all the way to absolute zero at low pressures.

The electrostatic attractions that hold ionic lattices together are much stronger than the forces between neutral molecules. Thus ionic crystals generally have high melting points. Most organic (carbon-containing) compounds that make up the bulk of biological materials are molecular species and have lower melting points than inorganic materials, which usually form ionic crystals.

For most substances, the atoms or molecules of the solid are packed together very efficiently. After a crystal melts, losing its rigid structure, the atomic or molecular units are usually farther apart than they were in the solid form, and so the density of the material is less. One major exception to this rule is water which, for reasons to be discussed in Chapter 8, has many unusual properties. Crystals of ice formed when water freezes at $0°C$ contain large cavities, and so its density is about 10% less than that of liquid water. This property is of great importance in our environment. If water were similar to most other substances, the more dense solid form would sink to the bottom. Lakes and rivers would freeze from the bottom up instead of from the surface downward and many life forms in the water would die without the protective insulation provided by the ice.

Ice cubes floating in a glass of water constitute a system at *equilibrium*. Liquid and solid water coexist only at a temperature of $0°C$ (unless one applies high pressure to the system). Then as long as both ice and liquid water are present, the temperature remains at $0°C$. Any heat added to the ice water causes ice to melt, but as long as some ice remains, the temperature stays constant at $0°C$. Only when the last of the ice melts does added heat cause the temperature of the water to rise.

The melting point of water can be changed by changing the pressure applied. High pressures favor the phase that is most compact, hence the densest. If a high pressure is applied to ice at a temperature below $0°C$, some of the ice will melt. This effect may help to make ice skating possible. Your weight concentrated the skate blade, which is a very small area, produces an enormous pressure on the ice. Because of the pressure and the heat generated by friction of the blade against the ice, a thin layer of ice melts, creating a thin layer of liquid water under the blade. The liquid water lubricates the motion of the blade across the ice.

The freezing (melting) point of water can also be lowered by dissolving salt in the water. When the temperature drops low enough for some of the water to freeze, most of the salt stays behind in the liquid solution, making it even more concentrated and the freezing point even lower. For this reason salt is often spread on highways and sidewalks in winter to lower the melting point and thus cause ice and packed snow to melt. Other substances dissolved in water also lower its freezing (melting) point, for example, the ethylene glycol that one adds to the coolant system of an automobile engine. It is essential to prevent freezing of water in engines or water pipes. Expansion of the water upon freezing can crack the engine block or split open the pipes— additional effects caused by this unusual property of water.

In the liquid phase of a substance, the molecules are not restricted to vibrate about a particular lattice position as in a solid. The molecules of a liquid move about, colliding with each other and the walls of the container. Although they are not restricted to particular positions, molecules of a liquid are on the average nearly as close to each other as in the solid phase. In a few substances such as water, the molecules are even closer together in the liquid phase. As the temperature is raised, the molecules of a liquid move around more and more rapidly. At any given temperature, some molecules of the liquid acquire enough velocity to escape from the attraction of the other molecules in the liquid and thus become gas molecules.

3–4 THE NATURE AND BEHAVIOR OF GASES

The gaseous form of a substance has a much lower density than the liquid or solid form. For example, the density of liquid water is approximately 1 g/cm^3, whereas that of water vapor (or steam) at $100°C$ is about 0.0006 g/cm^3. The density of a gas is lower because its molecules are spread out in space. A molecule in a gas bounces around, colliding with other molecules or the walls of the container. Collisions of molecules with the walls of the container give rise to the pressure of the gas.

The average velocity of molecules in a gas increases with temperature, being about 1000 mph for water molecules at room temperature. As molecules acquire higher velocities with increasing temperature, they collide harder with the container walls; i.e., the gas exerts a higher pressure. If a given quantity of gas is enclosed in a container of fixed volume, the pressure of the gas is proportional (to a very good approximation) to its temperature (see Fig. 3–3).

On the other hand, if the gas is placed in a cylinder beneath a piston that exerts a constant pressure, one finds that the volume of the gas is proportional to its temperature. As you probably know from experience, heating air causes it to expand. Warm air rises because it is less dense than surrounding cool air. If the temperature is held constant, the volume occupied by the gas is inversely proportional to the pressure applied via the piston—that is, the greater the pressure, the smaller the volume.

All of the above observations are summarized in the *ideal gas law*, a law which no real gas obeys exactly, but which all gases approach at low pressures and high temperatures:

$$pV = nRT, \qquad (3–1)$$

in which p, V and T are the pressure, volume and absolute (kelvin) temperature of the gas, n is the number of *moles* (or gram-atoms of atomic species) of gas present. The quantity R is a universal gas constant whose numerical value depends on the units used for pressure, volume, and temperature. If one measures volume in liters at standard atmospheric pressure, the appropriate value of R is 0.082 liter-atmosphere/mole K. Thus at 0°C (273 K), the volume occupied by one mole of any gas (e.g., 28 g of N_2, 32 g of O_2) at atmospheric pressure can be calculated from Eq. (3–1) as

$$V = \frac{nRT}{P}$$

$$= \frac{1 \text{ mole} \times 0.082 \text{ l-atm/mole K} \times 273 \text{ K}}{1 \text{ atm}} \qquad (3–2)$$

$$= 22.4 \text{ l.}$$

(A 22.4-liter cube has edges that are about 11 inches long.) The ideal gas law indicates that one mole or gram-atom of any gas occupies about the same volume under a given set of conditions, regardless of its atomic or molecular weight. Since a mole or gram-atom of any pure substance contains the same number of molecules or atoms, this means that the average volume occupied by any molecule in the gaseous state is about the same.

At room temperature, about 25°C, the volume of one mole of gas expands to 24.4 liters. Earth's atmosphere consists of about 78% N_2, 21% O_2, and 1% Ar, with molecular weights 28.01, 32.00 and 39.95, respectively, giving it an

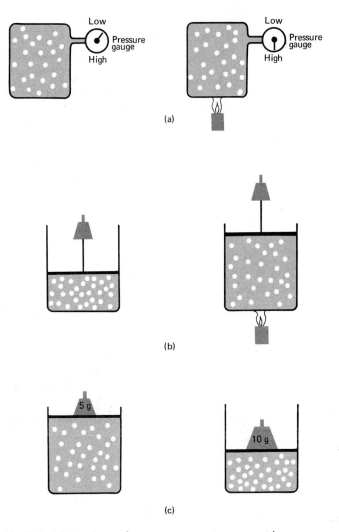

Fig. 3–3   (a) At a given volume, gas pressure increases with temperature. (b) At a given pressure, gas volume increases with temperature. (c) At a constant temperature, the volume of a gas decreases with higher pressure.

"average" molecular weight of 28.9. Therefore, one "mole" of air weighs 28.9 g and occupies about 24.4 l volume, giving air a density of approximately 28.9 g/24.4 l = 1.2 g/l at room temperature.

The weight of the air, from the outer reaches of Earth's atmosphere down to ground level, creates the atmospheric

Fig. 3–4   (a) Atmospheric pressure on a surface is equal to the weight of a column of air extending from the earth's surface to the upper ranges of the atmosphere. (b) Air pressure holds up a column of mercury 760 mm high (30 in.).

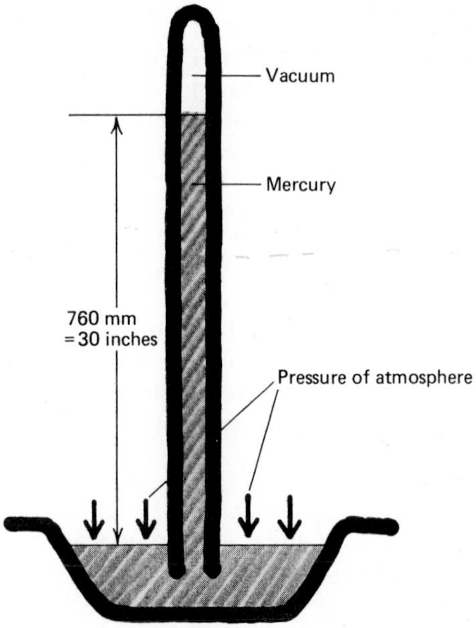

Vacuum

Mercury

760 mm
= 30 inches

Pressure of atmosphere

(b)

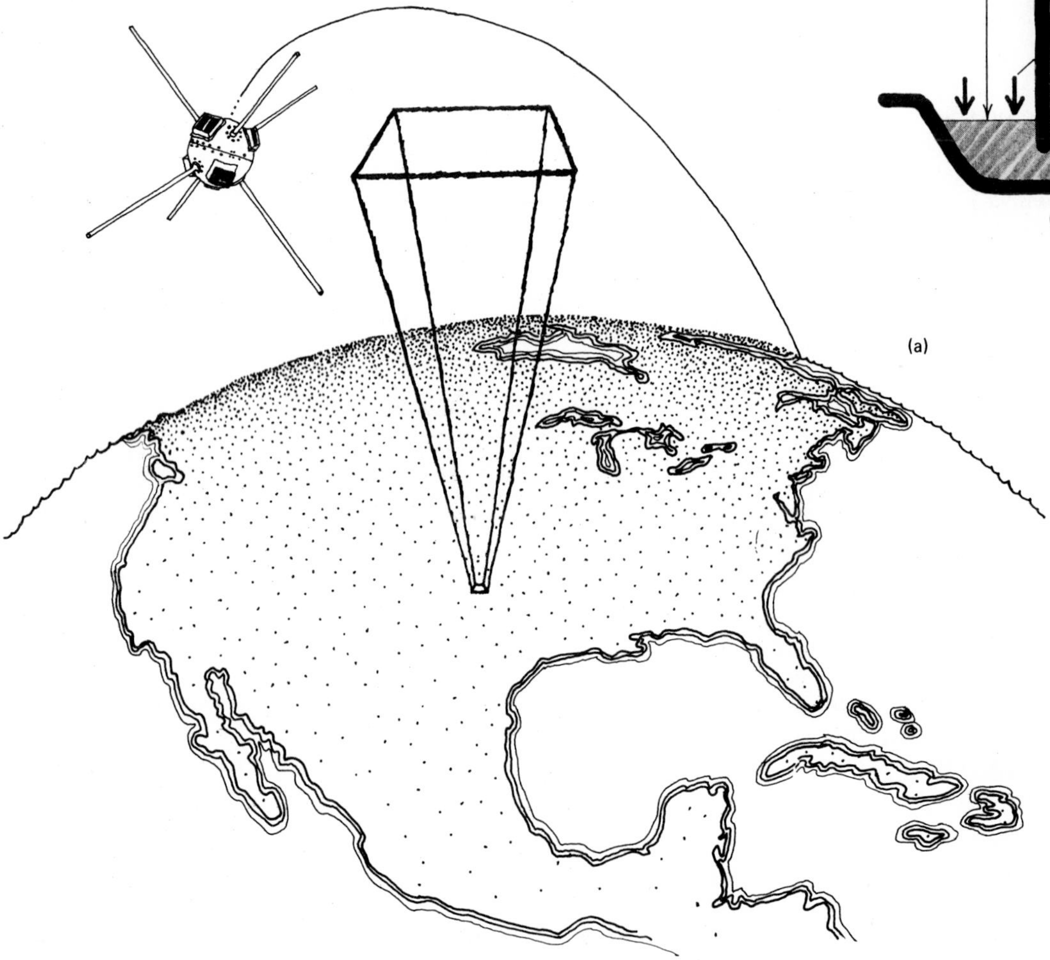

(a)

pressure. The force exerted by atmospheric pressure on an area of surface at the earth's crust is equal to the weight of the column of air resting on that area. Normal atmospheric pressure at sea level is about 1.03 kg/cm²—the weight of a column of air 1 cm square extending from the surface of the earth to the "top" of the atmosphere (see Fig. 3–4). In British units, atmospheric pressure is 14.7 lb/in².

If one dips a long piece of glass tubing into a vat of mercury and pumps all the air out of the top of the tube, the pressure of the air on the surface of the mercury pushes mercury up into the tube to a height of about 760 mm, or 30 inches.* The height of the column of mercury produced in a barometer is commonly used as a measure of the pressure of a gas. Thus, normal atmospheric pressure at sea level is called *760 mm of mercury, or 30 inches of mercury, or one atmosphere.* In weather reports on radio or TV, one frequently hears such statements as "the barometer is reading 29.85," meaning that the pressure of the local atmosphere is 29.85 inches of mercury. Barometric pressure and its direction of change are important indicators of climatic conditions and impending weather. Low pressures are often associated with warm weather and storms, high pressures with clear skies and cold air.

At very high altitudes, most of the mass of the atmosphere is below, so that atmospheric pressure becomes quite small. In Fig. 3–5 the pressure of the atmosphere is shown as a function of altitude. We see, for example, that atmospheric pressure drops to about one-fifth an atmosphere (150 mm Hg) at 35,000 ft (12.6 km) where commercial jet planes often fly.

At any given temperature, some of the molecules (or atoms) of a liquid acquire sufficient velocity to escape from the liquid and become part of the gas phase. The number of molecules with velocity sufficient to enter the gas phase increases with the temperature of the liquid. The rate at which molecules enter the gas phase and the pressure exerted by the gas, *the vapor pressure*, also increase with temperature of the liquid. We call the process *evaporation* or *vaporization*. In Fig. 3–6 we show the pressure of water vapor as a

---

* Note that a one-square-inch column of mercury (with a density of 13.54 g/cm³ or 0.49 lb/in³) that is 30 in. tall weighs 14.7 lb. Also, the atmosphere can support a column of water that is about 30 ft tall. The reason for the great difference between water and mercury is that the density of mercury is approximately 13.5 times that of water.

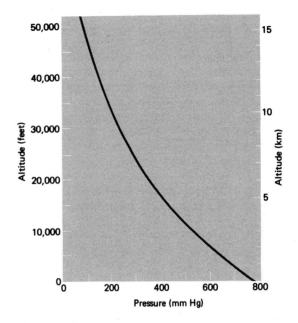

Fig. 3–5 Atmospheric pressure is much greater at sea level than at high altitudes.

function of temperature. When the temperature of a liquid is so high that the vapor pressure is equal to the pressure of the air (760 mm Hg at sea level), the liquid boils. In the case of water, boiling occurs at 100°C (212°F) at sea level.

Just like melting, the boiling of a liquid is an equilibrium process. Water in a pot on the stove will remain at 100°C as long as boiling water is present. Only after all the water in the pot has boiled away can the temperature of the pot go above 100°C. When you boil an egg you depend on the fact that the water keeps the egg at a specified temperature, the boiling point of the water, throughout the process. That temperature is 100°C near sea level. But if we go up on a high mountain, e.g., around 14,000 ft (4.3 km) on Pike's Peak in Colorado, the normal pressure of the air becomes much less, about 60% of its value at sea level. Up there the vapor pressure of water becomes equal to the air pressure and water boils at a temperature of about 86°C.

As in cooking an egg, the rate at which most chemical reactions take place decreases markedly with decreasing temperature. In fact, the rates of most chemical reactions de-

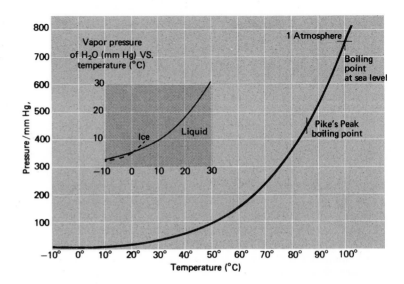

Fig. 3–6 The vapor pressure of solid and liquid water increases as the temperature rises.

crease by about a factor of two for each 10°C decrease in temperature. Thus on Pike's Peak we would need to boil the egg much longer than the usual three or eight minutes to make a soft- or hard-boiled egg. Alternatively, we can cook foods much more quickly by boiling them in a pressure cooker—a pressure vessel which can be adjusted to maintain pressures greater than that of the atmosphere, thus raising the boiling point of water.

There are several ways to interpret the vapor pressure curve of water. For example, at 20°C (68°F), the vapor pressure of water is 17 mm of mercury. If we were to put some liquid water into a vacuum chamber maintained at 20°C, water would evaporate into the vacuum until the pressure became equal to 17 mm Hg. At that point there would be equilibrium—water molecules would leave the liquid at the same rate at which they were being condensed out of the gas into the liquid phase.

The curve of Fig. 3–6 has important implications for meteorology and human comfort. The curve shows the capacity of the atmosphere to contain water as a function of temperature; for example, at 20°C the air can hold a maximum of 16.6 g of H_2O per cubic meter, giving a pressure of 17 mm Hg. The term *relative humidity*, given in weather reports, is the ratio of the actual amount of water vapor in the air to the maximum amount that could be held in the air at the particular temperature. If at 20°C the air actually contains

9.9 g of H_2O per cubic meter, a water vapor pressure of 10.2 mm Hg,* then the relative humidity (R.H.) would be

$$R.H. = \frac{\text{actual water vapor pressure}}{\text{max. water vapor pressure}}$$

$$= \frac{10.2 \text{ mm}}{17 \text{ mm}} = 60\%. \qquad (3-3)$$

Human comfort depends on both the temperature and the relative humidity of the atmosphere. Often in winter time, when there is not much liquid water to evaporate into the atmosphere, the relative humidity may drop to 30% or less. When cold outside air is brought inside and warmed, its relative humidity drops even lower, because the capacity of the air to contain water vapor increases with temperature. When the humidity is quite low, we say the air is "dry," having a large capacity to absorb water vapor evaporated from the surroundings. Dry air in a home tends to dry out mouth and nasal membranes, causing irritation. At low humidities the usual trapping capabilities of the nasal membranes may be decreased so that viruses can enter the lungs, resulting in a

_____

*   The pressure of the water vapor is strictly called the "partial" pressure of water in the atmosphere. The observed barometric pressure is the sum of the pressures of the other gases (N_2, O_2, Ar, etc.) plus the pressure of the water vapor.

higher number of colds and similar ailments for people living in dry environments. The wood and glue in furniture and woodwork may dry out and shrink. Under very dry conditions, even shuffling across a rug can build up charges of static electricity that are discharged when you touch a metal object or another person. Installing a humidifier in the heating system can alleviate these conditions. The humidifier adds water vapor to the hot air coming out of the furnace.

Unless you live in the West or Southwest, you are probably more annoyed by the discomfort of high relative humidity, a "muggy" feeling in the air. A large amount of heat is needed to evaporate H_2O molecules and place them into the gas phase. When water evaporates, the required heat is extracted from the surface from which the water evaporates, thus cooling the surface. We see this principle at work in the evaporative cooling towers on power plants. The same cooling effect takes place when our bodies perspire: when the perspiration evaporates, the skin is cooled. On a hot summer day in a humid area, the relative humidity may reach 80 to 90%. The air contains nearly as much water as it can hold, and so evaporation is very slow. Then the perspiration stays on the skin, making one uncomfortable instead of cool. In Chapter 9, we further discuss water in the atmosphere and some of its climatic effects.

Solids as well as liquids evaporate. The evaporation of a solid is referred to by the special name *sublimation*. Figure 3–6 shows the sublimation pressure of ice as a function of temperature. The curve indicates that in winter, when it is too cold for ice to melt, ice can sublime into the atmosphere without first becoming liquid. Another solid that sublimes is dry ice—frozen CO_2. At atmospheric pressure it goes directly from solid to gas form without going through the liquid phase—hence the name "dry ice."

3–5 SOLUTIONS

In addition to the simple phases of pure compounds and elements, i.e., the solid, liquid, and gas phases, we will also be concerned with more complex phases containing more than one component. The most important of these in laboratory chemistry are solutions in which one or more components, the *solutes*, are dissolved in another component, the *solvent*. When a non-ionic (or "molecular") solute is dissolved in a liquid solvent, e.g., sugar ($C_{12}H_{22}O_{11}$) in water or ethyl alcohol (C_2H_5OH) in benzene, individual molecules

of the solute compound move around randomly in the solvent. When ionic substances are dissolved, e.g., NaCl in water, the positive and negative ions move about in the solvent, usually surrounded by a sheath of solvent molecules.

A true solution is homogeneous; that is, if examined with the best optical microscopes, it appears completely uniform throughout. Or if we chemically analyze portions of the solution from various places in the container, we find the chemical composition to be the same throughout. Since the dissolved species are about the same size as the solvent molecules, we usually can't filter out the dissolved material except with certain kinds of biological membranes.

In the laboratory, many chemical studies are performed with solutions, usually with water as the solvent; we designate such a solution an *aqueous solution*. In many of the chemical reactions written in this book, we shall designate the physical states of reactants and products with the symbols s, l, g, and aq, meaning solid, liquid, gas, and aqueous solution, respectively.

When aqueous solutions of chemicals are used, it is usually important to know the *concentration* of the species, that is, the amount of dissolved material in a given volume of solution. The concentration can be stated in a variety of ways, e.g., grams of solute per liter of solution or kilogram of solvent. The most common unit for chemists is *molarity M, the number of moles of solute per liter of solution* (*not* per liter of solvent). A one-half *molar* aqueous solution, 0.5 *M*, of NaCl for example, contains one-half mole (29.22 g) of sodium chloride per liter of water solution.

When very small amounts of the dissolved species are present, it may be more convenient to work with parts per thousand (‰) or parts per million (ppm), i.e., the number of parts by weight of the dissolved species per thousand or million parts by weight of solution. These units are generally used to quote the concentrations of species present in sea water. For example, sodium ions have a normal sea-water concentration of 10.6‰, meaning 10.6 g of Na^+ per kg of sea water. The minor constituent Rb^+ (rubidium) has a concentration of 0.2 ppm, that is, 0.2 μg per gram of sea water (see also Appendix D).

Most solutes can dissolve in a given solvent only up to some maximum concentration, called its *solubility*, at which point the solution will have become *saturated*. Solubilities of most compounds in water increase with temperature. If one saturates a solution at high temperature and then cools the solution, some of the solute *precipitates* from the cooler

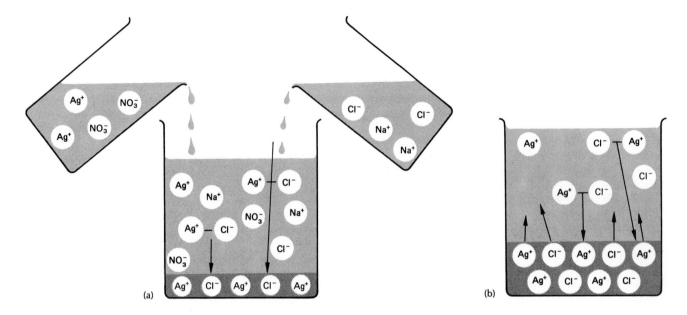

Fig. 3–7   (a) Precipitation of silver chloride by mixing solutions of AgNO₃ and NaCl. (b) Equilibrium of AgCl precipitate with water solution.

solution. If the solution is quite pure and its container is clean, one may be able to cool the solution very slowly without precipitation of the solute in excess of the solubility. Such a *supersaturated* solution is unstable; the excess solute usually precipitates quickly if we scratch the side of the beaker with a stirring rod or add a seed crystal of the solute to the solution.

Solubilities of compounds in water vary over wide ranges. At room temperature the solubility of NaCl is 6.2 M (360 g/l), whereas that of mercuric sulfide is 4.3×10^{-8} M or about 0.01 ppm (suggesting that one could remove Hg^{2+} ions from industrial waste water by adding sulfide ions, S^{2-}). Certain pairs of ions cannot exist together at high concentrations in water because they form very insoluble compounds. For example, both silver nitrate ($AgNO_3$) and sodium chloride are quite soluble in water. But if NaCl is mixed with an $AgNO_3$ solution, the Ag^+ ions are strongly attracted to the Cl^- ions to form silver chloride precipitate, which is white, crystalline, and almost insoluble:

$$Ag^+ (aq) + Cl^- (aq) \rightleftharpoons AgCl (s). \qquad (3-4)$$

Equation 3–4 is written with arrows going both directions because this is a *reversible*, *equilibrium* reaction. When insoluble AgCl is in contact with a solution containing Ag^+ and Cl^- ions, some of the ions in solution occasionally come together and form additional AgCl precipitate. Simultaneously, some of the ions of the precipitate go back into the solution. At equilibrium, the rates of the forward and backward reactions are equal, so that the amounts of precipitate and dissolved ions remain constant with time despite the exchange of ions between the solid and the solution (see Fig. 3–7).

Gases also dissolve in solvents to form solutions. The amount of oxygen dissolved in the water of a river or lake is critical in determining the ability of the water to support various types of aquatic life. When water is in good contact with the atmosphere, oxygen dissolves readily until the solution becomes saturated at about 0.01 g O_2/l (10 ppm) at 20°C. A major pollution problem in water bodies is that decomposition of biological materials often uses up most of the dissolved oxygen in bottom waters, making them uninhabitable by desirable forms of life (see Chapter 11). The oxygen can be replaced by contact between the water and the air, but there is often poor mixing between bottom waters and the surface. The solubility of most gases, including oxygen, decreases at higher temperature. Power plants that dump waste heat into bodies of water cause thermal pollution that increases water temperature and decreases the solubility of oxygen (see Section 4–5).

3–6 COLLOIDS, SUSPENSIONS AND AEROSOLS

In addition to forming true solutions, liquids and gases can use physical methods to contain other materials. For example, water can contain *colloids*, gelatinous clusters of hundreds of molecules, or suspended sediments, very small bits of soil or rock fragments. Unlike the dissolved material in a true solution, the particles of a suspension or colloid are often large enough to scatter light, causing the solution to appear cloudy. The particles stay suspended because of currents in the solvent or because of affinity of the particles for solvent molecules. If allowed to stand in a container, the suspended particles often settle to the bottom and can be removed by filtration.

The gases present in the atmosphere, including such pollutant gases as sulfur dioxide (SO_2) and carbon monoxide (CO), form a true gaseous solution. However, the atmosphere also contains immense numbers of suspended particles, called *aerosols*. Since most pollutant gases are colorless, the hazy appearance of polluted air is caused by suspended particles that scatter light. The aerosol particles range in size from radii less than 10^{-6} cm to about 10^{-3} cm. The larger particles settle to earth within hours or days, but the smaller ones may be kept aloft by air currents for months. Aerosols range in type from salty droplets of sea water to bits of tarry material and rock fragments released in the combustion of coal. Hair sprays and deodorants dispensed from aerosol cans consist of small liquid droplets or solid particles that are forced out of the can and dispensed into the air by a gaseous propellant (generally a Freon gas, for example, $C_2Cl_2F_2$) contained in the can under high pressure.

3–7 SUBSTANCES IN NATURE

In Chapter 2 we saw that pure compounds are made up of fixed proportions of elements. Pure sodium chloride consists of 0.649 g of sodium for every 1.00 g of chlorine. If we tried to vary the proportions by adding more sodium, we would obtain

the same amount of sodium chloride and the extra sodium would be left over unused. Furthermore, pure elements and compounds have fixed properties under specified conditions of temperature and/or pressure. That is, regardless of their origins, pure compounds have characteristic melting points, boiling points, densities, colors, etc. For example, pure water always melts at 0°C and boils at 100°C (at a pressure of one atmosphere) and has a density of 1 g/cm^3 at 4°C.

Most substances with which we are familiar in nature or in the household are mixtures of many compounds. Often we can see by inspection that the substance is not homogeneous, e.g., a leaf, a fragment of rock, a piece of wood do not appear uniform. In other cases classification is more subtle—to the naked eye, butter, cloth, brown sugar, and many other materials may appear to be homogeneous. However, if we examine these substances more carefully we find that their properties are not those of a pure substance. They may melt over a wide range of temperatures, their densities, colors, and other properties depend on the amounts of various ingredients present in the mixture. In a mixture, we can vary the proportions of the ingredients almost at will to achieve desired properties. For example, special types of steels are made by adding small amounts of other elements, e.g., carbon, chromium, or vanadium, to the basic iron. These steels have the properties needed for certain applications.

There are a few substances with which we have frequent contact that are nearly pure compounds, such as water, table salt, white sugar, rubbing alcohol, and baking soda. Since most elements tend to react with other elements, especially with oxygen in the atmosphere, not many elements exist pure in our daily surroundings. A few that do are: gold, silver, sulfur, carbon (graphite or diamonds), aluminum, copper, tin, iron, mercury, nickel, lead, and zinc. Most of these elements have been made pure by man's efforts. Of the elements mentioned, only carbon, sulfur, gold, and occasionally iron (usually in the form of iron meteorites) occur in almost pure form in nature.

In regard to purity, we should note that *no material is absolutely pure*. If we have sufficiently sensitive analytical techniques, we can detect impurity atoms of all naturally occurring elements in samples of so-called "pure" materials. A bottle labeled "NaCl" on the laboratory shelf will contain a fraction of 1% of KCl and smaller amounts of other elements. Tap water contains several ppm of such ions as Ca^{2+}, Mg^{2+}, Na^+, Cl^-, and CO_3^{2-}. Distillation of the water, i.e., boiling it and then condensing the steam in another, cooler container, removes major fractions of each impurity;

but with highly sensitive methods, one can still detect re-maining impurities.

Over the past several years, techniques for detecting and measuring chemical impurities have been vastly improved. These improvements have been extremely important in making us aware of certain subtle problems of the environ-ment and in providing tools for their observation and study. Until we could detect small amounts of DDT residues in birds and fish, mercury in water and fish, and lead in the atmos-phere and human tissue, we could not gauge the scope of those problems.

The quest for superior analytical methods continues to-day. More sensitive and convenient methods are especially needed by doctors and clinical chemists for diagnosing un-familiar diseases caused by toxic substances introduced into the environment by man's activities. Many children un-doubtedly died from ingestion of paint chips containing lead, before the toxicity of small quantities of lead was known and adequate analytical methods were available. When we are able to identify and measure trace quantities of many other elements and compounds in biological tissue, we may find evidence for toxicity of many substances of which we are unaware today. Simple, standard analytical methods for known pollutants are needed to write and enforce clean air and water standards. Laws restricting effluent discharges can-not be written until there are adequate testing procedures for enforcing the law. Thus the improvement of methods for chemical analysis is a key factor in enhancing our awareness of environmental problems and in helping us solve them.

SUGGESTED READING

K. J. Laidler and M. H. Ford-Smith, *The Chemical Elements* (Bogden and Quigley, Tarrytown-on-Hudson, N.Y., 1970). Paperback.

B. H. Mahan, *University Chemistry*, 2nd ed. (Addison-Wesley, Reading, Mass., 1969).

Linus Pauling, *General Chemistry*, 3rd ed. (W. H. Freeman, San Francisco, 1970).

PROBLEMS AND QUESTIONS

1. Briefly distinguish between members of each of the following pairs of terms.
   a) element and compound   b) compound and mixture
   c) solution and suspension   d) atom and molecule

2. Briefly define the following terms.
   a) equilibrium   b) freezing point
   c) thermal expansion   d) molarity of an aqueous
                                      solution
   e) density   f) atmospheric pressure
   g) vapor pressure   h) boiling point
   i sublimation   j) relative humidity
   k) ideal gas

3. Suppose you are given a glass that is filled to the brim with water, with part of an ice cube sticking out above the water. What will happen when the ice melts? Will the water overflow? Will it remain at the brim? Or will the water level fall below the brim? Explain.

4. a) Why are you able to see your breath on a cold day?
   b) Why do car windows fog over on a cold day when several people are inside?
   c) Why does frost gradually build up inside a food freezer?

5. In the following list of materials, indicate which are essentially pure elements, which are pure compounds, and which are neither:
   a) pepper   b) cream of tartar
   c) "lead" in a lead pencil d) milk
   e) air   f) baking soda
   g) permanent antifreeze   h) diamond
   i) glass

   Note: Do not be surprised if you miss some of these. The question is designed to ascertain your chemical back-ground at this stage.

6. Refer to Problem 5 of Chapter 2.
   a) If the density of gold is 19.3 g/cm³, how much volume does the 5-g sample occupy?

b) What is the volume occupied by a single atom of gold?

7. Suppose that a sample of ideal gas occupies a volume of 10 l at 0°C at a pressure of one atmosphere. What is the volume of the gas if

a) temperature is held constant and the pressure is raised to 3 atm?

b) pressure of the gas is held constant and the temperature is raised to 150°C?

c) the pressure is reduced to 0.2 atm and the temperature is lowered to −40°C?

d) What would be the volume occupied by one mole of an ideal gas under the conditions of part (c)?

8. What is the molecular weight of a gas if 500 ml of the gas has a mass of 0.210 g at standard temperature and pressure (STP)?

9. The density of metallic nickel ($_{28}$Ni) is 8.9 g/cm³.

a) How many grams of water would be displaced by a 35-g piece of nickel?

b) How many grams of mercury ($_{80}$Hg, density of 13.6 g/cm³) would be displaced by this same piece of nickel?

c) Does nickel float in water? in mercury?

d) Repeat parts (a) through (c) for a 50-g sample of tantalum ($_{73}$Ta, density 16.6 g/cm³).

10. Calculate the molarity M of each of the following solutions:

a) 10 g of NaCl dissolved in enough water to make 200 ml of solution.

b) 3 g of sucrose ($C_{12}H_{22}O_{11}$) dissolved in enough water to make 500 ml of solution.

11. Suppose the air temperature is 32°C and the relative humidity is 40%.

a) What is the pressure of water vapor in the atmosphere?

b) What weight of H_2O is contained in one liter of air under those conditions?

c) What is the dew point for the air, i.e., at what lower temperature would the relative humidity become 100%? (Ignore the small pressure drop of the air and water vapor that would result from cooling the air to that temperature.)

12. The air pollutant SO_2 can be produced by the reaction of elemental sulfur with oxygen.

a) Balance the equation below:

$$S + O_2 \longrightarrow SO_2.$$

b) If 16 g of sulfur react completely with oxygen, how many moles of S, O_2, and SO_2 are involved?

c) How many grams of O_2 are needed for the complete conversion of the sulfur to sulfur dioxide?

d) At standard temperature and pressure, what volume of SO_2 is produced?

13. Why does it take longer to boil an egg on a high mountain than at sea level?

4
ENERGY

Electricity, the final form of much of the energy that we need to
sustain our way of life, is carried through transmission lines like these.
(Photo by De Wys, Inc.)

4-1 THE IMPORTANCE OF ENERGY

Energy is usually considered the major subject of the field of physics. However, the transfer of energy is such an important factor in chemical transformations, including biochemical reactions in plants, animals, and man, that our discussions of chemistry would be incomplete without considering the energies involved.

From a practical standpoint, energy is a basic commodity needed by man and society. For his very survival, man needs energy in the form of food and heat. The developing nations of the world desperately need supplies of cheap, abundant energy in order to improve the quality of the lives of their citizens. Energy is needed to make fertilizer and fresh water and to propel trucks and tractors in order to produce food. Much of the burden of tedious human labor can be lifted by the use of other forms of energy. Vast supplies of energy are needed to run a modern industrial nation—to process ores and manufacture products, to generate electricity, to move airplanes, automobiles, trucks, and trains.

Some of our sources of energy appear to be running out. In the United States, we are already encountering shortages of electricity, natural gas, and gasoline during periods of peak usage. If we are to keep the industrial nations running and improve the lot of persons in the developing nations, we must expand the world supply of available energy.

On the other side of the coin, the production* and use of energy cause some of the world's most serious environmental problems: air pollution from the combustion of coal, oil and gasoline; land laid waste by strip mining; acidic waters draining from mines; oil slicks on water; and thermal pollution of lakes and rivers, to note some of the most serious forms of environmental degradation related to the use of energy. Thus we must not only find new sources of energy, but must insure that they cause minimum damage to the environment. In the meantime, we must find methods of cleaning up or replacing our present sources of energy.

4-2 FORMS OF ENERGY

The word "energy" is commonly used to denote the urge to get up and move around, to engage in athletics or perform physical work. Many physics texts give a formal definition of energy as "the ability to do work," but this does not really tell us much about what energy is.

There are two basic kinds of energy—*kinetic* and *potential*. Kinetic energy is energy "in action"—as possessed by a train running down the track, a car moving along the highway, or a flywheel revolving. The equation for the kinetic energy of a moving object is:†

$$E_k = \tfrac{1}{2}mv^2. \qquad (4-1)$$

In words, we could say that kinetic energy E_k is equal to one-half of the product of mass m multiplied by the square of velocity v^2. If the mass m is in grams and the velocity v is in cm/s, the energy is in *ergs*. The erg is a very small unit of energy—about the amount possessed by an aspirin tablet (about 1 g) moving along at 1 cm/s. A two-ton car moving down the highway at 60 mph has a kinetic energy of about 6.5×10^{12} ergs.

The other basic kind of energy is *potential* energy—energy that is not presently acting, but is in a form that can be converted to kinetic energy. The simplest example of potential energy is that of a boulder high up on the side of a mountain. It possesses considerable potential energy, which we can easily demonstrate by giving it a little push. The boulder will start rolling down the mountain, picking up speed as it is accelerated by the earth's gravity. As it goes down the mountain, picking up speed, it loses potential energy but gains kinetic energy. At the bottom of the mountain, all its potential energy has been converted to kinetic energy (see Fig. 4-1).

We can also write an equation for the potential energy of an object at height h above the ground:

$$E_p = mgh. \qquad (4-2)$$

The energy is again in ergs if the mass m is in grams, the height h is in cm, and g, the acceleration due to gravity, is expressed in cm/s². We cannot give a universal value for g because it depends on how far we are from the center of the earth (i.e., on the altitude above sea level) and on "local" variations—density of rocks underlying the area, presence of large mountains nearby, and so on. The average value of g is about 980 cm/s² at sea level on the earth.

---

*   Here, of course, we are using the word "production" in its industrial sense, because we cannot really "produce" energy; we can only transform it from one form to another.

†   This equation is not valid if the object is moving with a velocity close to the speed of light; in that case one must take into account an increase in the mass of the object due to relativistic effects (see Section 4-3).

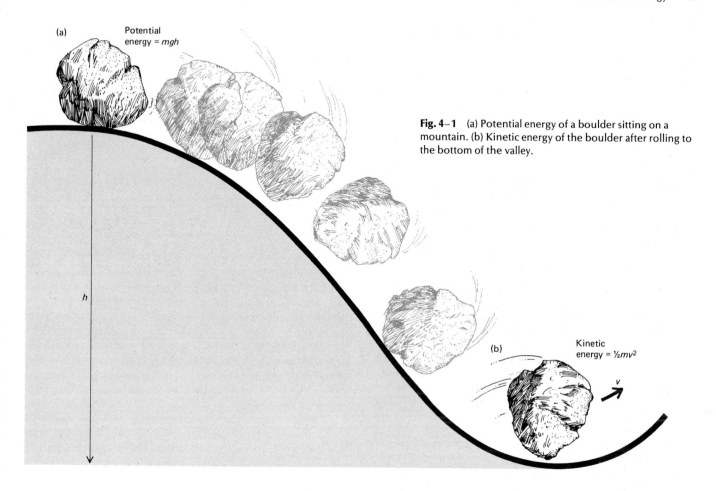

(a) Potential energy = mgh

h

Fig. 4–1   (a) Potential energy of a boulder sitting on a mountain. (b) Kinetic energy of the boulder after rolling to the bottom of the valley.

(b) Kinetic energy = $\frac{1}{2}mv^2$

v

Both kinetic and potential energy have many different forms. In the examples given above—an object in motion or at rest on a mountain—we have illustrated mechanical energy. Other types of energy are electrical, thermal (i.e., heat), chemical, radiant (light), and nuclear energy. One form of energy can be converted into another. Chemical potential energy in the form of lead (Pb), lead peroxide (PbO_2), and sulfuric acid (H_2SO_4) in a storage battery can become electrical energy which can in turn be converted into mechanical energy by passing the current through an electric motor. If we apply brakes (mechanical energy) to the motor, the friction slows the motor down and heats up the brakes, thus converting some of the mechanical energy into heat. The large lake behind a dam represents potential mechanical energy. When water flows down through the pipes to a hydroelectric plant, it acquires kinetic energy which it transfers to the blades of a turbine, causing it to turn. The mechanical energy of the turbine turns the rotor of a generator, producing electrical energy. If we pass the electricity through a light bulb, the energy appears partly as light and partly as heat. We could also use the electricity to charge a lead storage battery, i.e., convert electrical energy into chemical energy.

The energy unit that we are perhaps most familiar with is that used to measure electricity, the *kilowatt-hour* (kWh). The *watt* is a unit of *power* (as is horsepower), and power is energy per unit of time. A 100-watt bulb burning for 1 hr uses 100 watt-hours of energy, or 0.1 kWh. A typical household uses about 100 to 200 kWh of electrical energy each month. When you pay the electricity bill, you are buying energy. A kilowatt-hour is the same amount of energy as 3.6×10^{13} ergs.

Another unit of energy with which many people are familiar is the *calorie*, which is a unit of thermal energy. It is defined as the amount of heat needed to raise the temperature of one gram of liquid water by 1°C. Mechanical energy can be converted to heat. The conversion can be demonstrated, for example, by sliding down a rope. Friction of the hands on the rope converts some of the mechanical energy of the falling weight into heat. A given amount of energy can be added to a known amount of water in a thermally insulated container by operating a mechanical stirrer or electrical heater. The number of calories of heat produced is determined by measuring the temperature change of the water. From experiments of this type, it is known that one calorie of heat represents the same amount of energy as 4.18×10^7 ergs. Relationships among these and other units of energy are summarized in Appendix C. Although mechanical energy can be converted completely to heat, the converse is not true. Heat cannot be converted completely to mechanical energy, a fact we discuss in more detail in connection with the inefficiencies of steam-electric plants (Section 4–5).

Although we generally think of energy in connection with machines, our bodies also require energy. Every biological entity must take in energy to sustain growth and, in the case of humans and most animals, to facilitate muscular movements and maintain body temperatures higher than that of their surroundings. Essentially all the energy that passes through the food chain originates in sunlight. When sunlight falls on green plants, they carry out the process of *photosynthesis*; that is, the energy of sunlight is used to convert carbon dioxide, water, and other nutrients into oxygen and chemical compounds of high energy content. About 0.1% of the total energy from the sun that falls on Earth is converted, by means of photosynthesis, to chemical potential energy.*

Since photosynthesis does not occur in animals or humans, they must eat plants or other animals in order to obtain the energy and chemical nutrients they require. The transfer of energy through the food chain is rather inefficient—only about 10 to 15% of the energy is transferred in each step, e.g., from plants to small herbivores (plant-eating animals), from small animals to larger ones, and from animals to man. Therefore, one could optimize the use of energy from photosynthesis by eating plants directly instead of eating animals several steps removed.

A measure of the energy content of foods is the amount of heat that is released (called the *heat of combustion*) when the food is oxidized (burned) completely to form the oxides of the various elements. For example, the heat of combustion of common white sugar (sucrose) is the heat released in the following reaction:

$$C_{12}H_{22}O_{11}(s) + 12\ O_2(g) \longrightarrow$$

sucrose (solid)   oxygen (gas)

$$12\ CO_2(g) + 11\ H_2O(g), \quad (4\text{–}3)$$

carbon dioxide   water
(gas)   (gas)

which is about 3940 cal/g. This is approximately what the body does in metabolizing food; that is, we breathe in O_2 and exhale CO_2 and H_2O. However, the heat of combustion has to be modified for computing the nutritional energy of many foods because our bodies are not able to completely metabolize (change) them into the fully oxidized end products. (In the case of sugar, the nutritional energy is the same as the heat of combustion.) Nutritional energy is measured in *Calories* (note capital C), each having the same value as one kilocalorie (1000 calories).

The average adult requires about 2400 Cal/day from his food to make up for the energy he uses. If he eats more than he needs, the surplus is stored in his body in the form of fat. If he takes in less energy than he needs, the balance is obtained from the stored fat.

4–3 CONSERVATION OF ENERGY AND MASS

Until about the eighteenth century, men thought that mass could be created and destroyed. It was natural to suppose that when a stick of wood burned, much of its mass was destroyed and converted into heat. Other processes appeared to create matter. However, when careful measurements of the masses of all of the reactants and products were made (including the gases which could not be seen), it was found that mass is *conserved* in chemical reactions. That is, in chemical reactions, matter may change from one form to another, but the total mass remains the same. This fact is called the Law of Conservation of Mass.

---

* G. M. Woodwell, "The Energy Cycle of the Biosphere," *Scientific American* **223** (3), 64–74 (Sept. 1970).

Superficially, it also appears that energy can be created or destroyed—heat is released by burning things, the kinetic energy of a car appears to be destroyed when one applies the brakes, and so on. However, upon examining these processes more carefully, we find that energy is neither created nor destroyed, but simply changed from one form to another. This fact is the Law of Conservation of Energy. The energy released in the burning of wood was originally present as chemical potential energy of the wood and oxygen. The final products, mostly carbon dioxide and water vapor, represent lower states of chemical potential energy. The difference in chemical potential energy between the reactants and products is given off as heat released in the reaction. When one applies the brakes on a moving car, friction transforms the kinetic energy of the car into heat—in the brake drums and linings, in the tires, and on the surface of the road.

For practical purposes in the chemistry laboratory, the energies released per molecule in chemical reactions are small enough that the laws of conservation of energy and conservation of mass can be considered valid. However, nuclear reactions (discussed in Chapter 5), which release about one million times as much energy per atom as chemical reactions, can be understood only in terms of the *theory of relativity* proposed in 1905 by Albert Einstein, probably the outstanding mathematician and theoretical physicist of the first half of the twentieth century. Among other things Einstein's theory states that mass can be converted into energy and energy into mass. Thus in terms of relativity theory, energy and mass are not separately conserved. The separate laws of conservation of energy and of mass can be replaced by a single law of *conservation of energy and mass*; i.e., although energy and mass can be interconverted, the *sum of energy plus mass* in the universe remains constant.

The amount of energy E released when a mass m is converted to energy is given by

$$E = mc^2, \qquad\qquad (4–4)$$

where E is in ergs and m is in grams. The symbol c stands for the velocity of light in vacuum, which has the value 3×10^{10} cm/s. According to Eq. (4–4), the transformation of 1 g of matter to energy releases $(3 \times 10^{10})^2 = 9 \times 10^{20}$ ergs of energy. This is an enormous amount of energy: as shown in Appendix C, it's about the same amount of energy as is released by the burning of 3000 tons of coal! This result illustrates the fact that, in chemical reactions, it is generally safe to assume that energy and mass are separately conserved. According to the theory of relativity the carbon dioxide (CO_2) produced by burning the coal (mostly carbon) would weigh one gram less than the 3000 tons (2.7×10^9 g) of coal and 8000 tons (7×10^9 g) of oxygen consumed. However, in order to observe the mass change, one would have to weigh the coal, oxygen, and CO_2 to an accuracy of better than 1 g in ten billion!

The only processes that transform easily measurable amounts of mass into energy are nuclear reactions. When this is done, one does not take a 1-g sample of material and convert its mass entirely into energy. Instead, one takes a much larger amount of material and converts its atoms into other atoms that weigh slightly less than the original ones. Transformations of this sort provide the enormous energies that are released by nuclear weapons and nuclear reactors (see Chapter 5). In the first nuclear weapon used in warfare at Hiroshima in 1945, 1 kg of ^{235}U atoms were split (fissioned) to form products weighing about 999 g, the enormous explosion being produced by the conversion of about 1 g of mass to energy! Nuclear reactions also provide the vast energies released by stars, including the sun (see Chapter 6).

The examples given above illustrate the conversion of mass to energy, but according to the theory of relativity, energy can also be converted to mass. This phenomenon is most often encountered when scientists accelerate protons or other projectiles to very high velocities in cyclotrons. According to the theory of relativity, when the projectile reaches a velocity comparable to that of light, its mass increases; i.e., some of the energy used to accelerate the particle is converted into mass. For example, high-energy protons (with an energy of 140 million electron volts, MeV) from the University of Maryland cyclotron travel at about half the velocity of light and have masses about 15% greater than protons at rest (the so-called "rest mass" of protons). Protons from the world's highest energy accelerator, the National Accelerator Laboratory at Batavia, Illinois, may ultimately reach energies of 500 GeV (gigaelectron volts = billion electron volts) and masses more than 500 times those of protons at rest.

The conversion of energy to mass is the main reason for building very high-energy accelerators. To obtain information about the fundamental nature of matter, high-energy physicists search for and study the properties of a wide variety of mesons and "strange particles"—particles that exist for only a fraction of a second before disintegrating. These particles have to be made in nuclear reactions in order to be studied during their brief lifetimes. In order to make them, the projectile, usually a proton, must possess an energy at least equivalent to the rest mass of the particle. Thus higher

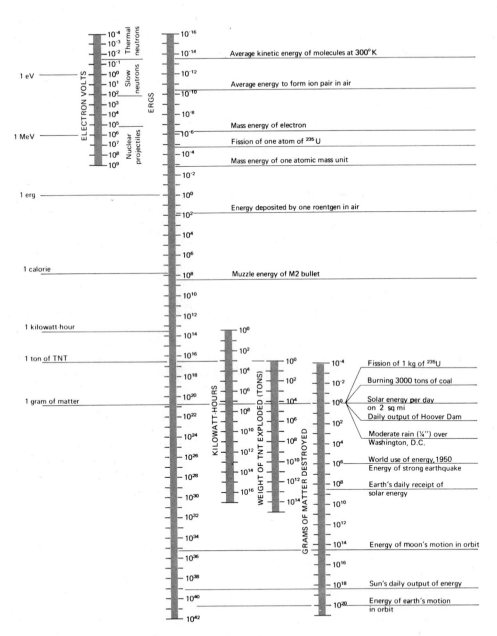

Fig. 4–2 The energy level of things. (Adapted from J. R. Williams, *Missiles and Rockets*, Oct. 3, 1960.)

and higher energy particle accelerators are needed to extend the search to heavier particles.

The relativistic increase of mass apparently occurs with all moving objects but becomes large enough to observe only when the velocity approaches the speed of light. It can be observed only with nuclei or electrons because massive objects are never accelerated to such high energy. We think of rocket ships as traveling very fast, but at 25,000 mph, their velocity is only 10^6 cm/s, thirty thousand times smaller than the velocity of light. According to the theory of relativity, it is impossible for any object to travel exactly at the speed of light, as it would have infinite mass and require an infinite amount of energy to bring it up to that speed. Thus as far as we know, the speed of light (in vacuum) is an upper limit on the velocity of objects.

Although the predictions of the theory of relativity are quite different from our experience with large objects, they are in agreement with the results of all relevant experiments. This does not, however, prove that the theory is correct. Scientists must remain skeptical of all theories and alert to any experimental results that disprove the theory. When that happens a new theory must be constructed to explain the new observations as well as all the older results that were correctly predicted by previous theories. For example, Newton's laws of motion correctly predict the behavior of objects traveling at low speeds, but they break down when applied to particles with extremely high velocities. Einstein's theory of relativity correctly predicts the behavior of high-velocity particles and gives the same predictions as Newton's laws for low velocities.

4–4 SCALES OF ENERGY

The energies involved in natural and man-made processes vary in magnitude from the very small energies of individual molecules in a gas at room temperature (about 10^{-14} erg) to the vast energies of the sun's daily output (10^{39} ergs) and the energy of Earth's motion about the sun (10^{41} ergs). Figure 4–2 compares the magnitudes of energy involved in a variety of familiar processes. The energy of all processes can be expressed in the fundamental unit of ergs. But, for very small-scale processes, such as those involving individual atoms and molecules, it is more convenient to measure energy in the much smaller units of *electron volts* (eV). Suppose we had a one-volt battery (flashlight batteries are about 1.5 volts) hooked up to two flat metal plates as shown in Fig. 4–3. An electron placed near the surface of the negative plate would be repelled by the negative plate and attracted by the positive

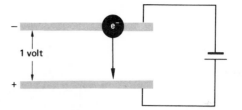

Fig. 4–3 An electron gains one electron volt (eV) of energy by jumping a gap of one volt.

plate. If we performed this experiment in vacuum, so that the electron would not bump into molecules of air, the electron would accelerate toward the positive plate and bang into it. Its kinetic energy just before striking the plate is defined as one *electron volt*. The electron volt is a very small energy unit: the electron would be traveling quite fast, 6×10^7 cm/s, but it has a very small mass, about 9×10^{-28} g. We can relate it to the erg by the conversion factor (see Appendix C).

$$1 \text{ eV} = 1.602 \times 10^{-12} \text{ erg.} \qquad (4-5)$$

As shown in Fig. 4–2, the average kinetic energy of a molecule in a gas at 300° K (27° C, a little higher than room temperature) is about 0.03 eV. The most weakly attached electrons on atoms or molecules can be knocked loose by projectiles having a few eV of energy. The energies that hold atoms together in molecules are typically in the range of a few eV. Thus most chemical transformations involve energies of one to several eV per atom or molecule.

Radioactive decay and other nuclear transformations discussed in Chapter 5, typically involve energies about one million times greater than those of atomic and molecular processes. Thus, for nuclear processes, it is convenient to use million electron volts (MeV) to express energies. Nuclear fission of one atom of uranium, for example, releases about 200 MeV of energy. We can calculate the energy released when one atomic mass unit (one amu is about 1.66×10^{-24} g) is transformed to energy by substitution into the Einstein equation, Eq. (4–4):

$$E = mc^2 = 1.660 \times 10^{-24} \text{ g} \times (2.998 \times 10^{10} \text{ cm/sec})^2$$

$$= 1.492 \times 10^{-3} \text{ erg}$$

$$= \frac{1.492 \times 10^{-3} \text{ erg}}{1.602 \times 10^{-12} \text{ erg/eV}} = 9.315 \times 10^8 \text{ eV} = 931.5 \text{ MeV.}$$

A fossil-fuel burning electric plant emits invisible SO_2 and particles (smoke) into the surrounding air. (Photo courtesy of Billy Davis, Louisville *Courier-Journal*, and EPA.)

Thus in Chapter 5, when we wish to convert the mass change (in amu) of a nuclear transformation into MeV, we can use the conversion factor

$$1 \text{ amu} = 931.5 \text{ MeV} \tag{4-6}$$

Although the energy released in the nuclear transformation of a single atom is much smaller than an erg, so many atoms (e.g., 10^{24} atoms) are typically involved in producing a nuclear explosion or running a nuclear reactor, that the total energy released is quite large.

For very large-scale processes, such as those shown near the bottom of Fig. 4–2, it is appropriate to use such energy units as kilowatt-hours, the weight of TNT (trinitrotoluene) exploded, or grams of matter converted to energy. The size

of nuclear weapons, for example, is typically quoted in terms of the tons of TNT that would have to be set off to release the same energy as the weapon. The conversion factors relating the various units of energy are summarized in Appendix C.

Although we think of such man-made energy sources as large hydroelectric dams and nuclear weapons as being quite vast, we see in Fig. 4–2 that their energies are small compared with those involved in many natural processes. For example, a moderate rainfall over Washington, D.C., releases the same amount of energy as a Hiroshima-size weapon, and a strong earthquake releases as much energy in a minute or two as is used by the world population over an entire year.

4–5 ENERGY AND THE ENVIRONMENT

A nation's per capita use of energy is a rather accurate indicator both of the standard of living of its people and of the potential for damage to the nation's environment. In the United States, we have satisfied most citizens' basic needs for energy and have long been making extensive luxury use of energy—the second car, air conditioners in the home, office, and car, electric ice crushers and can openers, etc. However, the developing nations of the world desperately need energy to improve the quality of the lives of their people: they need it to heat homes, propel trucks and tractors, manufacture fertilizer, refine ores, cast metals, generate electricity, and desalinate sea water to produce fresh water for irrigation and drinking.

Many of the most serious environmental problems of the technological nations result from the use of energy. Every form of energy production, with the possible exception of direct utilization of solar radiation, causes some damage to the surroundings. A large part of urban air pollution is caused by emissions from internal combustion engines. Other forms of urban air pollution, especially particulate material and sulfur dioxide, result from the combustion of coal and low-grade oil in steam electric plants or central heating plants. The use of coal often causes additional forms of environmental degradation, such as the destruction of large land areas and the flow of acidic water into rivers that often accompany the mining of coal. Off-shore pumping of oil and marine transport of it often create large oil slicks on water. Both fossil-fueled and nuclear steam electric plants dump from half to two-thirds of their energy into flowing or standing water as waste heat—the so-called "thermal pollution" that may cause desirable fish to die or leave the area to "trash"

In a search for more coal to supply energy for electric power plants, large areas of the Appalachian region have been laid bare by strip mining techniques. Above, an abandoned strip mine near Coalton, West Virginia. (Photo courtesy of the Bureau of Mines, U.S. Department of the Interior.) Below, strip mine reclamation in Morgan County, Tennessee. (Photo courtesy of the Tennessee Valley Authority.)

fish that can survive at elevated temperatures. Nuclear plants and their associated fuel reprocessing plants are capable of polluting the environment with radioactive atoms of various elements. Fossil-fueled power plants also contaminate the environment with naturally occurring radioactive elements that are released to the environment when the fuel is burned.

Hydroelectric Power

The environmental damage caused by some forms of energy production are rather subtle. It is probably obvious to you that coal- or oil-fired power plants emit large quantities of air pollutants. However, it is probably not so obvious that hydroelectric plants also cause some serious problems. After all, they consist simply of a dam across a river that holds back the water which tumbles through large turbines that power electric generators. It is perhaps surprising to learn that this "clean" form of power generation can cause severe problems in the environment.

One major problem of hydroelectric plants is the enormous weight of the water that fills the lake behind the dam rather quickly after the dam is constructed. The added weight places severe stresses on the geological formations in the vicinity of the lake. These stresses may even cause earthquakes in the area as the lake fills if there are instabilities in the geological formations.* During the filling of Lake Mead behind Hoover Dam in Arizona, many small earthquakes up to magnitude five on the Richter scale were observed.† As the lake behind the Kariba Dam in Zambia, Africa, filled, many earthquakes up to magnitude 6.1 occurred. The most severe earthquakes happened as the lake behind a dam in Kogna, India, filled. The worst of the earthquakes there registered 6.5 on the Richter scale (about the same as that of the major shock in the Los Angeles area in February 1971) and caused about 200 deaths.

Examples of most of the other types of problems caused by dams and hydroelectric plants are very much in evidence in the case of the Aswan High Dam on the Nile River in lower

---

*   G. F. MacDonald, "The Modification of the Planet Earth by Man," *Technology Review* **72** (1), 26–35 (Oct./Nov. 1969).

†   The Richter scale is logarithmic, i.e., an increase of one unit corresponds to a tenfold increase in the magnitude of the earthquake. An earthquake of magnitude 6, for example, causes a ten-fold greater deflection of seismograph recorders than one of magnitude 5.

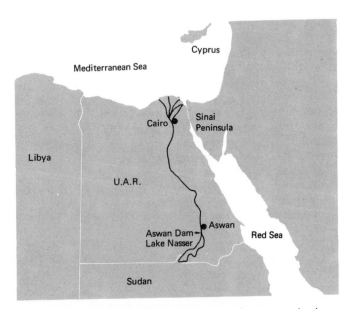

Fig. 4–4 Photo shows the Aswan High Dam under construction in 1964, with the waters of Lake Nasser in the foreground (Wide World Photos). The map shows the location of the dam.

Egypt.* The dam, which was begun in 1960, was intended to control flooding, serve as a reservoir for irrigation of vast areas of previously arid land, and produce electricity (see Fig. 4–4).

The Nile was sealed off in 1964 and, although Lake Nasser behind it is now only about half full, it is already clear that the dam has created many unfortunate problems. The lake was supposed to have filled by 1970, but large amounts of water have been lost via evaporation from the large surface of the lake and seepage of water through permeable strata of the lake bed. The water is badly needed for reclamation of lands now too dry for agriculture, but it now seems doubtful that the lake will fill to much more than half of its designed volume for many years to come. The dam has controlled flooding, but the sediments containing the vital nutrients that once made the flood plains extremely fertile are now trapped behind the dam instead of being deposited on the flood plains. This will make it necessary to apply expensive chemical fertilizers to the plains to maintain their fertility. The decreased flushing action of the fresh-water floods from the Nile has allowed salt deposits to build up on some lands along the river through-out upper and middle Egypt, making them unfit for crops. The decreased flow of nutrients from the Nile into the Mediterranean Sea has drastically reduced the catch of sardines and other fish vital to the economy and food supply of Egypt.

Perhaps the most tragic problem created by the Aswan High Dam is the increase of the parasitic disease *bilharzia* (also called *schistosomiasis*) among the people of Egypt. The still waters behind the dam create an excellent habitat for snails that harbor the larvae of the blood flukes that later attach themselves to humans and enter their bloodstreams. Bilharzia can be fatal, but more commonly it causes a great deal of pain to the afflicted person, weakening him and making him more susceptible to a variety of other diseases. Almost any contact with the infested water, such as working barefoot in irrigated fields, exposes one to bilharzia, for which there is no known, lasting cure. An estimated 35% of Egypt's population suffered from the disease before con-struction of the dam, and the figure will surely rise dramatic-ally in the future. Still water also provides a good breeding ground for insects carrying diseases, such as mosquitoes bear-

---

*   Claire Sterling, "Aswan Dam: Predictions Came True with a Vengeance," *Washington Post*, Feb. 15–24, 1971 (four-part series).

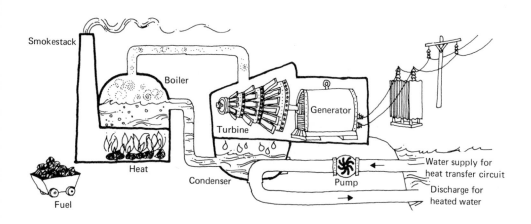

Smokestack

Boiler

Generator

Turbine

Heat

Condenser

Pump

Fuel

Water supply for
heat transfer circuit

Discharge for
heated water

Fig. 4–5 Major components
of a steam electric plant.
[From Joseph Priest,
*Problems of Our Physical
Environment* (Addison-
Wesley, Reading, Mass.,
1973), p. 51.]

ing fatal malaria. As yet this problem has not arisen in Lake
Nasser, but chances are high that it will in the future.

There are other problems caused by dams such as the
filling in of scenic canyons and valleys and the growth of
extensive algae "blooms" and mats of aquatic weeds in the
lakes behind them. Probably no single dam causes all of the
undesirable effects we have mentioned. However, in every
case there are enough serious potential problems that they
should be weighed carefully against the possible benefits
before a decision is made to construct a new dam.

Thermal Pollution

Another form of environmental degradation common to
several types of electric power generation is *thermal pollution*—the dumping of wasted heat into streams of water or
the atmosphere. Steam electric plants produce electricity by
heating water to make steam and allowing the steam to
expand in a turbine. Expansion of the steam against the
blades of the turbine causes the central shaft to rotate; it in
turn is connected to the rotor of a generator that produces
the electricity. The steam condenses as liquid water and is
cooled before it is recycled to the heat source. Most steam
electric plants flow large volumes of water from a river or lake
through long coils of tubing to cool the steam water.

Even if all the equipment works perfectly (that is, completely free of friction), it is impossible to completely convert
a given quantity of heat into mechanical energy (and then to
electricity) unless the steam is condensed to water at absolute
zero (0 K), an impossible achievement in practice. As a con-

sequence of a fundamental law of nature, the Second Law of
Thermodynamics, the theoretical efficiency ϵ of a heat engine
(i.e., the ratio of work produced to heat expended) is given
by

$$\epsilon = \frac{\text{work}}{\text{heat}} = \frac{T_1 - T_2}{T_1}. \tag{4-7}$$

In this equation, T_1 is the *absolute* (or kelvin) temperature of
the gas (e.g., steam) used in the engine, and T_2 is that of the gas
or liquid after expanding (and cooling) in the engine. We see
that, unless $T_2 = 0$ K, the efficiency ϵ is less than one. The
application of Eq. (4–7) to a plant operating with steam at
300°C and cooling water at 20°C is shown schematically in
Fig. 4–6.

Equation (4–7) shows that the efficiency of a plant is
maximized when it uses steam at the highest possible temperature T_1 that can be handled in the plant and uses a coolant
at the lowest possible temperature T_2. These facts explain why
power plants are designed for the maximum practical temperature of steam and why they are usually located near a large
supply of cold water. Since one can never find a coolant in the
environment with a temperature close to absolute zero, a
fraction of the heat, T_2/T_1, is transferred to the coolant instead of being used. If the coolant is water from a stream, the
wasted heat is added to the cooling water before it flows back
to the lake or stream. The wasted heat must go somewhere in
the surroundings, e.g., to a water body or the atmosphere.
The heat cannot simply disappear, because as we have seen
above, energy must be conserved.

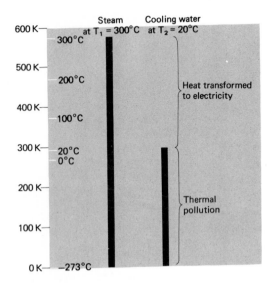

Fig. 4–6 Theoretical efficiency of a steam electric plant.

Looking again at Fig. 4–6, we can calculate the theoretical efficiency of a typical steam electric plant by using the appropriate *absolute* (kelvin) temperatures:

$$\epsilon = \frac{573 - 293}{573} = \frac{280}{573} = 0.49 = 49\%.$$

The actual efficiency of a plant would be somewhat less than this value, say about 35%, because of frictional losses and other inefficiencies. Although one most often hears thermal pollution mentioned in connection with nuclear power plants, *all* steam electric plants obey the Second Law of Thermodynamics (that is, Eq. 4–7). However, fossil-fuel plants operate with higher steam temperatures (about 600°C) than nuclear plants (typically 315°C). Because of the difference in steam temperatures, fossil-fuel plants normally have overall efficiencies of 38 to 40%, as against 30 to 32% for present-day nuclear power plants.* Thus, fossil-fuel plants dump a somewhat smaller fraction of the heat they generate into the surroundings. Nuclear power plants are a much younger application of technology than coal- and oil-fired plants. As experience in the use of nuclear energy is gained, more efficient plants will undoubtedly be designed and operated.

The total amount of waste heat given off by a modern power plant is almost unimaginable. For example, the Baltimore Gas and Electric Company is building two large nuclear units at Calvert Cliffs, Maryland. Each unit will produce about 2450 megawatts of heat, which at 31% efficiency will yield about 800 megawatts of electricity.† Most of the balance of the heat, 1650 megawatts per unit, will be dumped into Chesapeake Bay as thermal pollution. About 420,000 gal/min of Chesapeake Bay water will be drawn into the plant as cooling water, and its temperature will be raised by 10°F before it returns to the bay. The effluent "river" flowing back into the bay will be the fourth largest river feeding the bay, ranking only behind the Susquehanna, Potomac, and James Rivers.‡

The warmed water will mix rather quickly with the other water, but the average surface temperature of the bay will be raised about 3°F over about 23 acres (1/30 of a square mile).§ Compared with the approximately 50°F variation of surface water temperature over the year, the change caused by the nuclear plants will be rather small in this case, but the effects are often much greater for power plants located on smaller bodies of water.

Why should slightly warmed water cause so much concern, especially when one considers that homeowners pay a fair sum of money for water heaters and electricity to obtain warm water? In a sense the warm water from a power plant is a resource out of place. It could be a desirable commodity, but it may have a deleterious effect upon the ecological balance of a stream or lake when dumped there indiscriminately.

An increase of temperature speeds up most chemical reactions. This speed-up occurs in biochemical reactions just as in any other type of reaction. The time sequence for organisms in the water can be greatly changed by the addition of heat. Fish eggs may hatch abnormally early in the season and be ready for growth before the nutrients they need come down the stream, and thus they may starve. Fish are much more active at elevated temperatures and, because of enhanced activity, their life spans are often shortened. Their heightened activity demands greater than normal amounts of oxygen, but there is less oxygen available in the water because of its decreased solubility at high temperatures.

---

* D. E. Abrahamson, *Environmental Cost of Electric Power*, A Scientists' Institute for Public Information Workbook, 1970.

---

† J. W. Foerster, H. Laster, H. Seliger, and O. M. Phillips, "Nuclear Power Plants and Our Environment," A Report to the Maryland Academy of Sciences by the Study Panel on Nuclear Power Plants, Jan. 1970.

‡ E. P. Radford et al., "Statement of Concern," *Environment* **11** (7), 18 (1969).

§ "Nuclear Power Plants in Maryland," Governor's Task Force on Nuclear Power Plants, Annapolis, Md., Dec. 1969.

This infrared photo shows the Indian Point power plant on the Lower Hudson River. Warm water from the plant (shown by the lighter shades in the river) is traveling south with the tide. (Photo by Texas Instruments, Inc., courtesy of Consolidated Edison Co.)

Above a certain temperature, different for each species, fish die.* The reasons for their death are not clear—perhaps the lack of oxygen, strain on the nervous system, or breakdown of critical enzymes in their bodies (see Chapter 14). Usually, the more desirable game fish are the most sensitive to increased temperature. Brown trout, for example, thrive at temperatures between 55 and 65°F and die at about 78°F, whereas "trash" fish such as carp can survive at much higher temperatures. If the temperature of water where desirable fish spawn or mature is raised, they may move to other areas of the water body that have the right temperature. However, at the new location, other conditions such as salinity or availability of nutrients may not be adequate, so that species may completely disappear from the water body, leaving only undesirable species.

Besides heating the water, power plants frequently add chlorine (Cl_2) or other biocides to the intake water to prevent organisms from clogging the condenser coils. The biocides kill many of the organisms in the water passing through the plant and near the exit channel of the cooling water. Also, corrosion of the coils causes metals such as copper to pollute the effluent water. In some regions of Chesapeake Bay, the copper concentration in the water is so high that oysters and clams are contaminated with it to the point that they are unfit for human consumption. Thus some of the effects that

* J. R. Clark, "Thermal Pollution and Aquatic Life," *Scientific American* **220** (3), 19–27 (March 1969).

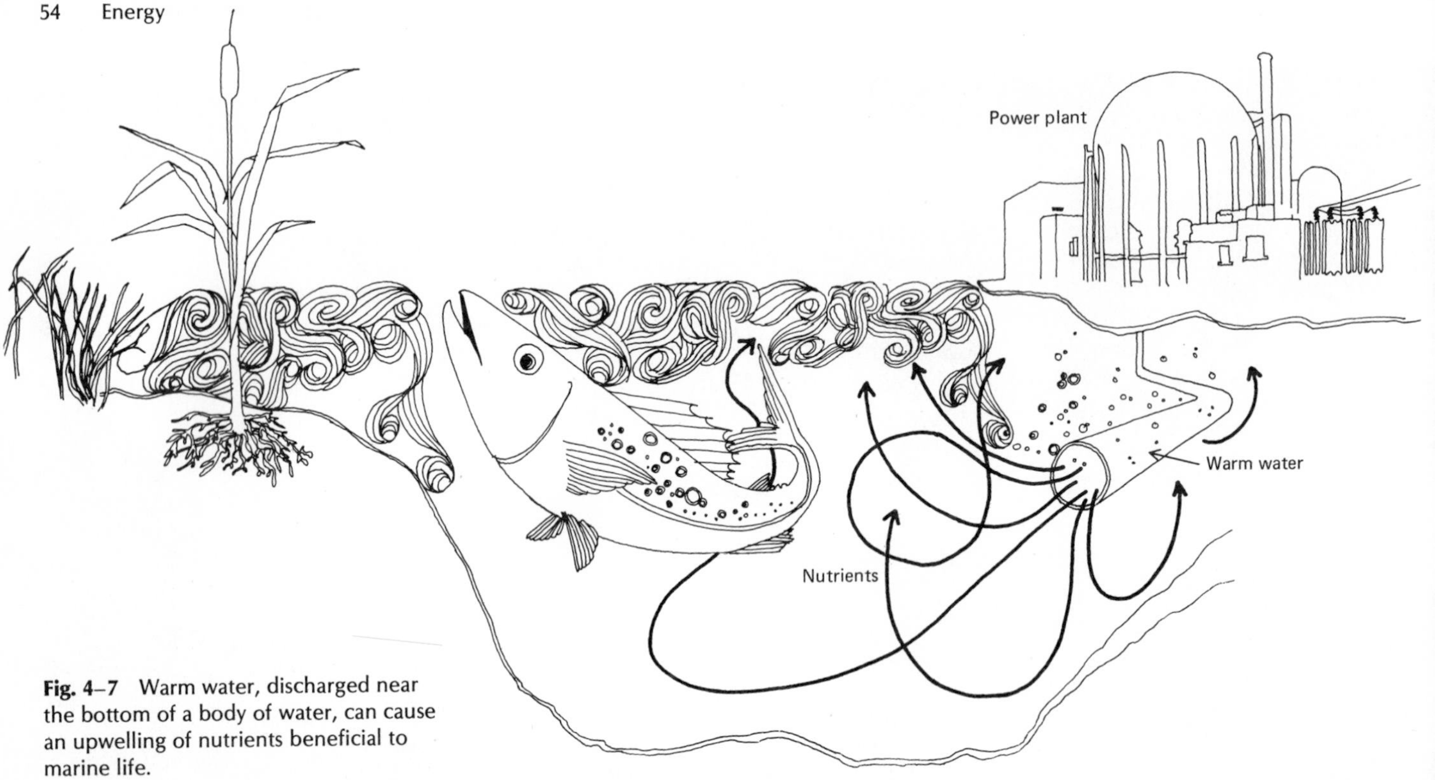

Fig. 4–7 Warm water, discharged near the bottom of a body of water, can cause an upwelling of nutrients beneficial to marine life.

people attribute to thermal pollution are, in fact, caused by biocides or trace metal contamination rather than the elevated temperature of the water.

Under some circumstances, heated water may cause no problem or even be beneficial. When a plant is located on a large body of water with considerable current or wind action, the warmed water spreads out quickly over a large area and gives up the excess heat to the atmosphere. Properly designed plants located at appropriate sites on such large bodies of water as the Great Lakes of Chesapeake Bay should cause little if any noticeable thermal effect. Even when the plant is located on a small stream, so that it uses a major fraction of the water flow, the effects of calefaction are not always harmful. Daniel Merriman has examined various indicators of water quality and aquatic life in the Connecticut River before and after the building of a large nuclear power plant at Haddam Neck. So far he has found no serious deleterious effects of the warmed water.*

The size of the water body is not the only criterion. A proposed plant of the Florida Power and Light Co. at Turkey Point on the calm, salty water of Biscayne Bay has drawn considerable opposition because of probable harmful effects on some tropical marine organisms that die at temperatures above 92°F.†

Warmed water can be beneficial. It may, for example, make the water ideal for fish that require warmer water than normally exists in the area. By introducing an appropriate species, we may be able to do "fish farming" in the heated water. Other beneficial uses for the heat should be sought. Warm water could be dumped at the bottoms of lakes or seas in which the surface water is deficient in nutrients. The heated water would rise, bringing some nutrients to the surface waters, where most of the biological life needing nutrients exists (see Fig. 4–7). This might be very useful in the Mediterranean, where the surface waters are impoverished of nutrients. One could envision a series of nuclear power plants

---

* Daniel Merriman, "The Calefaction of a River," *Scientific American* **222** (5), 42–52 (May 1970).

† See Gershon Fishbein, "Storm Boils Up over Future of Fish in Biscayne Bay," *Washington Post*, Feb. 22, 1970.

ringing the eastern part of the Mediterranean to provide energy and fresh water to the coastal countries and to bring nutrients to the surface to enhance the growth of fish. Heated water could also be used beneficially for irrigation in fairly cool climates. Use of the water in the spring would promote earlier growth and perhaps increase crop yields. Some heated water could be used in industry or in home heating. With sufficient effort, it should be possible to devise other imaginative, beneficial uses of thermal pollution.

In areas where the warmed water is harmful, other cooling methods can be used. Near the power plant a cooling pond is sometimes constructed, typically with an area of several square miles. Warm water is dumped into the pond,

where it gradually cools by contact with the air and evaporation before it is circulated back through the plant and warmed again. Or the plant can employ cooling towers that transmit the heat to the air. "Dry" cooling towers transfer the heat directly to the air, just like the radiator of an automobile engine. In "wet" cooling towers, the warmed water is sprayed into the air and cooled by evaporation. Both types of cooling towers have drawbacks. The dry towers are expensive because they must draw in enormous quantities of air to achieve the necessary cooling. Evaporation of vast quantities of water in the wet towers increases the air humidity in the region and may cause fogs if the air cools downwind from the plant. (See Fig. 4–8.)

Fig. 4–8 Cooling towers at the Calder Hall nuclear plant in England. [From D. R. Inglis, *Nuclear Energy: Its Physics and Social Challenge* (Addison-Wesley, Reading, Mass., 1973), p. 34. Reprinted by permission.]

The choice of the best cooling method for a power plant thus depends on local conditions. Near a large river or lake, use of the water may be quite acceptable. In dry climates, wet cooling towers or ponds may be best. In very humid climates, dry towers may be needed.

In the future, it may be possible to greatly improve the efficiency of power plants and reduce thermal pollution with the use of magnetohydrodynamics (MHD), a process that is still in the experimental stage. Instead of heating steam, the heat from a nuclear reactor or fossil-fuel burner would produce a very hot, moving stream of gas to which some easily ionized material such as potassium would be added. Movement of the charged species would generate electricity directly. After passing through the MHD cycle, the gas would still be hot enough to generate additional electricity by conventional steam-electric methods.

4–6 ENERGY SOURCES—PRESENT AND FUTURE

Perhaps the most important questions considered in this book are those related to energy use and production: Can the developing nations find abundant, clean, and cheap sources of the energy that they so desperately need? Are there ways in which technological nations can economize on the use of energy? Failing that, how can they produce the additional energy they require with minimum damage to the environment and minimum expenditure of natural resources?

During this century, the average per capita use of energy in the United States has risen from 25 billion calories per year to about 86 billion calories.* The latter figure is about 70 times the average human intake of calories in food; by the use of other energy sources, each of us has, in effect, 70 "slaves" working for us. Our per capita energy use is about seven times the world average.† Looking at it in another way, the U.S. has only about 6% of the world's population, but we use nearly one-third of the world's energy! Our appetite for energy has so far shown no sign of becoming sated. As shown in Tables 4–1 and 4–2, total U.S. use of energy is expected to more than double, from 13.3 to 30.6 quintillion (10^{15}) cal/yr between 1968 and 2000, although the population will rise by only about 50%.

We can make some interesting observations regarding the data in Tables 4–1 and 4–2. The "gross input" total represents energy generated by all methods, including production of electricity. The "net input" total represents the energy actually used by various types of consumers. The total net input is somewhat less than gross because of losses (mainly as thermal pollution) in the conversion of heat to electricity. In 1968, 3.53 units of energy were used in the generation of electricity but only 1.13 units of electrical energy were produced, the remaining 2.7 units being lost to bodies of water or the atmosphere.

Nearly all our energy today comes from fossil fuels—15.1 units out of the gross input of 15.7 units. Despite the expected growth of nuclear energy (from 0.03 to 9.6 units), fossil fuels are expected still to be the major energy source in the year 2000–30.1 out of 40.95 units of gross input. Hydroelectric energy will probably double but still remain a minor source. It cannot be expanded greatly beyond that point as we will then be utilizing most of the potential energy of rain water that falls on ground above sea level.

The projections both of sources and uses of energy for the year 2000 may prove to be inaccurate. If nuclear fusion can be developed as a practical, economical energy source, nuclear energy may become far more important than indicated in Table 4–2. If control of coal and oil combustion emissions proves to be highly expensive, then new energy sources, such as geothermal and solar energy (discussed below), as well as nuclear energy, may become more competitive economically. Natural gas is a big question mark: it is a very clean fuel, producing mainly just carbon dioxide and water vapor, but its supply is uncertain. The use of gas has expanded rapidly in recent years, causing critical shortages of delivery capacity during severe winter cold periods. We have recently been using gas faster than new supplies have been developed. A 1970 survey by the Potential Gas Committee indicated that the U.S. may have ultimate gas supplies amounting to about seventy times our current annual use; however, much of it may be deposited at such great depths (more than 15,000 ft) that it will be very expensive to tap.‡ Processes for conversion of coal and oil to gas are now under extensive study and may someday become important.

One particularly disturbing aspect of Table 4–2 is the projected large increase in use of electricity—from 1.13

---

*   The Energy Industry: Progress—Into Doom?" *Technology Review* **73** (4), 55–56 (Feb. 1971).

†   A. M. Weinberg and R. P. Hammond, "Limits to the Use of Energy," *American Scientist* **58**, 412–418 (1970).

‡   "Plenty of Gas, But Costly," *Chemical and Engineering News*, June 7, 1971, p. 11.

Table 4–1   Sources and Uses of Energy in the United States—1968

| Type of consumer | Energy (quintillions of calories) | | | | | | | |
| | Fossil fuels | | | Hydro-electric | Nuclear | Gross input | Electricity used | Net input |
	Coal	Gas	Petroleum					
Household + commercial	0.15	1.64	1.66			3.45	0.63	4.08
Industrial	1.41	2.34	1.13			4.88	0.50	5.38
Transportation	0.03	0.15	3.65			3.83		3.83
Electricity generation	1.79	0.81	0.30	0.60	0.03	3.53	(1.13)	
Total	3.38	4.94	6.74	0.60	0.03	15.69		13.29

Data from U.S. Bureau of Mines; see G. A. Mills, H. R. Johnson, and H. Perry, "Fuels Management in an Environmental Age," *Environmental Science and Technology* 5, 30–38 (1971).

Table 4–2   Projected Sources and Uses of Energy in the United States—2000

| Type of consumer | Energy (quintillions of calories) | | | | | | | |
| | Fossil fuels | | | Hydro-electric | Nuclear | Gross input | Electricity used | Net input |
	Coal	Gas	Petroleum					
Household + commercial		4.81	0.50			5.31	5.04	10.35
Industrial	0.50	4.41	3.30			8.21	2.77	10.98
Transportation		0.25	8.96			9.21	0.03	9.24
Electricity generation	6.09	1.03	0.23	1.28	9.59	18.22	(7.84)	
Total	6.59	10.50	12.99	1.28	9.59	40.95		30.57

Same source as Table 4–1.

units in 1968 to 7.84 units in 2000. Because of the inefficiency of steam electric plants, only one-third to one-half of the input energy is converted to electricity, the balance being wasted mainly as thermal pollution. In A.D. 2000, an energy input of 18.2 units will be needed for the production of 7.84 units of electrical energy, which indicates a waste of 10.4 units, about 80% of our total net use of energy in 1968! One unfortunate trend is the increased use of electrical resistance heating in homes. Electrical resistance heating is the generation of heat simply by passage of electricity through a wire, as in the heating coils of a toaster or iron. Contractors fre-

quently install electrical heating units in new houses because of low initial cost, the small amount of labor required, and because of encouragement from the power industry. This is a very inefficient way to heat space—it would be more economical of energy to burn the fuel directly in the home and use nearly all of it rather than wasting half or more of the heat at the steam electric plant! A much more efficient form of electrical heating would be the use of heat "pumps"—reversible air conditioners that would pump heat indoors in winter. The theoretical efficiency of this process is given by Eq. (4–7). If the inside temperature T_1 is 72°F (295 K) and the outside temperature T_2 is 20°F (265 K), then every kilowatt-hour of work (electricity) would pump nearly ten kilowatt-hours of heat into the house! So far, heat pumps have been practical only in mild climates because of heat-transfer problems when it is quite cold outside, but the method deserves more study.

Who are the big users of energy and what steps might be taken to reduce or limit energy consumption? Although the homeowner flicking on a switch is sometimes described as "demanding" more energy, direct household use of energy is rather small. "Household and commercial" consumption is less than one-third of "net input"—household use alone amounts to about 20% of net input. About 40% of our energy is used by industry to mine and refine ores and make various products. Much of this energy (as well as natural resources such as iron ore) is wasted because our economy is based on planned obsolescence—the manufacture of throwaway automobiles and disposable TV sets. The industrial demand for energy could be greatly reduced if we built automobiles and appliances to last longer. The wastage of energy and resources is enhanced because we presently recycle very little of the materials in consumer products. Less energy would be required to recover some materials from refuse than is now required to obtain the materials from ores (see Chapter 12).

Another major use of energy, which also entails considerable wastage, is transportation, which uses about 30% of U.S. energy. The private automobile is an important convenience of modern society, but it is an inefficient mode of transportation. Most of the gasoline is used to move the two tons of steel in the car and very little to transport the 180-lb person riding in it! Mass transit systems are much more efficient for moving people. A large direct saving of energy could be made by building improved urban transit systems.

Even if the economies of energy suggested above are made, factors such as population growth will doubtless require increases in energy in the technological nations. And to raise per capita energy supplies in the underdeveloped nations to U.S. standards, a several-fold increase of the worldwide energy supply would be needed. What will be the sources of the additional energy? Of the fossil fuels, only coal is sufficiently abundant to have a major long-range impact, but it is one of our most polluting fuels. In order to use coal without environmental degradation, new methods for mining it and controlling emissions from its combustion will be needed.

Nuclear reactors of the types now being built will provide increasing amounts of electricity over the next several years, but will not be a major future source because of the scarcity of the isotope ^{235}U, which is used as fuel. In order to obtain enormous amounts of energy from nuclear fission it will be necessary to develop practical, safe, breeder reactors that create fuel faster than they burn it (see Chapter 5). The largest potential source of nuclear energy is nuclear fusion, by which the nuclei of small atoms are combined to form larger nuclei. The development of devices for controlled release of nuclear fusion energy is still in a very early stage. It is not certain that the work will be ultimately successful, but recent advances in the technology are hopeful (see Chapter 6).

Two other potentially large sources of energy may be widely used in the future—geothermal energy and solar energy (i.e., energy of sunlight).

Geothermal energy is derived from heat contained in rocks beneath the earth's surface. Particularly in areas of recent or active volcanoes, there are pockets of molten rocks called *magma chambers*. When water from the surface trickles down to these chambers, contact with the molten rocks transforms the water into steam of high temperature and pressure. The high pressure forces the steam through cracks and channels in the rocks, back towards the earth's surface, where it often comes blasting out of the surface as a *fumarole*, like Old Faithful in Yellowstone National Park. By running pipes down to the chambers that contain the steam, one can make use of the steam's energy, either directly or indirectly by using it to generate electricity. (Refer to Fig. 4–9.)

Only in a few areas of the world is use being made of geothermal energy today. In Iceland, an area of considerable active volcanism, homes are heated by the hot water or steam from fumaroles. New Zealand and Italy have been using geothermal energy for some time. The only geothermal development in the United States today is the Geysers power plant in Northern California (see Fig. 4–10). The plant now produces relatively little power, but by 1975 it is to be expanded to 600 megawatts (thermal) power, comparable in size with modern power plants of conventional design. There are many other locations in the United States where such geothermal power plants could be built, particularly in the deserts of the Southwest. Just in the region of the Imperial Valley in Southern California, an estimated 20,000 to 30,000 megawatts of electrical generating capacity could be

Fig. 4–9 A geothermal electric plant utilizes heat from rocks beneath the surface of the earth.

Fig. 4–10   Part of the Geysers power plant in Northern California. Note the steam pipes with expansion loops in the foreground. (Photo courtesy of the Pacific Gas and Electric Co.)

developed!* As an added side benefit, the steam can be condensed to form pure water, which could be used by municipal water systems or for irrigation of desert lands.

Geothermal power plants are not completely free of environmental problems. Accompanying the steam from underground, there may be sulfurous gases and water droplets containing salts. Control measures have to be taken to prevent the release of these potentially harmful effluents to the surroundings. Probably it will be necessary to pump the salty water back into the ground to replace some of that withdrawn in order to prevent settling of the earth's crust in the region.

The sun is the source of all of our energy at the earth's surface except nuclear energy. The energy of fossil fuels was obtained by photosynthesis in plants, which in turn depended on sunshine. The energy of wind and rain is de- rived from the warming of the earth's surface by the sun. Today we make little direct use of sunlight as an energy source, although the energy of sunlight falling on the earth's surface is enormous. In warm, cloudless areas (e.g., the South- west U.S.) sunlight deposits nearly one kilowatt per square meter during the middle hours of the day! Solar batteries, thin wafers of semiconductor material such as silicon, have been developed to convert the energy of sunlight directly into electricity. Their major use has been in unmanned satellites, where they power cameras and transmitters for years without deterioration. At present, solar batteries are too expensive for use as large power sources.

Recently, scientists and engineers have begun to think seriously about using the vast power of sunlight to produce large amounts of electricity.† In large facilities similar to greenhouses, sunlight would be absorbed on thin, dark-

*   R. H. Gilluly, "The Earth's Heat: A New Power Source," *Science News* **98**, 415 (1970).

†   A. L. Hammond, "Solar Energy: A Feasible Source of Power," *Science* **172**, 660 (1971).

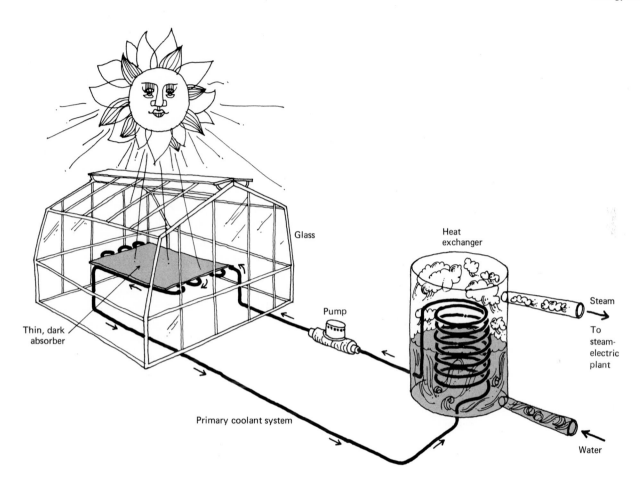

Glass

Heat
exchanger

Steam

To
steam-
electric
plant

Pump

Thin, dark
absorber

Primary coolant system

Water

Fig. 4–11   A possible type of solar electric plant.

colored, nonreflective surfaces. These surfaces would become quite hot. A fluid with good heat-conducting properties, e.g., liquid sodium, would be pumped to the surfaces to remove the heat to make steam and generate electricity (Fig. 4–11). Vast amounts of electricity could be obtained from solar energy, and by condensation of the steam, one could also obtain fresh water. Because of the large light collection areas needed (about 8 square kilometers for a 1000-megawatt plant), the cost of the electricity would not be very competitive with that from fossil-fuel plants in the U.S. today. However, as fossil fuels become more scarce, solar power may become more economically attractive, especially if fossil fuel plants are required to spend large sums to clean up their emissions. Solar power plants would be almost totally free of pollutant emissions, although they would have the problem of thermal pollution common to all steam electric plants.

SUGGESTED READING

D. E. Abrahamson, *Environmental Cost of Electric Power*, A Scientists' Institute for Public Information Workbook, 1970.

D. K. Anthrop, "Environmental Side Effects of Energy Production," *Science and Public Affairs* **26** (8), 39–41 (1970).

H. Brown, "Human Materials Production as Process in the Biosphere," *Scientific American* **223** (3), 194–208 (September 1970).

"Energy Technology to the Year 2000," a three-part series, *Technology Review* **74**, No. 1–3 (1971–1972).

D. Fabun, "Energy: Transactions in Time," in *The Dimensions of Change* (Glencoe Press, Beverly Hills, Calif., 1970).

L. P. Gaucher, "Energy in Perspective," *Chemical Technology*, March 1971, pp. 153–158.

Henry Jarrett, ed., *Environmental Quality in a Growing Economy* (Johns Hopkins University Press, Baltimore, 1966).

Joint Committee on Atomic Energy, Congress of the United States, "Selected Materials on Environmental Effects of Producing Electric Power," Government Printing Office, Washington, D.C., 1970.

G. J. F. MacDonald, "The Modification of Planet Earth by Man," *Technology Review* **72** (1), 27–32 (1969).

G. A. Mills, H. R. Johnson, and H. Perry, "Fuels Management in an Environmental Age," *Environmental Science and Technology* **5**, 30–38 (1971).

G. T. Seaborg and W. R. Corliss, *Man and Atom* (Dutton, New York, 1971).

Scientific American **225** (3), (September 1971). Entire issue devoted to energy.

S. F. Singer, "Human Energy Production as a Process in the Biosphere," *Scientific American* **223** (3), 173–193 (September 1970).

"The Energy Crisis," two-part series, *Science and Public Affairs* **27**, No. 7 and 8 (1971).

A. M. Weinberg and R. P. Hammond, "Limits to the Use of Energy," *American Scientist* **58**, 412–418 (1970).

A. M. Weinberg, *Reflections on Big Science* (M.I.T. Press, Cambridge, Mass., 1967). Paperback.

Mitchell Wilson and the Editors of Life, *Energy* (Time, Inc., New York, 1968).

PROBLEMS AND QUESTIONS

Note: Conversion factors that may be needed for working the problems are summarized in Appendix C.

1. Give practical examples (e.g., "A chemical battery produces electricity that runs a motor") of the following energy transformations:
   a) mechanical to electrical to chemical
   b) radiant to heat to electrical
   c) mechanical to electrical to nuclear
   d) nuclear to heat to mechanical to electrical to mechanical

2. What steps might humans take to make more efficient use of energy in the food chain?

3. It would appear that the U.S. philosophy of urban transportation is to build freeways for the movement of each person in a private automobile rather than construct extensive mass-transit systems. In later chapters we will discuss the major contributions of auto exhaust to urban air pollution, but for now, discuss the implications of the present system vs. rapid transit *in regard to economy in the use of energy and mineral resources.*

4. Briefly list the ways in which hydroelectric plants can cause environmental degradation.

5. A spacecraft must achieve a velocity of about 25,000 mi/hr (about 11,000 m/s) to escape from the earth's gravitational field.
   a) How much energy (in ergs) would a 10,000 kg spacecraft have at the escape velocity?
   b) A typical large modern electric plant generates 500,000 kw. How long would such a plant have to operate in order to generate the amount of energy equivalent to that of the spacecraft at escape velocity?
   c) If an automobile at 60 mi/hr has an energy of about 6.5×10^{12} ergs, how many such automobiles would be needed to equal the energy of the spacecraft? (When energy is considered in this way, one can see why a spacecraft must burn fuel at such an enormous rate, i.e., about 3 tons/s, to achieve its great velocity!)

6. Suppose we have rain clouds at a height of 2000 m over a 10 × 10 km area at sea level. How much energy in ergs is released of water falls from the clouds to form a layer 2 cm deep on the ground? Use Fig. 4–2 to decide how many tons of TNT would be equivalent to the released energy in the rainfall.

7. A suburbanite has a square 20 × 20 m lawn that is heavily shaded by trees, so that it is very difficult to get grass to grow. He decides to attempt to get grass by providing about 12 hr of artificial light per day of an intensity (energy) equal to that of the noontime sun. If electricity costs 5¢/kwh, how much will his electricity bill for the grass amount to over the three months of summer?

8. a) If an average person takes in 2400 Calories per day (note Calories), what is his average power consumption in watts? By comparison, an incandescent bulb in a typical floor lamp typically uses about 150 watts.

   b) If people give off heat at the rate calculated in part (a), how big an air conditioner (in BTU/hr) is needed to remove the heat given off by 100 people in a room?

9. How much mass is converted to energy when 100,000 kg of sugar (sucrose) is burned, as in Eq. (4–3)?

10. We are going to design a coal-fired steam electric plant.

    a) If the plant employs steam at 500°C and cooling water from a river at an average temperature of 20°C, what is the theoretical efficiency ϵ of the plant?

    b) Suppose the actual overall efficiency of the plant, i.e., the ratio of electrical energy produced to heat generated, is 44%. If the plant is to generate 500,000 kw of electrical power, how many tons (1 ton = 909 kg) of coal must be burned each hour, if each ton releases about 6.6×10^9 cal?

    c) Assume that all the heat not converted to electrical energy is dumped into the river as thermal pollution. If laws permit the temperature of the water to be raised by only 5°C, how many liters of water must flow through the cooling channel each minute to remove the wasted heat? (By comparison, the average flow of the Connecticut River, where the Haddam Neck plant is located, is about 3.5×10^7 l/min.)

    d) If instead of dumping the excess heat into a river we were to use a "wet" cooling tower in which most of the wasted energy was used to evaporate water, how many liters of water would be evaporated per minute? (To evaporate 1 ml of water requires about 540 cal.)

5 NUCLEI AND NUCLEAR ENERGY

Fireball from a nuclear explosion rises above Frenchman Flat in 1957.
(Photo courtesy of Lookout Mountain Laboratory, U.S. Air Force.)

5–1 THE SIGNIFICANCE OF NUCLEAR ENERGY

In this chapter and in Chapter 6, we discuss nuclear processes and nuclear energy in more detail than is normally done in introductory chemistry texts. We feel the subject is of overriding importance now and will be so for many decades, probably even centuries, in the future. Since 1945, when the first nuclear weapons were exploded, our civilization has lived under the threat of nuclear annihilation. The United States (with British and Canadian cooperation) and then the Soviet Union, France, and China have acquired nuclear weapons, and several other nations are capable of developing them. During the Cuban missile crisis of October 1962 it seemed that we were at the brink of a massive nuclear war. Since that time, the nuclear threat has receded somewhat, with hopeful signs such as the Limited Nuclear Test-Ban Treaty of 1963, the Nuclear Non-Proliferation Treaty, and the Strategic Arms Limitations agreements of 1972. However, as long as nuclear weapons exist, there is always the dire possibility that a conflict in Viet Nam, the Middle East, or elsewhere will escalate into nuclear war.

Nuclear energy is both a "sword" and a "plowshare." As the latter it holds the promise of unlimited supplies of energy that are needed to improve the lot of mankind throughout the world. However, even the peaceful application of nuclear energy is not an unmixed blessing—severe environmental degradation could result from improper use of nuclear energy.

We are now at a crossroads in the development of new sources of energy. Many conventional sources are in limited supply and, furthermore, have created many of our most serious environmental problems. With proper use, nuclear energy can serve as a solution to the problem. Power companies are increasingly turning to nuclear fission for production of electricity. However, the construction of nuclear power plants has been encountering strong opposition from conservationists and other citizen groups and from some state and local regulating bodies. Much of the criticism of nuclear energy has been highly emotional. Some, we feel, is so outlandish that it may destroy the credibility of reasoned criticism. In this chapter, we attempt to explain the principles and problems of nuclear fission reactors in sufficient detail so that you will be able to understand the issues involved in the use of nuclear energy. In Chapter 6, we discuss nuclear fusion, a potentially much larger energy source which may be safer than nuclear fission.

5–2 NUCLEAR BINDING ENERGIES

Why are only certain isotopes of an element stable and the others unstable? For example, why is the element chlorine made up of the isotopes $^{35}_{17}Cl$ and $^{37}_{17}Cl$, whereas other isotopes such as $^{36}_{17}Cl$ and $^{38}_{17}Cl$ are unstable and transform themselves into atoms of other elements? To answer these questions we must consider the relationships between atomic masses, energies, and stability.

One cannot obtain the atomic mass of an atom by adding up the masses of the neutrons, protons, and electrons present in the atom. For example, we could "make" an atom of $^{4}_{2}He$ from two neutrons and two $^{1}_{1}H$ atoms:

$$2\ ^{1}_{1}H + 2\ ^{1}_{0}n \longrightarrow\ ^{4}_{2}He. \tag{5–1}$$

Consider the atomic masses:

$2\ ^{1}_{1}H$:	$2 \times 1.007825 =$	2.015650 amu
$2\ ^{1}_{0}n$:	$2 \times 1.008665 =$	2.017330 amu
	Total	4.032980 amu

Actual atomic mass of ^{4}He: 4.002604 amu

Difference: 0.030376 amu

This shows that if we were to make an $^{4}_{2}He$ atom from its constituents, we would have a small amount of mass left over. What happens to that mass? As seen in Section 4–3, it is converted into energy. Processes similar to this one provide most of the energy of stars, including our sun.

With the use of Eq. (4–6), we find that the energy released per atom of ^{4}He formed is

$$0.03038\ \text{amu} \times 931.5\ \frac{\text{MeV}}{\text{amu}} = 28.30\ \text{MeV} \tag{5–2}$$

or

$$28.30\ \text{MeV} \times 1.602 \times 10^{-6}\ \frac{\text{erg}}{\text{MeV}} = 4.54 \times 10^{-5}\ \text{erg}.$$

Now 4.54×10^{-5} erg is a very small amount of energy, but there are so many atoms in a gram of ^{4}He (about 1.5×10^{23}) that formation of a gram of it would yield 6.8×10^{18} ergs of energy, about 30 million times as much energy as one would obtain by burning a gram of coal!

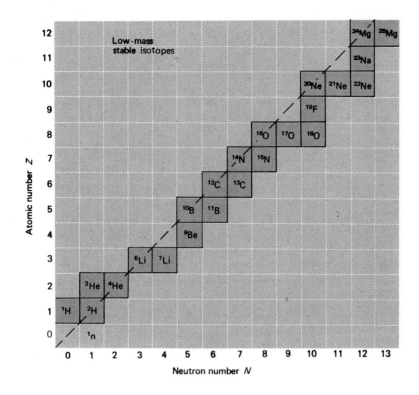

Fig. 5–1 Pattern of stable isotopes among the very light elements.

If we were to perform the same calculation for any type of atom that exists in nature, we would obtain a similar result: the mass of the atom is always less than the sum of the masses of the "parts." This means that the separated parts have more nuclear "potential energy" than the atom. Formation of an atom is analogous to a rock rolling down a mountain side. Just as a rock is in a more stable position at the bottom of the mountain, the electrons, neutrons, and protons are in a more stable configuration when they are brought together to form an atom.

The energy released when two neutrons and two ^{1}H atoms are brought together to form a ^{4}He atom is called the *binding energy* of ^{4}He. This is the amount of energy that holds ^{4}He together, since that much energy would have to be expended to reverse the process and break ^{4}He into its parts. The binding energy of a nucleus results from the strong, attractive forces that exist between the various particles in the nucleus.

5–3 RADIOACTIVITY

Pattern of Stable and Radioactive Isotopes

In Fig. 5–1 we show the stable isotopes among the very light elements on a plot of atomic number Z versus neutron number N. Stable isotopes have lower energies than their unstable, radioactive neighbors. Among the light elements, the stable isotopes have about equal numbers of neutrons and protons; e.g., along the line of stability are the important isotopes with $N = Z$: $^2_1$H, $^4_2$He, $^{12}_6$C, $^{14}_7$N, $^{16}_8$O, $^{40}_{20}$Ca, and others.

Among the heavier elements, stable isotopes tend to have more neutrons than protons, and thus the line of stability tilts somewhat (see Fig. 5–2). For example, the stable isotope of element 83, bismuth, is $^{209}_{83}$Bi$_{126}$, which has 43 more neutrons than protons. The reason for this excess of neutrons is that protons, all being positively charged, repel each other. The added neutrons hold the nucleus together in spite of the pro-

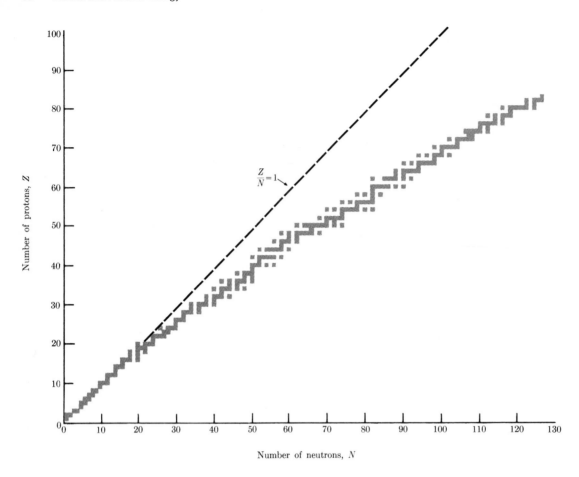

Fig. 5–2   Pattern of stable isotopes over the entire mass range. [From Bruce H. Mahan, *University Chemistry*, 2nd ed. (Addison-Wesley, Reading, Mass., 1969), p. 793. Reprinted by permission.]

tons' "pushing" each other by the repulsion between their positive charges. Above bismuth, even the added neutrons are not sufficient to keep the protons from "exploding" (fissioning) the nucleus.

☓   What happens to isotopes that have too many neutrons to be stable? One of the neutrons in the nucleus transforms to a proton and an electron. The proton stays in the nucleus and the electron is emitted with an energy up to the difference in the nuclear potential energies between the parent and daughter atoms, i.e., the difference between their atomic masses. The process

$$n \rightarrow p + \beta^- + energy$$

is called *beta decay* (β decay), and the electrons emitted are called β-rays or particles.*

---

*   We are ignoring the small particles, the "neutrino" (v) and the "antineutrino" ($\bar{v}$), that are also emitted in β decay processes. They have no mass or charge, are almost impossible to observe, and have negligible interaction with the surroundings; however, they carry off various amounts of the available decay energy, and only the balance appears as energy of the β^- particle.

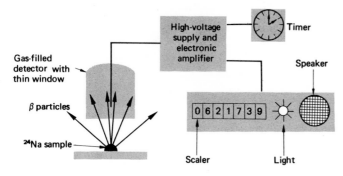

Fig. 5–3 Experimental arrangement for observing β^- particles from a sample of ^{24}Na.

An example of a β^- emitting radioactivity is ^{24}Na. If ordinary sodium, ^{23}Na, is bombarded with neutrons in a nuclear reactor, some of its atoms undergo neutron captu.e to form ^{24}Na:

$$^{23}_{11}\text{Na}_{12} + {}^{1}_{0}\text{n} \longrightarrow {}^{24}_{11}\text{Na}_{13}. \qquad (5\text{–}3)$$

Sodium-24 has a greater mass (thus greater nuclear potential energy) than $^{24}_{12}\text{Mg}_{12}$, the only stable isotope that has 24 nucleons in its nucleus. Thus ^{24}Na atoms spontaneously transform into ^{24}Mg atoms by β^- decay in which a neutron in the nucleus is converted to a proton, with the emission of a β^- particle:

$$^{24}_{11}\text{Na}_{13} \longrightarrow {}^{24}_{12}\text{Mg}_{12} + \beta^-. \qquad (5\text{–}4)$$

One can place a sample of radioactive material near a detector which measures the nuclear radiation given out by clicking, flashing a light, or recording a count every time one of the β^- particles from the decay of a ^{24}Na atom strikes it (Fig. 5–3). Thus we measure the *activity* of the ^{24}Na sample, i.e., the number of ^{24}Na atoms that decay to ^{24}Mg per unit of time. If we measured the activity of a ^{24}Na sample several times a day over a period of several days, we would obtain data similar to those plotted in Fig. 5–4a. We see that the ^{24}Na activity decreases sharply during the period. The data can be interpreted more easily if we plot the *logarithm* of activity (see Appendix A), instead of activity itself, as a function of time, as in Fig. 5–4b. Mathematically, the straight line

Fig. 5–4 Radioactivity of a sample of ^{24}Na over a period of several days. (a) Linear plot of the count rate of β-particles. (b) Logarithmic plot of the same data. On the logarithmic scale, the 15-hr half-life of ^{24}Na can be readily observed.

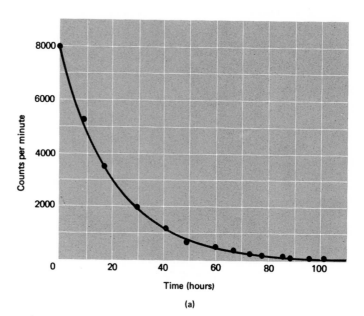

(a)

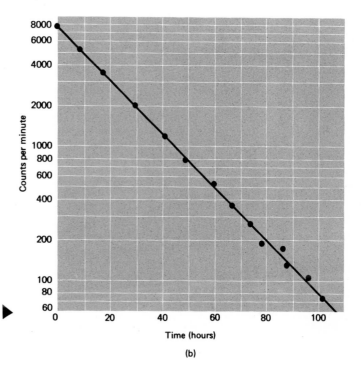

(b)

obtained on such a plot means that the activity is proportional to the number of atoms remaining as ^{24}Na at any time. This, of course, makes sense—as soon as an atom has decayed to stable ^{24}Mg, it can no longer decay. (Note in Fig. 5–4 that the experimental points do not fall exactly on the curve. This is typical of experimental data, since measurements always have some error.)

Every radioactive species is characterized by a *half-life*, the amount of time it takes for the activity of a sample to decrease to half of its initial value. The half-life of ^{24}Na is 15 hours, as we can observe from Fig. 5–4b. The activity at time zero is about 8000 counts per minute. Fifteen hours later, the activity has dropped to half the initial value, to 4000 c/min. In the following 15-hr period, the activity again drops by half, to 2000 c/min. Over any 15-hr period, the radioactivity of ^{24}Na drops by one-half. Since, as noted above, the activity is proportional to the number of atoms present, the number of atoms of the radioactive species also drops to half its initial value over a period of one half-life. Half-lives of radioactive species range from a fraction of one second to periods greater than the age of the solar system.

Another example of a β^- emitting radioactivity is the familiar ^{90}Sr, present in fallout from nuclear weapons explosions, which decays with a half-life of 28 years to ^{90}Y, which in turn decays rather quickly to stable ^{90}Zr (see Fig. 5–5):

28-yr $^{90}_{38}\text{Sr}_{52} \longrightarrow \,^{90}_{39}\text{Y}_{51} + \beta^-,$ (5–5)

64-hr $^{90}_{39}\text{Y}_{51} \longrightarrow \text{stable } ^{90}_{40}\text{Zr}_{50} + \beta^-.$ (5–6)

The 28-yr half-life of ^{90}Sr means that if we have 1000 atoms of it today, 28 years from now one-half of them, 500, will have decayed and the other 500 will remain. During the succeeding 28 years, half of the remaining 500 atoms will decay, and so forth. This is one of the problems of radioactivity in the environment—some species have such long half-lives that they remain in existence for many years after being formed. Unlike chemical "poisons" in the environment, radioactive atoms cannot be rendered harmless by chemical or any other known practical means. Strontium-90 behaves chemically the same as ordinary stable strontium atoms. Stable strontium is chemically similar to the element calcium, which is deposited primarily in bones and teeth. In Section 5–4, we discuss some of the biological effects caused by the radioactive decay of nuclides such as ^{90}Sr inside the body.

When isotopes have too many protons to be stable, a proton in the nucleus can transform into a neutron by one of

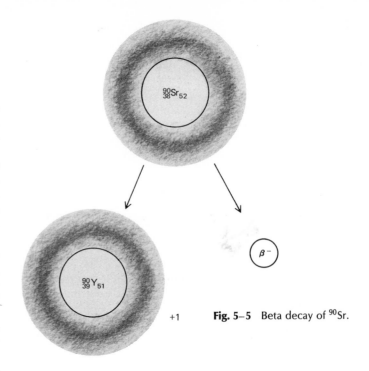

+1   **Fig. 5–5** Beta decay of ^{90}Sr.

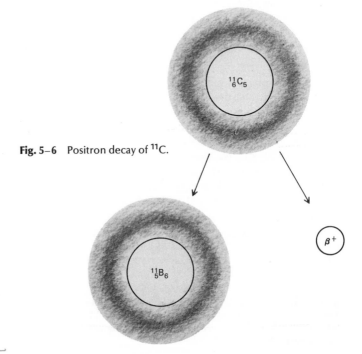

Fig. 5–6 Positron decay of ^{11}C.

two possible mechanisms. On the one hand, the proton can transform into a neutron and a *positron* (β^+), i.e., a particle having the same mass and amount of charge as an electron, but with positive instead of negative charge. The artificial nuclide ^{11}C, with a half-life of 20 min, is an example of a β^+ emitter (see Fig. 5–6):

$$20\text{-min} \quad {}^{11}_{6}\text{C}_5 \longrightarrow {}^{11}_{5}\text{B}_6 + \beta^+. \tag{5–7}$$

The positron is the *antiparticle* of the electron. From science fiction you may have heard of vast explosions occurring when a chunk of *antimatter* collides with matter. This is what happens on a small scale, since the positron is antimatter and unstable in our world of "matter." It is ejected from the nucleus with an energy of a fraction of an MeV or more. When it has passed through enough material to slow down almost to rest it comes in contact with an electron and the two *annihilate*. Their masses completely disappear and the energy appears as γ rays (see Fig. 5–7):

$$\beta^+ + e^- \longrightarrow 2\gamma. \tag{5–8}$$

Each of the γ rays has an energy equivalent to the mass of an electron, 0.51 MeV.*

The other mechanism by which protons in the nucleus can transform into neutrons is *electron capture* (EC). In species such as ^{7}Be, in which the nucleus has too many protons to be stable, one of the electrons around the nucleus may be "captured" by the nucleus, where it converts a proton into a neutron (see Fig. 5–8):

$$^{7}_{4}\text{Be}_3 \xrightarrow{\text{EC}} {}^{7}_{3}\text{Li}_4 \tag{5–9}$$

In general, isotopes above and to the left of the line of stable isotopes in Figs. 5–1 and 5–2 have an excess of protons and thus decay toward the stable isotopes by electron capture or β^+ emission. Isotopes below and to the right of the line of stability have an excess of neutrons and thus decay by β^- emission.

Most naturally occurring isotopes are stable in the sense that they cannot undergo β^-, EC/β^+, or alpha decay (see

---

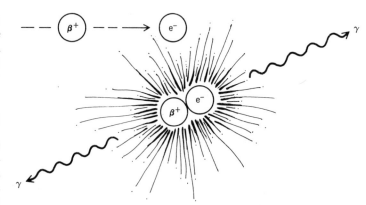

Fig. 5–7 Annihilation of a positron and an electron.

below). However, a few radioactive isotopes exist in nature simply because their half-lives are comparable with or greater than the time that has elapsed since the most recent formation of the elements of the earth (about 4.5×10^9 years). Prominent examples of naturally occurring radioactive isotopes are the following:

Isotope	Half-life
^{235}U	7.1×10^8 yr
^{238}U	4.5×10^9 yr
^{232}Th	1.4×10^{10} yr

Fig. 5–8 Electron-capture decay of ^{7}Be.

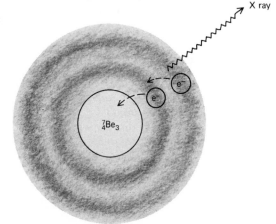

---

*   So far as we know, our entire galaxy consists of matter. However, there may be galaxies made up of antimatter. If material from an anti-matter galaxy were to enter our galaxy and collide with some matter, all the antimatter and an equal amount of matter would be destroyed and the mass converted to energy!

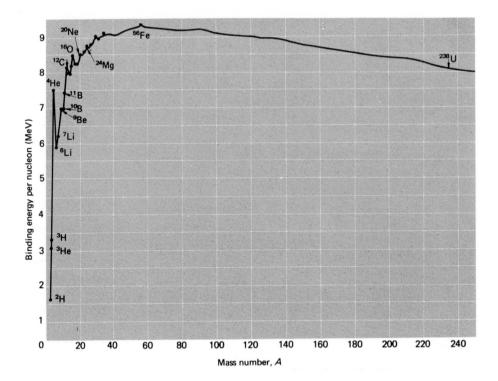

Fig. 5–9 Binding energy per nucleon for isotopes that are not subject to β^- decay.

All these decay by alpha emission (see below). The other major natural activity is that of ^{40}K, which has a half-life of 1.27×10^9 yr; it decays partly by β^- emission to ^{40}Ca and partly by EC/β^+ decay to ^{40}Ar. The latter process accounts for the large abundance of argon (Ar) in the earth's atmosphere (about 1%).

Alpha Decay and Spontaneous Fission

So far we have been considering radioactive decay between species with the same number of *nucleons* (neutrons and protons) and thus the same mass number A. Simply by looking at the atomic masses, one can determine which species decay to others by conversion of neutrons to protons (or vice versa), while leaving the number of nucleons unchanged.

In order to determine the relative stability of species with different mass numbers, we may consider the *binding energy per nucleon* of each species. This is a measure of the average strength with which each nucleon is attracted to the others present in the nucleus. As shown in Section 5–2, the binding energy of ^{4}He is 28.3 MeV, giving a binding energy per nucleon of 28.3/4, or about 7.1 MeV. Figure 5–9 shows the binding energy per nucleon for β-stable species, i.e., those that are not subject to β^- or EC decay. The greatest binding energy occurs in nuclei having mass numbers in the region around $A = 60$. Iron-56 is the most stable configuration of neutrons, protons and electrons known to us on earth. (There may be other more stable forms of nuclear matter existing in such bodies as "neutron stars"; see Chapter 6.)

Above the mass $A = 60$ region, binding energy per nucleon decreases because of mutual electrostatic repulsion of all the positively charged protons. Each proton repels each of the other protons in the nucleus. Even the excess neutrons are not enough to stabilize nuclei above $_{83}$Bi, and two new decay modes, both of which reduce nuclear charge, become prominent. One of these is *alpha decay* (α decay), in which the nucleus of ^{4}He (called an α particle) is emitted from a heavy nucleus. For example,

$$^{238}_{92}U_{146} \longrightarrow {}^{234}_{90}Th_{144} + {}^4_2He_2. \qquad (5-10)$$

The half-life of ^{238}U is 4.5×10^9 yr, close to the age of the solar system. Since all nuclei above Pb and Bi are unstable, ^{234}Th is unstable, as are its "daughter," ^{234}Pa, and "granddaughter," ^{234}U, etc., so that if one waits long enough one finds that the original ^{238}U atom has decayed through a series of eight α decays and six β^- decays to $^{206}_{82}$Pb.

The energies of α particles emitted from nuclei are usually in the range of 4 to 8 MeV, which is somewhat higher than the energies typically released in β^- decay. Since the heaviest elements have the greatest internal electrostatic repulsion, they tend to have the shortest half-lives. This is one of the limitations encountered in attempts to make new, artificial, heavy elements.

The other method by which very heavy nuclei reduce their charge is *spontaneous fissions*; the nucleus spontaneously divides into two nuclei of roughly half the mass of the original nucleus and a few "left-over" neutrons, e.g.,

$$^{252}_{98}Cf_{154} \longrightarrow {}^{140}_{56}Ba_{84} + {}^{108}_{42}Mo_{66} + 4\,{}^1_0n. \qquad (5-11)$$

This is only one of the hundreds of ways in which ^{252}Cf can divide. Nuclei with mass numbers between about $A = 85$ and $A = 165$ are formed, with the most likely products having masses of about $A = 106$ and $A = 144$. Tremendous energy is released in fission (about 220 MeV), because of the large amount of electrostatic repulsion in the fissioning nucleus. Since the original nucleus has more neutrons per proton than stable nuclei in the fission product region, nearly all the fission products are neutron-rich and therefore subsequently decay by β^- and γ emissions. Each fission that occurs thus produces two radioactive atoms that may decay to other radioactive atoms before stability is attained. In Eq. (5–11), both the ^{140}Ba (with a half-life of 12.8 days) and its daughter, ^{140}La (half-life, 40.2 hr), are unstable, decaying eventually to stable ^{140}Ce:

$$\text{12.8-day} \quad {}^{140}_{56}Ba \xrightarrow{\beta^-}$$

$$\text{40.2-hr } {}^{140}_{57}La \xrightarrow{\beta^-} {}^{140}_{58}Ce \text{ (stable).} \qquad (5-12)$$

The ^{108}Mo in Eq. (5–11) is so far from stability that its half-life has never been measured, but it is probably quite short.

Another important fact about fission is that it has a greater tendency to occur as we move to heavier nuclei. Shorter half-lives in the heavier nuclei, due to fission and α-decay, make it increasingly difficult to synthesize heavy elements.

Gamma Rays and X Rays

Gamma rays (γ rays) and x rays are high-energy forms of ordinary light (see Chapter 7). Whereas a bundle or "quantum" of ordinary visible light typically has an energy of about 2 eV, an x-ray quantum has an energy of about 10,000 eV (or 10 keV, kilo electron volts), and a γ ray has about 1 MeV. X rays are produced in x-ray tubes by allowing electrons of several tens of keV to strike a target. They are also produced in radioactive decay. In electron capture, for example, electrons that are closest to the nucleus are the most likely to be captured, leaving an electron "vacancy" in the structure of the atom. The vacancy may be filled by an electron of a higher energy, which "drops" into the vacant position. The atom then generally emits an x ray with an energy that is characteristic of the element. Gamma rays are emitted from nuclei left in excited states following α or β decay.

5–4 THE EFFECTS OF RADIATION

Interactions of Radiation with Matter

Nuclear radiations initially have energies (typically 1 MeV) that are about a million times as great as chemical binding energies (a few eV). As α particles go through material, their 2+ electronic charge causes them to interact with the atoms of the medium. They give up their energy to the medium by knocking electrons off atoms or molecules (*ionization*) or by breaking molecules apart (*dissociation*). Although α particles have a lot of energy, their large mass causes them to move rather slowly and they give up their energy to the medium in a very short distance (or *range*). They leave behind a dense trail of ionization and disrupted molecules. They can travel only a few centimeters in air and cannot penetrate a sheet of aluminum foil (see Fig. 5–10). Since they can be stopped by a thin layer of skin, α activity presents little danger to the body (except possibly for skin burns if in close contact) provided it is kept outside of the body. However, if α activity is ingested into the body, the chemical characteristics of most α-emitting elements cause them to be deposited in bones. Later we will discuss the effects of radioactivity on human tissue.

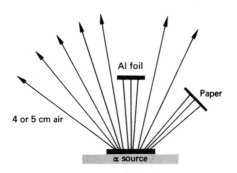

Fig. 5-10 Penetration of α particles from radioactive decay. They travel a short distance through the air, but are stopped by paper or household aluminum foil.

Beta particles interact with matter in the same way as α particles. However being much smaller than α particles (about 1/7000 the weight of an α), a β particle travels much faster than an α particle of the same energy and leaves a much less dense trail of ionization. Consequently, a β particle has a longer range than an α particle of the same energy. Beta particles can pass through a meter or more of air and some can penetrate the glass of a test tube. They can penetrate about a centimeter or less through tissue.

Gamma and x rays lose their energy in "one-shot" processes. Instead of loosing energy continuously as it moves through material, a γ ray looses no energy until it collides with an electron and gives up all or a major part of its energy to the electron. Thus the γ or x ray itself causes little direct ionization and dissociation along its path. However, the one or two struck electrons go through matter interacting with it in the same way as a β particle (see Fig. 5-11).

Since γ rays interact via "one-shot" processes, it is meaningless to speak of the range of a γ ray in matter. Instead one speaks of a *half-thickness* of a given material for γ rays of a specific energy. The half-thickness is the distance over which half of the γ rays entering the material interact with electrons. The half thickness of aluminum for 1-MeV γ rays is about 4 cm. This means that if 100 γ rays start through a piece of aluminum 4 cm thick, about half of them would interact with the aluminum and the other 50 would pass on through (see Fig. 5-12).

Lead is a very good absorber of γ and x rays because it has a high atomic number and very high density. A great deal of it is used in laboratories that handle large amounts of γ ray–emitting radioactivity, because it protects personnel from harmful radiation and shields radiation detectors from any radioactivity in the vicinity other than the sample whose radiations are being counted. Lead shielding is also used around medical and dental x-ray equipment to ensure that the patient is exposed to x rays only in the desired area and that the x-ray technicians are shielded from radiation.

Gamma and x rays penetrate human tissue very effectively—the higher the energy, the greater the penetrating power. Gamma rays of 1 MeV have a half-thickness of about 11 cm in tissue, and x rays of 10 keV have a half-thickness of about 0.5 cm. X rays of somewhat higher energy are useful for diagnostic purposes, because a fraction of the original x-ray beam can pass through the body and expose an x-ray film. The intensity of x rays passing through depends on the density of tissue, and particularly on the atomic number of the tissue material. For example, bones and teeth absorb x rays quite readily because of their high content of calcium (Ca), which has a higher atomic number than most soft-tissue atoms. The rather small half-thickness of x rays in body tissue

Fig. 5-11 Interaction of γ rays with electrons.

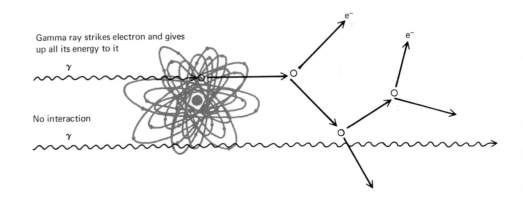

Gamma ray strikes electron and gives up all its energy to it

γ

No interaction

γ

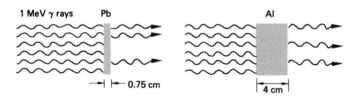

Fig. 5–12 Gamma rays passing through aluminum and lead. Note that the greater density and higher atomic number of the lead give it a smaller half-thickness than the aluminum.

Some materials, such as crystals of sodium iodide (NaI) and zinc sulfide (ZnS), give off light when struck by nuclear radiations. The latter substance was used by Rutherford to detect α particles in his fundamental experiments on the structures of atoms (see Chapter 2). Radiations also release silver from its salts on photographic film. Nuclear radiations (γ and x rays, and high-energy β rays) can pass through paper wrapped around the film and "expose" it even before it is unwrapped. Radiation workers carry film badges with them to record their exposure to penetrating radiations.

means that they are not very useful for irradiation of deep-seated tumors, because more radiation is deposited in the skin than in the tumor. Deep therapy is normally performed with γ rays from the radioactive nuclide ^{60}Co (half-life, 5.2 yr), which emits γ rays of 1.173 and 1.333 MeV energies.

Radioactivity is detected by the effects of the radiation on its surroundings. When an α or β particle or a γ or x ray passes through a gas, electrons are knocked loose from their molecules, leaving free negative electrons and positively charged ions in the gas. If this occurs between plates connected to a battery, as in Fig. 4–3, the electrons in the gas are attracted to the positive plate, and positive ions to the negative plate, causing a flow of current in the circuit that can be measured with an ammeter or more sensitive device (see Fig. 5–13). Such effects are the basis of radiation detectors such as Geiger counters, which are used in detecting radioactive contamination and in prospecting for uranium deposits.

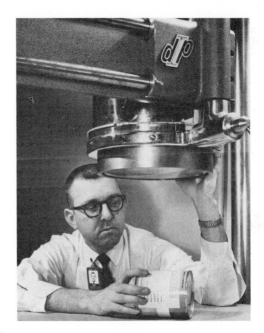

A sample can of paint being placed under a gamma scintillation counter at Oak Ridge, Tennessee. Radiation from the sample will then be counted to determine its radioactivity content. The scientist is wearing a radiation badge on his shirt. (Photo courtesy of Union Carbide Corp.)

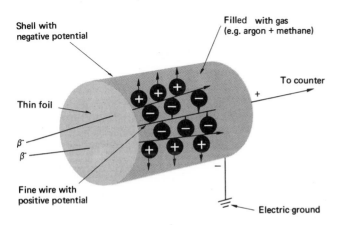

Fig. 5–13 Radiations are typically detected by collecting the ions they produce in a gas-filled tube.

Biological Effects of Radiation

When nuclear radiations pass through matter, they disrupt atoms and molecules by ionizing them or dissociating the molecules. A single nuclear radiation typically has enough energy to affect nearly a million molecules! The harmful effects of radioactivity on human beings and other forms of life result from the disruption of molecules in the tissue after the radiation passes through.

Radiation effects on living systems fall into two general categories: *somatic* and *genetic*. Somatic effects cause damage to a living organism during its lifetime. For example, disruption of the molecules of a healthy cell may damage the cell's growth-regulation mechanism, causing it to reproduce itself in an uncontrolled manner—a condition called "cancer." Another common somatic effect of radiation is leukemia, a cancer-related disease of the blood-forming areas of bone marrow that causes production of excessive numbers of white blood cells.

Since radiation can cause cancer, you may wonder why radiation therapy is used as a treatment for cancer. Radiation "kills" both healthy and cancerous cells, but, probably because of their higher rate of division, cancer cells are generally more susceptible to radiation than healthy cells.

Genetic effects of radiation result from molecular damage in the genes and chromosomes of the reproductive material of the body (see Chapter 15). The various features of a child's body, e.g., height, eye color, and hair color, are determined by the genes derived from the father's and mother's reproductive material. Radiation passing through the material can cause changes in the genes that will govern the characteristics of children not yet conceived. When a child is born, such changes in the parents' genes may cause it to have, for instance, one eye brown and the other blue. These changes in genes are called *mutations*.

Mutations caused by natural radiation (cosmic rays or naturally occurring radioactivity—see Section 5–3) have probably been very important in the process of evolution, although other factors such as excessive heat can also cause mutations. To see how evolution comes about, suppose that a mutation causes the offspring of an animal to have two heads. Suppose further that two heads increase the chances of survival of the offspring by enabling it to search for food more effectively and see its predators more quickly than normal members of its species. If its chances of surviving to an age at which it can reproduce two-headed offspring are greater than those of one-headed members of the species, the two-headed variety will increase in number faster than the one-headed, so that eventually, after many generations, the entire species may consist of the two-headed type. The process by which characteristics that enhance the power of survival eventually become normal characteristics of a species is called *natural selection* or *survival of the fittest*.

From the foregoing example, one might conclude that mutations are desirable, so that for the good health of future generations, we should subject ourselves to considerable radiation in order to accelerate the process. Unfortunately, most mutations are undesirable; living systems, especially humans, are so highly developed and complex that most mutations weaken the system rather than improve it. If a mutation is sufficiently degrading, the offspring will not survive long enough to reproduce and pass the genetic defect on to future generations, and so the defect will quickly disappear from the species. The ratio of good to bad mutations in humans is probably of the order of one to a thousand or ten thousand. It's as if you blindfolded yourself and stuck a screwdriver into the works of a computer. There is a small chance that whatever change you made would improve the computer's operation. However, the chances are vastly greater that you would damage the computer.

Because of the great differences in their penetrating powers, γ rays and high-energy β^- and x rays on the one hand, and α particles and low-energy β particles on the other, pose very different problems of potential harm to people. The latter radiations have such short ranges that they are almost harmless if one prevents skin contact with the source and does not ingest (by breathing or swallowing) the source of the radiation. Of course, if the α or low-energy β^- emitter also gives off penetrating radiations, one must take the precautions noted below for γ- and x-ray sources.) Large amounts of α activity or low-energy β^- activity are generally handled in air-tight "glove boxes" which prevent the material from being spread around the room or becoming airborne.

Most α-emitting species have chemical properties similar to those of the elements present in bones and teeth. If α-emitting material is ingested, it may find its way to the bones or teeth and be incorporated there. When this happens, it is difficult or impossible to remove more than a fraction of the radioactive material. An important naturally occurring α-emitter is 1600-yr ^{226}Ra. The chemical behavior of radium is very similar to that of calcium, an important constituent of bone and teeth. If a large amount of radium is incorporated into the bone, constant bombardment of the bone and its marrow (where blood formation occurs) by the densely ionizing α particles often causes bone cancer or leukemia.

The worst example of radium "poisoning" occurred in the early 1930s in the United States. A number of young women died as a result of painting radium-dial watches—watches on which the numerals and hands glow in the dark. Some ^{226}Ra and zinc sulfide were added to the paint to make it glow. The painters ingested large amounts of α activity as a result of licking the tips of their brushes in order to produce fine points for painting the small numbers.

In the case of radioactive samples that emit x-rays, γ rays or high-energy β particles, we must consider both internal and external irradiation of humans. Not only is it important to prevent ingestion of these radioactive materials, but it is also necessary to shield personnel from the radiations, when the sources are outside the body. This is generally done by surrounding the samples with lead, iron, or high-density concrete. Large stationary sources of radiation such as x-ray tubes, nuclear reactors, and nuclear accelerators must have carefully designed shielding to protect workers.

Measuring Radiation Doses

Radiation doses are commonly measured in units of *rads* and *rems*. One rad is the amount of radiation that deposits 100 ergs of energy per gram of tissue. A rad of densely ionizing radiations (for example, α particles) does more damage than a rad of x rays, γ, or β^- rays. Thus to assess the biological damage caused by radiations one uses the *rem* (roentgen equivalent man). This unit is obtained by multiplying the dose in rads by the "relative biological effectiveness" (RBE), which has the value of one for β^-, γ, and x rays, but ten for α particles.

In Table 5–1 we list some of the effects observed when people are exposed to very large doses of radiation over a short time. No effect can be observed below about 25-rem doses, and 450 rem is the LD$_{50}$, that is, the dose that is lethal to about 50% of the population. The effects listed are for irradiation of the entire body with the indicated dose. One can receive much higher doses before observing the effects if the dose is limited to certain portions of the body, particularly the extremities, which are rather insensitive to radiation. The doses listed are assumed to be received over a period of a few days or less. If the dose is spread out over several weeks or

Table 5–1 Radiological Data

Expected effects of acute radiation exposure (whole body)

Death of half the exposed population (LD$_{50}$)	450 rem
Nausea and fatigue	100–200 rem
Slight temporary blood change	25–50 rem
No detectable clinical effects	0–25 rem

Radiation protection standards

Occupational exposure	5 rem/yr
Individual in general population	0.5 rem/yr (500 mrem)
Sizable sample population group	0.17 rem/yr (170 mrem)

Typical Nonoccupational Exposure

Background from natural sources in U.S.	0.08–0.2 rem (80–200 mrem)
Medical application to portions of the body	
Average chest x ray	0.2 rem (200 mrem)
Average GI tract examination	22 rem
Range for fluoroscopic examination	5–400 rem
Average yearly exposure to diagnostic x rays	0.055 rem (55 mrem)

Adapted from statement of T. J. Thompson, Hearings before the Joint Committee on Atomic Energy, Part 1, Oct. 28–31 and Nov. 4–7, 1969, p. 190.

months, the body's defense mechanisms repair much of the damage caused. Cancer patients often receive several thousand rem on small portions of the body over several months. This produces extensive damage to the irradiated area and perhaps to the blood, but the patients can survive and recover.

All of us receive radiation from natural causes. The radiation from natural background ranges from 80 to 200 mrem/yr (where 1000 millirem = 1 rem) for U.S. residents, with an average of about 170 mrem/yr. About 70% of the background comes from naturally occurring radioactive species in our environment—about one-third of this from ^{40}K present in our bodies among the atoms of the essential element potassium, and the remainder from radioactivity in bricks, rocks, concrete, and other materials around us. The other 30% of the background comes from cosmic rays—protons and other high-energy nuclear projectiles that bombard the earth from the sun and outer space. Some people receive higher background doses than others because of variations in radioactivity in the construction materials of their houses, offices, etc., and because of the change of cosmic-ray intensity with altitude (an increase by a factor of three from sea level to the "mile-high" city of Denver, Colorado).

The Federal Radiation Council and the Atomic Energy Commission have established regulations on the use of radioactive materials that are designed to limit the exposure of any major population group to radiation from man's activities (exclusive of medical uses) to less than 170 mrem/yr in addition to that of the natural background. There is considerable controversy concerning this guideline.* Regulatory bodies are under intense pressure from environmental groups and various critics to lower the guidelines on permissible radiation doses from man's activities, especially from nuclear power plants. Much of the criticism is highly emotional, with little basis in fact—perhaps because of the horrors of nuclear weapons or people's unfamiliarity with nuclear radiation.

Perhaps the most severe criticism has been that by Drs. John Gofman and Arthur Tamplin of Lawrence Livermore Laboratory, Livermore, California.† They estimate that if every U.S. resident received 170 mrem/yr from man-made

sources, the radiation would cause about 32,000 cancer-related deaths, among the 200 million Americans in addition to the 300,000 that now occur from natural causes.

Other radiation scientists question the validity of the assumptions upon which the Gofman-Tamplin calculations are based. The problem is that we have very little knowledge about the effects of very low radiation doses—below about 50 rem the incidence of radiation-caused afflictions is almost impossible to distinguish from natural and other causes. In rebuttal, it is also noted that the nuclear industry's activities today cause the average person to receive only about 1 mrem/yr—less than 1% of the guideline value. Over the next few decades, nuclear energy will come into much wider use, but even then it is most unlikely that the radiation exposures will average more than one-fiftieth of the present guideline.

It is impossible to prove or disprove the conclusions reached by Gofman and Tamplin. In the absence of definitive data, it would seem advisable to set the radiation guidelines at the lowest practical value compatible with the development of nuclear energy. In 1971, the AEC proposed new regulations applicable to all reactors using ordinary water as the coolant (see Section 5–6). The new regulations would limit exposure of persons at the plant boundary to 5 mrem/yr.

When one realizes that a person receives an average of about 55 mrem/yr from medical diagnostic x rays, one wonders whether we should be so concerned about the 1 mrem/yr we receive from the operations of the nuclear energy industry. The dose from medical x rays has dropped considerably over the years, as machines have been better shielded, more sensitive film has become available, and operators have improved practices (Figs. 5–14 and 5–15). But many people receive chest x rays every two years and the dose for each is about 200 mrem with the best equipment (and up to 700 mrem on some older equipment still in use), which is greater than the yearly exposure permitted from nuclear energy activities. One receives an enormous dose of radiation from a fluoroscope, a camera in which the physician observes the picture instantaneously on a fluorescent screen.‡ Despite the improvements in medical x-ray technology and practice, Dr. Karl Morgan, Director of the Oak

---

* For a more detailed discussion, see R. W. Holcomb, "Radiation Risk: A Scientific Problem?" *Science* **167**, 853 (1970).
† Many of their critical papers, and rebuttals of them, are collected in the Hearings before the Joint Committee of Atomic Energy, "Environmental Effects of Producing Electric Power" (see Suggested Reading).

‡ During the early 1940s, many shoestores had fluoroscopes in which the customer inserted his feet to see how well his shoes fitted. These devices were quite fascinating, especially to children, and we shudder to think what doses the children must have received! Fortunately, as noted above, the extremities are the least sensitive parts of the body to radiation.

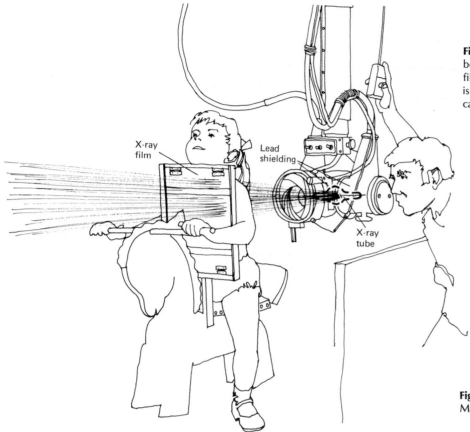

X-ray
film

Lead
shielding

X-ray
tube

Fig. 5–14   X rays penetrate the body and produce a picture on the film on the other side. The toy horse is a device used by some doctors to calm children's fears.

Fig. 5–15   Typical medical x ray (courtesy of Massachusetts General Hospital).

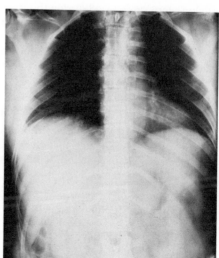

Ridge National Laboratory Health Physics Division, has suggested* that the exposure from medical diagnostic x rays could be reduced to about 10% of its present level without reduction of medically important information.†

The foregoing comments on radiation doses apply to somatic effects. Much less is known about genetic effects. Many experiments have been performed on radiation-induced mutations in fruitflies and mice, and the results suggest the rate of human mutation induced by radiation,

---

*   K. Z. Morgan, in Hearings on "Environmental Effects of Producing Electric Powers" p. 1271 (see Suggested Reading).
†   It is perhaps disturbing to note that, despite our poor knowledge of the effects of very low-level exposure to radiation, we know far less about the long-term effects of many common substances to which we are frequently exposed, e.g., caffeine, drugs and medicines, and food additives.

but direct evidence in man is almost impossible to obtain. Mutations take so long to manifest themselves and can arise from such a variety of sources that their attribution to particular causes is almost impossible. Assuming that Hiroshima survivors received about 200 rem, Muller has estimated that the dose probably caused each of their offspring to inherit an average of one mutation produced by the radiation exposure, in addition to several natural mutations, most derived from past generations.* Detailed studies of children conceived by survivors of the Japanese nuclear explosions have turned up no evidence for radiation-caused birth malformations, stillbirths, decreased birth weights, or child mortality. The only effect observed was a small shift of the male/female sex ratio over the first ten years after the explosions, but not since.†

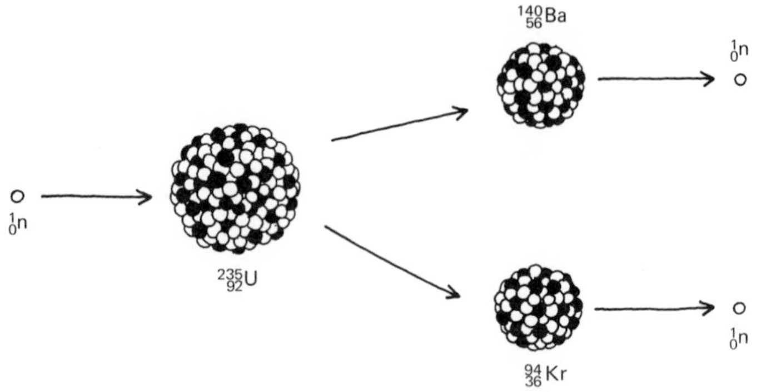

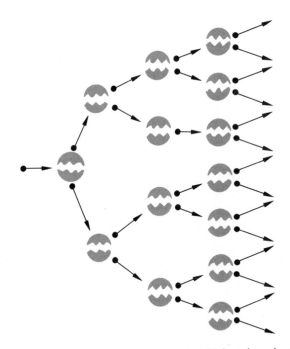

Fig. 5–16   A typical fission reaction of ^{235}U.

5–5 NUCLEAR WEAPONS

Thermal Neutron Fission and Chain Reactions

In Section 5–3 we saw that spontaneous fission is an important decay mode among heavy, man-made elements such as californium. Spontaneous fission releases enormous amounts of energy. However, it is not a practical source of energy because it occurs so infrequently among elements that are present naturally on Earth. The heaviest naturally occurring nuclide, $^{238}_{92}$U, undergoes spontaneous fission in only $10^{-4}\%$ of its decays.

The keys that unlocked the vast energies of nuclear fission were discovered during the late 1930s and early 1940s on the threshold of World War II. It was found that if the rare uranium isotope ^{235}U (0.7 % of natural uranium) is struck by a neutron, the energy deposited in the nucleus usually causes it to undergo fission (see Fig. 5–16):

$$^{235}_{92}U + ^{1}_{0}n \longrightarrow \text{2 fission fragments + 2 or 3 neutrons.} \quad (5-13)$$

The significance of the finding is that this reaction, which is initiated by a neutron, releases more neutrons than the one it uses up! This indicates that the energy of fission can be released by a *chain reaction*. Suppose that a neutron strikes

Fig. 5–17   A simple chain reaction in which each nucleus fissions when struck by a neutron, and each fission reaction releases two neutrons.

---

* H. J. Muller, "The Genetic Damage Produced by Radiation," *Bulletin of Atomic Scientists* **11**, 210 (1955).

† R. W. Miller, "Delayed Radiation Effects in Atomic-Bomb Survivors," *Science* **166**, 569–574 (1969).

one nucleus in a chunk of pure ^{235}U, causing it to fission with the release of two neutrons. If each of these neutrons strikes another ^{235}U nucleus, causing it to fission, four or more neutrons would be released in the second "generation." In Fig. 5–17, we show how the fission rate would double in each generation if neutrons from each fission event caused two new fissions. (Actually, an average of 2.5 neutrons are released per fission.) The fission rate would build up very quickly, spreading throughout the chunk of ^{235}U like a fire. However, before all the ^{235}U atoms were consumed, the enormous energy released would cause an explosion that would blow the uranium atoms apart, shutting off the reaction. This is what happens in a nuclear fission bomb.

Although the fission chain reaction in ^{235}U sounds very simple, several problems must be overcome to make it happen. First, about 99.3% of natural uranium is the isotope ^{238}U, which usually does not undergo fission when struck by a neutron. Instead, the neutron is captured to produce ^{239}U:

$$^{238}_{92}U + {}^{1}_{0}n \longrightarrow {}^{239}_{92}U. \tag{5–14}$$

The ^{239}U undergoes two β decays to form the "artificial" elements neptunium and then plutonium:

$$\text{23-min } {}^{239}_{92}U \longrightarrow {}^{239}_{93}Np + \beta^-,$$

$$\text{2.3-day } {}^{239}_{93}Np \longrightarrow {}^{239}_{94}Pu + \beta^-. \tag{5–15}$$

Although the ^{239}Pu produced is quite valuable, these reactions do not produce the neutrons needed to sustain the chain reaction. By clever tricks involving the slowing down of the neutrons emitted in fission, it is barely possible to sustain chain reactions in natural uranium. The *fast neutrons* (about 1 MeV) emitted in fission are allowed to have many collisions with atoms of the *moderator* (generally water or carbon) to give up most of their energy to become *thermal neutrons*, i.e., neutrons having the same energies ($\sim$0.03 eV) as molecules of a gas at the temperature of the reactor. Moreover, chain reactions can be sustained only in large reactors containing many tons of uranium, with very little power output. It is much easier to construct a nuclear reactor if one removes most of the ^{238}U from natural uranium to produce material that is enriched in the ^{235}U isotope. In order to produce fission weapons that can be transported by airplane and then release enormous energy in a short time, it is essential that highly enriched ^{235}U be used. However, since chemical behavior is determined almost solely by the atomic number Z and only slightly by the mass number A, it is extremely difficult to separate the small amount of ^{235}U from the ^{238}U of natural uranium.

The second problem of sustaining the chain reaction is that of losing neutrons. If the reaction is initiated in a very small piece of ^{235}U, most of the neutrons will escape to the surroundings and not cause new fission events. Thus a minimum amount of material is needed, termed the *critical mass*, in order to reduce losses at the surface of the material to the point where enough neutrons cause fissions to sustain the chain reaction. The size of the critical mass depends on several factors, including the purity of the ^{235}U. The critical mass for uranium containing a lot of ^{238}U is much greater than that of a pure sample of ^{235}U. For pure ^{235}U, the critical mass is several kilograms.

Nuclear Fission Weapons

The facts about fission noted in the preceding section were known to scientists in the United States at the start of World War II. Since fission had been discovered by German scientists in 1938, it was presumed that scientists in Nazi Germany had realized the possibility of making a bomb of enormous destructive power released by a chain reaction. Thus it was considered imperative that the United States and her allies investigate the possibility of making such weapons. The most distinguished collection of scientists and engineers ever brought together for a single project were assembled to work on the Manhattan Project—the highly secret quest to make nuclear weapons. The project was led by J. Robert Oppenheimer and included many of America's top scientists, such as Ernest O. Lawrence (inventor of the cyclotron), Glenn T. Seaborg (who has participated in the discovery of about a dozen synthetic elements), Hans Bethe, Charles D. Coryell, Harold C. Urey, and Arthur H. Compton, as well as many outstanding European scientists, including Enrico Fermi, Niels Bohr, Leo Szilard, Emilio Segrè, Edward Teller, and Eugene Wigner.

There were two possible routes to the bomb, both of them quite difficult. On the one hand, one could separate enough ^{235}U from natural uranium to make a critical mass. The second possibility was that of making ^{239}Pu (which is also quite fissionable) in reactors fueled with natural uranium (see Eq. 5–15). Because of the high stakes involved, the United States pursued both methods.

Smoke billowing 20,000 feet above Hiroshima after a ^{235}U bomb was dropped there in 1945. (Photo courtesy of the Atomic Energy Commission.)

Although the various isotopes of an element have the same chemical behavior, there are very slight differences in the velocity of gaseous molecules having different masses. Uranium was converted to a gas, uranium hexafluoride (UF$_6$), and passed through a series of porous barriers. The slightly lighter molecules of ^{235}UF$_6$ diffuse through barriers a little faster than those of ^{238}UF$_6$, so that after passage through thousands of barriers, the gas emerging is somewhat enriched in ^{235}UF$_6$. Huge diffusion plants were built to perform this enrichment process in a rural area of eastern Tennessee (now called Oak Ridge) in order to take advantage of electrical energy from the Tennessee Valley Authority. Additional enrichment of the ^{235}U was done with electromagnetic separators in which ions of uranium compounds were passed through a magnetic field that bent the path of ions containing ^{235}U in a slightly different direction from those with ^{238}U atoms.

The second path to the bomb involved the production of ^{239}Pu in large reactors built at Hanford, Washington, on the Columbia River. (Again refer to Eq. 5–15.) The uranium fuel was periodically removed from the reactor in order to separate the ^{239}Pu from the uranium. Since $_{94}$Pu and $_{92}$U are different elements, they can be separated by chemical means. However, this was an enormously difficult task. Plutonium is a man-made element that was discovered only in 1940. Because of the time pressure, the entire chemical process for removing a few grams of plutonium from tons of uranium had to be designed and studied at a time when far less than one gram of plutonium was available. Normally when a new industrial chemical process is introduced, it is first tried in a laboratory and then in a pilot plant before the full-scale plant is designed. But there was no time for that during the Manhattan Project—it had to work the first time!

As it turned out, both the ^{235}U enrichment process and the ^{239}Pu production method succeeded in producing enough material for weapons in 1945. When a bomb is constructed, it must contain more than enough ^{235}U or ^{239}Pu to make a critical mass, but the supercritical mass must not be assembled until the moment of detonation. In the bomb dropped over Hiroshima in August 1945, this was achieved by placing two pieces of ^{235}U in the bomb. To set the bomb off, the two pieces were "shot" together to form a supercritical mass. This type of bomb was never tested before it was dropped on Hiroshima, because there was enough ^{235}U for only one bomb and the scientists were confident that it would work on the basis of their calculations! (See Fig. 5–18.)

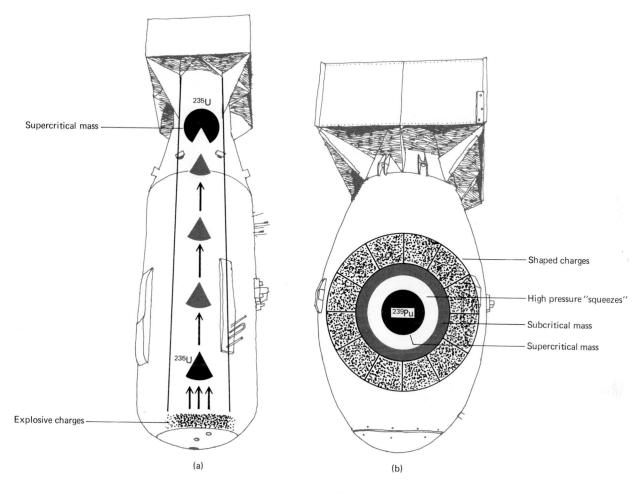

Supercritical mass

²³⁵U

²³⁵U

Explosive charges

(a)

Shaped charges

High pressure "squeezes"

Subcritical mass

Supercritical mass

²³⁹Pu

(b)

Fig. 5–18   (a) The "gun" type nuclear fission bomb with ^{235}U material. (b) The "implosion" type weapon with ^{239}Pu material.

The bomb dropped on Nagasaki was made of ^{239}Pu. For complicated reasons, it was necessary to "implode" the ^{239}Pu bomb. A nearly critical sphere of ^{239}Pu was surrounded with shaped charges of explosives. When the charges were set off, the ^{239}Pu was imploded (squeezed), making it super-critical and holding it together long enough to cause a con-siderable amount of fission. Since the implosion technique was much more tricky than that involved in the ^{235}U weapons,

a ^{239}Pu device was tested at Almagordo, New Mexico, on July 16, 1945.

The Japanese government surrendered a few days after the bombs were dropped on Hiroshima and Nagasaki. They probably thought that these bombs were just the start of a continuing series. However, no more bombs were ready for use at that time and it would have probably taken several months to prepare more.

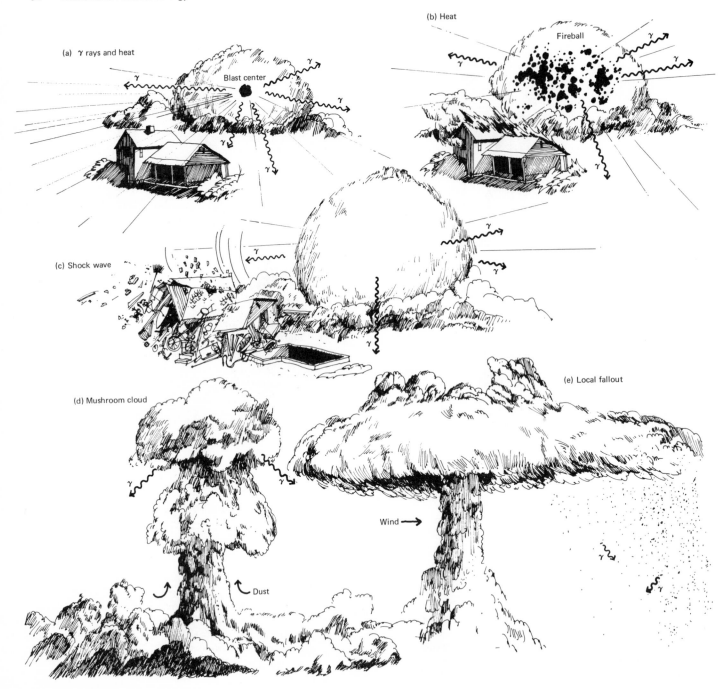

(a) γ rays and heat

Blast center

γ

(b) Heat

Fireball

γ

(c) Shock wave

γ

(d) Mushroom cloud

γ

Dust

(e) Local fallout

Wind →

γ

Fig. 5–19 Sequence of events in a nuclear weapon explosion.

Each bomb dropped on Japan released energy equal to the explosion of about 20 kilotons (20,000 tons) of the conventional explosive, trinitrotoluene (TNT). An explosion of this size corresponds to the fission of about 1 kilogram (2.2 lb) of ^{235}U or ^{239}Pu, a very small amount of material (see Fig. 4–2). The bombs undoubtedly contained more fissionable material than this, since the material flies apart before all of it fissions, but these details are still classified.

Although the fission bombs were much larger and more terrifying than any weapons previously used, they were small by comparison with the thermonuclear weapons now available, which we discuss in Chapter 6. During a series of weapons tests in 1962, the Soviet Union exploded a weapon that released an energy equivalent to about 100 megatons of TNT, i.e., about 5000 times as large as the bombs dropped on Japan.*

Rubble near ground zero in Hiroshima, 1945. (Photo courtesy of the Atomic Energy Commission.)

The Effects of Nuclear Weapons Explosions

From a technical point of view, what happens after a nuclear weapon is set off? First, about 35% of its energy is radiated as heat, which travels away from the blast with the speed of light. So much energy is released so quickly in a small area that the air in the vicinity becomes hot enough to glow. This "fireball" has a temperature of several thousand degrees Centigrade until it starts to cool off a few seconds after the explosion. The fireball expands to a diameter of about 450 ft within a millisecond and to about 7200 feet within ten seconds. The intense heat radiated from the fireball can set combustible material afire several miles away. Fires are started over an area 15 to 20 miles in diameter following the explosion of a one-megaton weapon.† The shock wave (a wave of high air pressure) travels away from the blast center with the speed of sound, about 1000 ft/s. The shock wave is the "noise" of the blast that breaks windows and knocks buildings down. (See Fig. 5–19.)

The unique feature of nuclear weapons is their nuclear radiation. Gamma rays are produced instantaneously by nuclear reactions in the bomb itself and are radiated in all directions with the speed of light. A one-megaton weapon exploded at several thousand feet (an "air burst") would cause people within several miles of the blast to receive lethal doses of radiation if exposed directly to the bomb's rays. At later times, "fallout" settles to earth, bringing with it radioactive fission products made in the explosion.

Many of the fission products have such short lives (e.g., less than 15 minutes) that they decay before they can settle to earth. Some fission products are produced with high yields and have such long half-lives that they can settle back to earth and cause extreme radioactive contamination of the area. Prominent among these are such fission products as 28-yr $^{90}_{38}$Sr, 65-day $^{95}_{40}$Zr, 67-hr $^{99}_{42}$Mo, 8-day $^{131}_{53}$I, 30-yr $^{137}_{55}$Cs, and 12.8-day $^{140}_{56}$Ba.

There are several kinds of fallout to consider. If the bomb is small (less than about 200 kilotons) and is set off on or close to the ground or water surface, large particles of dust and water droplets are drawn up into the "mushroom cloud." Nearly all the fission products become attached to the particles and settle to the ground within an area of a few thousand square miles around the target and downwind from it. This *local fallout* occurs within hours after the blast, and coats everything in the area with intensely radioactive particles. In this area, danger would arise not only from ingestion of radioactivity in food and water, but even from simply being outdoors. A calculation by Pauling‡ gives some idea of

---

*   Herbert Scoville, Jr., "The Limitation of Offensive Weapons," *Scientific American* **224** (1), 15–25 (Jan. 1971).
†   Most U.S. and Soviet intercontinental ballistic missiles carry one-megaton warheads. The small number of U.S. Titans carry five-megaton weapons and the Soviet SS-9 carries 25-megaton weapons. American B-52 airplanes originally carried 25-megaton bombs but now use larger numbers of lower-yield weapons. It is not known whether the Soviet Union has deployed its 100-megaton weapons in bombers.

‡   Linus Pauling, *No More War* (Dodd, Mead, New York, 1958), pp. **47** ff.

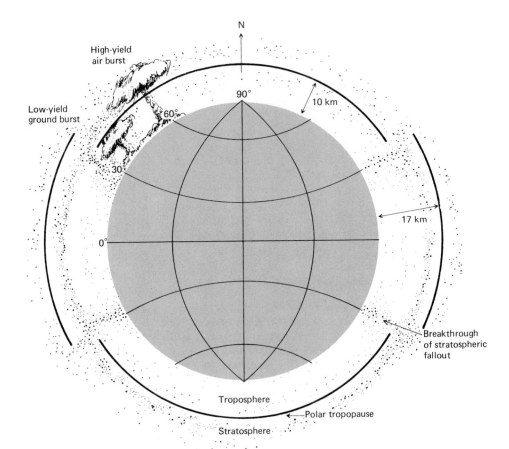

Fig. 5–20 Structure and typical temperature profile of the atmosphere.

the intensity of the radiation. He assumes that the products from an eight-megaton fission explosion uniformly cover a circle 100 miles in diameter. He estimates that the total dose of radiation at three feet off the ground would be about 3000 rem over the period from one hour after the blast to the end of the first day, with an additional 1100 rem throughout the following week. The radiation level would decrease quickly with time, because of rapid decay of the shorter-lived fission products, but since the lethal dose is about 500 rem (see Table 5–1), no one could survive many hours of exposure to local fallout. The purpose of fallout shelters is to protect people from the radiation of local fallout. It would be necessary to stay in the shelter almost constantly for the first week after a blast. During the second week, it would probably be safe to leave the shelter briefly, and at the end of two weeks, to leave it for much of the time. Of course, it is to be expected that most food and water supplies would be heavily contaminated.

If the weapon is large (one megaton or more) or is set off at an altitude of several thousand feet, local fallout may be rather small, with most of the radioactivity eventually settling to earth as *tropospheric* or *stratospheric* fallout. Near the earth, the atmosphere normally gets colder with increasing altitude (see Chapter 9). In the temperate zone, this trend continues to an altitude of about 10 km, where the temperature is typically about −55°C; at this stage, the *tropopause* intervenes, dividing the atmosphere into the *troposphere* below and the *stratosphere* above. Temperature in the stratosphere remains rather constant and then begins to increase with altitude. Radioactive clouds from large weapons set off high in the atmosphere penetrate the tropopause and deposit most of the fission products on very small particles in the stratosphere, leaving the remaining activity, generally about 5 to 25% of the total, in the troposphere (Fig. 5–20).

Transfer of material across the tropopause is slow. The fission products trapped below it in the troposphere usually

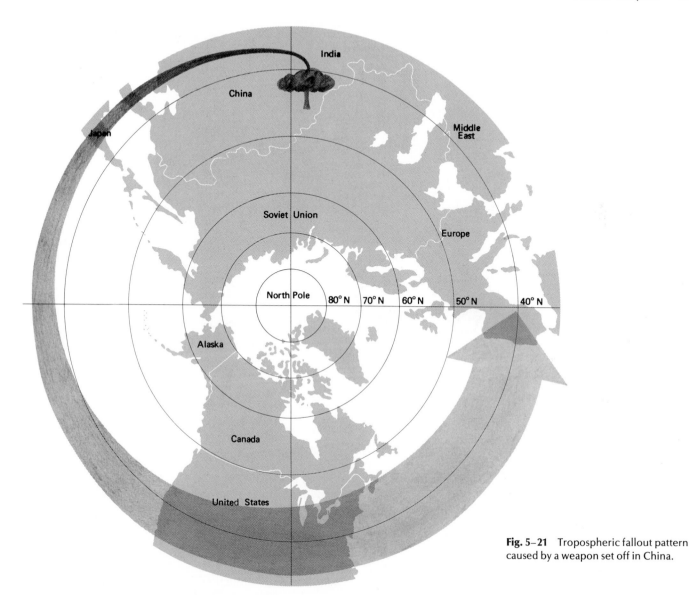

Fig. 5–21 Tropospheric fallout pattern caused by a weapon set off in China.

travel around the world from west to east, staying in a fairly narrow band of latitude. The radioactive particles slowly settle to the ground and are carried down, especially with rainfall. Thus several days after a nuclear weapons test in Central Asia, rains in the United States may bring down radioactive debris from the troposphere. The small radioactive particles in the stratosphere come down through the tropopause very slowly, so that debris from a given test shot may be re-entering the troposphere for five years or more, although for most strato-

spheric products, the average residence time is one year. Much of the stratospheric activity falls out between latitudes 30° and 40°, probably because of breaks in the tropopause at about latitude 30°. (For an illustration of this refer to Fig. 5–21.)

High-yield shots set off near the equator deposit comparable amounts of radioactivity in the stratospheres of both the northern and southern hemispheres. However, mixing across the equator is slow, so that explosions farther from

Fig. 5–22 Pathways of fallout ^{131}I to man.

the equator deposit most of their activity in the hemisphere in which they were set off.

Since tropospheric fallout does not occur until many days after the blast, species with half-lives of several days or longer produce the most contamination. Stratospheric fallout occurs on a time scale of years, so that only long-lived products such as 28-year ^{90}Sr are of importance. From various sources we estimate that U.S. citizens have received average doses of 200 to 300 mrem from fallout from weapons tests.

One particularly troublesome product in local and tropospheric fallout is 8-day ^{131}I, which may be deposited on vegetables and eaten directly by people. Or it may fall on grass

eaten by cows, the ^{131}I subsequently appearing in the cows' milk (Fig. 5–22). Much of the iodine consumed is deposited in the thyroid gland. The ^{131}I bombards the thyroid with β^- and γ rays as it decays, possibly causing cancer or other injury of the gland. Children are particularly susceptible because they drink a lot of milk, because they have small thyroid glands (causing greater concentration of the radioactivity), and because their thyroids take up more iodine than an adult's.* As a result of U.S. weapons tests in Nevada in the

---

* If a child's diet consists mostly of milk, one can avoid the radio-active material by using canned or powdered milk that has been on

Fig. 5–23 Pathways of fallout ^{90}Sr to man.

early 1950s, an estimated 430,000 children (mainly in Western states) received radiation doses of more than 20 rads to their thyroids.[†] The St. Louis Committee for Environmental Infor-

mation has estimated that the radiation will ultimately cause between 59 and 596 cases of cancer in the affected children.[‡]

One of the most prominent stratospheric fallout products is 28-year ^{90}Sr. It is concentrated strongly in cows' milk along with the chemically similar element, calcium. Children are the ones most affected by this product also, because they drink a lot of milk and because their bones and teeth, where most of the strontium and calcium are deposited, are just being formed (see Fig. 5–23). Because of the long life of ^{90}Sr, there is no simple way to prevent its consumption, as there is for ^{131}I. The best solution is to avoid further testing and use of nuclear weapons.

---

the shelf for several weeks. The eight-day half-life of ^{131}I means that its activity is cut in half every eight-days, so that in two months (about eight half-lives) it drops by a factor of more than 200.

[†]   A. R. Tamplin and H. L. Fisher, "Estimation of Dosage to Thyroids of Children in the U. S. from Nuclear Tests Conducted in Nevada during 1952 through 1955," Univ. of Calif. Lawrence Radiation Laboratory UCRL-14707, May 1966.

[‡]   "From Rads to Cancers—CNI Estimates," *Scientist and Citizen* **8** (9), 6 (1966).

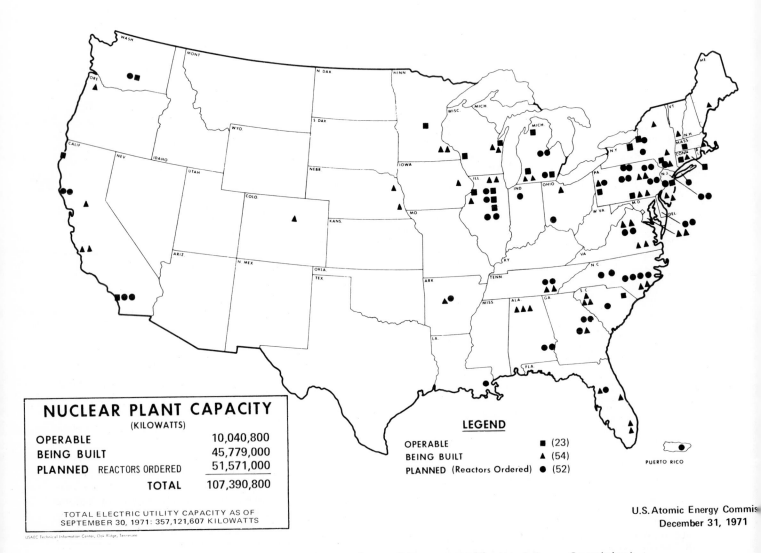

NUCLEAR POWER REACTORS IN THE UNITED STATES

NUCLEAR PLANT CAPACITY
(KILOWATTS)

OPERABLE	10,040,800
BEING BUILT	45,779,000
PLANNED REACTORS ORDERED	51,571,000
TOTAL	107,390,800

TOTAL ELECTRIC UTILITY CAPACITY AS OF
SEPTEMBER 30, 1971: 357,121,607 KILOWATTS

USAEC Technical Information Center, Oak Ridge, Tennessee

LEGEND

OPERABLE	■	(23)
BEING BUILT	▲	(54)
PLANNED (Reactors Ordered)	●	(52)

PUERTO RICO

U.S. Atomic Energy Commis
December 31, 1971

Fig. 5–24 Nuclear power plants in the United States. (Map courtesy of the Atomic Energy Commission.)

5–6 NUCLEAR POWER PLANTS

The Basic Design of Nuclear Reactors

Nuclear fission can also be used in a beneficial way as a source of enormous energy. In 1971 there were 23 operating nuclear power plants in the United States, with a total generating capacity of 10,000 megawatts (MW) of electricity. This is a small fraction of the total U.S. electrical generating capacity, (about 360,000 MW), but 54 additional nuclear plants were under construction and another 52 were in planning stages (see Fig. 5–24). By 1980 nuclear plants will probably have a generating capacity of 120,000 to 170,000 MW, about 25% of the U.S. total. By the year 2000, more than half the electricity in the United States may be generated by nuclear power plants.

The components of reactors include ^{235}U or other *fissile fuel* to sustain a chain reaction, a *moderator* to slow down the neutrons (so they will react more readily with ^{235}U), a *coolant* to extract heat from the reactor, *control rods* to limit the energy production, *shielding* to protect personnel near the reactor, and a *containment shell* to prevent escape of radioactivity to the outside (see Fig. 5–25).

In power reactors the fuel is typically in the form of pellets of uranium oxide encased in tubes of zirconium or stainless steel. There are about 35 to 40 thousand such fuel rods in a reactor core, each of them about 12 ft long. The fuel rods are surrounded by circulating water under high pressure that acts as the coolant, to remove heat from the reactor, and as the moderator.

The reactor is started up by insertion of a neutron source. Gradually the fission rate and the number of neutrons released per second increase. When power production reaches the desired level, further increase of the fission rate is stopped by inserting control rods, made of materials like cadmium (Cd) or boron (B), which have a very high probability for capturing neutrons. The control rods act as a "sponge" for neutrons, and they are inserted far enough to keep the number of neutrons and fissions constant from one generation to the next. The reactor then operates at a constant power level until it is shut down by further insertion of the control rods. While a reactor is operating, the control rods are poised ready to be dropped full length into the reactor immediately in case of any emergency.

Fission releases a tremendous amount of energy with a very small expenditure of fuel. A 500 MW power plant consumes only about 1 kg of ^{235}U per day. The energy initially appears as kinetic energy of the fragments and energies of the

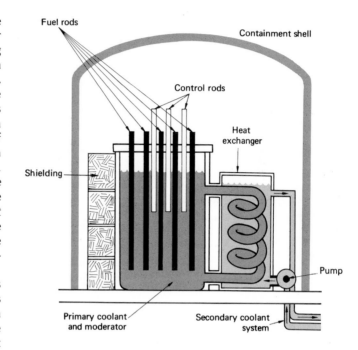

Fig. 5–25 Components of a modern nuclear reactor.

β^- and γ rays from the radioactive products. The fragments and radiations pass through the material of the reactor, depositing energy by ionizing and dissociating atoms and molecules as discussed in Section 5–4. As the material recovers from passage of the radiations, the energy is converted to heat. The water circulating through the reactor removes the heat from the reactor, to produce electricity.

There are two principal types of power reactors in operation today: pressurized water reactors (PWR) and boiling water reactors (BWR). In a PWR, the primary coolant water is at about 315°C and is kept under a pressure of about 2250 psi (pounds per square inch) to prevent boiling. This very hot water is circulated through a heat exchanger, where it gives up heat to a secondary water stream, converting it to steam at about 265°C. The superheated steam then expands into a

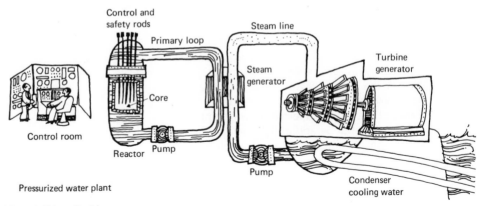

Fig. 5–26   Pressurized water reactor. [From Joseph Priest, *Problems of Our Physical Environment* (Addison-Wesley, Reading, Mass., 1973, p. 165. Reprinted by permission.]

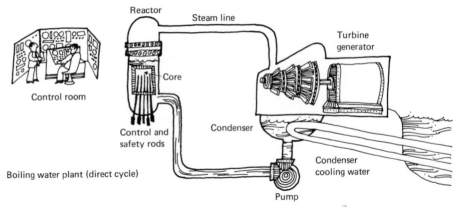

Fig. 5–27   Boiling water reactor. (From Priest, *Problems of Our Physical Environment*, p. 165. Reprinted by permission.)

turbine connected to an electrical generator to make electricity (see Fig. 5–26). A BWR typically operates at about 280°C and the pressure (1020 psi) is low enough to allow the water to boil, forming steam in the reactor, which then passes directly to the turbine (see Fig. 5–27). In both cases, water from a large source, such as lake or river, is normally used as a secondary coolant and then passed back to the source. The water is used only to cool the turbine condenser and is *not* passed through the reactor. The major problem caused by effluent cooling water is *thermal* pollution (see Section 4–5), not radioactive pollution.

Breeder Reactors

The only fissile nuclide that occurs in nature is ^{235}U. Since its abundance in natural uranium is only 0.7%, the future of nuclear fission as an important energy source would not be bright if this were the only possible fuel. Fortunately, there are *fertile* materials which, while not themselves fissile, can be converted into fissile materials. One example is ^{238}U which can be converted to 24,000-year ^{239}Pu, as in Eq. (5–15). Similarly, the principal isotope of thorium, $^{232}_{90}$Th, can be converted to 1.6×10^5-year $^{233}_{92}$U. Both ^{233}U and ^{239}Pu are like ^{235}U in that

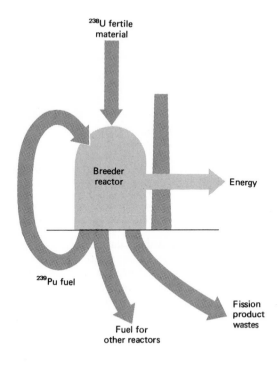

Fig. 5–28 A breeder reactor produces new fuel for itself and for other reactors at the same time that it is producing electricity.

they readily capture low-energy neutrons and undergo fission. They can be used instead of ^{235}U in nuclear reactors or weapons.

A breeder reactor converts fertile material, i.e., ^{238}U or ^{232}Th, into fissile material while simultaneously producing electricity. In an ordinary reactor, which contains some ^{238}U along with the ^{235}U fuel, some of the ^{238}U is converted to ^{239}Pu. However, a conventional reactor is not called a breeder because it uses up more atoms of ^{235}U than it produces of ^{239}Pu, whereas a breeder produces more fuel than it "burns." (See Fig. 5–28.)

One of the greatest technological challenges of the present age is that of developing power breeder reactors. By making the fertile ^{238}U into fuel, we can increase our nuclear energy reserves at least thirty-fold, assuring the world of an adequate energy supply for the next several centuries. However, breeder reactors are still in the prototype stage, and

many engineering and safety problems need to be solved before they can be a reliable source of electricity.

A breeder reactor operates best with "fast" rather than thermal neutrons. In order to keep the neutrons from losing energy in collisions with nonfissile or nonfertile atoms, the reactor must be quite small. The small size requires a coolant that is able to transfer heat out of the reactor very efficiently. Materials such as liquid sodium metal, molten salts, and helium gas have been used as coolants in experimental breeders and are the most likely candidates for use in commercial breeders.

Breeder reactors are potentially more hazardous than conventional reactors because of the dangers in handling those coolants. As most chemistry students know, a chunk of sodium dropped into water produces hydrogen gas so vigorously that it often explodes with the oxygen of the air! Furthermore, sodium becomes intensely radioactive when exposed to neutrons. On the other hand, the sodium has a safety advantage in that it does not have to be kept under the high pressure of the water used in conventional reactors.

The only commercial breeder and power reactor to have operated in the U. S., the Enrico Fermi reactor near Detroit, suffered a severe accident in October 1966 when a metal part of the reactor became detached from its position and blocked the flow of coolant in part of the reactor, causing extensive damage to the reactor, including the melting of two fuel rods. Radiation levels became quite high in the reactor building, but little if any contamination outside the building occurred. The damage in the surrounding area might have been serious had not the reactor been shut down by a reactor operator after the reactor building was automatically sealed off and alarms went off, indicating abnormal levels of radioactivity near the reactor.*

Clearly, a great deal of research and development is needed to ensure the safety of breeder reactors before they can come into wide use. Because of the severe need for worldwide energy sources, it is important that the development be done before readily available supplies of ^{235}U are "burned up" in conventional nuclear reactors. The AEC, the Tennessee Valley Authority, and the Commonwealth Edison Co. of Chicago are jointly planning the construction of a demonstration breeder reactor that will start up by 1980. The plutonium-fueled reactor will be located somewhere in eastern Tennessee.

The Hazards of Nuclear Reactors

There is little danger that a nuclear reactor out of control would undergo a vast, nuclear explosion. If the control rods were pulled out and the fission rate allowed to increase without limit, the heat generated would eventually exceed the capacity of the coolant to remove heat. The reactor would expand at the higher temperature and eventually parts of it would melt. Either the expansion or the melting would probably make the reactor subcritical and shut it off. Although a reactor will not go off like a bomb, it is potentially dangerous because of the intense radioactivity contained in it. After a typical power reactor has been running for several months, the fission products present in it produce about 10^{20} β^- and γ decays per second!

---

*   S. Novick, *The Careless Atom* (Houghton Mifflin, Boston, 1969).

Several possible dangers of such large reactors must be considered:

1. Radiation from the reactor may affect operating personnel and other people in the vicinity.

2. Small amounts of radioactivity are released to the atmosphere and water during normal operation of the reactor.

3. Much greater amounts of radioactivity may be released to the surroundings in case of an accident (fire, earthquake, etc.).

4. Radioactivity may be released from nuclear fuel reprocessing plants.

5. Radioactivity may be released from waste disposal systems.

Problem 1 is rather easily solved. Reactors are surrounded with blocks of concrete several feet thick, in order to shield personnel and nearby residents from radiations of the reactor. Radiation detectors in the building and around the area continuously monitor radioactivity. If unusually high levels are detected, the reactor is automatically shut down and warnings are sounded (although the failure of the Fermi reactor to shut down is a matter of some concern). The worker safety record of the nuclear industry is considerably better than that of most conventional industries, probably because of the extreme precautions taken in handling radioactive materials.

Concerning problem 2, federal government regulations permit reactors to release only extremely minute fractions of the enormous radioactivity they contain. Guidelines on permitted radioactive effluents from most reactors are based on the premise that no major population group should receive more than 5 mrem/yr of radioactive exposure from reactor operations (see Section 5–4). The primary coolant of a reactor is in a closed system, and the secondary coolant (e.g., stream water) never passes through the reactor at all. Most radioactive products that get into water bodies do so through small, pin-hole leaks by which a little of the primary coolant may mix with the secondary coolant. The major radioactive water pollutant is tritium, 12-yr ^3H, made by neutron irradiation of deuterium (^2H) present in all water and from neutron reactions with lithium and boron. Tritium is not a serious pollutant as it emits only low-energy β^- particles and is not concentrated in the body.

The nuclear power plant at Haddam Neck, Connecticut. (Photo courtesy of the Connecticut Yankee Atomic Power Co.)

Problem 3 involves a potentially serious threat. An almost inconceivably large amount of radioactivity is present in a reactor that has been operating continuously for some time. The amounts of 12.8-day ^{140}Ba and other products of moderate half-life that are present after several weeks of operation are almost the same as would result from the explosion of a 20-kiloton fission bomb. If an earthquake or explosion were to break open a reactor and spread its fission products over several hundred square miles, the danger would be about the same as that from local fallout from a small weapon. Thousands of deaths by radiation would occur if such an event took place near a large city. There has been much concern regarding possible "loss of coolant" accidents, i.e., episodes in which a portion of the reactor loses its coolant because of a leak. Provisions are made for adding emergency cooling water if this should happen, but if the affected portion of the reactor should get too hot, steam formed in the cooling channels could block the entry of the emergency water.* If this happened, a portion of the reactor might well melt. The melting could bring together reactive chemical species to cause a fire or explosion that could break open the containment shell and allow escape of radioactive material.

---

* R. Gillette, "Nuclear Reactor Safety: A New Dilemma for the AEC," *Science* **173**, 126–130 (1971).

There are several lines of defense against massive reactor accidents. First, the design, construction, and operating procedures of all reactors are carefully reviewed and regulated by the Atomic Energy Commission (AEC) and state regulatory bodies. Second, the air-tight containment shells that surround reactors are designed to withstand substantial explosions. Finally, not much of the fission product activity would be released even if the reactor were broken open, since most of the radioactive atoms are embedded in the fuel rods.

Problem 4, the reprocessing of fuel, is perhaps the most hazardous of routine procedures in the nuclear industry.

Control-rod mechanism of a pressurized water reactor, with the pressure vessel partially disassembled. [From D. R. Inglis, *Nuclear Energy: Its Physics and Its Social Challenge* (Addison-Wesley, Reading, Mass., 1973), p. 93. Reprinted by permission.]

After fuel rods have been in a reactor for several months or years, they must be reprocessed to replace spent fuel, to separate the ^{239}Pu or ^{233}U that has formed, and to remove the fission products. Fuel rods are removed from the reactor and stored for several months to allow much of the radioactivity to decay. Then they are transported in shielded containers to a fuel reprocessing plant where they are dissolved in acids; then the uranium is chemically purified and recast into new fuel rods. When the rods are dissolved, gaseous fission products trapped in the rod are released. Under current practice much of the volatile fission-product activity is released to the atmosphere. Much of the activity results from noble gas products, which are not very dangerous. However, methods for trapping the gaseous effluents are known and will probably be put into use as the nuclear industry grows. (See Fig. 5–29.)

The final hazard of the nuclear energy industry, problem 5, involves the "left-overs" from fuel reprocessing, many gallons of liquids containing vast concentrations of long-lived radioactive species such as 28-year ^{90}Sr and 30-year ^{137}Cs. There is no known way of rendering these activities harmless. The best that can be done is simply to keep the materials out of environmental circulation while they decay. At present, the liquid wastes (about 75 million gallons to date) are stored in large underground tanks.* In some of them, the radioactivity is so intense that the solution would boil if not cooled. There is the constant danger that a tank will develop a leak, allowing the radioactive liquid to contaminate the local groundwater. The AEC has developed a safer waste storage method by which wastes are converted to an impervious ceramic solid. The AEC had planned to store the solid wastes in a salt mine near Lyons, Kansas. However, the plan was blocked by the State of Kansas and environmental groups. Since nearby salt mines contain some water, it was feared that water might get into the radioactive storage area.† The Kansas Geological Survey feared that heat released by the radioactive wastes might cause earthquakes or other disturbances of the overlying ground. The AEC is continuing its search for appropriate underground sites. In the meantime, they will store the wastes in ceramic form in above-ground concrete-shielded tanks that can be monitored continuously for evidence of any leakage.

*   J. A. Snow, "Radioactive Wastes from Reactors," *Scientist and Citizen* **9**, 89 (1967).
†   Thomas O'Toole, "AEC to Store Atomic Waste on Surface," Washington *Post*, 19 May 72.

Laura phonecall 9:00 P.M Feb 24

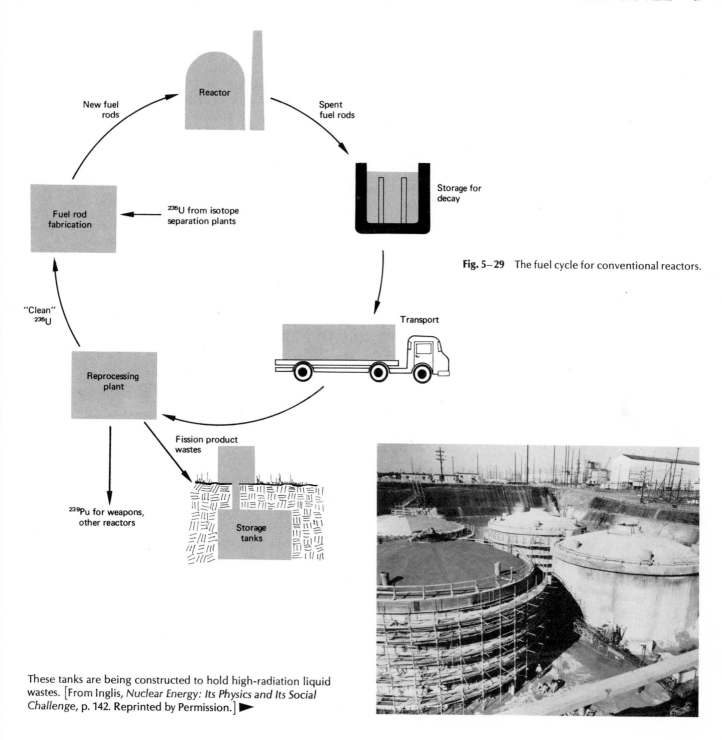

New fuel
rods

Reactor

Spent
fuel rods

Storage for
decay

Fuel rod
fabrication

^{235}U from isotope
separation plants

Fig. 5–29   The fuel cycle for conventional reactors.

"Clean"
^{235}U

Transport

Reprocessing
plant

Fission product
wastes

^{239}Pu for weapons,
other reactors

Storage
tanks

These tanks are being constructed to hold high-radiation liquid
wastes. [From Inglis, *Nuclear Energy: Its Physics and Its Social
Challenge*, p. 142. Reprinted by Permission.] ▶

In summary, nuclear reactors, both conventional and breeders, offer the promise of abundant energy supplies. There are some hazards associated with the use of nuclear energy. However, unlike fossil-fuel plants, nuclear plants emit no sulfur dioxide, nitrogen oxides, or particulate material, which now cause so much of our urban air pollution. The air quality of our cities would be greatly enhanced if we could replace fossil-fuel plants with nuclear reactors. If the clean-up of emissions from fossil-fuel plants proves to be impractical or excessively expensive, nuclear energy may be the most economical solution. With proper development, it should be both possible and not unreasonably expensive to operate nuclear reactors safely. However, officials and citizens must maintain vigilance to ensure that potentially hazardous reactors are designed and operated safely.

SUGGESTED READING

D. E. Abrahamson, "Environmental Cost of Electric Power," A Scientists' Institute for Public Information Workbook, 1970 (SIPI, 30 East 68th St., New York, N.Y. 10021, 1970).

G. R. Choppin, *Nuclei and Radioactivity* (W. A. Benjamin, New York, 1964). Paperback.

N. P. Davis, *Lawrence and Oppenheimer* (Simon and Schuster, New York, 1968). Paperback.

L. Fermi, *Atoms in the Family* (University of Chicago Press, Chicago, 1954). Paperback.

G. Friedlander, J. W. Kennedy, and J. M. Miller, *Nuclear and Radiochemistry*, 2nd ed. (John Wiley, New York, 1964).

S. S. Glasstone, *Sourcebook on Atomic Energy*, 3rd ed. (Van Nostrand, New York, 1967).

S. S. Glasstone, ed., *The Effects of Nuclear Weapons*, rev. ed. (U.S. Government Printing Office, Washington, D.C., 1962).

B. G. Harvey, *Nuclear Chemistry* (Prentice-Hall, Englewood Cliffs, N.J., 1965).

Hearings on "Environmental Effects of Producing Electric Power," Joint Committee on Atomic Energy, U.S. Congress, Oct. 1969–Feb. 1970 (U.S. Government Printing Office, Washington, D.C.).

R. G. Hewlett and O. E. Anderson, Jr., *The New World 1939/1946. Volume I of a History of the U.S. Atomic Energy Commission* (Pennsylvania State University Press, University Park, Pa., 1962).

D. R. Inglis, *Nuclear Energy: Its Physics and Its Social Challenge* (Addison-Wesley, Reading, Mass., 1973). Paperback. An excellent, more detailed account of material covered in this chapter.

G. W. Rathjens and G. B. Kistiakowsky, "The Limitations of Strategic Arms," *Scientific American* **222** (1), 19–29 (Jan. 1970).

Herbert Scoville, Jr., "The Limitation of Offensive Weapons," *Scientific American* **224** (1), 15–25 (Jan. 1971).

G. T. Seaborg and J. L. Bloom, "Fast Breeder Reactors," *Scientific American* **223** (5), 13–21 (Nov. 1970).

G. T. Seaborg and W. R. Corliss, *Man and Atom* (Dutton, New York, 1971).

E. C. Tsivoglou, "Nuclear Power: The Social Conflict," *Environmental Science and Technology* **5**, 404–410 (1971).

"Underground Nuclear Testing," *Environment* **11** (6), 2–53 (1969).

A. M. Weinberg, "Social Institutions and Nuclear Energy," *Science* **177**, 27–34 (1972).

A. M. Weinberg and R. P. Hammond, "Limits to the Use of Energy," *American Scientist* **58**, 412–418 (1970).

PROBLEMS AND QUESTIONS

1. Give a one- or two-sentence definition or explanation of each of the following terms.

   a) chain reaction   b) critical mass

   c) control rods   d) moderator

   e) implosion   f) tropopause

   g) primary coolant   h) breeder reactor

2. Complete the following nuclear reactions.

   a) beta decay: $^{60}_{27}\text{Co}_{33} \longrightarrow \beta^- +$

b) positron emission: $^{22}_{11}Na_{11} \longrightarrow \beta^+ +$

c) electron capture: $^{41}_{20}Ca_{21} \longrightarrow$

d) alpha decay: $^{239}_{94}Pu_{145} \longrightarrow {}^4_2He_2 +$

e) nuclear fission: $^{236}_{92}U_{144} \longrightarrow {}^{99}_{42}Mo_{57} + \qquad + 3{}^1_0n_1$

3. One nuclear reaction that may be involved in thermo-nuclear weapons is

$$^7_3Li_4 + {}^2_1H_1 \longrightarrow 2\,{}^4_2He_2 + {}^1_0n_1$$

How many MeV of energy is released per reaction? Take into consideration the following data on the atomic mass of each of the reactants.

^7Li: 7.016005 amu ^1n: 1.0086654

^2H: 2.0141022 amu ^4He: 4.0026036

Recall that 1 amu is equivalent to 931.5 MeV.

4. If the atomic mass of ^1H is 1.0074825, what is the binding energy (in MeV) of ^7Li? (Use any data from Problem 3 that you may need.)

5. A species of odd mass number A and a given Z and N has more neutrons and fewer protons than the most stable species having that mass number A. In order to become more stable, does the species emit an electron, emit a positron, or capture an electron?

6. In the very heavy elements, the stable atoms have considerably more neutrons than protons (see the periodic table inside the front cover). Why?

7. Arrange the following in order of their increasing ability to penetrate long distances through material: 1-MeV β^-, 5-MeV α particle, 1-MeV γ ray.

8. Which is, in general, more dangerous *inside* the body, an emitter of 5-MeV α particles (with no γ rays) or an emitter of 1-MeV γ rays (with no α's) of the same activity (disintegrations per second)? Which would be more dangerous at a distance of 1 m from a person? Relate your answers to those of Question 7.

9. Why are very heavy nuclei unstable with respect to fission and alpha decay?

10. a) One of the most important radioactive products from a nuclear fission explosion is ^{131}I, which has a half-life of eight days. A bomb of the size that exploded over Hiroshima produces about 20 grams of ^{131}I. How many grams of ^{131}I would still remain 16 days after such an explosion?

b) If ^{131}I were ingested into the body, where would it be concentrated?

11. a) We know that $^{137}_{56}Ba$ is a stable nuclide; therefore $^{137}_{54}Xe$ is unstable. By what process, β^- or EC/β^+, does ^{137}Xe decay?

b) Write an equation for the decay of ^{137}Xe, indicating the products that would be formed.

12. Helium is a very rare gas in the earth's atmosphere and crust, but it can be found in appreciable quantities in uranium mines. Why?

13. a) What is meant by the "somatic" effects of radiation? Name a few somatic effects.

b) When nuclear radiations strike human reproductive material, the genetic code can be altered, producing mutations in the offspring. Since mutations cause evolution, a strengthening of the species, one might suppose that we should increase the irradiation of our reproductive organs. What are the fallacies in this line of reasoning?

c) Some people have suggested that we need not be concerned about environmental problems, because if we do nothing about them, we will evolve into creatures with a high tolerance for air and water pollution, etc., just as bugs develop resistance to DDT. What's wrong with this very reassuring point of view?

14. Below are listed several high-yield fission products and their half-lives. Indicate which would be important constituents of (a) local, (b) tropospheric, and (c) stratospheric fallout following an atmospheric fission blast.

a) 15-min $^{89}_{37}Rb$

b) 52-day $^{89}_{38}Sr$

c) 28-yr $^{90}_{38}Sr \longrightarrow$ 64-hr $^{90}_{39}Y$

d) 10-hr $^{93}_{39}Y$

e) 65-day $^{95}_{40}Zr \longrightarrow$ 35-day $^{95}_{41}Nb$

f) 14.6-min $^{101}_{42}Mo$

g) 8.05-day $^{131}_{53}I$

h) 30-yr $^{137}_{55}Cs$

15. How does one control a reactor, i.e., turn it on and off?

16. List five principal ways in which the operation of nuclear power reactors can degrade the environment.

17. In addition to fission products, fallout from nuclear weapons contains the unused ^{235}U or ^{239}Pu from which the bomb was made. Both decay by α emission accompanied by very low-energy γ rays. Are these nuclides a hazard mainly by means of external irradiation of the body or mainly if they are ingested? Explain.

18. From your knowledge of the nature of the large amounts of radioactive waste storage from nuclear fuel reprocessing, can you suggest any uses for that "resource out of place"?

19. In the book and movie *On the Beach*, the last survivors of a major nuclear war were in Australia and New Zealand. Can you explain the logic of this scenario in terms of likely targets in a nuclear war and the patterns of fallout that would result?

20. Antiballistic missiles (ABM's) are designed to intercept incoming missiles at least several kilometers above the earth by exploding nuclear weapons near them. (Whether or not they will actually work is another question!) What kind of fallout (local, tropospheric, or stratospheric) would be most prominent after such an interception and what population would be the most seriously affected by fallout: (a) people living at 30 to 40° latitude in the hemisphere of the explosion, (b) people in the immediate area beneath the explosion, or (c) people in the same latitude band as the explosion? Would you expect the fallout to be most prominent (a) for just a few days after the blast, (b) spread out over a one- to five-year period after the blast, or (c) spread over several weeks after the blast?

21. Explain why the development of the breeder reactor is so important for the world's supply of energy. Why should it be done before conventional reactors use up most of the readily available ^{235}U?

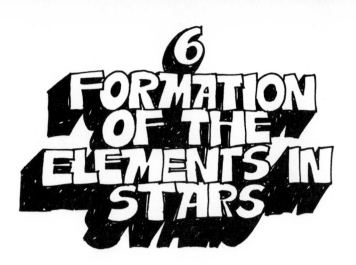

6 FORMATION OF THE ELEMENTS IN STARS

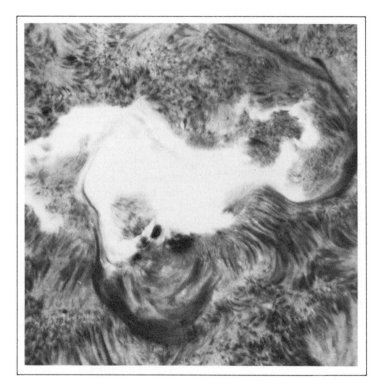

Solar flare. (Courtesy of the Hale Observatories.)

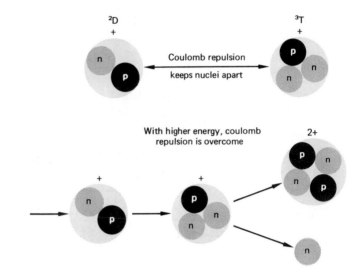

Fig. 6–1 In order to initiate fusion reactions between deuterons and tritons, the nuclei must collide with enough energy to overcome the mutual repulsion of their positive charges.

6–1 NUCLEAR FUSION

Nuclear fusion, the amalgamation of small nuclei to form larger ones, is a source of enormous energy. In the universe, it provides most of the energy radiated by stars. Here on Earth we have known fusion only in the form of awesome weapons of terror—the *thermonuclear weapons* ("hydrogen bombs") that have been tested but never used in warfare. In contrast with the great destructive threat of thermonuclear weapons, controlled nuclear fusion may in the future serve as an almost limitless source of energy for beneficial purposes. Because of the world's needs for new energy sources (see Chapter 4), the quest to control nuclear fusion is one of the most important scientific and engineering projects now in progress. Although in recent years newspapers have reported breakthroughs in the search for control of fusion, as we discuss below, there are still very large problems that must be overcome before we can tap this abundant source of energy. Although there is some justification for optimism, as yet it is not certain that nuclear fusion reactors will ever become practical sources of energy.

In trying to understand the broader aspects of man's environment, there are many questions that come to one's mind: What is the origin of the elements in the universe? How much of each element is present in the world? What is the life history of a star? What is the chemical composition and structure of Earth? We take up these questions, which depend in large part on nuclear fusion reactions, in the latter sections of this chapter.

In Chapter 5 we saw that nuclei having mass numbers A of about 50 to 60 are the most stable that occur in nature, i.e., they have the greatest amount of binding energy per nucleon (see Fig. 5–9). The binding energy per nucleon rises sharply from very small nuclei such as ^2H (deuterium, also designated ^2D), ^3H (radioactive tritium, ^3T), and ^3He to more stable nuclei such as ^4He, ^{12}C and ^{16}O. The rapid increase in binding energy shows that if the very small nuclei can be combined to form larger ones, in a process called nuclear fusion, large amounts of energy can be released.

An example of a fusion reaction is that between the hydrogen isotopes deuterium and tritium to form the very stable ^4He and a neutron:

$$^2_1D + ^3_1T \longrightarrow ^4_2He + ^1_0n + 17.6 \text{ MeV.} \qquad (6-1)$$

The energy released per reaction is small compared to the more than 200 MeV released in fission of uranium. But since uranium atoms are so much heavier than those of deuterium and tritium, the energy released *per gram of reactant* in this fusion reaction is 3.5 times as much as that from fission per gram of uranium!

In order for reactions such as (6–1) to occur, the reactants must collide with high energies to overcome repulsion between the positive charges of the two nuclei (see Fig. 6–1). The deuteron ($^2D^+$) must have at least 10 keV of energy to come close enough to the tritium nucleus to cause the nuclear reaction. Deuterons can be accelerated to that energy with high-voltage equipment, but much more energy is used in the process than is given off by the reaction because most of the accelerated deuterons miss the tritium nuclei. The only practical way in which to obtain energy from the reaction is to heat a mixture of deuterium and tritium gases to extremely high temperatures. As discussed in Chapter 2, the higher the temperature of a gas, the higher is the average energy of the atoms or molecules in the gas. At temperatures above about ten million degrees, a few of the deuterium nuclei will collide hard enough with tritium nuclei to cause reaction (6–1) to occur.*

6–2 THERMONUCLEAR WEAPONS

Temperatures high enough to cause fusion occur at the center of stars, including our sun, but are difficult to achieve on earth. The explosion of nuclear fission bombs produces extremely high temperatures of the sort needed. Soon after the explosion of the first fission bombs, attention shifted to the possible development of thermonuclear weapons: Could the heat developed by fission bombs initiate fusion reactions of small nuclei, and should the United States attempt to develop such a "superbomb"? Debate on the massive weapon raged throughout the late 1940s, but an affirmative decision was made when the Soviet Union exploded their first nuclear weapon in 1949.

Until 1951, most American plans envisioned a fission bomb surrounded by layers of solid D_2 and T_2, which would undergo reactions such as Eq. (6–1) upon initiation by the fission explosion. Since hydrogen becomes frozen only below $-259°C$, devices of this sort had to be very large and complicated in order to maintain the low temperature necessary to keep the solid T_2 and D_2 from melting and then boiling

away. These cumbersome devices could not be delivered by airplanes. The U.S. government tested various devices on islands in the Pacific as early as 1952, but probably didn't have deliverable weapons until about 1954.

The "secret of success" in making thermonuclear weapons was the use of compounds of lithium and deuterium (LiD) or tritium (LiT). These compounds are solids up to 445°C, and so they can be packed into deliverable weapons. Another advantage was that the lithium itself can undergo nuclear reactions that release energy, for example,

$$^6_3Li_3 + ^1_0n \longrightarrow ^4_2He_2 + ^3_1T_2 + 4.8 \text{ MeV.} \qquad (6–2)$$

The ^3T atoms emerge from the reaction with so much energy that they can initiate other reactions, such as (6–1).

Details of modern thermonuclear weapons are, of course, highly classified. One can guess that there is a fission "trigger" at the center, as shown in Fig. 6–2, that is fired to produce a high temperature and a burst of neutrons. The neutrons and the heat then set off thermonuclear reactions like Eq. (6–1) and Eq. (6–2) in the lithium compounds around the trigger. If there were nothing else around the bomb, it would fly apart before much of the material had reacted. To hold it together, the whole package is surrounded by a dense *tamper* material that holds the bomb together as long as possible. Uranium, the densest known element, was used as the tamper in early

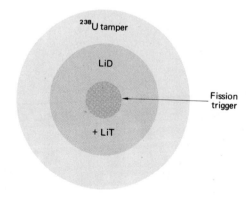

Fig. 6–2 A fission-fusion-fission thermonuclear weapon.

---

*   At such high temperatures, the molecules T_2 and D_2 (which are the normal form of these types of hydrogen at room temperature) are completely broken into atoms, and the atoms are ionized into electrons and deuterium and tritium nuclei.

Thermonuclear blast. This photo was taken at a height of approximately 12,000 feet—
50 miles from the detonation site. Two minutes after the detonation, the cloud rose
to 40,000 feet. Ten minutes later, as it neared its zenith, the cloud stem pushed up to
about 25 miles, deep into the stratosphere. The mushroom portion went up to 10
miles and spread for 100 miles. (Courtesy of U.S. Air Force.)

thermonuclear weapons. Although low-energy neutrons don't cause ^{238}U to fission, the higher-energy neutrons from the thermonuclear weapons do. Thus the explosive force of the weapon is greatly increased by nuclear fission of the uranium tamper!

The thermonuclear part of a large weapon is relatively "clean," in that it doesn't produce fission-product radioactivity that returns to earth as fallout. However, the uranium tamper creates an enormous amount of fission-product activity, making a very "dirty" weapon. In 1957, the U.S. government announced the development of a so-called "clean" bomb. Probably some nonfissioning material was substituted for uranium in the tamper, and the size of the fission trigger may have been reduced. No nuclear weapon is truly "clean," however, as all have fission triggers and the thermonuclear reactions produce radioactive tritium and neutrons. Many of the neutrons react with nitrogen of the atmosphere to produce radioactive ^{14}C:

$$^{14}_{7}N + ^{1}_{0}n \longrightarrow ^{14}_{6}C + ^{1}_{1}H, \tag{6–3}$$

a 5730-year activity that behaves like ordinary carbon, an important constituent of all living matter.

6–3 NUCLEAR FUSION REACTORS

If the high temperature of nuclear weapons could be achieved under controlled conditions in nuclear fusion reactors, fusion could supply all the energy needed by man for many centuries. The deuterium present in the hydrogen of water in the

seas could be used as a cheap, inexhaustible fuel. Although the isotopic abundance of ^{2}D is only 0.015% (one ^{2}D atom for every 6500 ^{1}H atoms), fusion of the deuterium atoms in a gallon of water would release about the same amount of energy as the combustion of 300 gallons of gasoline![*] The deuterium could be extracted from a gallon of water at a cost of less than ten cents.

Unfortunately, enormous practical problems must be overcome before useful energy can be extracted from nuclear fusion reactions. To derive energy from reaction (6–1) it would be necessary to inject deuterium and tritium gases into a container, heat the mixture to about forty million K and contain the resulting plasma (electrons and bare nuclei) while the thermonuclear reaction occurred. Since no material can withstand the enormous temperatures required, most schemes for fusion reactors rely on the use of high magnetic fields to keep the charged particles of the hot plasma from striking the container walls (see Fig. 6–3).

The United States and the Soviet Union are working on the development of fusion reactors, and recent developments are encouraging.[†] However, many problems remain to be solved before this type of reactor can produce practical, economical power.[‡] Prototype fusion reactors are able to contain the plasma for only a few thousandths of a second

---

* G. J. Mischke, "The Search for Fusion Power," *Naval Research Reviews*, April 1971, pp. 1–16.
† W. C. Gough and B. J. Eastland, "The Prospects of Fusion Power," *Scientific American* **224** (2), 50–64 (Feb. 1971).
‡ D. J. Rose, "Controlled Nuclear Fusion: Status and Outlook," *Science* **172**, 797–808 (1971).

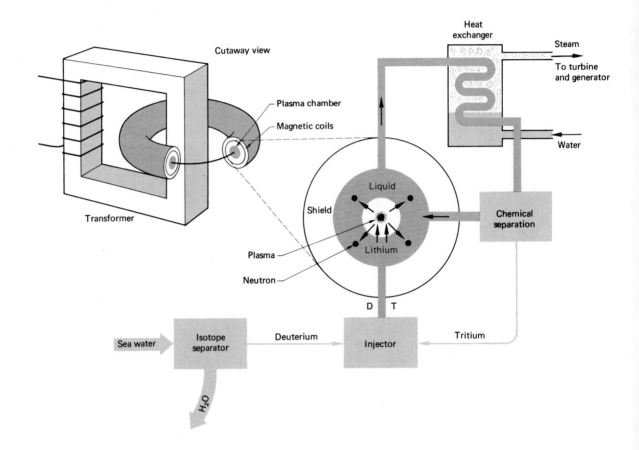

before it "leaks" out of the magnetic container—much too short a time to produce more energy than it uses. The density and temperature of the plasma are still well below the required values. Each of the three parameters—temperature, density, and duration of plasma confinement—is about a factor of ten below the value needed.

Another type of fusion reactor would require no magnetic containment. Pellets of frozen D_2 and T_2 would be dropped into the middle of a large vat of swirling liquid lithium and then be struck by an immense burst of light from a laser (see Fig. 6–4). The laser light would almost instantaneously heat the pellet to the temperatures required for fusion and much of the material would react before it had time to blow apart—

a sort of "mini" thermonuclear bomb that would go off, giving up its energy to the lithium.* The development of this type of fusion reactor appears to be closer at hand than that of the "magnetic container" reactor.

Fusion reactors would be safer than fission reactors as they produce no fission products. If all the neutrons are absorbed by the lithium, then the only radioactive product is 12-yr ^3T, which can be reinjected into the reactor as fuel. Even if some ^3T escapes, it is less hazardous than most fission products.

---

*   M. T. Lubin and A. P. Fraas, "Fusion by Laser," *Scientific American* **224** (6), 21–33 (June 1971).

Fig. 6–3 A possible fusion reactor based on magnetic containment of plasma in a toroidal (doughnut-shaped) chamber. The reaction $D + T \rightarrow$ $^4He + n$ occurs in the chamber, and the neutrons escape with most of the energy into the liquid lithium.

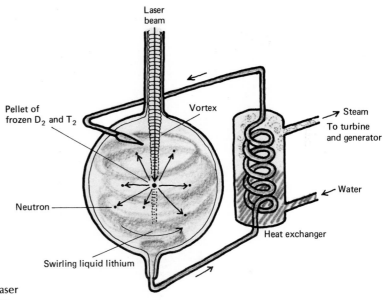

Fig. 6–4 A possible fusion reactor in which a high-powered laser beam initiates fusion in a pellet of frozen deuterium and tritium.

6–4 ABUNDANCE OF ELEMENTS IN THE SOLAR SYSTEM

In this and the following section we wish to consider two major related questions: How did the elements of our environment originate? And how do stars generate their enormous energies?

How do we learn about the distribution of elements in the solar system? First, we know something about the elements that make up the earth. Geochemists have made chemical analyses of the gases in our atmosphere, the materials dissolved in the oceans, and rocks at the earth's surface. With the use of seismographs, geophysicists observe the velocity of shock waves that pass through Earth following earthquakes or

explosions. From this information, we know approximately the density of rocks in the earth at various depths. Most important, we know that the earth has a dense solid inner core surrounded by a molten outer core that extends out to about half of Earth's 6400-km radius (see Fig. 6–5).

Earth's core is probably made up largely of iron (Fe) and nickel (Ni), with a few percent of sulfur (S) or silicon (Si). Since we can dig into the earth only a few kilometers, chemical data about Earth's interior is obtained from analyses of *meteorites*.

Meteorites are objects that fall onto the earth from somewhere in space. On a clear night you may have observed "falling" or "shooting" stars—"stars" that suddenly move across the sky. Most "shooting stars" are *meteors*—small

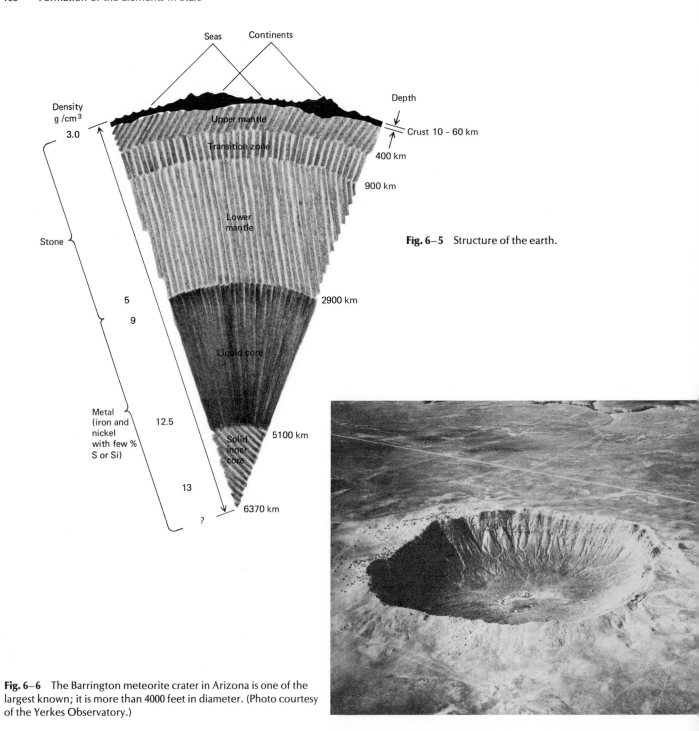

Seas   Continents

Density
g /cm³

3.0

Depth

Crust 10 – 60 km

Upper mantle

400 km

Transition zone

900 km

Lower
mantle

Stone

5

9

2900 km

Liquid core

Metal
(iron and
nickel
with few %
S or Si)

12.5

5100 km

Solid
inner
core

13

6370 km

?

Fig. 6–5   Structure of the earth.

Fig. 6–6   The Barrington meteorite crater in Arizona is one of the
largest known; it is more than 4000 feet in diameter. (Photo courtesy
of the Yerkes Observatory.)

Fig. 6–7　Orbits and relative sizes of the planets of the solar system. Note the asteroid belt between Mars and Jupiter.

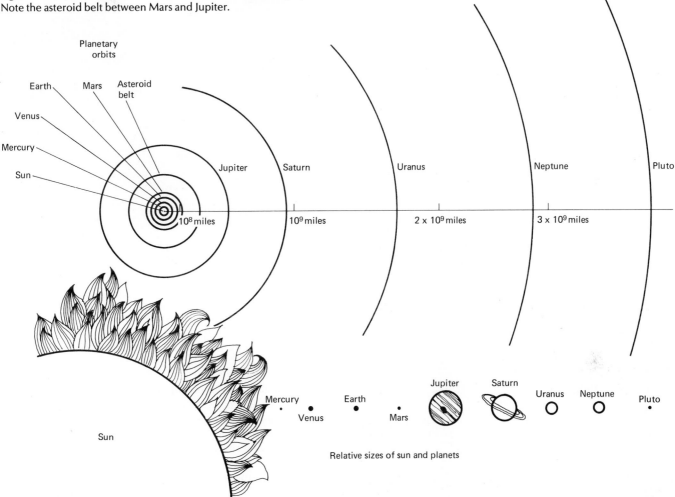

Relative sizes of sun and planets

objects from space that are attracted by the earth's gravity. The friction of Earth's atmosphere causes them to get hot enough to glow like stars, and most of them completely burn up before they hit the ground.* The few that are large enough to survive and strike the ground are called meteorites (see Fig. 6–6).

---

* All of us have become familiar with this effect via flights of manned spacecraft. Upon re-entry into the atmosphere, the leading surface of the spacecraft, the "heat shield," becomes so hot that pieces of the fiery material burn off ("ablate"), carrying much of the heat with them, thus protecting the astronauts.

There are three major classes of meteorites: the "irons" made up largely of iron and nickel, "stones" which consist of rocky material, and "stony-irons" composed of a mixture of stone and iron-nickel. Most weigh only a few kg or less, but some of the iron-nickel meteorites are immense. The largest one known was found at Hoba West Farm, near Grootfontein, South West Africa, and weighs about sixty tons.

The origin of meteorites is not certain, but they are thought to be fragments of a planet or planets that once existed. By comparing the orbits of the planets of the solar system, one would expect a planet between the orbits of Mars and Jupiter. There is no planet at that distance from the sun, but there is a group of *asteroids* there, i.e., many objects much smaller than planets, the largest of which (Ceres) has a diameter of about 800 km (see Fig. 6–7). One or more planets may once have existed in that orbit and later have broken apart into the asteroids. Many scientists believe that the meteorites are pieces of material from the asteroid belt.

Meteorites provide clues about the interior of the earth. The core of the earth is probably similar in composition to the iron-nickel meteorites. Stony meteorites contain minerals that are rare at Earth's surface but that are probably predominant at the high temperatures and pressures in the stony mantle of the earth. From analyses of meteorites we can tell where most of the elements went when the earth separated into core, mantle, and crust. Elements such as gold ($_{79}$Au), platinum ($_{78}$Pt), iridium ($_{77}$Ir), and osmium ($_{76}$Os), for example, would have had a strong affinity for molten iron-nickel. They have probably gone mostly into Earth's core, accounting for their scarcity (and thus their high value) at the crust of the earth.

Moon rocks being collected by scientist-astronaut Harrison H. Schmitt during Apollo 17 expedition to the moon in December 1972. (Courtesy of NASA.)

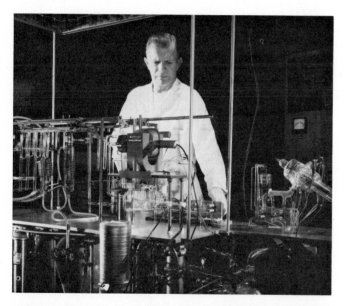

A high-sensitivity rare-gas mass spectrometer used in chemical research. This instrument is capable of detecting as little as one billionth of a cubic millimeter of gas. It is used in studies of the age of meteorites and investigations of the bombardment of satellites in space by helium atoms. (Courtesy of Brookhaven National Laboratory.)

Since 1969, samples from the surface of the moon have been brought to Earth for study. Analysis of these samples will provide some information about the composition of the solar system, but the major findings will probably have a stronger bearing on our understanding of the formation of the moon and its relationship to the earth.

A major source of information about the composition of the solar system is the light emitted by the sun and the planets. The sun is so hot that we obtain light from atoms near the sun's surface. As described in Chapter 7, atoms of a given element give off light of particular well-defined colors when heated to a very high temperature. By analyzing the colors in the light that reaches Earth, we can learn about the composition of material near the surface of the sun (and other stars). The planets are not large enough to have the high temperature and pressure at their centers to generate their own energy, as the sun does. However, when sunlight strikes the planets, it is either reflected or absorbed by them and re-emitted into space. By analyzing the light from planets, we can obtain information about their chemical compositions. In this way, for example, it is found that the very dense atmosphere of Venus contains a large amount of carbon dioxide. The atmosphere of the largest planet, Jupiter (318 times Earth's mass), contains large amounts of methane (CH_4) and ammonia (NH_3).

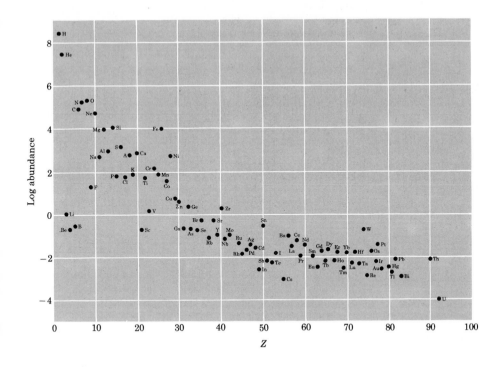

Fig. 6–8 Relative abundance of the elements in the solar system plotted on a logarithmic scale. [From Gregory R. Choppin and Russell H. Johnsen, *Introductory Chemistry* (Addison-Wesley, Reading, Mass., 1972), p. 24. Reprinted by permission.]

Data from these diverse sources have been pieced together, with guidance from chemical and nuclear theory, to estimate the abundances of all the elements in the solar system.* Table 6–1 lists a few of the important elements and groups of elements and their relative abundance. In Fig. 6–8, the abundances of the elements are plotted versus their atomic numbers. The relative abundance of the 14 most prominent elements in the earth's crust, the whole earth, and meteorites is listed in Table 6–2.

Note in Table 6–1 that the lightest two elements, $_1$H and $_2$He, make up 98% of the weight of the solar system despite the fact that neither element is among the top fourteen elements of the whole earth. The reason for this difference is that the sun comprises about 99.8% of the mass of the solar system; thus, the relative abundance of elements in the solar system is

Table 6–1. Abundance (by Weight) of Elements in the Solar System

Element		%
Hydrogen:	^{1}H	71%
	^{2}H	0.01%
Helium:	^{4}He	27%
	^{3}He	0.006%
Masses $A = 5$ and 8		0%
$_3$Li, $_4$Be and $_5$B		10^{-6}%
$_6$C, $_7$N, $_8$O, $_{10}$Ne ($_9$F)		1.8%
$_{11}$Na through $_{22}$Ti		0.2%
Iron group ($50 < A < 62$)		0.02%
$63 < A < 100$		10^{-4}%
$A > 100$		10^{-5}%

*   For more details, see L. H. Ahrens, *Distribution of the Elements in Our Planet* (McGraw-Hill, New York, 1965); also see H. E. Suess and H. C. Urey, "Abundances of the Elements," *Review of Modern Physics* **28**, 53–74 (1956).

Table 6–2.   Abundance of the Most Prominent Elements (by Weight) in the Earth and Meteorites

Earth's crust		Whole earth		Stony meteorites*	
Element	%	Element	%	Element	%
O	46.6	Fe	34.6	O	33.2
Si	27.7	O	29.5	Fe	27.2
Al	8.1	Si	15.2	Si	17.1
Fe	5.0	Mg	12.7	Mg	14.3
Ca	3.6	Ni	2.4	S	1.9
Na	2.8	S	1.9	Ni	1.6
K	2.6	Ca	1.1	Ca	1.3
Mg	2.1	Al	1.1	Al	1.2
Ti	0.44	Na	0.57	Na	0.64
H	0.14	Cr	0.26	Cr	0.29
P	0.11	Mn	0.22	Mn	0.25
Mn	0.095	Co	0.13	P	0.11
F	0.063	P	0.10	Co	0.09
Ba	0.043	K	0.07	K	0.08

Data from B. Mason, *Principles of Geochemistry*, 3rd ed. (John Wiley, New York, 1966).

*   Data are for "chondritic" meteorites, the major class of stony meteorites. The major elements of iron meteorites are: 90.8 % Fe, 8.6 % Ni, and 0.6% Co.

nearly the same as in the sun alone which consists mainly of those two elements. When it was being formed, the earth may have had considerably more H and He than it does today. However, the earth is such a small planet that its gravitational force is not sufficient to prevent the escape of lightweight H_2 molecules and He atoms from its gravitational field, and so most of those gases would have been lost. The composition of the larger planets such as Saturn and Jupiter, which have much stronger gravitational fields than the earth, is more similar to the sun's.

The composition of the earth's crust is not the same as that of the entire earth. Elements such as Fe, Ni, Co, and Cr are more prominent in the whole earth because of the large amounts of them present in the earth's core. The rocks of the mantle consist mainly of the first four elements of the whole earth, Fe, O, Si, and Mg, as do stony meteorites, whereas crustal rocks contain increased amounts of Al, Ca, Na, K, Ti, and Ba. Hydrogen is more prominent in the crust because large amounts are combined with oxygen in the water of the oceans and in crustal minerals.

6–5 NUCLEAR REACTIONS IN STARS: FORMATION OF THE ELEMENTS

Now that we have an idea of the composition of the solar system, can we explain the formation of these elements by nuclear reactions in stars?

The most widely accepted theory of the origin of the universe is the *big bang theory*.* Present evidence suggests that the universe started out about 12 to 13 billion years ago with a so-called "big bang"—a hot, primordial ball of neutrons or protons and electrons. Nuclear reactions in the primordial "fireball" may have produced much of the helium now observed in the sun and stars before it blew up, expanding to fill much of the universe with a very diffuse concentration of matter, mostly hydrogen originally present or from decay of neutrons (see Fig. 6–9).

Just by chance, regions of high-density matter would occasionally be formed in the diffuse gas. Once a region of high density is formed, it grows like a snowball: the region of higher density has a gravitational attraction for other material in the vicinity and a star may be formed.

The embryonic star grows by collecting more material, mostly hydrogen. The center of the star grows hot from the energy released when distant molecules fall into the star. High pressures are developed as the mass of material outside the center grows. When the temperature reaches about ten million degrees celsius and the central density of the star is about 100 g/cm³ (about five times as great as lead or uranium), the star is able to initiate nuclear reactions that produce tremendous energy.

One of the most important initial reactions in a hot mass of hydrogen is

$$_{1}^{1}H + {_{1}^{1}H} \longrightarrow {_{1}^{2}D} + {_{1}^{0}\beta^{+}} + 1.4 \text{ MeV.} \tag{6–3}$$

Once reactions such as this are "turned on," they release enough energy to prevent further collapse of the star, which maintains about the same size during this period of "hydrogen burning."

Deuterons formed in Eq. (6–3) can react with additional hydrogen:

$$_{1}^{2}D + {_{1}^{1}H} \longrightarrow {_{2}^{3}He} + 5.5 \text{ MeV,} \tag{6–4}$$

*   For a more detailed, but understandable discussion of the topics covered in this section, see I. Iben, Jr., "Globular-Cluster Stars," *Scientific American* **223**(1), 26–39 (July 1970).

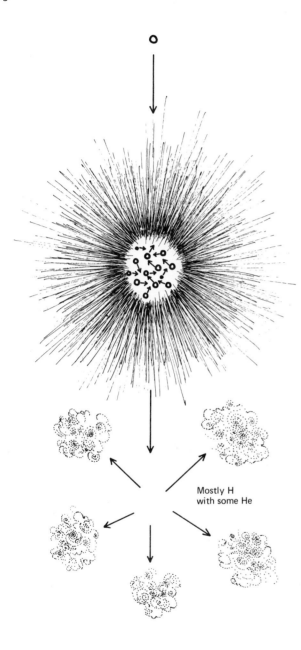

Mostly H
with some He

◀ **Fig. 6–9**   The "big-bang" hypothesis of the origin of the universe.

and, finally, two ^{3}He nuclei from reaction (6–4) can react to form ^{4}He:

$$^3_2\text{He} + {}^3_2\text{He} \longrightarrow {}^4_2\text{He} + {}^1_1\text{H} + {}^1_1\text{H} + 12.9 \text{ MeV}. \qquad (6\text{–}5)$$

Combining reactions (6–3) through (6–5), we see that the net reaction is the combination of four ^{1}H atoms to form one ^{4}He atom:

$$4\,{}^1_1\text{H} \longrightarrow {}^4_2\text{He} + 2\,{}^0_1\beta^+ + 26.7 \text{ MeV}. \qquad (6\text{–}6)$$

This is the main reaction that produces the energy of the sun and other "main sequence" stars. As the star gets hotter, it expands to form a *red giant*, and hydrogen burning takes place in a shell some distance out from the center.

Several problems are encountered when one tries to imagine ways in which heavier elements are built up. One difficulty is that there are no stable atoms of mass $A = 5$ and 8. It is difficult to get past those masses up to the heavier elements that we know exist. When hydrogen burning is complete, the core collapses somewhat, heats up, and then helium burning is initiated. One might suppose that two ^{4}He nuclei would react to form ^{8}Be nuclei; however, ^{4}He nuclei are so stable that ^{8}Be would almost immediately break apart into two ^{4}He nuclei again. But if a third ^{4}He nucleus strikes the ^{8}Be before it has broken up, then the stable nucleus ^{12}C can be formed:

$$3\,{}^4_2\text{He} \longrightarrow {}^{12}_6\text{C} + 7.6 \text{ MeV}. \qquad (6\text{–}7)$$

This reaction occurs at a density of 10,000 g/cm^3 and a temperature of 100 million degrees celsius (see Fig. 6–10). Stars in which these conditions for "helium burning" exist are called *horizontal-branch stars*. By bringing three ^{4}He nuclei together to form a ^{12}C nucleus, we have leap-frogged the region of masses $A = 5$ to 11. And we have accounted for the very small abundance of the elements $_3$Li, $_4$Be, and $_5$B that lie in that region (see Table 6–1 and Fig. 6–8).

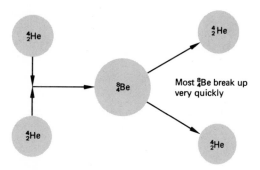

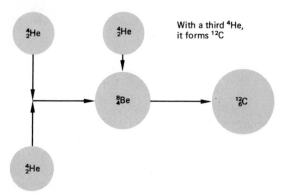

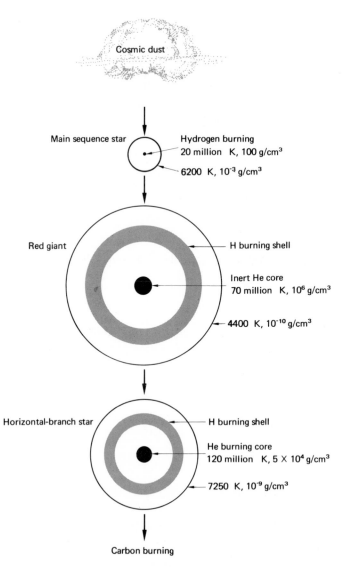

Fig. 6–10 An atom of ^{12}C is occasionally formed when three ^{4}He nuclei come together.

Fig. 6–11 A hydrogen-burning star. When it burns out, the core collapses, pressure and temperature increase, and ^{4}He burning is initiated. ▶

When a helium-burning star contains a lot of ^{12}C, other reactions become possible (see Fig. 6–11). For example,

$$^{12}_{6}C + ^{4}_{2}He \longrightarrow ^{14}_{7}N + ^{2}H + energy, \qquad (6-8)$$

$$^{12}_{6}C + ^{4}_{2}He \longrightarrow ^{16}_{8}O + energy. \qquad (6-9)$$

After most of the helium is consumed, the star collapses somewhat, converting some of its potential gravitational energy into heat. At about 600 million degrees celsius, reactions of heavier nuclei become possible, for example,

$$^{12}_{6}C + ^{12}_{6}C \longrightarrow ^{24}_{12}Mg + energy,$$

$$^{12}_{6}C + ^{16}_{8}O \longrightarrow ^{28}_{14}Si + energy, \qquad (6-10)$$

$$^{12}_{6}C + ^{12}_{6}C \longrightarrow ^{23}_{11}Na + ^{1}_{1}H + energy.$$

These carbon-burning reactions yield isotopes of many elements up to about iron. Among the most prominent are the *alpha-particle nuclei*—species made from an integral number of ^{4}He nuclei, such as $^{12}_{6}C$, $^{16}_{8}O$, $^{20}_{10}Ne$, $^{24}_{12}Mg$, and so on up to $^{40}_{20}Ca$. As shown in Fig. 6–8, elements that have one of these species as an isotope are abnormally abundant.

The reactions discussed above release energy, which heats up the star, keeps it from collapsing, and causes it to radiate light. There is a fundamental change of this situation in the vicinity of $^{56}_{26}Fe$, the most stable nucleus that exists in nature (see Section 5–2). Above the region of ^{56}Fe, most stellar reactions *do not release energy*, rather they require energy in order to occur at all. Thus additions of ^{4}He nuclei, reactions of ^{12}C nuclei, and similar reactions do not build up nuclei much beyond ^{56}Fe. Note in Fig. 6–8 that the relative abundance of elements with atomic number greater than $_{26}$Fe drops off rather sharply.

There are two principal routes by which nuclides above ^{56}Fe can be built up, both involving addition of neutrons. Some reactions involving light nuclei release neutrons, for example,

$$^{13}_{6}C + ^{4}_{2}He \longrightarrow ^{16}_{8}O + ^{1}_{0}n + energy. \qquad (6-11)$$

The neutrons released strike other nuclei in the star, including those of ^{56}Fe and heavier species.

Neutron capture by a stable nucleus often produces a nucleus that has too many neutrons to be stable and undergoes β^{-} decay to an isotope of the next higher element. For example, when a neutron strikes an atom of stable ^{58}Fe, the ^{59}Fe produced decays to ^{59}Co:

$$^{58}_{26}Fe + ^{1}_{0}n \longrightarrow ^{59}_{26}Fe,$$

$$\qquad (6-12)$$

$$45\text{-day } ^{59}_{26}Fe \longrightarrow ^{59}_{27}Co + \beta^{-}.$$

This is the so-called *s* process—neutron capture on a *slow* time scale. It may be many years between one neutron capture and the next for a given nucleus, and so there is time for β^{-} decay between captures. Neutron capture followed by β^{-} decay slowly produces heavier and heavier nuclei along the line of β stability and account for the existence of a large fraction of the naturally occurring species above ^{56}Fe (as well as some below it). Some process other than the *s* process is needed to explain the formation of other nuclides. Figure 6–12 shows some of the nuclei near the line of β stability in the region of Zn to Sr. The arrows show the pathways by which nuclei heavier than ^{68}Zn are built up by the *s* process. The isotopes $^{70}_{30}Zn$, $^{76}_{32}Ge$, $^{82}_{34}Se$, $^{86}_{36}Kr$, and $^{87}_{37}Rb$ are stable and occur in nature in reasonable isotopic abundance, yet they are bypassed in the *s* process. For example, consider ^{70}Zn: neutron capture by ^{68}Zn produces ^{69}Zn, a 14-hour activity that decays to form ^{69}Ga. Since ^{69}Zn exists for only a few hours before decaying, there is little chance that it will be struck by a neutron during its brief existence and form ^{70}Zn.

The clearest evidence for a more rapid process is the existence of $^{232}_{90}Th$, $^{235}_{92}U$, and $^{238}_{92}U$ in nature. There is no way to get from $^{209}_{83}Bi$ to those nuclei by a slow process, because most of the intermediate nuclei are too short-lived for α and/or β^{-} decay. Thus a very rapid burst of neutrons is needed to build up nuclei through that region of very short half-lives.

The most dramatic example of the *r* process, neutron capture on a *rapid* time scale, occurs in very large stars after completion of nucleosynthesis up to ^{56}Fe. Once such a star contains mostly ^{56}Fe, there is no further source of nuclear energy and the inner part of the star collapses. If the star is considerably larger than our sun, the collapse is most spectacular—the great pressures and temperatures (about four billion degrees) at the core initiate many nuclear reactions that release neutrons, for example,

$$^{56}_{26}Fe + energy \longrightarrow 13\, ^{4}_{2}He + 4\, ^{1}_{0}n. \qquad (6-13)$$

The energy released by collapse, which occurs in times as short as 10 to 1000 seconds, blows material from the outer

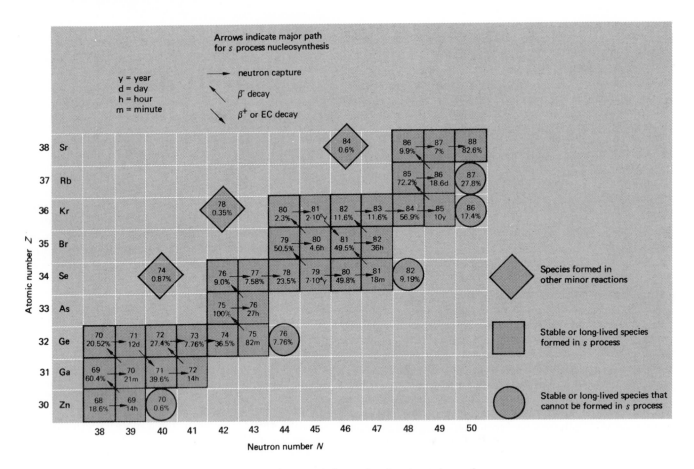

Fig. 6–12 Chart of the isotopes in the region of $_{30}$Zn and above, showing the pathways for build-up of elements via the *s* process and indicating isotopes that require other reactions for their formation.

parts of the star into space (see Fig. 6–13). Stars that undergo such spectacular explosions are called *supernovas*. Supernovas occur about once every hundred years per galaxy. About one star in a hundred ever explodes as a supernova. For two weeks after the explosion, the remnants of the star radiate more energy than a billion suns.* Material from the star's surface is ejected at nearly the speed of light. The most dramatic example of a supernova now observable in the sky is the Crab Nebula, the remnants of a star that exploded in A.D. 1054 and has continued to expand in the sky ever since.

Neutrons released in reactions in the core bombard the outer portion of the star with such an intense flux of neutrons that large numbers of neutrons can be absorbed by a nucleus before α and β^- decay can occur. For example, ^{56}Fe may quickly capture 23 neutrons to form ^{79}Fe which has far too many neutrons to be stable and decays by a rapid series of β^- decays, eventually to form stable $_{35}^{79}$Br. During the few seconds of intense neutron bombardment, nuclei with mass numbers all the way up to $A = 238$ and beyond are made. Since our solar system contains elements in the uranium-thorium region, it is clear that some of the cosmic dust from which the solar system was made was once in stars that became supernovas. This is not surprising, as the life cycles for

*   P. Gorenstein and W. Tucker, "Supernova Remnants," *Scientific American* **225** (1), 74–85 (July, 1971).

Fig. 6–13 Supernova: explosion of a large star.

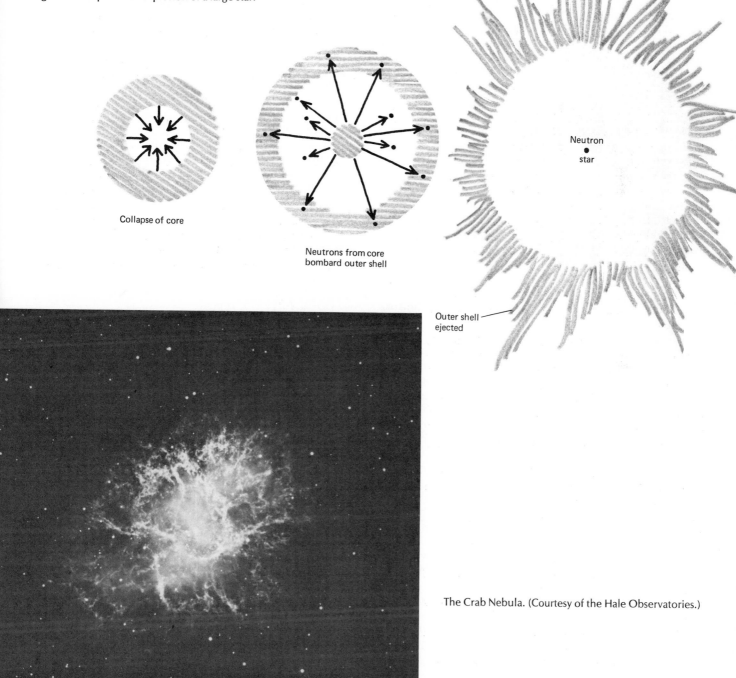

Collapse of core

Neutrons from core
bombard outer shell

Neutron
star

Outer shell
ejected

The Crab Nebula. (Courtesy of the Hale Observatories.)

large stars capable of becoming supernovas are much shorter than those of small stars. Since the solar system was formed only 4.5 to 5 billion years ago (the universe is probably about 12 to 13 billion years old), there would have been sufficient time for massive stars to form, pass through their various stages and then explode before the solar system was formed.

Artificial supernova explosions have been performed on Earth. In the test of a thermonuclear device at Eniwetok in November 1952, the device was apparently surrounded by ^{238}U which was subjected to intense neutron bombardment in the explosion. After the explosion, airplanes were flown through the radioactive cloud to collect products of the explosion on filter papers. Analysis of the filters revealed products all the way up to $^{255}_{100}Fm$ in the debris. (The weapons test was one of the methods used in the discovery of the new artificial elements with atomic numbers $Z = 99$ and 100, later named einsteinium and fermium in honor of Einstein and Fermi.)

These results showed that the flux of neutrons was so intense that ^{238}U nuclei absorbed as many as 17 neutrons before undergoing β^- decay toward nuclides on the line of β stability.

Only a minority of stars are massive enough to die in a "blaze of glory" as a supernova. When stars the size of our sun or smaller exhaust their nuclear fuels, they collapse to form stars known as *white dwarfs*. Although these stars have about the same mass as our sun (over 300,000 times that of Earth), their radii are comparable with that of Earth. Thus white dwarfs are extremely dense—in the range of 10^4 to 10^8 g/cm^3 (by comparison, the densest material known on each is uranium, 19.1 g/cm^3).

The cores of most stars left over from supernova explosions become *neutron stars*, whose radii are thought to be only about 10 km, with a density of about 3×10^{14} g/cm^3, nearly the same as that of atomic nuclei. At the high density and pressure near the center of the star, electrons acquire such high velocities that they strike nuclei and become absorbed, converting nuclear protons to neutrons which are given off as free neutrons. We say that the electrons "dissolve" nuclei in the star, converting them to free neutrons. The star becomes a huge nucleus consisting almost entirely of neutrons!

In 1967, astronomers using radio telescopes observed pulsating radio waves coming from outer space—signals that came at regular intervals that ranged from 0.03 to 4 s. At first it was speculated that these signals were being transmitted by intelligent beings on planets near other stars.* However, it is now thought that the more than 60 known "pulsars" are neutron stars, whose rapid rotation cause the pulsation of the radio signals. Evidence supporting this explanation is the fact that there is a pulsar in the Crab Nebula, as one would expect following a supernova explosion.

More massive stars are thought to collapse to even greater densities than those of nuclei (to about 10^{16} g/cm^3) to form *black holes*, so called because no light can escape from them. Nearly all our knowledge of black holes, their properties and even their existence, is based on theoretical predictions. According to Einstein's theory of general relativity, the gravitational force of a massive object deflects light rays passing near it. The gravitational field of a black hole would be so strong that light rays emitted by it would be curved back into it. Thus no light or matter could escape from the black hole (see Fig. 6–14). Since black holes give no light or other radiation, it is impossible to observe them directly with telescopes.

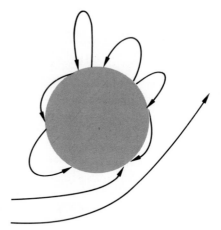

Fig. 6–14   The strong gravitational field of a black hole prevents the escape of light rays.

However, there are indirect ways in which it may be possible to confirm their existence and infer their properties. These methods depend on the fact that astronomers occasionally observe binary star systems—two stars close together that

*   A. L. Hammond, "Stellar Old Age," *Science* **171**, 994, 1133, and 1228 (1971).

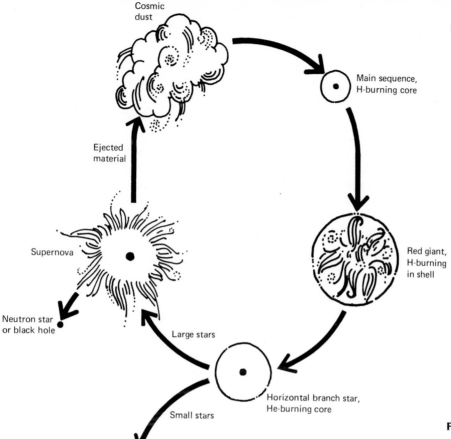

Cosmic dust

Main sequence, H-burning core

Ejected material

Red giant, H-burning in shell

Supernova

Neutron star or black hole

Large stars

Horizontal branch star, He-burning core

Small stars

White dwarf

Fig. 6–15   The life cycle of stars.

rotate about their mutual center of mass. If such a binary system includes a black hole and an ordinary star, the presence of the black hole can perhaps be determined by observation of light from its visible partner. Also, material ejected by the ordinary star will be strongly attracted to the black hole. Collisions between atoms approaching the black hole would cause emission of light which could escape into space if far enough away from the center of the body. Thus, material approaching the black hole may cause a visible halo around the object which emits no light itself. Indirect observations of two possible black holes by the above methods have been reported, but the evidence is not yet universally accepted as proof of the existence of black holes.*

* D. E. Thomson, "Black Holes: No Longer Hypothetical," *Science News* **103**, 28–29 (1973).

Figure 6–15 summarizes the life cycles of stars. They start giving off light, as "main sequence" stars, in their hydrogen-burning phase. When hydrogen in the core is exhausted, the density of the core increases and hydrogen-burning takes place in shells farther out from the core. The outer portions of the star expand to about fifty times the original radius of the star and become somewhat cooler in the "red-giant" phase. Continued collapse of the core produces densities and temperatures high enough for helium-burning to occur in "horizontal-branch" stars.

Beyond that point, small stars collapse to form white dwarfs. More massive stars become supernovas, throwing off much of their mass, with their cores collapsing to form neutron stars or black holes, depending on the amount of material in the core. Heavy elements and other debris from supernovas can be incorporated into new stars that form later, as in the case of our sun and the solar system. The more massive the star is, the more rapidly it passes through its life cycle. Stars massive enough to become supernovas probably have lifetimes of 100 million to one billion years. Thus there were probably many more supernova explosions occurring early in the history of the universe than there are today.

Stars of mass about the same as or less than that of our sun live for many billions of years. The relatively small stars in globular clusters at the edges of our galaxy were probably formed very early in the 12 to 13 billion-year life of the universe, as they are quite deficient in the heavy elements that would have been formed in pre-existing stars. The light we receive from the most distant observable stars indicates that the universe is continuing to expand from an initial, primordial fireball. If the universe contains sufficient mass (especially in the unobserved black holes), gravity will halt the expansion at some stage and cause the collapse of the entire universe back into a massive black hole. Or will it be like the original fireball and start the universe on a new cycle of expansion? Has the universe been through previous cycles prior to the one that seems to have started about 12 to 13 billion years ago?* We cannot answer these questions, which border on philosophy and religion.

SUGGESTED READING

L. H. Ahrens, *Distribution of the Elements in Our Planet* (McGraw-Hill, New York, 1965).

B. J. Bok, "The Birth of Stars," *Scientific American* **227** (2), 49–61 (Aug. 1972).

C. D. Coryell, "The Chemistry of Creation of the Heavy Elements," *Journal of Chemical Education* **38**, 67–72 (1961).

N. P. Davis, *Lawrence and Oppenheimer* (Simon and Schuster, New York, 1968).

W. A. Fowler, *Nuclear Astrophysics* (American Philosophical Society, Philadelphia, 1967).

P. Gorenstein and W. Tucker, "Supernova Remnants," *Scientific American* **225** (1), 74–85 (July 1971).

F. Heide, *Meteorites*, translated by E. Anders and E. R. DuFresne (University of Chicago Press, Chicago, 1964). Paperback.

Brian Mason, *Principles of Geochemistry*, 3rd ed. (John Wiley, New York, 1966).

J. P. Ostriker, "The Nature of Pulsars," *Scientific American* **224** (1), 48–63 (Jan. 1971).

Roger Penrose, "Black Holes," *Scientific American* **226** (5), 38–55 (May 1972),

M. J. Rees and J. Silk, "The Origin of Galaxies," *Scientific American* **222** (6), 26–35 (June 1970).

M. A. Ruderman, "Solid Stars," *Scientific American* **224** (2), 24–31 (Feb. 1971).

PROBLEMS AND QUESTIONS

1. Why is it necessary to have very high temperatures to initiate fusion reactions (e.g., in thermonuclear weapons or in the center of stars)?

2. a) If the isotopic abundance of ^2D is 0.015% of natural hydrogen, what is the weight of ^2D contained in 1 g of water?

   b) If fusion of 1 g of deuterium (employing ^6Li in secondary reactions) releases about 1.1×10^{19} ergs, how much energy would be released by fusion of the deuterium contained in one cubic kilometer of water?

   c) In Table 3–2 we saw that gross input of energy in the U.S. in the year 2000 is expected to be about 4×10^{19} calories per year. If we were to continue using energy at that rate in future years, for how many years could we supply all the energy by fusion of the deuterium contained in the waters of the Great Lakes (see Fig. 11–10)?

3. From what four sources do we get information about the cosmic abundance of the elements?

4. a) The earth has a mass of 5.98×10^{27} g and a radius of 6371 km. What is its average density?

   b) The sun has a mass of 2×10^{33} g and a radius of 6.95×10^5 km. What is its average density?

   Note that you have calculated an *average* value of the density. In any planetary body, the density is greater than

---

*   J. A. Wheeler, "Our Universe: The Known and the Unknown," *American Scientist* **56**, 1–20 (1968).

average near the center and decreases to much smaller values at the surface and in the atmosphere.

c) If a black hole has an average density of about 10^{16} g/cm^3 and a mass twice that of the sun, what is its approximate radius?

5. Let us assume that a planet similar to earth in size and structure broke up to form the meteorites that now fall to earth. What would be the expected ratio of stony to iron meteorites falling to earth, assuming that both types are of about the same average size and have about the same chance of surviving burn-up upon entry into the earth's atmosphere? Meteorites observed to fall on Earth show a stones/irons ratio of about 20/1.

6. How does the existence of thorium and uranium here on earth demonstrate that the material of our solar system was made in part by neutron bombardment on a rapid time scale?

7. The number of atoms of a radioactive species decreases to half its initial value over a period of time known as its *half-life*. Since the half-life of ^{238}U is 4.5 billion years, which is about equal to the age of the solar system, we know that there was twice as much ^{238}U present in the solar system when it was first formed as there is now.

a) The half-life of ^{235}U is 7×10^8 years. In present naturally occurring U, the ratio ^{235}U/^{238}U is 0.72/99.27 = 0.00725. What would have been the ratio ^{235}U/^{238}U when the solar system was first formed? Just give an approximate answer.

b) Would it have been easier or more difficult to operate reactors with natural uranium as a fuel at that time than it is now? Explain.

8. What is meant by a fission-fusion-fission nuclear weapon?

9. Why are Li, Be, and B in such low abundance in the solar system?

10. What is the difference between the *r* and *s* processes in a star?

7 LIGHT AND ITS RELATION TO ATOMIC STRUCTURE

Intense radiation emitted by a used fuel element from a nuclear reactor causes water of the cooling basin to give off visible light. (Courtesy of the Savannah River Operations Office of the Atomic Energy Commission.)

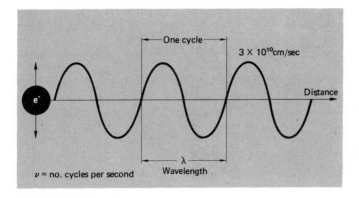

Fig. 7–1 A light wave (or "electromagnetic radiation") of frequency v cycles/s and wavelength λ travels a distance c (3 × 10^{10} cm) per second.

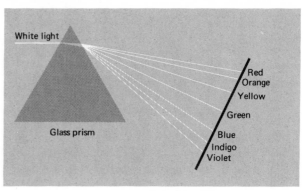

Fig. 7–2 One can spread out the colors of the sun's white visible spectrum by passing the light through a prism.

7–1 THE NATURE OF LIGHT

Visible light is a form of *electromagnetic radiation*—an oscillating electric and magnetic field. One could make a light wave by shaking a charged object such as an electron up and down, but one would have to do this about 10^{15} times per second to make visible light. That would be a physical impossibility for a person, but atoms are able to accomplish it. Electrons in a radio aerial also oscillate back and forth to produce radio waves, which are closely related to light.

When light is produced by oscillation of a charge, the light travels away from the source with the speed of light, which travels a distance of c = 3 × 10^{10} cm/s (in vacuum). The *frequency* of the light, v, is equal to the number of oscillations (or *cycles*) per second.* Since light goes through a distance c in one second and makes v oscillations over that distance, the length for each oscillation, called the *wavelength* λ, is equal to c divided by the frequency v:

$$\lambda \text{ (wavelength)} = \frac{c \text{ (speed of light)}}{v \text{ (frequency)}}, \qquad (7-1)$$

as shown in Fig. 7–1.

---

* Frequencies are now often given in the unit of *hertz* (Hz), in honor of H. R. Hertz, who made important discoveries in the field of radio waves. One Hz is the same as one cycle per second. Radio stations announce their assigned broadcast frequencies in terms of "megahertz."

The different colors of light have different frequencies and hence different wavelengths. Sunlight contains essentially all the possible colors of visible light, as we can demonstrate by passing sunlight through a prism (Fig. 7–2). The light emerging from the prism is spread out into the various colors that make up the "white light" of the sun. The passage of sunlight through droplets of water similarly spreads out light into its colors to produce rainbows.

A beam of white light can also be spread out into the spectrum of its component colors by allowing it to strike a *diffraction grating*, a piece of glass ruled with many closely spaced lines (about 1000/cm). The diffraction of light scattered from or passed through a grating demonstrates the wave nature of light. Consider light waves emerging from neighboring slits of the grating (see Fig. 7–3). Light rays A and B travel different distances to reach point C. If the difference in path lengths is equal to a whole number of wavelengths of light, rays A and B will be "in phase" when they reach point C; i.e., both will have crests and valleys at the same points. This *constructive interference* between the different waves causes a bright line of light of that particular wavelength at point C. At any given angle, there will be some wavelength of light that meets the condition of constructive interference.

You have probably seen constructive interference when sunlight is reflected from oil floating on a puddle of water. In this case, there is a difference in path length between rays reflected from the oil and water surfaces. Since the thickness

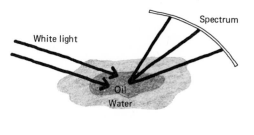

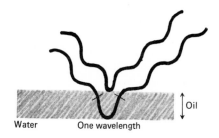

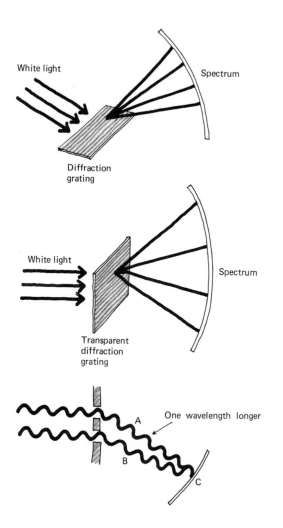

Fig. 7–4 Diffraction of light from an oil slick.

Fig. 7–3 Passage of light through a diffraction grating or scattering from a grating causes constructive interference when path lengths differ by a whole number of wavelengths.

of the oil slick is comparable to the wavelengths of light, we can see the "rainbow" of constructive interferences at various angles to the sunlight (see Fig. 7–4).

Visible light represents only a very limited range of frequencies or wavelengths of the *electromagnetic spectrum*. As shown in Fig. 7–5, the spectrum ranges from γ rays with wavelengths of about 10^{-11} cm or less emitted by nuclei (see Section 5–3) to very long-wave radio waves with $\lambda = 10^{10}$ cm. There is nothing particularly unusual about visible light. It just happens that radiations with wavelengths between about 4000 Å (violet) and 7500 Å (red) cause chemical reactions to occur in the receptors of the retina of the eye. These biochemical reactions are eventually translated in the brain into the signals that we perceive as light.

At frequencies just below that of visible red light are the infrared (IR) waves. They are not visible but produce heat which can be felt. At even lower frequencies are microwaves (important in radar and microwave ovens), and various kinds of radio waves employed by FM radio, television, "shortwave" radio, and AM radio. At frequencies above the visible

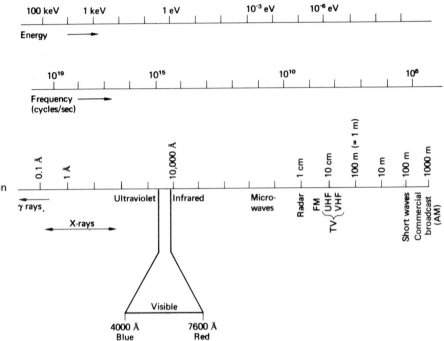

Fig. 7–5 Visible light is only a very small portion of the total electromagnetic spectrum. Scales showing the frequencies and energies (in eV) of the photons are shown below the wavelength scale.

spectrum, there is ultraviolet (UV) radiation, which tans light-colored skin and forms vitamin D in the body. At still higher frequencies are x rays and γ rays, whose properties were discussed in Chapter 5. All electromagnetic radiations have the same nature as visible light, but with different frequencies and wavelengths.

The phenomena discussed above relate to the wave nature of light. But light has aspects of both particles and waves. The wavelength and frequency of light relate to its wave nature. However, light is produced in the form of individual units or "particles" called *photons*. The energy E of each photon of light is related to the frequency v by

$$E = hv, \qquad (7-2)$$

where

h = Planck's constant

= 6.62×10^{-27} erg-s or 4.14×10^{-15} eV-s.

We have already encountered this phenomenon in Section 5–3. We saw that when a single nucleus drops from one energy state to a lower one, the energy difference appears as a single γ-ray photon of that energy. A person who has stayed in the dark for a half-hour or more, so that his eyes are dark-adapted, can perceive a flash of light when the retina absorbs as few as five photons (about 10 eV of energy).

When individual atoms of a particular element are excited, they give off photons of light of particular colors (frequencies) characteristic of the element. For example, if

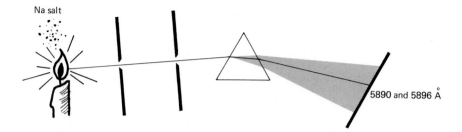

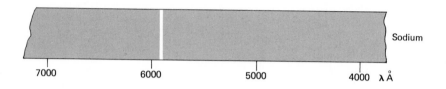

Fig. 7–6   Spectrum emitted by sodium atoms in a flame.

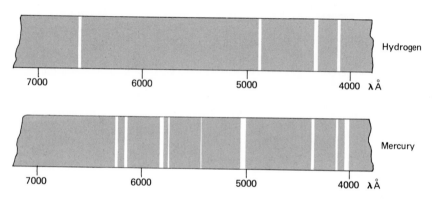

Fig. 7–7   Spectra of excited hydrogen and mercury atoms.

a salt or solution containing sodium atoms is dropped into a hot flame, the sodium atoms give off a brilliant yellow light. If the light is passed through a prism, the spectrum obtained consists of two closely spaced yellow lines with wavelengths $\lambda = 5890$ and 5896 Å (see Fig. 7–6). Using Eqs. (7–1) and (7–2), we can calculate the frequency and the energy of a single photon of light of one of these wavelengths, say the 5890 Å light (5.890×10^{-5} cm). The frequency is

$$\nu = \frac{c}{\lambda} = \frac{3.00 \times 10^{10} \text{ cm/s}}{5.890 \times 10^{-5} \text{ cm}} = 5.10 \times 10^{14}/\text{s}.$$

Each photon of that wavelength and frequency has an energy

$$E = 4.14 \times 10^{-15} \text{ eV-s} \times 5.10 \times 10^{14}/\text{s} = 2.11 \text{ eV}.$$

Thus a photon of 2.11 eV is emitted when an electron in a sodium atom drops from one energy level to another one that is 2.11 eV lower.

When mercury vapor or hydrogen gas is placed in a sealed tube and a high voltage is applied, the elements give off the more complicated spectra shown in Fig. 7–7. Every element gives a characteristic spectrum of lines different from that of all other elements. The spectrum of an element is always the same, whether the element is made in the laboratory or excited by the high temperature on the surface of the sun. These spectra are called bright-line or emission spectra—just a few bright lines are observed on a dark background when the light emitted by the atoms is passed through a prism.

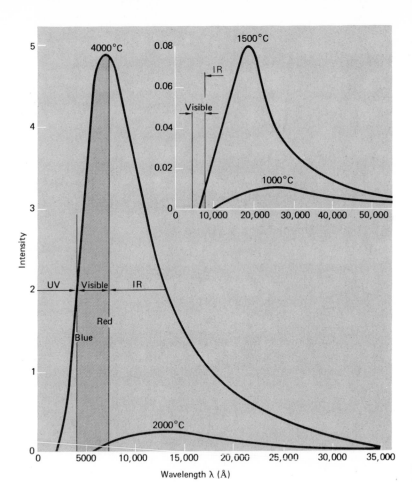

Fig. 7-8 Continuous spectra of objects at different temperatures.

There are also *continuous spectra*. For example, if a piece of iron is heated to "red hot," agitation of the atoms and electrons becomes so vigorous that light of every frequency over a wide spectrum is emitted, similar to that of the sun. The intensities of the various colors of the continuous spectrum depend almost solely upon the material's temperature and little upon its chemical composition.

Every object that has a temperature above absolute zero emits a continuous spectrum of light. In Fig. 7-8, we show the continuous spectra emitted by objects in several different temperature ranges. We do not see this light except from very hot objects because our eyes are sensitive only to frequencies in the visible range, which are not emitted strongly until the temperature gets quite high. A piece of tungsten (W) at room temperature reflects a gray light. If heated with a torch, it emits more and more photons of high frequencies as it warms

up. Long before we can see the light, we can feel the heat produced by the invisible IR light it gives off.

With continued heating, the first light appears in the low frequency (low energy, long wavelength) red region of the visible. Upon further heating, the temperature becomes high enough to cause the intensity of the blue part of the spectrum to build up, so that the light becomes more "white." The tungsten has changed from "red hot" to "white hot." The tungsten filament of a light bulb is heated to white hot by the passage of electricity and gives off light.

Even our bodies give off light. The temperature of our bodies is so low that we see each other only by reflected light. However, if we turned off the lights in the classroom and took a picture with a camera containing film sensitive to IR light, persons in the room would show up clearly, because their body temperature is about 98.6°F (310 K), whereas objects in the room are only about 72°F (295 K). Infrared photography has many important applications. For example, in an IR picture one can clearly observe the mixing of warm water from a power plant with the cooler ambient water of a river or lake (see p. 53). There are also many military applications of IR sensitive detectors, such as devices to guide a missile towards the hot tailpipe of a jet plane.

Both the bright-line and continuous spectra discussed above are emission spectra, light given off. On the other hand, if white light is passed through a gas, certain frequencies of the light are absorbed—those whose energies are just right to raise electrons of the gas atoms from one state of energy to another. Those frequencies of light are missing in the *dark-line spectrum* emerging from the gas (see Fig. 7–9). Such spectra are also called *absorption spectra*.

Examples of all these types of spectra are provided by observation of the sun's light. The 6000 K temperature of the outer portions of the sun produces a continuous spectrum in which about 46% of the energy is emitted in the visible portion of the electromagnetic spectrum, 9% in the UV region, and 45% in the IR region (see Fig. 7–10). Some of the elements

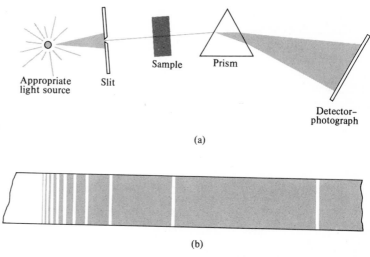

(a)

(b)

Fig. 7–9 Dark-line spectra produced when certain frequencies are absorbed as a continuous spectrum passes through a vapor. [From Paul F. Weller and Jerome H. Supple, *Chemistry-Elementary Principles* (Addison-Wesley, Reading, Mass., 1971), p. 106. Reprinted by permission.]

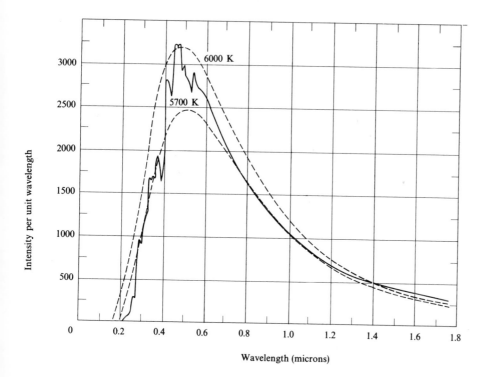

Fig. 7–10 The continuous spectrum actually given off by sunlight (solid black line), compared with two theoretical spectra at 5700 K and at 6000 K. The term "micron" is another name for micrometer. [After R. G. Fleagle and J. A. Businger, *An Introduction to Atmospheric Physics* (Academic Press, New York, 1963).]

in that hot portion of the sun's surface give off characteristic bright-line spectra. Light from the hot surface passes through the cooler atmosphere of the sun, which absorbs certain lines, thus producing dark-line spectra. These complications make it difficult to analyze the sun's spectrum and compute its chemical composition, as discussed in Chapter 6. Observation of light from planets, the sun, and stars is further complicated by absorption of some frequencies by Earth's atmosphere—the reason why space scientists would like to place an observatory in orbit beyond the earth's atmosphere.

7–2 QUANTUM THEORY OF ATOMIC STRUCTURE

By 1913 it was known that atoms consist of positively charged nuclei surrounded by negatively charged electrons. It was thought that each electron orbits the nucleus in the same way that a satellite orbits the earth. Electrons in orbits farthest from the nucleus would have the highest energy and those in closer would have lower energies. When an electron drops from an outer orbit to an inner one, the energy difference ΔE between the two orbits is given off as a photon of light with frequency $v = \Delta E/h$ (recall that h is Planck's constant). See Fig. 7–11.

Two important questions could not be answered by theories then in existence. First, since the electron is strongly attracted to the positively charged nucleus, why doesn't the electron fall into the nucleus? Second, why do atoms of a given element emit photons of just certain frequencies of light, instead of a wide variety of frequencies, when their electrons drop from high-energy to low-energy orbits? This question was based on the observation that only electron orbits of certain energies occur in atoms of a given element.

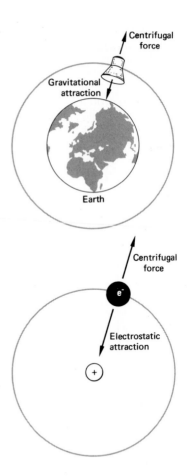

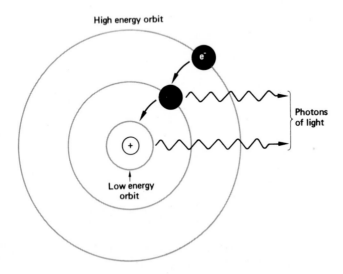

Fig. 7–11 When an electron drops from a high energy level in an atom to one of lower energy, the energy is given off as a photon of light.

Fig. 7–12 In both satellite and electron orbits (in the Bohr model), the attractive force inward is balanced by the outward centrifugal force.

Electron orbits of the type envisioned early in this century were similar to those of a satellite orbiting the earth. A satellite in a stable orbit at a given altitude must have a particular velocity so that the pull of the earth's gravity is exactly balanced by the outward-directed *centrifugal force*—the force you feel if you tie a rock on a string and twirl it around, or the force you feel when you go around a sharp curve in an automobile. Similarly, an electron in a stable orbit would have a velocity just right to cause its centrifugal force to exactly balance the attraction of the positive nucleus for the negative electron (see Fig. 7–12). A satellite can achieve stable orbit at any desired altitude (if high enough to avoid "drag" by the atmosphere, which would brake the satellite) by traveling at the required velocity—the lower the altitude, the higher must be the velocity to provide enough centrifugal force to overcome the increased gravitational pull.

Similar considerations govern electron orbits with one important difference—only a few of the infinite number of conceivable electron orbits close to the nucleus actually exist. This fact is demonstrated by the limited number of spectral lines of light emitted by the atoms of a particular element.

In 1913, Niels Bohr, a Danish theoretical physicist, made a very important breakthrough in the understanding of atomic structure. Although Bohr's theory contained some radical new ideas, it was rather quickly accepted because his model of the hydrogen atom accurately predicted the energies of lines observed in the hydrogen spectrum. Figure 7–13 shows the circular orbits of the electron predicted by the theory and the energy of an electron in each of the various orbits. When an electron drops from an outer orbit to one closer to the nucleus, the energy difference between the two states is given off as a photon. At the top of Fig. 7–13 we show how the various groups of spectral lines in the UV, visible, and IR regions are produced when the electron falls from the outermost orbit to the inner orbits.

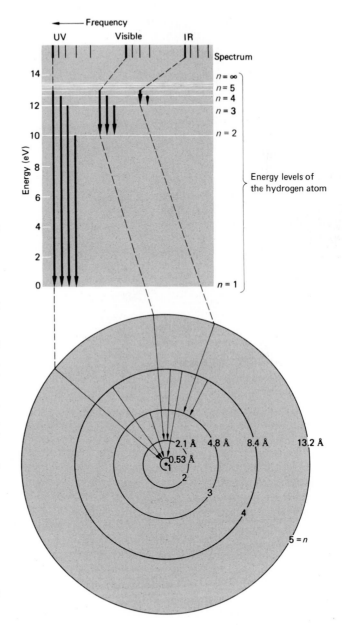

Fig. 7–13 The Bohr model of the hydrogen atom. Radii of the various orbits are shown at the bottom and the energy of the electron when in the various orbits is shown in the center of the diagram. When an electron drops from one level to a lower one, the energy difference is given off as a photon of light. These transitions, depicted by arrows, produces the hydrogen spectrum shown at the top.

Niels Bohr, 1885–1962, whose 1913 model for atomic structure correctly predicted the observed spectrum of light emitted by hydrogen atoms.

Despite the success of the Bohr model of the hydrogen atom, it became clear by the early 1920s that it could not explain a number of observations on more complicated atoms. During the middle 1920s, the great German theoretical physicists, particularly Werner Heisenberg and Erwin Schrödinger, extended the ideas of Bohr to construct new and fundamental concepts of physics called *quantum mechanics*, or *wave mechanics*. These developments, along with Einstein's theories of relativity, represent the greatest advances in physics of this century, equal in importance to Newton's Laws of Motion of the seventeenth century.

The details of quantum theory are so mathematically complex that we cannot present them here. However, we can discuss some of the implications of the theory for very small objects. These conclusions are quite different from those we arrive at by observing objects large enough to be seen and weighed. Some of the more unusual results are the following.

1. Wave nature of particles. Just as light has both wave and particle properties, objects that we normally think of as particles, e.g., electrons and protons, *also have wave properties.* Except at velocities near the speed of light, the wavelength of a particle is given by

$$\lambda = \frac{h}{mv},$$

(handwritten annotation: h = Planck's constant, m = mass of the particle, v = velocity)

(7–3)

where h is Planck's constant ⌊used to obtain photon energies in Eq. (7–2)⌋, m is the mass of the particle, and v is its velocity. A 100-eV electron, for example, has a wavelength of 1.2 Å. Since this wavelength is in the range of the spacings between atoms of a crystal lattice, it is possible to test the theory by scattering electrons from crystals (see Fig. 7–14). The waves of electrons exhibit constructive interference at certain angles, just as light rays do when they pass through a diffraction grating, thus demonstrating the wave nature of particles.

2. Quantization. One of the most unusual features of quantum mechanics is that some variables can have only certain discrete values and none in between. One of these variables is angular momentum. When an object of mass m has a velocity v in a circular orbit of radius r, its angular momentum is

$$\text{A.M.} = mvr.$$

(7–4)

According to quantum mechanics, electrons and other particles can have only discrete amounts of angular momentum:

$$\text{A.M.} = \frac{nh}{2\pi},$$

(7–5)

where n is an integer and h is Planck's constant. If the theory is correct, the angular momentum of anything that rotates is restricted to values given by Eq. (7–5)—even a bicycle wheel. You don't notice the quantization of bicycle wheels because the value of h is so small compared to the angular momentum of the wheel that the jump from one value of n to a value $n + 1$ produces an extremely minute change in its speed of rotation. But if the value of h were about 10^{33} times greater than it is, the bicycle could move only at certain speeds such as 1, 2, or 3 km/hr, but not at any intermediate speed such as 1.3 km/hr. It would be a jerky ride!

Although according to modern quantum theory, the electrons do not move in circular orbits as in Bohr theory, angular momentum is still quantized, being restricted to the values of Eq. (7–5).

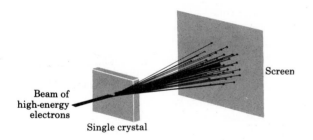

Fig. 7–14 The wave nature of electrons is demonstrated by scattering electrons of a particular energy from a crystal to produce strong scattered beams at certain angles, somewhat like a rainbow. [From Gregory R. Choppin and Russell H. Johnsen, *Introductory Chemistry* (Addison-Wesley, Reading, Mass., 1972), p. 98. Reprinted by permission.]

3. Probability distributions. Prior to the advent of quantum mechanics, it was in principle possible to make exact predictions of the outcome of certain experiments if one knew the initial conditions. For example, if one knew the initial position, mass, and velocity of an electron in the vicinity of a positively charged particle, it was thought to be possible to calculate what those conditions would be at some later time, from one's knowledge of the electrostatic force and its effects on the motion of the electron.

According to the quantum theory, we are not able to specify exactly where the electron will be and in what direction it will be moving at a later time. The best we can do is to indicate the *probability* of finding the electron in a particular region. In Fig. 7–15 we depict a probability distribution by

darkening the region where one would be most likely to find the particle. A calculation might show, for example, that there is a 50% probability of finding the particle inside the dark ring and an equal probability of its being outside the ring. This is like a weather report: the weatherman generally can't say with certainty whether or not it will rain tomorrow. Instead he indicates the likelihood of rain by saying, "There's a 60% chance of rain." If his predictions are accurate, it will rain on six out of ten such days.

In Fig. 7–16 we compare the Bohr model for a hydrogen atom with the quantum mechanics model, in which the electron occupies the *orbital* closest to the nucleus. We use the term "orbital," specifying a diffuse electron probability distribution of particular size and shape, to distinguish it from

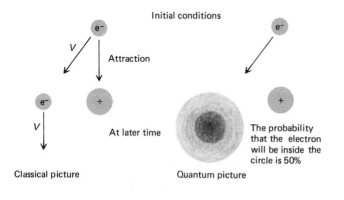

Fig. 7–15 With quantum mechanics, we can indicate only the probability of finding an electron in a particular region at a later time. The probability may be only 50%, as shown in the indicated area in this figure.

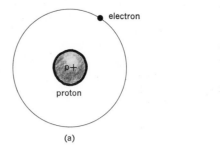
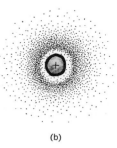

Fig. 7–16 The innermost orbital of the hydrogen atom: (a) according to the Bohr model, and (b) according to quantum mechanics. [From Arthur L. Williams, Harland D. Embree, and Harold J. Debey, *Introduction to Chemistry*, 2nd ed. (Addison-Wesley, Reading, Mass., 1973), p. 54. Reprinted by permission.]

the well-defined "orbits" of the Bohr model. In Fig. 7–16b, the darker region around the nucleus indicates a greater probability of finding the electron there.

We would obtain this picture if we could photograph the atom many times and record the position of the electron each time by a dot, as on Fig. 7–16b. Although we might take many pictures of the atom, we would not be able to say exactly where the electron would be found the next time a picture was taken. All we can do is indicate the probability of finding it in various regions. As an analogy, consider a dart board after a reasonably good player has thrown many darts aimed at the bull's-eye. We cannot say exactly where his next dart will land, but from the density of marks left by previous throws, we can state the probability that the dart will land in certain regions (see Fig. 7–17.

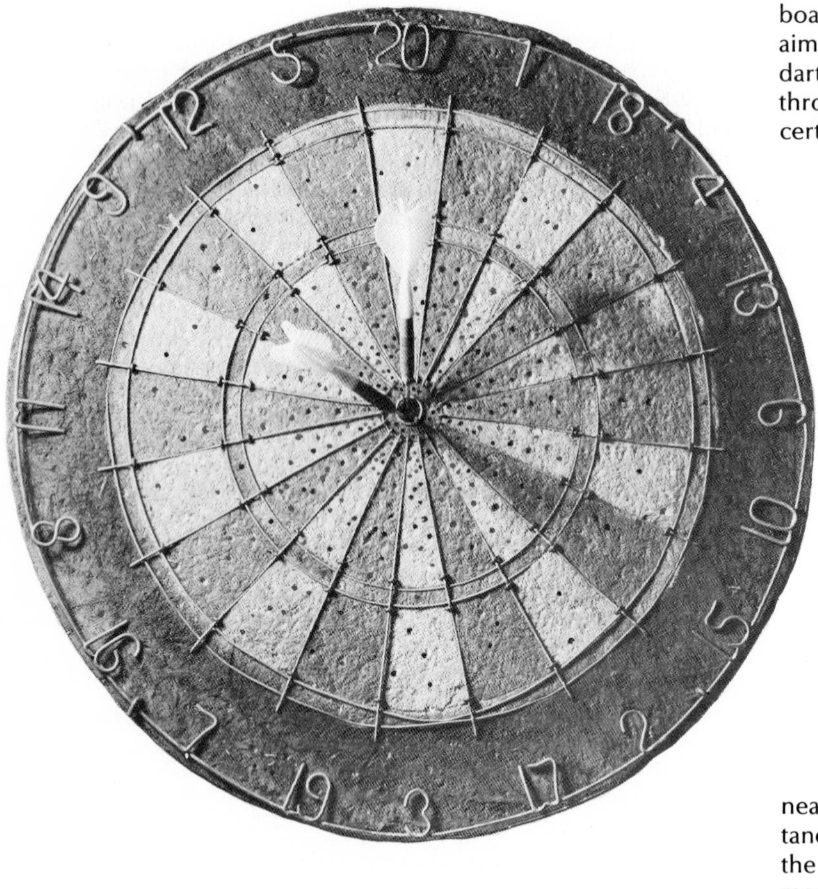

Fig. 7–17 The density of marks on a dart board allow one to determine the probability that the dart will land in a given location on the next throw. (Photograph by Bruce Anderson.)

In Fig. 7–16b we see that the electron density is highest near the nucleus and drops off smoothly with increasing distance from the nucleus, with a reasonably high probability of the electron's being outside the orbit predicted by the Bohr model. In that model, the electron's circular orbit has a radius of 0.529 Å. According to the new model, the average distance of the electron from the nucleus is 0.793 Å. Although Fig. 7–16b is a flat, two-dimensional representation, the probability distribution for this orbital is the same in all directions in space, i.e., the orbital is spherically symmetric (see Fig. 7–18).

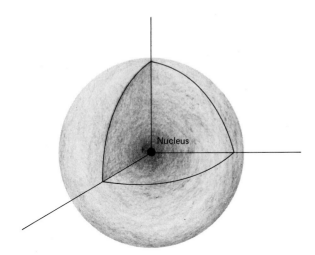

Fig. 7–18 The innermost orbital (1s) of the hydrogen atom is spherically symmetric.

4. The Heisenberg uncertainty principle. Many of the concepts of quantum mechanics discussed above are embodied in the *Heisenberg uncertainty principle*, which states that one cannot make precise determinations of certain pairs of quantities simultaneously. One cannot, for example, make accurate measurements of both the position and the momentum (mass × velocity) of a small particle. The best one can do is to determine that a particle is in a certain region of space and that its momentum falls within a given range. The Heisenberg uncertainty principle has nothing to do with the quality of the instruments used to measure these quantities—even with perfect equipment it would not be possible because of the "fuzziness" of the probabilities described above.

Prior to quantum mechanics, if one knew the initial conditions of all particles in the universe, it was supposedly possible to compute the future course of any portion of the universe. Quantum mechanics replaced this earlier concept, *determinism*, with uncertainty and probability, causing quite a stir among philosophers.

Despite the unusual predictions of quantum mechanics for very small objects, all tests of the theory to date have yielded results in agreement with the predictions. As the objects treated by quantum mechanics grow larger, the predictions gradually become the same as those with which we are familiar, i.e., those predicted by Newton's laws of motion.

As we observed in Chapter 4, however, there is no way to prove that any theory, including quantum theory, is correct. New experiments may someday yield results that disagree with its predictions. It will then be necessary to devise a theory that explains both the results correctly predicted by quantum mechanics and the new conflicting observations.

7–3 STRUCTURE OF THE HYDROGEN ATOM

So far we have discussed only the hydrogen electron orbital that is closest to the nucleus—the lowest energy orbital. If energy is added to the hydrogen atom (as by striking it with a light photon), the electron can be raised to one of the many higher energy orbitals farther from the nucleus. Each orbital is characterized by a set of *quantum numbers* that describe the properties of the orbital. The *principal quantum number* n is a measure of the distance from the nucleus to the orbital and the energy of the electron in that orbital. With increasing n, the orbitals are increasingly distant from the nucleus and have increasing energy, much like the Bohr model orbits in Fig. 7–13. All the orbitals having the same value of n make up an electron *shell*.

In the first shell, the only orbital available is the spherically symmetric one shown in Figs. 7–16 and 7–18. But if the hydrogen electron is displaced to the second shell ($n = 2$), two kinds of orbitals are possible. One is spherically symmetric, just as in Fig. 7–16b, but located farther out from the nucleus. The second type has a dumbbell-shaped probability distribution. The former is called an *s* orbital, and the latter, a *p* orbital. In each shell (except the first) there are three *p* orbitals arranged at right angles to each other.

The third shell has one *s* orbital, three *p* orbitals (both kinds located farther away from the nucleus than the $n = 2$ orbitals), and five *d* orbitals, whose complex shapes are shown in Fig. 7–19. The $n = 4$ and higher shells have all these orbitals, plus seven *f* orbitals whose shapes are so complex that they are rarely shown.*

All the orbitals of a given shape, that is, *s*, *p*, *d*, or *f*, in a particular shell are referred to as a *subshell*. Each is designated by the shorthand notation 1s, 2s, 2p, etc. In the hydrogen atom, the single electron in a particular shell has the same

---

* Electrons in the different types of orbitals have different amounts of orbital angular momentum. Electrons in the *s*, *p*, *d*, and *f* orbitals have 0, 1, 2, and 3 units of angular momentum ($h/2\pi$), respectively (see preceding section).

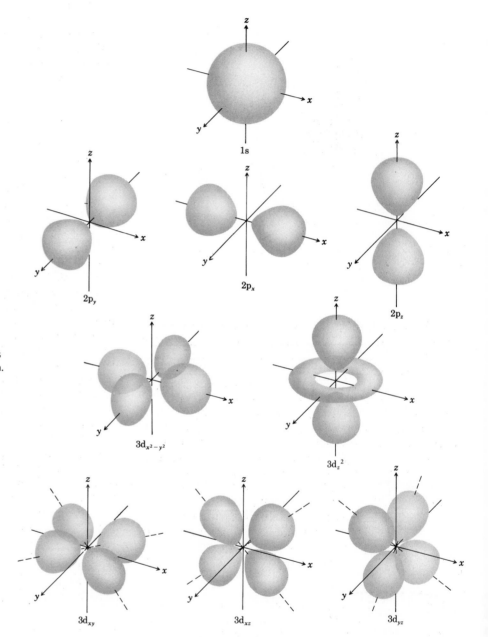

Fig. 7–19 Shapes of several types of orbitals available to an electron in a hydrogen atom. (Adapted from Gregory R. Choppin and Russell H. Johnsen, *Introductory Chemistry* (Addison-Wesley, Reading, Mass., 1972), p. 121. Reprinted by permission.]

energy regardless of which subshell it is in. For example, it would have the same energy whether in the 3s, 3p, or 3d subshell. However, it would have more energy in the $n = 4$ electron shell. In hydrogen atoms, there is only one electron to move around, but in more complex atoms, more than one electron will occupy many of the subshells. The capacities of the first five shells, and their subshells are listed in Table 7–1 and their energies for the hydrogen atom are shown schematically in Fig. 7–20.

Table 7–1. Summary of Orbitals of the First Five Electron Shells and their Capacities

Principal quantum number n	Subshell designation	No. of orbitals	Electron capacity (2 per orbital)	Total electron capacity of shell	Order of filling†
1	1s	1	2	2	1
2	2s	1	2	8	2
	2p	3	6		3
3	3s	1	2		4
	3p	3	6	18	5
	3d	5	10		7
4	4s	1	2		6
	4p	3	6	32	8
	4d	5	10		10
	4f	7	14		13
5	5s	1	2		9
	5p	3	6	32	11
	5d	5	10		14
	5f	7	14	(50)	17
	(5g)*	(9)	(18)		20

*   Not necessary to use this in elements below $Z = 121$, which has not been discovered.

†   Missing numbers are assigned to orbitals of higher shells, e.g., the 6s orbital is the 12th to be filled.

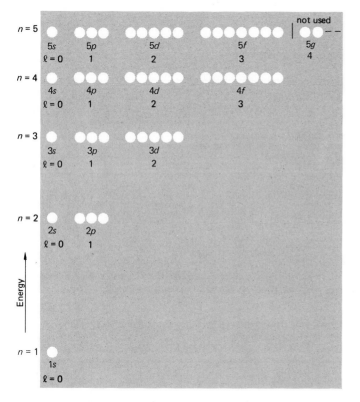

Fig. 7–20 Energies and designations of the orbitals in the first few shells of the hydrogen atom. Electrons in s, p, d and f orbitals have orbital angular momentum l of 0, 1, 2, and 3 units, respectively.

7–4 MANY-ELECTRON ATOMS

When we consider elements above hydrogen that have many electrons, how do we determine which orbitals are occupied by the electrons? The lowest state of energy of an atom, the *ground state*, is that in which the electrons occupy orbitals of the lowest possible energy. If this were the only considera-tion, all the electrons in a large atom would occupy the $n = 1$ electron shell closest to the nucleus, which has the lowest energy. However, it is found experimentally that the place-ment of electrons is governed by the *Pauli exclusion principle*, which in effect states that each orbital can hold only two electrons (one that spins clockwise on its own axis and another that spins counterclockwise). When many electrons are added to an atom, the Pauli principle means that orbitals of low energy get filled up first, so that additional electrons must be added to orbitals of successively higher energy.

In the ground state of the hydrogen atom, the single electron occupies the lowest energy orbital, the $1s$. We designate the electronic structure by the notation $1s^1$, where the superscript "1" means that one electron is present in the $1s$ subshell. When we move to $_2$He, we can also place the second electron in the $1s$ subshell. The He electronic structure is designated as $1s^2$.

At this point we have filled the $n = 1$ electron shell. It has only the single $1s$ orbital, and it can hold only two electrons, with spins in opposite directions. The third electron in $_3$Li must be placed in the $n = 2$ shell. Although in the hydrogen atom a $2p$ electron has the same energy as a $2s$, the presence of the two electrons in the $1s$ subshell causes an electron in the $2s$ to be lower than in the $2p$. Thus in the ground state of $_3$Li, the third electron occupies the $2s$ subshell since it is slightly lower in energy than the $2p$ subshell. The electron configuration of $_3$Li is $1s^22s^1$.

The $2s$ subshell is filled by the fourth electron in $_4$Be. The six electrons added from $_5$B to $_{10}$Ne enter the three $2p$ orbitals. At that point the $n = 2$ shell is full and the eleventh electron in $_{11}$Na, is added to the $3s$ orbital. The $3s$ subshell is filled in $_{12}$Mg and the six electrons added between $_{13}$Al and $_{18}$Ar go into the $3p$ subshell.

At the next element, $_{19}$K, we encounter a new phenomenon—the $3d$ orbitals lie so much higher in energy than the $3p$ that they are above the $4s$ orbital. Thus the nineteenth electron of $_{19}$K goes into the $4s$ orbital, which is filled by the twentieth electron in $_{20}$Ca. Only then is the $3d$ subshell filled by the ten electrons added between $_{21}$Sc and $_{30}$Zn. Similarly, the $5s$ subshell is filled before the $4d$. The fourth electron shell also has $4f$ orbitals. They lie so high in energy that they are not filled until after the $6s$ subshell. The last column of Table 7–1 shows the order in which the orbitals are filled.

Table 7–2 shows the electronic configurations of a number of atoms. You may have found the discussion of the order of filling somewhat confusing, especially above $3p$, where $4s$ is filled before $3d$. However, there is a simple memory device shown in Fig. 7–21 that gives a fairly accurate order for filling the orbitals. By starting at the top and working down each succeeding diagonal arrow we see that $3d$ is correctly indicated to be filled after $4s$, and $4f$ after $6s$, etc. The actual

Table 7–2. Electronic Configurations of Neutral Atoms of Some of the Elements

Atom	Configuration
$_1$H	$1s^1$
$_2$He	$1s^2$
$_3$Li	$1s^22s^1$
$_4$Be	$1s^22s^2$
$_5$B	$1s^22s^22p^1$
$_{10}$Ne	$1s^22s^22p^6$ = neon core
$_{11}$Na	Neon core + $3s^1$
$_{13}$Al	Neon core + $3s^23p^1$
$_{18}$Ar	Neon core + $3s^23p^6$ = argon core
$_{19}$K	Argon core + $4s^1$
$_{20}$Ca	Argon core + $4s^2$
$_{21}$Sc	Argon core + $4s^23d^1$
$_{30}$Zn	Argon core + $4s^23d^{10}$
$_{31}$Ga	Argon core + $4s^23d^{10}4p^1$
$_{36}$Kr	Argon core + $4s^23d^{10}4p^6$ = krypton core
$_{37}$Rb	Krypton core + $5s^1$
$_{38}$Sr	Krypton core + $5s^2$
$_{39}$Y	Krypton core + $5s^24d^1$
$_{48}$Cd	Krypton core + $5s^24d^{10}$
$_{49}$In	Krypton core + $5s^24d^{10}5p^1$
$_{54}$Xe	Krypton core + $5s^24d^{10}5p^6$ = xenon core
$_{55}$Cs	Xenon core + $6s^1$
$_{56}$Ba	Xenon core + $6s^2$
$_{57}$La	Xenon core + $6s^24f^1$*
$_{70}$Yb	Xenon core + $6s^24f^{14}$
$_{71}$Lu	Xenon core + $6s^24f^{14}5d^1$

*   La may have the last electron in $5d$ instead of $4f$. Many of the rare earth elements have one electron in $5d$.

electronic configurations of the atoms occasionally differ in the placement of one or two electrons from our general predictions in this section. Such discrepancies arise because some electron orbitals have very nearly the same amount of energy, so that subtle effects can cause changes in the placement of one or two electrons.

Several important points about electronic structure should be made before we consider chemical behavior of the elements. First, there can never be more than eight electrons

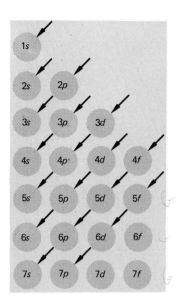

Fig. 7–21 Device for remembering order in which electrons are added to atoms.

Li $1s^2 2s^1$

Fig. 7–22 The two electrons of the 1s orbital shield the outermost electron from about two of the three positive charges in the $_3$Li nucleus. [From Paul F. Weller and Jerome H. Supple, *Chemistry: Elementary Principles* (Addison-Wesley, Reading, Mass., 1971), p. 159. Reprinted by permission.]

in the outermost electronic shell, i.e., those which fill the *s* and *p* subshells. The capacity of the third and higher shells is greater than eight, but electrons are not added to the *d* and *f* subshells until some electrons are added to shells farther from the nucleus.

Second, there is a big energy jump from the *p* subshell of one shell to the *s* subshell of the next higher shell. This means that electrons in the filled *p* subshells that occur in each of the noble gas elements (the last column on the right in the periodic table), except $_2$He, are very tightly bound to the atom. This structure has great stability. As a result, the noble gas elements have little tendency to gain or lose electrons by reacting with atoms of other elements. In the next higher elements, the alkali metals in the first column of the periodic table, there is a single electron in the *s* orbital of the outermost shell. It is shielded from most of the attractive nuclear charge by electrons in the inner shells (see Fig. 7–22). As a result, the electron is weakly bound to the atom and can easily be removed in chemical reactions. Jumps of this single electron from one outer orbital to another produce the spectral lines that yield the characteristic colors when these elements are dropped into a flame. The strong yellow color produced by sodium (see Fig. 7–23) is produced when the outermost electron drops from the 3*p* to the 3*s* subshell.

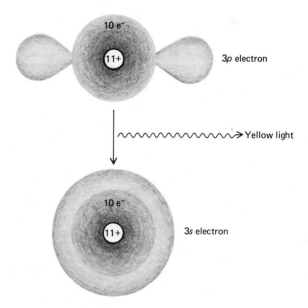

Fig. 7–23 The 3*p* to 3*s* transition in Na causes emission of the characteristic yellow light.

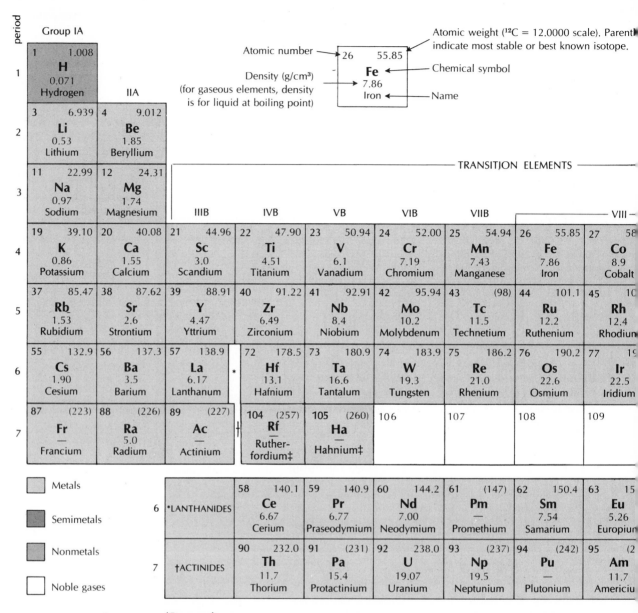

Group IA / IIA / TRANSITION ELEMENTS periodic table

Legend:
- Atomic number → 26
- Atomic weight (¹²C = 12.0000 scale). Parenthesis indicate most stable or best known isotope. → 55.85
- Density (g/cm³) (for gaseous elements, density is for liquid at boiling point) → 7.86
- Chemical symbol → Fe
- Name → Iron

	Group IA	IIA	IIIB	IVB	VB	VIB	VIIB	VIII	
1	1 1.008 **H** 0.071 Hydrogen								
2	3 6.939 **Li** 0.53 Lithium	4 9.012 **Be** 1.85 Beryllium							
3	11 22.99 **Na** 0.97 Sodium	12 24.31 **Mg** 1.74 Magnesium							
4	19 39.10 **K** 0.86 Potassium	20 40.08 **Ca** 1.55 Calcium	21 44.96 **Sc** 3.0 Scandium	22 47.90 **Ti** 4.51 Titanium	23 50.94 **V** 6.1 Vanadium	24 52.00 **Cr** 7.19 Chromium	25 54.94 **Mn** 7.43 Manganese	26 55.85 **Fe** 7.86 Iron	27 58 **Co** 8.9 Cobalt
5	37 85.47 **Rb** 1.53 Rubidium	38 87.62 **Sr** 2.6 Strontium	39 88.91 **Y** 4.47 Yttrium	40 91.22 **Zr** 6.49 Zirconium	41 92.91 **Nb** 8.4 Niobium	42 95.94 **Mo** 10.2 Molybdenum	43 (98) **Tc** 11.5 Technetium	44 101.1 **Ru** 12.2 Ruthenium	45 10 **Rh** 12.4 Rhodium
6	55 132.9 **Cs** 1.90 Cesium	56 137.3 **Ba** 3.5 Barium	57 138.9 **La** 6.17 Lanthanum *	72 178.5 **Hf** 13.1 Hafnium	73 180.9 **Ta** 16.6 Tantalum	74 183.9 **W** 19.3 Tungsten	75 186.2 **Re** 21.0 Rhenium	76 190.2 **Os** 22.6 Osmium	77 19 **Ir** 22.5 Iridium
7	87 (223) **Fr** — Francium	88 (226) **Ra** 5.0 Radium	89 (227) **Ac** — Actinium	104 (257) **Rf** Rutherfordium‡	105 (260) **Ha** Hahnium‡	106	107	108	109

Legend:
- ▨ Metals
- ▨ Semimetals
- ▨ Nonmetals
- ☐ Noble gases

		IIIB	IVB	VB	VIB	VIIB	VIII	
6	*LANTHANIDES	58 140.1 **Ce** 6.67 Cerium	59 140.9 **Pr** 6.77 Praseodymium	60 144.2 **Nd** 7.00 Neodymium	61 (147) **Pm** — Promethium	62 150.4 **Sm** 7.54 Samarium	63 15 **Eu** 5.26 Europium
7	†ACTINIDES	90 232.0 **Th** 11.7 Thorium	91 (231) **Pa** 15.4 Protactinium	92 238.0 **U** 19.07 Uranium	93 (237) **Np** 19.5 Neptunium	94 (242) **Pu** — Plutonium	95 (2 **Am** 11.7 Americium

‡Proposed names

Third, the chemical behavior of an element is largely determined by the number of electrons present in the outermost, or *valence shell*. Thus all the elements that have the same number of valence electrons (the alkali metals, for example, have one) generally have similar chemical behavior. In some series of elements, e.g., the *transition* elements between ₂₁Sc and ₃₀Zn, electrons are filling an inner shell of electrons; in this example, it is the 3d electron subshell. Since the outermost shell has nearly the same structure throughout the series, elements of the series are chemically similar to each other. In the *rare earth* series, between ₅₇La and ₇₁Lu, electrons are filling the 4f subshell, two shells in from the valence shell. These elements are so similar to each other that it is very difficult to separate them chemically.

NOBLE GASES

			IIIA	IVA	VA	VIA	VIIA	He
								2 — 4.003 — **He** — 0.126 — Helium
			5 — 10.81 — **B** — 2.34 — Boron	6 — 12.01 — **C** — 2.26 — Carbon	7 — 14.01 — **N** — 0.81 — Nitrogen	8 — 16.00 — **O** — 1.14 — Oxygen	9 — 19.00 — **F** — 1.505 — Fluorine	10 — 20.18 — **Ne** — 1.20 — Neon
	IB	IIB	13 — 26.98 — **Al** — 2.70 — Aluminum	14 — 28.09 — **Si** — 2.33 — Silicon	15 — 30.97 — **P** — 1.82 — Phosphorus	16 — 32.06 — **S** — 2.07 — Sulfur	17 — 35.45 — **Cl** — 1.56 — Chlorine	18 — 39.95 — **Ar** — 1.40 — Argon
58.71 — **Ni** — 8.9 — Nickel	29 — 63.54 — **Cu** — 8.96 — Copper	30 — 65.37 — **Zn** — 7.14 — Zinc	31 — 69.72 — **Ga** — 5.91 — Gallium	32 — 72.59 — **Ge** — 5.32 — Germanium	33 — 74.92 — **As** — 5.72 — Arsenic	34 — 78.96 — **Se** — 4.79 — Selenium	35 — 79.91 — **Br** — 3.12 — Bromine	36 — 83.80 — **Kr** — 2.6 — Krypton
106.4 — **Pd** — 12.0 — Palladium	47 — 107.9 — **Ag** — 10.5 — Silver	48 — 112.4 — **Cd** — 8.65 — Cadmium	49 — 114.8 — **In** — 7.31 — Indium	50 — 118.7 — **Sn** — 7.30 — Tin	51 — 121.8 — **Sb** — 6.62 — Antimony	52 — 127.6 — **Te** — 6.24 — Tellurium	53 — 126.9 — **I** — 4.94 — Iodine	54 — 131.3 — **Xe** — 3.06 — Xenon
195.1 — **Pt** — 21.4 — Platinum	79 — 197.0 — **Au** — 19.3 — Gold	80 — 200.6 — **Hg** — 13.6 — Mercury	81 — 204.4 — **Tl** — 11.85 — Thallium	82 — 207.2 — **Pb** — 11.4 — Lead	83 — 209.0 — **Bi** — 9.8 — Bismuth	84 — (210) — **Po** — (9.2) — Polonium	85 — (210) — **At** — — — Astatine	86 — (222) — **Rn** — — — Radon
	111	112	113	114	115	116	117	118

157.3 — **Gd** — 7.89 — Gadolinium	65 — 158.9 — **Tb** — 8.27 — Terbium	66 — 162.5 — **Dy** — 8.54 — Dysprosium	67 — 164.9 — **Ho** — 8.80 — Holmium	68 — 167.3 — **Er** — 9.05 — Erbium	69 — 168.9 — **Tm** — 9.33 — Thulium	70 — 173.0 — **Yb** — 6.98 — Ytterbium	71 — 175.0 — **Lu** — 9.84 — Lutetium
(247) — **Cm** — — — Curium	97 — (247) — **Bk** — — — Berkelium	98 — (249) — **Cf** — — — Californium	99 — (254) — **Es** — — — Einsteinium	100 — (253) — **Fm** — — — Fermium	101 — (256) — **Md** — — — Mendelevium	102 — (254) — **No** — — — Nobelium	103 — (257) — **Lr** — — — Lawrencium

Fig. 7–24
Periodic table of elements.

7–5 THE PERIODIC TABLE OF THE ELEMENTS

In Section 7–4 we saw that there are various repeating patterns of electron configurations as we add electrons to build up higher elements. These repeating patterns are closely related to the position of each of the elements in the periodic table of the elements (see Fig. 7–24). For example, in all the elements of the last six columns except $_2$He, the *p* orbitals of the outermost shell are being filled to their capacity of six electrons. The *p* orbitals become filled in the elements of the last column, the *noble* (or *rare*) gases. In the next higher elements, the *alkali metals* of the first column, a single electron has been added to an otherwise empty electron shell having a higher principal quantum number.

In the second column are the *alkaline earth* elements, in which the *s* orbital of the outermost shell has its full complement of two electrons. In the *transition elements* of the ten columns in the middle of the table, electrons are being added to the *d* orbitals of the electron shell one level in from the outermost shell. In the transition elements of the fourth row of elements, for example, where some electrons are present in the $n = 4$ shell, electrons are being added to the $3d$ orbitals between $_{21}$Sc and $_{30}$Zn. Finally, in the sixth and seventh rows of elements, the table should be stretched out and the elements $_{57}$La through $_{71}$Lu (the *lanthanides*) and $_{89}$Ac through $_{103}$Lw (the *actinides*) placed in the locations occupied by $_{57}$La and $_{89}$Ac respectively; however, for convenience in display of the table on a normal-sized piece of paper, these elements are usually placed by themselves at the bottom. In these elements, the *rare earths* (or lanthanides) and actinides, electrons are mostly being added to the *f* orbitals two shells in from the outermost. In the rare earths, for example, which contain electrons in the $6s$ orbital, electrons are being added to the $4f$ orbitals.

The position of an element in the table, which tells us which orbitals are being filled and how many electrons are present in the *valence shell*, also tells us a lot about the chemical characteristics to be expected for the element. In particular, elements in a given column, or *chemical group*, tend to have similar chemical properties. Thus the periodic table represents a useful summary of systematic chemical characteristics. Once you learn how to use the table, you can predict the properties of a given element on the basis of the known properties of other elements in the same column.

In this chapter we have developed the rules of electronic structure and used them to find the electronic configuration of the elements; then we placed the elements in the periodic table. However, the development of the table, which was so important for the systematic understanding of chemical behavior, occurred long before anything was known about atomic structure.

The first reasonably correct periodic table was constructed by a Russian chemist, Dimitri Mendeleev, in 1869. By the middle of the nineteenth century, a number of chemically similar families of elements were known. For example, it was known that Li, Na, K, Rb, and Cs were chemically similar. Another known group was Be, Mg, Ca, Sr, and Ba; another was Cl, Br, and I; still another was O, S, Se, and Te. We know now that the order of elements in the table is determined by the atomic number Z, the number of protons in the nucleus.

Dmitri Mendeleev, 1834–1907, who developed the first successful periodic table of the elements in 1869.

But in 1869, nothing was known about protons or atomic numbers. However, rather accurate atomic weights were known for about 55 elements, so Mendeleev arranged the elements in order of increasing atomic weight, which with two exceptions (Co and Ni, Te and I) is the same order as the atomic numbers.

Mendeleev's job was difficult—rather like putting together a jigsaw puzzle with many of the pieces missing. Only about 63 elements were known and the atomic weights for eight of them were wrong (for example, La = 92, U = 116).* None of the noble gas elements had been discovered, and only a few of the rare earths. The elements above bismuth, except uranium and thorium, were also missing, as well as a few other rare elements.

When Mendeleev arranged the known elements in order of increasing atomic weight, he noticed that there were repeating patterns of chemically similar elements. Among the light elements he had:

H Li Be B C N O ? Na Mg Al Si P S Cl K Ca ? Ti

*   J. R. Partington, *A Short History of Chemistry*, 3rd ed. (St. Martin's Press, New York, 1957).

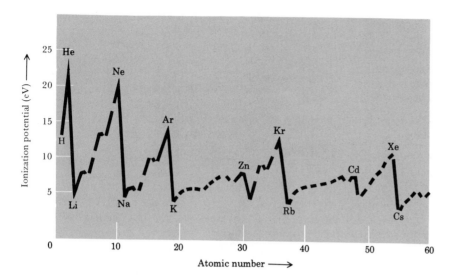

Fig 7–25 Ionization potentials of the elements. [From Gregory R. Choppin and Russell H. Johnsen, *Introductory Chemistry* (Addison-Wesley, Reading, Mass., 1972), p. 86. Reprinted by permission.]

As indicated for Li-Na-K and Be-Mg-Ca, Mendeleev saw that chemically similar elements repeated every seven elements much like the notes of a piano scale.

The brilliance of Mendeleev lay in his realization that some of the elements were missing. In constructing the table, he noticed that when he placed chemically similar elements below one another, there were some gaps. Not only did they indicate which elements were missing, but by studying the properties of the known elements in the same group, one could predict those of the undiscovered member. For example, in the sequence shown above, the question mark between Ca and Ti indicates a missing element with atomic weight of about 45. In 1879, the element scandium (atomic weight, 44.96) was discovered and placed there.

The periodic table has been used extensively over the past thirty years in the discovery of the unstable, radioactive elements $_{43}$Tc, $_{61}$Pm, $_{85}$At, and others, especially the artificial elements, $_{93}$Np through $_{105}$Ha (hahnium, after Otto Hahn, the co-discoverer of fission). Even today, scientists designing experiments to make or discover new elements use the periodic table in this way. As new heavy elements are discovered, we find that the half-lives of their isotopes are shorter and shorter, on the order of seconds or less for $_{105}$Ha, the heaviest element now known. One might suppose that the half-lives

of heavier elements would simply keep on getting shorter; however, there are theoretical reasons for expecting a region of more stable nuclei, with half-lives perhaps as great as many years, in the vicinity of element 114. If some of these elements were made in nucleosynthesis, they might be found deposited with chemically similar elements on Earth. Looking at the periodic table, we see that elements 110, 112, and 114 are probably chemically similar to $_{78}$Pt, $_{80}$Hg, and $_{82}$Pb. Scientists have looked for evidence of the undiscovered elements in deposits of platinum, mercury, and lead, but so far without success.

7–6 THE VALENCE SHELL AND SIMPLE IONS

Let us see how the periodic properties of electronic structure affect the behavior of atoms. Consider the energy needed to remove a single electron from a neutral atom of each of the elements—*the ionization potential.* Figure 7–25 shows that ionization potential is a regular, repetitive function of the atomic number Z. The values gradually rise to a maximum, then there is a precipitous drop, followed by another slow rise. The maximum values occur for the noble gas elements: $_2$He, $_{10}$Ne, $_{18}$Ar, $_{36}$Kr, $_{54}$Xe, and $_{86}$Rn. Ionization potential of the succeeding elements, the alkali metals, $_3$Li, $_{11}$Na, $_{19}$K, $_{37}$Rb and

$_{55}$Cs are very low. What is the meaning of these observations?

First we see that it requires a great deal of energy to remove one electron from a noble gas atom. In each noble gas except $_2$He, the p orbitals of the valence shell are completed. Their valence shells contain eight electrons, two in the s orbital and six in the p orbitals. Atoms having eight electrons in the valence shell have exceptional stability, as shown by their large ionization potential.

Further evidence for the special stability of a valence shell with eight electrons is shown by the ionization potential of the elements just above the noble gases, the alkali elements. The single electron in the outermost shell of each of these elements is loosely attached and can easily be removed by atoms of other elements that have a stronger attraction for electrons. The size of atoms also reflects the strength of its attachment to its electrons. In Fig. 7–26 we see that the alkali metal atoms are much larger than succeeding elements of the same row. In sodium, for example, the ten electrons of the $n = 1$ and 2 shells effectively cancel nearly ten of the positive charges of the 11+ nucleus. Thus the single electron in the $n = 3$ shell is held to the atom by a net charge of only about one positive charge. The orbital of this weakly bound electron is therefore quite large.*

In $_{12}$Mg, the nuclear charge has increased to 12+. With the 10 electrons of the inner two shells cancelling about ten of the positive charges, the $3s$ electrons are bound by a net charge of about 2+ units. The increased attraction (compared to sodium) causes the valence shell orbitals to contract appreciably and accounts for the higher ionization potential of magnesium. This trend continues through the period to $_{17}$Cl, which has the smallest atoms in the third row of the periodic table. In Fig. 7–27 we see that atomic radii increase gradually as we go down the column of elements in a chemical group. The largest known atomic radius, that of $_{55}$Cs, is 2.25 Å. (That of the radioactive element $_{87}$Fr is undoubtedly larger, but it has not been measured.)

The strong tendency of atoms to achieve an outer shell of eight electrons explains the general chemical behavior of

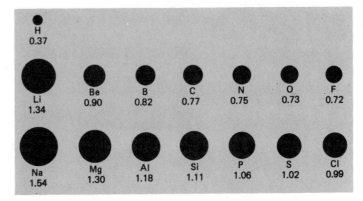

Fig. 7–26 The radii of neutral atoms decrease across a row of the periodic table.

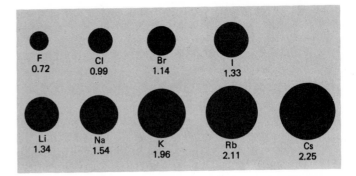

Fig. 7–27 The radii of neutral atoms increase as we move down the columns of the periodic table.

several groups of elements. The fact that neutral atoms of the noble gases already have eight electrons in their outer shells (except helium) explains their behavior—they rarely react with any elements at all. Before 1960 we would have said that the noble gases form no compounds. However, during the early 1960s a few compounds of the noble gases were synthesized, mainly with xenon and fluorine, but it is still accurate to say that the noble gases are almost chemically inert.†

---

*   Since electron density decreases gradually with distance from the nucleus, rather than suddenly dropping to zero, it is difficult to define the radius of an atom. The radii shown in Fig. 7–26 are equal to half the distance between the nuclei of two atoms of the element when held together by covalent bonding (see Chapter 8), as in Cl_2, H_2, etc. Since noble gas atoms do not form such bonds with each other, comparable radii for that family of elements cannot be determined.

†   A review of the noble gas compounds by H. H. Hyman (ed.), *Noble-Gas Compounds* (University of Chicago Press, Chicago, 1963), lists some of the moderately stable compounds. They include XeF_2, XeF_4, XeF_6, XeF_8, $XeOF_4$, KrF_2, KrF_4, XeO_3, and ArF_4.

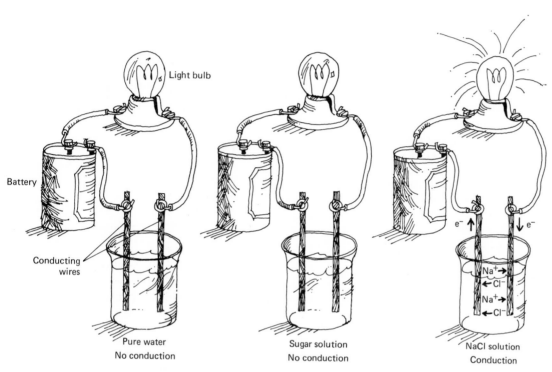

Light bulb

Battery

Conducting
wires

e^- ↑   ↓ e^-

Na$^+$→
← Cl$^-$
Na$^+$→
← Cl$^-$

Pure water
No conduction

Sugar solution
No conduction

NaCl solution
Conduction

Fig. 7–28 Solutions of ionic compounds such as NaCl are good conductors of electricity, whereas those of non-ionic compounds such as sugar ($C_{12}H_{22}O_{11}$) are not.

The alkali metal atoms readily give up their outside electrons to become *ions* with net 1+ charges, provided there are other atoms available that can accept the electrons. Examples of the latter are the *halogens*—the elements in the next to last column (VIIA) of the periodic table. With seven electrons in the valence shells of the neutral atoms, the halogens have a strong tendency to acquire an eighth electron to become *halide* ions with 1− charges. If metallic sodium, Na, is placed in contact with chlorine gas, Cl_2, electrons given up by the Na atoms are "grabbed" by Cl atoms, producing Na$^+$ and Cl$^-$ ions:

$$2Na\ (s)\ +\ Cl_2\ (g)\ \longrightarrow\ 2Na^+\ +\ 2Cl^-\text{.} \qquad (7-6)$$

The product formed in reaction (7–6) is sodium chloride, or common table salt. As may be observed with a magnifying glass or simple microscope, sodium chloride formed by evaporation of a salt-water solution occurs as very regular crystals. From studies of NaCl crystals with x rays, we know that the crystals are made up of a three-dimensional lattice of alternating Na$^+$ and Cl$^-$ ions (see Fig. 2–12). The electron is completely detached from the sodium atom and attached to the chlorine atom. The transfer of the electron can be depicted by the use of electron-dot formulas, in which electrons of the valence shell are represented by dots. The combination of a sodium and a chlorine atom is shown as

$$Na\cdot\ +\ \overset{\cdot\cdot}{\underset{\cdot\cdot}{:Cl}}\cdot\ \longrightarrow Na^+\ +\ \overset{\cdot\cdot}{\underset{\cdot\cdot}{:Cl}}\cdot^{-} \qquad (7-7)$$

The lattice of the sodium chloride crystal is held firmly in place by the strong electrostatic attractions between neighboring Na^+ and Cl^- ions. Since there are six Cl^- ions around each Na^+ ion, we cannot say that the Na^+ ion "belongs" to a particular Cl^- ion; however, we can say that the ions are held together in the crystal lattice by *ionic bonds*, the attraction between positive and negative ions. When an ionic crystal is dissolved in water, the individual ions, in this example, positive Na^+ *cations* and negative Cl^- *anions*, move around independently in the water. One can demonstrate the presence of the charged species in such an ionic solution by passing an electric current through the solution, as shown in Fig. 7–28. Very pure water contains so few ions that insufficient current flows through it to light a bulb. When a strongly ionized compound such as NaCl is added to the water, the ions can carry enough current through the solution to light the bulb. Solid NaCl crystals do not conduct electrical current since the ions cannot move without the water. Non-ionic substances such as sugar do not increase the conductivity of the water and will not light the bulb.

Alkali metals and, in general, other elements on the left side of the periodic table are *electropositive*, i.e., they tend to give up electrons rather easily and become positive ions. The halogens and other elements toward the right side of the table are generally *electronegative*, i.e., they accept electrons and become negative ions. Linus Pauling has established a numerical scale of electronegativities in order to give some quantitative indication of these tendencies.* Figure 7–29 shows that elements such as fluorine (F), oxygen, nitrogen (N), and chlorine have the greatest electronegativity. As one would expect, alkali metals and alkaline earths, the electropositive elements, have the least electronegativity.

The greater the difference between elements of a compound (in terms of electronegativity), the greater is the compound's ionic nature. Although many compounds form separated ions in crystals and aqueous solutions, distinct molecules are present in the gas phase of the compounds. The ionic character of these molecules increases with increasing difference in the electronegativity of the elements. For example, in LiF, the difference is $4 - 1 = 3$, and the bond between the elements has an ionic character of 92% because the F atom has a much greater attraction for electrons than the Li atom. On the other hand, when the difference in

* L. C. Pauling, *General Chemistry*, 3rd ed. (W. H. Freeman, San Francisco, 1970).

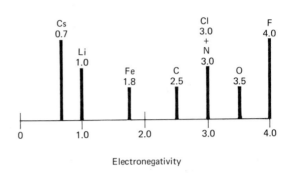

Fig. 7–29 Electronegativity of several important elements, according to the scale developed by Linus Pauling.

electronegativity is small, the ionic character is small. In that case, the atoms share the electrons about equally in a covalent bond, as described in more detail in Chapter 8.

Ionic compounds are formed between many pairs of elements in which one member is a strongly electropositive element from the left side of the periodic table and the other is an electronegative element from the right side. Elements of the second column (IIA) of the periodic table, the alkaline earths, form 2+ ions by giving up the two electrons from their valence shells. The charge of the ions is referred to as the *valence* or *oxidation number* of the species (for reasons discussed in Chapter 10). For example, Ba^{2+} has a valence or oxidation number of 2+. Since atoms of elements at the right-hand side of the periodic table have valence shells that are nearly filled, they often acquire enough electrons to complete the shell, thus becoming negative ions. The halogens of column VIIA have seven electrons in their valence shells, and so they often pick up an additional electron to form 1− ions. Similarly, elements such as oxygen, in column VIA, have six electrons in their valence shells and often pick up two more, giving them 2− oxidation states.

Since a compound must have a net charge of zero, the number of positive charges of the cations must be balanced by an equal number of negative charges of the anions. Thus we combine one K^+ with one Br^- to form KBr, one Ba^{2+} with

Table 7–3. Some Simple Compounds Formed between Electropositive and Electronegative Elements

Electropositive element	Oxidation state	Electronegative element	Oxidation state	Compound formed	Name
Na	+1	Cl	−1	NaCl	Sodium chloride
Ca	+2	Br	−1	$CaBr_2$	Calcium bromide
Ba	+2	O	−2	BaO	Barium oxide
Al	+3	Cl	−1	$AlCl_3$	Aluminum chloride
Al	+3	O	−2	Al_2O_3	Aluminum oxide
C	+4	O	−2	CO_2	Carbon dioxide
H	+1	N	−3	NH_3	Ammonia
Fe	+2	O	−2	FeO	Ferrous oxide
Fe	+3	O	−2	Fe_2O_3	Ferric oxide

Table 7–4. Oxidation States of the Transition Metals of the Fourth Period

Element	$_{21}Sc$	$_{22}Ti$	$_{23}V$	$_{24}Cr$	$_{25}Mn$	$_{26}Fe$	$_{27}Co$	$_{28}Ni$	$_{29}Cu$	$_{30}Zn$
Electron configuration	$3d^14s^2$	$3d^24s^2$	$3d^34s^2$	$3d^54s^1$	$3d^54s^2$	$3d^64s^2$	$3d^74s^2$	$3d^84s^2$	$3d^{10}4s^1$	$3d^{10}4s^2$
Oxidation states	3+ *	3,4+	2,3,4,5+	2,3,6+	2,3,4,6,7+	2,3+	2,3+	2,3+	1,2+	2+

*   Most common oxidation state in inorganic compounds is underlined.

two Cl^- ions to form $BaCl_2$, two Rb^+ ions with one O^{2-} to form Rb_2O, and one Ca^{2+} with one O^{2-}, forming CaO. From the column of the periodic table in which the element occurs, we can often predict the oxidation state that it will acquire in compounds formed with other elements. By balancing the positive and negative oxidation numbers, we can predict the formula of the compound formed by combinations of two different elements, as shown by several examples in Table 7–3. The names of these binary compounds (i.e., those formed by just two different elements) are given in the last column. Most such compounds have the name of the more electropositive element followed by that of the more electronegative element with the suffix "ide." The common name "ammonia" is an exception to this general rule. In some cases, the Latin root is used for an element, e.g., "ferric" or "ferrous" for "iron."

Chemistry would be very simple if one could predict all the formulas of the compounds that could be formed just from the number of electrons present in the valence shell. However, many elements can have many more oxidation states than the one which we would expect. In Table 7–3 we see that iron exhibits both 2+ and 3+ oxidation states. The 2+ state is formed, as expected, by the loss of the two s electrons in the $n = 4$ shell. But an iron atom can also lose a $3d$ electron to form the oxidation state 3+. As with iron, the lower and higher oxidation states are often indicated by the suffixes "-ous" and "-ic", respectively. Most of the transition metals in columns IVB through IIB in the periodic table have 2+ oxidation states, but nearly all have additional states as shown in Table 7–4. Those in column IIIB, the elements Sc, Y, La, and Ac, have only 3+ oxidation states, formed by the loss of two s electrons and the single d electron.

SUGGESTED READING

I. D. Garard, *Invitation to Chemistry* (Doubleday, Garden City, N.Y., 1969). Paperback.

G. Gamow, *Mr. Tompkins in Wonderland* (Macmillan, New York, 1940).

B. H. Mahan, *University Chemistry*, 2nd ed. (Addison-Wesley, Reading, Mass., 1969).

J. R. Partington, *A Short History of Chemistry*, 3rd ed. (St. Martin's Press, New York, 1957). Paperback.

L. Pauling, *General Chemistry*, 3rd ed. (Freeman, San Francisco, 1970).

G. C. Pimentel and R. D. Spratley, *Chemical Bonding Clarified through Quantum Mechanics* (Holden-Day, San Francisco, 1969). Paperback.

A. C. Wahl, "Chemistry by Computer," *Scientific American* **222** (4), 54–71 (April 1970).

PROBLEMS AND QUESTIONS

1. Arrange the following types of electromagnetic radiation in order of: (a) increasing energy per photon, (b) increasing wavelength, and (c) decreasing frequency.

blue light	γ rays
red light	microwaves
UV	radiowaves
IR	x rays

2. What are: (a) continuous spectra, (b) bright-line spectra, and (c) dark-line spectra?

3. Why would IR film be able to record a truck driving along a tree-covered road?

4. How many subshells are to be found in the $n = 4$ and $n = 5$ shells? What is the maximum number of electrons that can be placed in each of these subshells?

5. a) State the Pauli exclusion principle and indicate how it governs the filling of electrons in shells around an atom?

   b) If the Pauli principle were not valid, in which shell would the electrons of all atoms be found? Why?

6. What are the electronic configurations of Na, Sc, Ne, K^+, Cl^-, and Li^+?

7. Why do the rare earth elements react chemically very much like each other?

8. a) When we put lithium (Li) in a flame, an electron on a Li atom can become excited to a higher energy level. When the electron drops back down to its original level, red light with a wavelength of about 6708 Å is emitted. To how much energy does this correspond?

   b) What is the wavelength of a 1-MeV γ ray?

9. a) What is the wavelength of an electron after passage through a potential difference (increase of voltage) of 20,000 volts?

   b) What is the wavelength of a proton after passage through the same potential (in the opposite direction)?

   c) Using the criterion of wavelength alone, which would be superior for investigating atomic and subatomic structure, 20-keV electrons or 20-keV protons?

10. The properties of He^+, that is, a 2+ nucleus with just one electron circulating around it, are rather similar to those of the hydrogen atom; however, the 2+ charge on the nucleus of the helium causes some quantitative differences.

    a) Would the average position of an electron in the 1s orbital of He^+ be closer to or farther from the nucleus than in H? Explain.

    b) The energy of an electron in orbitals of such one-electron atoms is

    $$E_n = -2.175 \times 10^{-11} \frac{Z^2}{n^2} \text{erg,}$$

    where Z is the nuclear charge and n is the principal quantum number. (Note that the energy is measured as though the electron were infinitely far from the nucleus, and so all such values are negative.) Would the energy of the $2p \rightarrow 1s$ transition be greater in He^+ or in H?

    c) Calculate the frequency and wavelength of the light emitted in the $2p \rightarrow 1s$ transition in He^+.

11. What kinds of orbitals (i.e., s, p, d, f, g, etc.) are permitted in the $n = 6$ shell?

12. How many electrons an be placed in an s subshell, e.g., the 2s subshell? How many can be placed in the p, d, and f subshells?

13. a) Using Table 7–2 and Fig. 7–21, predict the electronic configuration of $_{24}Cr$. Is this the configuration that is observed experimentally?

b) Predict the electronic configuration of the undiscovered element that has $Z = 116$.

14. In 1869, the atomic weight of lanthanum was thought to be 92. This value was probably determined by weighing the amount of lanthanum that combined with a given weight of oxygen. We now know that the formula for lanthanum oxide is La_2O_3. Show how the atomic weight could have been obtained as 92 with an incorrect formula for lanthanum oxide.

15. a) What is the meaning of the term *ionization potential*?

b) In which group of elements are the ionization potentials generally the smallest, and in which are they the largest?

c) Can you suggest explanations for the observed small drop in ionization potential between $_4Be$ and $_5B$ and between $_{12}Mg$ and $_{13}Al$?

16. a) If we move across a row of the periodic table, e.g., from $_{11}Na$ to $_{18}Ar$, do the radii of the neutral atoms increase or decrease? Explain.

b) If we move down a column of the periodic table, e.g., from $_4Be$ to $_{88}Ra$, do the atomic radii increase or decrease? Explain.

17. What is the usual charge of ions formed by elements of Group IA? Why?

18. Why do the ionization potentials and atomic radii remain about the same from $_{21}Sc$ through $_{30}Zn$?

19. Why do noble gas elements occur as individual atoms in the gas phase (e.g., He, Ne, Ar, Xe), whereas for most other elements (e.g., O_2, N_2, Cl_2), the gas phase of the pure element usually consists of diatomic molecules?

20. What kind of bond (ionic or covalent) would you expect to be formed between:

a) Cs and Cl   b) C and Br

c) Sr and Cl   d) S and O

e) Ba and Br

21. Give the formula for the simplest compound likely to be formed from the following pairs of elements (for example, C and H form CH_4).

a) Sc and O   b) Ga and Te

c) Sb and H   d) Ge and O

e) C and Cl   f) Rb and Br

g) Ca and O   h) Y and S

i) H and Se   j) Mg and I

22. The species $_{17}Cl^-$, $_{18}Ar$, and $_{19}K^+$ all have 18 electrons.

a) Which species would you expect to have the smallest radius? Why?

b) Which would have the most weakly bound outer electron (i.e., the lowest ionization potential)? Why?

8 MOLECULAR STRUCTURE

Bottle of carbonated beverage showing bubbles of CO_2 coming out of solution. (Photograph by Bruce Anderson.)

8–1 COVALENT BONDING

In Chapter 7 we saw that formation of simple compounds between elements of widely differing electronegativity can be understood in terms of the formation of ions and ionic bonds. When atoms of two elements of about the same electronegativity react, they form bonds in which the electrons are shared about equally between the atoms. Bonds formed by the sharing of electrons are called *covalent bonds*.

Covalent bonding is particularly important for elements in which the valence shell is about half full. In order to form ions with outer shells of eight, these atoms would have to gain or lose several electrons. Large amounts of energy would be required to remove or add several electrons. When one or two electrons have been removed, electrostatic attraction makes it difficult to pull additional negative electrons away from the positive ion. Similarly, it is difficult to form highly charged negative ions; once a negative charge has been acquired, more energy is required to bring additional electrons up to the ion. Thus the atoms of many elements, especially those that would need to gain or lose three or four electrons to achieve a full outermost shell, form compounds by the sharing of electrons in a covalent bond.

A common element that forms covalent bonds is carbon, the principal element of organic chemistry and biochemistry. Every carbon atom has four electrons in its valence shell. It can acquire four more electrons by combining with four hydrogen atoms to form the compound methane, CH_4 (the major constituent of natural gas), whose structure we can depict as:

$$\cdot \overset{\cdot}{\underset{\cdot}{C}} \cdot \;+\; 4H\times \;\longrightarrow\; H \overset{\overset{H}{\times}}{\underset{\underset{H}{\times}}{\overset{\cdot}{\underset{\cdot}{C}}}} \overset{\cdot}{\underset{\cdot}{\times}} H$$

For identification, we have indicated the electrons originally from the H atoms as ×'s, although of course all electrons are identical. See also Figs. 8–1 and 8–2. By sharing electrons with hydrogen atoms, the carbon atom acquires the desired eight electrons in its valence shell. Likewise, each hydrogen atom acquires a second electron from the carbon atom, filling its $n = 1$ shell. Since the electrons in the C—H bonds repel each other, the four hydrogens are aligned as far apart as possible, forming a tetrahedron. Covalent compounds have little tendency to break up into ions if, for example, they are dissolved in water. Sugar is an example of this type of compound, and its non-ionic character is indicated by the fact

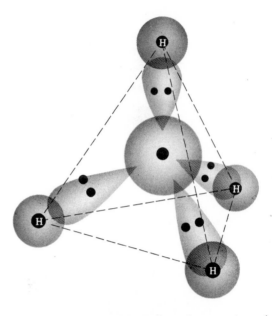

Fig. 8–1 By sharing electrons the carbon atom in methane achieves a complete valence shell of eight electrons, and each hydrogen atom achieves a full $n = 1$ shell of two electrons. [From Gregory R. Choppin and Russell H. Johnsen, *Introductory Chemistry* (Addison-Wesley, Reading, Mass., 1972), p. 246.]

that it does not conduct electrical current when dissolved in water.

The structure of many covalent compounds can be correctly predicted by the *octet rule*; that is, by arranging the electrons in bonds so that each atom has a total of eight outer electrons. (Of course, in the case of hydrogen, only two electrons are needed to fill its outermost and only shell, the 1s orbital.) We can, for example, apply the octet rule to determine the structure of water molecules. Oxygen has six electrons in its valence shell and acquires two others from hydrogen atoms:

$$H \overset{\cdot \cdot}{\underset{\underset{H}{\times}}{\overset{\times}{O}}} \!:\;$$

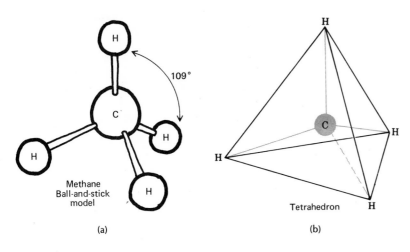

Fig. 8–2 The tetrahedral shape of the methane molecule. [Part (b) from Choppin and Johnsen, *Introductory Chemistry*, p. 245. Reprinted by permission.]

Nitrogen, which has five outer electrons, acquires three more from hydrogen atoms in the compound ammonia, NH_3:

$$H \overset{\times}{\underset{\underset{H}{\cdot\cdot}}{N}} \overset{\times}{{}} H$$

In most pure elements in the gaseous state, except the noble gases, two or more atoms of the element combine to form molecules in which the atoms generally have eight electrons, for example, the halogens:

$$:\!\overset{\cdot\cdot}{\underset{\cdot\cdot}{F}}\!\overset{\times\times}{\underset{\times\times}{F}}\!\times \qquad :\!\overset{\cdot\cdot}{\underset{\cdot\cdot}{Cl}}\!\overset{\times\times}{\underset{\times\times}{Cl}}\!\times \qquad :\!\overset{\cdot\cdot}{\underset{\cdot\cdot}{Br}}\!\overset{\times\times}{\underset{\times\times}{Br}}\!\times \qquad :\!\overset{\cdot\cdot}{\underset{\cdot\cdot}{I}}\!\overset{\times\times}{\underset{\times\times}{I}}\!\times$$

In bonds between two atoms of the same element, the electrons are attracted with equal strength by both nuclei. Thus the distribution of electrons between the two atoms is symmetrical. But in bonds between atoms of elements with greatly different electronegativity the electrons are more strongly attracted to the more electronegative atom of the bond. These are called *polar bonds*. An example is hydrogen chloride, HCl, which in the gas phase is a highly polar molecule:

$$\overset{\delta+}{H}\!\overset{\delta-}{:}\!\overset{\cdot\cdot}{\underset{\cdot\cdot}{Cl}}:$$

The shared electrons spend much more of their time near the Cl atom than the H atom. The molecule has an electric dipole moment, that is, a slight positive charge at the H end and a slight negative charge toward the Cl end. When placed in water, HCl breaks up completely into H^+ and Cl^- ions. Water molecules are also polar, having a small negative charge on the oxygen end of the bonds and a small positive charge on the hydrogen atoms.

The symmetrical molecules such as the halogens mentioned above are called *nonpolar*, since they have no separation of charge between the parts of the molecule. Any molecule with a separation of charge which is not symmetrical is called a *polar molecule*. Some examples of polar molecules are

$$H\!\overset{\cdot\cdot}{\underset{\times}{O}}: \qquad \overset{\times}{\underset{\times}{Br}}\!\overset{\cdot\cdot}{\underset{\cdot\cdot}{Cl}}: \qquad H:\!\overset{H}{\underset{H}{C}}\!\overset{\times\times}{\underset{\cdot\cdot}{Cl}}\!\times \qquad Na\!\overset{\cdot\cdot}{\underset{\cdot\cdot}{Cl}}:$$

Although sodium chloride is ionic in the solid phase or in water solution, it does form highly polar covalent molecules in the gaseous state. Except in solution or in crystals, no compound is completely ionic. One of the most nearly ionic compounds in the gas phase is that formed between lithium and fluorine, which have electronegativity values of 1.0 and 4.0, respectively. Even with such a large difference in electronegativity, the bond has only about 92% ionic character, i.e., the electrons of the bond are close to the fluorine atom 92% of the time and near the lithium atom 8% of the time. With decreasing difference in electronegativity between atoms, the bonds become less polar, so that the electrons are shared more equally.

Molecules that contain polar bonds such as C—F or C—Cl are nonpolar if the molecules are symmetrical, so that the polar characteristics of each bond are cancelled by those of

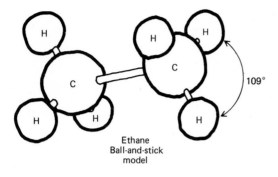

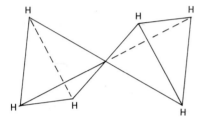

Fig. 8–3   The tetrahedral shape of the ethane molecule.

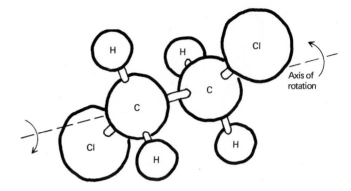

Fig. 8–4   Ball-and-stick model of $C_2H_4Cl_2$, in which there is free rotation around the carbon-carbon bond.

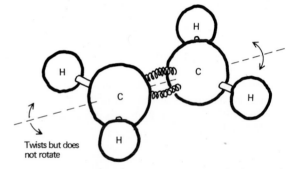

Fig. 8–5   Ball-and-stick model of C_2H_4 ethene, which twists back and forth but does not rotate about the carbon-carbon bond.

the other polar bonds. Examples are carbon tetrafluoride and carbon tetrachloride:

Now consider some more complicated molecules that can be made by bonding carbon atoms to each other, as in most organic and biochemical compounds (see Chapters 13 and 15). Let's start by removing a hydrogen atom from a methane molecule and forming a covalent bond to a second carbon atom:

The second carbon is left with three unpaired electrons and an outer shell of only five. This situation can be easily remedied by bringing in three other atoms, such as hydrogens, that can form bonds with the second carbon atom:

In this compound, called *ethane*, the octet rule is satisfied for the carbon atoms and each hydrogen has its two electrons.

Just as in methane, the four bonds around each carbon atom are directed toward the corners of a tetrahedron, giving the molecule the shape shown in Fig. 8–3. We could make a molecule of similar shape by replacing some of the hydrogen atoms with halogen atoms, as in dichloroethane:

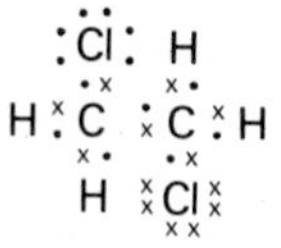

$$\begin{matrix} & :\!\ddot{C}\!l\!: & H \\ H\!:\!C & :C\!:\!H \\ H & \overset{\times\times}{\underset{\times\times}{\times}C l}\!\overset{\times}{\underset{\times}{}} \end{matrix}$$

You might suppose that the two forms written below are different compounds because of the different placement of the two chlorine atoms:

$$\begin{matrix} :\!\ddot{C}\!l\!: & H \\ H\!:\!C & :C\!:\!H \\ H & \times C l \times \end{matrix} \qquad \begin{matrix} :\!\ddot{C}\!l\!: & \times C l \times \\ H\!:\!C & :C\!:\!H \\ H & H \end{matrix}$$

However, one end of the molecule can rotate freely relative to the other end, as we see in Fig. 8–4, so that all three positions about either carbon atom are equivalent. Thus the two structures written above are just two ways of writing the same compound. Of course, if we were to place both chlorine atoms on the same carbon atom,

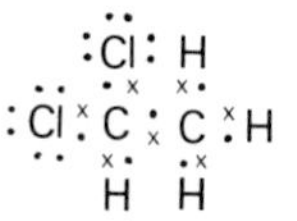

$$\begin{matrix} & :\!\ddot{C}\!l\!: & H \\ :\!C l\!:\!C & :C\!:\!H \\ & H & H \end{matrix}$$

we would have a different compound with somewhat different properties.

Another well-known compound of carbon and hydrogen is ethene, C_2H_4. How can we make an octet structure for it? Let's start by removing two hydrogen atoms from ethane:

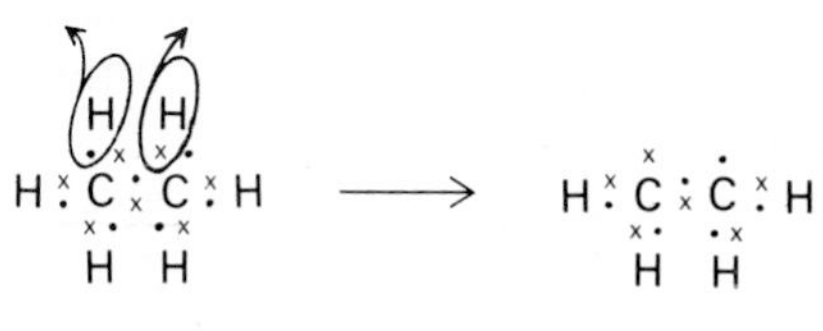

$$H\!:\!C\!:\!C\!:\!H \longrightarrow H\!:\!C\!:\!C\!:\!H$$

Each carbon atom now has only seven electrons in its valence shell. However, we can use the unpaired electron from each

carbon atom to form a second bond between the carbon atoms:

$$H\!:\!C\!:\!C\!:\!H \longrightarrow \begin{matrix} H & H \\ C\!:\!C \\ H & H \end{matrix}$$

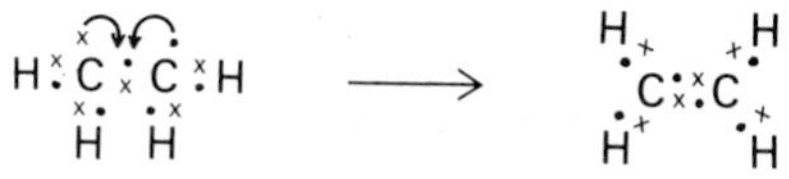

By sharing two pairs of electrons between the carbon atoms, the molecule is able to satisfy the octet rule. The four-electron bond is called a *double bond*, to distinguish it from the *single bonds* in which a single pair of electrons is shared. Instead of continuing to draw dots for all the electrons, we will use a shorthand notation in which single bonds are represented by single lines and double bonds by two lines, for example:

$$\begin{matrix} H & H \\ | & | \\ H-C-C-H \\ | & | \\ H & H \end{matrix} \qquad \begin{matrix} H \\ \diagdown \\ C=C \\ \diagup \\ H \end{matrix}\begin{matrix} H \\ \diagup \\ \diagdown \\ H \end{matrix}$$

ethane ethene

The actual shape of the ethene molecule is just as drawn above: all the atoms lie in the same plane and the angles between bonds are approximately 120°. The two ends of this molecule cannot rotate about the double bond (see Fig. 8–5). Therefore, the two structures

$$\begin{matrix} Cl & & Cl \\ \diagdown & & \diagup \\ & C=C & \\ \diagup & & \diagdown \\ H & & H \end{matrix} \quad \text{and} \quad \begin{matrix} Cl & & H \\ \diagdown & & \diagup \\ & C=C & \\ \diagup & & \diagdown \\ H & & Cl \end{matrix}$$

cis *trans*

are different compounds and have slightly different properties. They are called the *cis* and *trans* isomers of dichloroethene. *Isomers are compounds which have the same chemical formula, in this case $C_2H_2Cl_2$, but different molecular structures.*

Yet another well-known compound of carbon and hydrogen is ethyne (C_2H_2), more commonly known as "acetylene," the gas used in welders' torches. After removal of two hydrogen atoms from the ethene molecule, we have

$$\begin{matrix} H & & H \\ :\!C\!:\!C\!:\! \end{matrix}$$

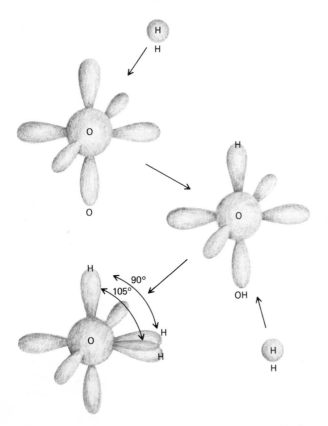

Fig. 8–6 Electron probability distribution in oxygen and hydrogen when they form a hydroxide and then a water molecule. Although we might expect the H—O bonds to form 90° angles, they are actually about 105°.

The structure of many molecules can be constructed by application of the octet rule. However, there are many molecules whose structures cannot be understood so simply. For example, one would expect the atoms of a molecule of ordinary oxygen, O_2, to be joined by a double bond; however, magnetic measurements show that there is an unpaired electron attached to each oxygen atom. When the total number of valence electrons in a molecule is odd, there is, of course, no way to completely satisfy the octet rule. One example is nitric oxide, NO, which has eleven valence electrons. Oxygen and nitric oxide are but two of the numerous compounds containing unpaired electrons. Many of these compounds are colored, as the unpaired electron can absorb photons of visible light and be excited to a higher energy state.

We can often make several different electronic structures that satisfy the octet rule, e.g., for sulfur trioxide SO_3:

$$\ddot{O}\overset{x}{\underset{xx}{x}}S\overset{x}{x}\ddot{O}\qquad :\ddot{O}\overset{x}{\underset{xx}{x}}S\overset{x}{x}\ddot{O}:\qquad \ddot{O}\overset{x}{\underset{xx}{x}}S\overset{x}{x}\ddot{O}:$$
$$:\ddot{O}:\qquad\qquad :\ddot{O}:\qquad\qquad :\ddot{O}:$$

If the SO_3 molecule actually contained one double bond and two single bonds as depicted by any of the above structures, experiments would show two different bond lengths (since double bonds are shorter than single bonds) and two different O—S—O bond angles, one between two single bonds and another between a double and single bond. Experimentally all the bond lengths are found to be equal, as well as the bond angles (all 120°). This apparent contradiction is resolved when we learn that when several equivalent structures can be made just by moving the electrons around, the actual structure is an average of all the possible structures. In SO_3, the true structure is an average of the three structures depicted above; i.e., instead of two single bonds and a double bond, we find that each bond is a $1\frac{1}{2}$ bond. This averaging of electronic structures is called *resonance*.

Many compounds have both ionic and covalent parts; for example, sodium sulfate dissociates completely into sodium and sulfate ions:

$$Na_2SO_4(s) \longrightarrow 2Na^+(aq) + SO_4^{2-}(aq). \qquad (8-1)$$

How can the octet rule be satisfied? Again we can move the two unpaired electrons into the carbon-carbon bond:

$$H\overset{x}{\underset{x}{C}}\overset{\cdot\,\cdot}{}\overset{x}{\underset{x}{C}}H \qquad \text{or} \qquad H-C\equiv C-H$$

This bond, containing three pairs of electrons, is a *triple bond*. The four atoms are all in line, just as drawn. Another common triple-bonded molecule is that of nitrogen gas, N_2, which makes up 80% of Earth's atmosphere:

$$:N\overset{x}{\underset{x}{}}\overset{\cdot\,\cdot}{}N\overset{x}{\underset{x}{}} \qquad \text{or} \qquad N\equiv N$$

The atoms of the sulfate ion are held together by covalent bonds:

$$\left[\begin{array}{c} :\ddot{O}: \\ :\ddot{O} \overset{xx}{\underset{xx}{\times\, \overset{\triangle}{\underset{\triangle}{S}} \,\times}} \ddot{O}: \\ :\ddot{O}: \end{array} \right]^{2-}$$

The two electrons depicted by triangles were obtained from the sodium atoms and cause the SO_4^{2-} ion to have its net charge of 2–. Some of these covalent-bonded ions have resonance structures, e.g., carbonate ion, CO_3^{2-}.

8–2 SPECTRA OF MOLECULES

In Chapter 7 we saw that one can learn a great deal about the electronic structure of atoms by observing the spectra of light emitted when individual electrons of atoms of a particular element are excited. We can also learn a great deal about the structure of molecules by determining the radiations that they emit or absorb. Most of the facts about molecular structure discussed in Section 8–1 were derived from spectroscopic observations. As we shall see in Chapter 9, the earth's climate is strongly affected by the absorption of radiations by certain molecular species present in the atmosphere.

When a molecule is formed by covalent bonding, the probability distribution of electrons in a bond is often similar to that of electrons in orbitals of the atoms. However, the probability distribution is changed somewhat because of the presence of the second atom of the bond. Consider the formation of an O— H bond in a water molecule when a single hydrogen atom is brought up to an oxygen atom (see Fig. 8–6). The hydrogen atom has a single electron in the spherical 1s orbital. The electronic configuration of the oxygen atom is $1s^2\, 2s^2\, 2p^4$; that is, it has two electrons in one 2p orbital and an unpaired electron in each of the other two 2p orbitals. When the bond is formed, the hydrogen atom shares its single electron with one of the unpaired electrons in an oxygen 2p orbital. The probability distributions for the two electrons of the bond are identical to each other and different from those of the original atomic orbitals. The electrons spend much of their time between the two positive nuclei, to which they are attracted. There is a higher probability of finding the electrons near the oxygen atom because of its greater electronegativity.

When the complete water molecule is formed, the second hydrogen forms a bond with the remaining unpaired 2p electron from the oxygen atom. Since the p orbitals are perpendicular to each other, one might suppose that the H—O—H bond angle would be 90°, but it is actually about 105°. The mutual repulsion of electrons of the bonds forces the bonds to a greater angle than 90°. Also, since the bonds are somewhat polar, the hydrogen atoms acquire a partial positive charge, causing them also to repel each other. In the analogous compounds of hydrogen with sulfur and selenium (which are in the same chemical group as oxygen), the bond angles are 92° and 91°, respectively. Repulsion effects are not as great in these molecules because of the larger size of the central atom and the smaller differences in electronegativity, which cause a decrease of the polar character of the bonds.

In Chapter 7 we saw that electrons of an atom could be raised from one orbital to another by absorption of a photon with an energy equal to the energy difference between orbitals. Many of these electronic excitations of atoms are caused by UV or visible light. Molecules, like atoms, can absorb visible or UV light that produces electronic excitation of the molecule. The electron probability distributions shown in Fig. 8–6 are for the lowest-energy electronic state of the molecule. Just as in atoms, the electrons of orbitals of a molecule can be raised to higher energy states that have probability distributions of different shapes. There is one major difference between electronic transitions in molecules and atoms: the spectra of atoms consist of series of very sharp lines, but because of certain complexities of molecules (discussed (below), the electronic transitions of molecules are spread over wide ranges of frequencies. As a result, we usually do not learn as much about molecules from their visible and UV spectra as we do in the case of atoms.

On the other hand, molecules can be excited in ways that are not possible for individual atoms. The bonds that hold atoms together in a molecule are rather flexible, so that they can be made to bend and stretch. It is as if the atoms were held together by springs. If radiation of the proper energy is absorbed by the molecule, the atoms vibrate about their average positions. The quantum theory permits only vibrational states of certain energies. Thus molecules of a particular compound can absorb only certain well-defined frequencies of radiation. There are two major classes of vibrations: "stretching" vibrations of the atoms along the axis of the bond and "bending" vibrations in which the angle between two bonds oscillates (see Fig. 8–7).

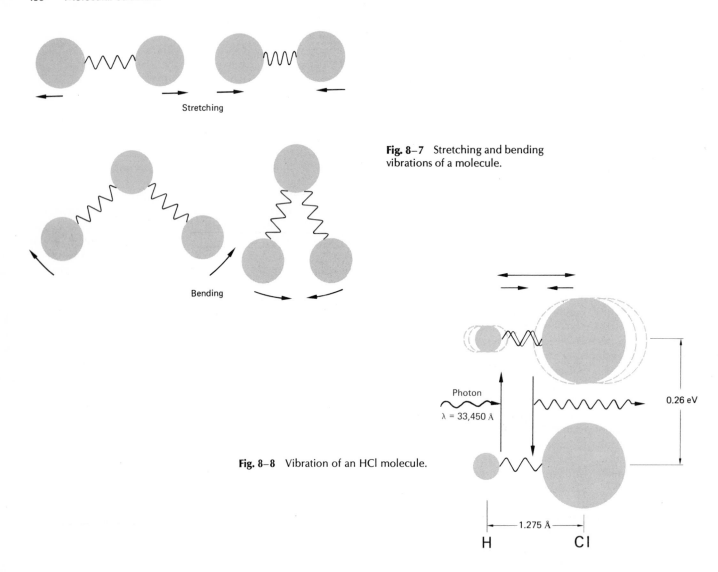

Fig. 8-7 Stretching and bending vibrations of a molecule.

Stretching

Bending

Fig. 8-8 Vibration of an HCl molecule.

Photon
λ = 33,450 Å

0.26 eV

1.275 Å

H   CI

The energies of vibrational transitions are much smaller than those of electronic transitions and fall mainly in the infrared (IR) region of the spectrum. For example, when a simple diatomic molecule such as HCl absorbs IR radiation of wavelength 33,450 Å, it vibrates rapidly back and forth as shown in Fig. 8–8. The vibration continues until the molecule gives off the energy, generally in the form of another photon of IR light of the same energy as the one originally absorbed. A photon of twice the energy will cause the HCl molecule to vibrate about twice as fast.

Molecules containing three or more atoms have more complicated vibrations. Figure 8–9 shows some of the vibra-

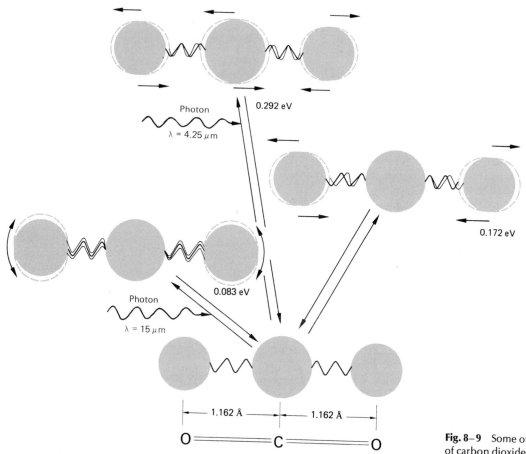

Photon
λ = 4.25 μm

0.292 eV

0.172 eV

0.083 eV

Photon
λ = 15 μm

|← 1.162 Å →|← 1.162 Å →|

O ═══ C ═══ O

Fig. 8–9 Some of the vibrational excited states of carbon dioxide molecules, most of them induced by IR radiation. (For rather complicated reasons, the vibration of CO_2 shown in the right center of the figure cannot be induced by absorption of IR light, only by being hit by higher energy radiations.)

tions of carbon dioxide, CO_2, and their energies. When a continuous spectrum of IR light is passed through CO_2 gas, photons of these and other energies are absorbed. As we shall see in Chapter 9, the absorption of IR light by atmospheric carbon dioxide has important implications for global climatic change.

Fig. 8–10 Some of the vibrational excited states of the water molecule.

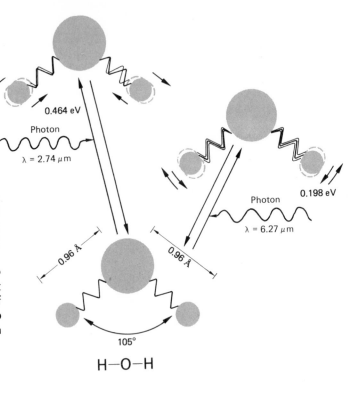

0.464 eV

Photon

$\lambda = 2.74\,\mu m$

0.198 eV

Photon

$\lambda = 6.27\,\mu m$

0.96 Å   0.96 Å

105°

H—O—H

The fact that water molecules are bent (see Fig. 8–10) causes their vibrations to be more complicated than those of CO_2, in which the atoms lie in a straight line. Water vapor in the atmosphere also absorbs IR light and in Chapter 9 we shall show the absorption spectra of both H_2O and CO_2.

Infrared spectrometry is used extensively by chemists to determine the structure of unknown molecules and to detect various compounds in mixtures. Certain configurations of atoms in molecules, called *functional groups*, strongly absorb IR light of particular frequencies. Thus we can identify such functional groups as

$$-C=C-, \quad -C-O-H, \quad -C\overset{\displaystyle O}{\underset{\displaystyle OH}{\Big\langle}}, \quad C=O, \text{ etc.}$$

Every compound that has a number of functional groups produces a characteristic absorption spectrum when IR light

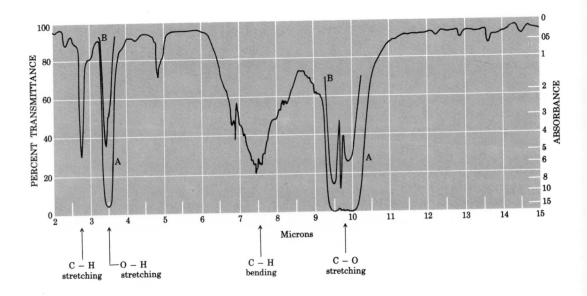

C – H stretching   O – H stretching   C – H bending   C – O stretching

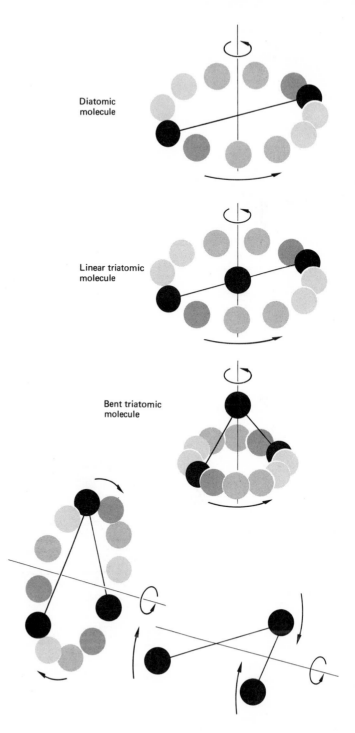

passes through it. A typical IR absorption spectrum is shown in Fig. 8–11. Infrared spectrometry is used extensively for the detection of various organic compounds in complex environmental samples, e.g., pesticide residues, breakdown products in motor oil, organic constituents of air pollutants, illicit drugs and narcotics, and many others.

At energies even lower than those of IR light, molecules have other forms of excitation, particularly rotations. According to quantum theory, molecules can rotate only at certain speeds. Molecules can absorb just certain frequencies of radiation to raise them from one state of rotation to another (see Fig. 8–12). The specific radiations which cause rotation are mostly in the microwave region, i.e., very short radio waves with wavelengths ranging from about 0.1 to 1000 cm, with corresponding frequencies from 300,000 down to 30 megahertz (1 megahertz = 1 million cycles per second).

Fig. 8–11   The infrared spectrum of a typical organic molecule. [From Gregory R. Choppin and Russell H. Johnsen, *Introductory Chemistry* (Addison-Wesley, Reading, Mass., 1972), p. 425. Reprinted by permission.]

Diatomic molecule

Linear triatomic molecule

Bent triatomic molecule

Fig. 8–12   Rotational excitation of molecules.

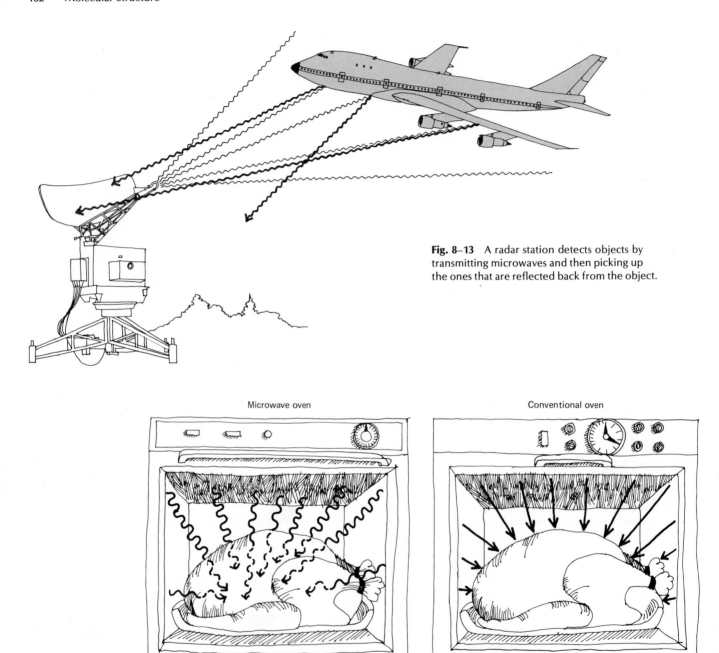

Fig. 8–13   A radar station detects objects by transmitting microwaves and then picking up the ones that are reflected back from the object.

Microwave oven

Conventional oven

Fig. 8–14   Food cooks rapidly in microwave ovens because the microwaves penetrate into the food, heating it throughout, unlike conventional ovens in which the heat is slowly conducted from the outside to the inner portions of the food.

Techniques for the use of microwaves were developed during World War II for use in radar (see Fig. 8–13). Microwaves are used by television networks for transmission of television pictures to local stations, because microwaves can be directed from one relay station to another with parabolic aerials. Microwave spectroscopy has been used by astronomers to detect chemical compounds that exist between the stars and in the outer reaches of the universe. Some of the observed compounds, such as formaldehyde (CH_2O) are believed to be the building blocks for the more complicated molecules of primitive life forms.

A growing practical use of microwaves is in ovens that cook foods very rapidly. In conventional ovens, it takes a long time for heat to be conducted from the surface of a piece of meat to the inside in order to cook it thoroughly. Microwave ovens bombard food with radiations of about 2450 megahertz that are transmitted deep into meat and other products (see Fig. 8–14). Radiations of this frequency are absorbed by water molecules throughout the food and cause localized heating wherever there are water molecules. The food cooks thoroughly within seconds or minutes, but there are serious potential hazards associated with microwaves. If they leak out of the oven, they can heat and damage various parts of the body, in extreme cases "cooking" them just like a piece of meat.* Microwaves can be contained safely by metal sheets or screens (whose openings must be smaller than the several-cm wavelength of the microwaves). However, if the oven door does not shut tightly or if a switch fails to shut off the oven when the door is opened, one can receive excessive doses of microwave radiation. One of the greatest microwave hazards is the formation of cataracts on the eyes.

8–3 PHYSICAL PROPERTIES OF WATER

Water is one of the most important substances in our environment. It is one of the essential requirements for all biological life. The oceans, which cover three-fourths of Earth's surface, have important moderative effects on Earth's climate. Water is an excellent solvent for many polar and ionic substances; thus chemists carry out many of their studies in water solutions.

Water has unusual properties when compared with other compounds. Most of these unusual features result from the polar nature of the O—H bonds. The oxygen end of the O—H bond acquires a slight negative charge and the hydrogen end, a slight positive charge. In liquid or solid water, the H_2O molecules tend to line up with the hydrogen of one molecule

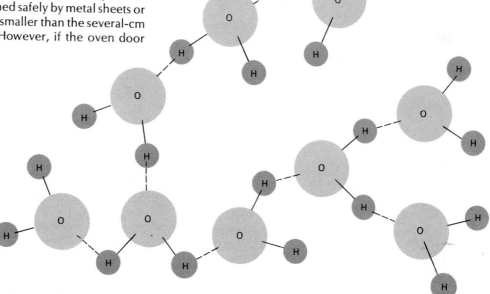

Fig. 8–15 Water molecules are attracted to each other rather strongly because of hydrogen bonding.

---

*   Terri Aaronson, "Out of the Frying Pan," *Environment* **12**(5), 26–31 (June 1970).

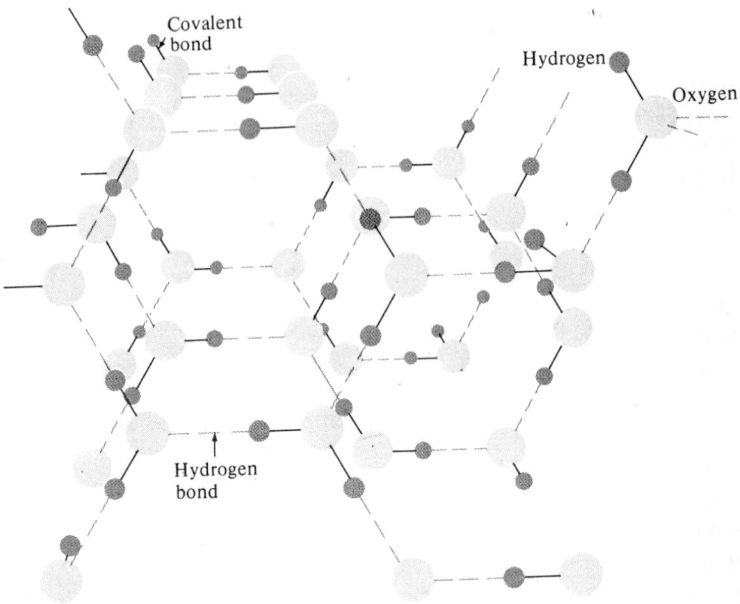

Fig. 8–16 The open structure of ice, caused by hydrogen bonding, causes it to have a lower density than liquid water. [From Paul. F. Weller and Jerome H. Supple, *Chemistry: Elementary Principles* (Addison-Wesley, Reading, Mass., 1971), p. 267].

attracted electrostatically to the oxygen of a neighboring molecule. This effect is called *hydrogen bonding*, which is not a true covalent or ionic bond, but an unusually large attraction between molecules (see Fig. 8–15).

For most substances, molecules are farther apart in the liquid state than in the solid, so that the solid is denser than the liquid. Just the opposite is true of water: because of hydrogen bonding, water molecules in ice form a very open structure with large cavities (see Fig. 8–16). As ice is warmed and approaches the melting point, the molecules vibrate more vigorously about their average positions in the crystal lattice, breaking many hydrogen bonds. At 0°C the rigid structure collapses as the ice melts to form liquid water, a process that requires 80 calories of heat per gram of ice. Upon melting, the H_2O molecules, no longer held rigidly in a lattice, come slightly closer together, so that the density increases about 0.91 g/cm^3 to nearly 1.0 g/cm^3. Because of its unusual change in density when it melts, ice floats on the surface of liquid water.

Some of the hydrogen-bonded structure remains in localized areas of the liquid water. Additional heating continues to break up the structure causing a further increase in density, which reaches a maximum of 1.00 g/cm^3 at 4°C.

Above that temperature, further breakup of the structure is more than counterbalanced by the increased agitation of the molecules. The average distance between molecules increases and the density decreases slightly with increasing temperature (see Fig. 8–17). The density of ice and of liquid water at various temperatures has important implications for temperature profiles in water bodies. When the air temperature is well below 0°C, ice floats on the surface of a lake and the densest liquid water, with a temperature of 4°C, goes to the bottom. When the temperature throughout a lake is well above 0°C, the warmest, least dense water stays near the surface and cooler, denser water sinks to the bottom.

Another unusual feature of water is its large *heat capacity* —the amount of heat (measured in calories) needed to raise the temperature of 1 g of a substance by 1°C. The heat capacity of liquid water is about 1 cal/g. As shown in Table 8–1, the heat capacity of water is among the highest known for common substances. Temperature is a measure of the velocity of agitation of the atoms and molecules of a substance. In part, the large heat capacity of water results from its hydrogen bonding. When molecules stick together tightly, a great deal of energy must be added to the substance to cause the agitation necessary to raise the temperature.

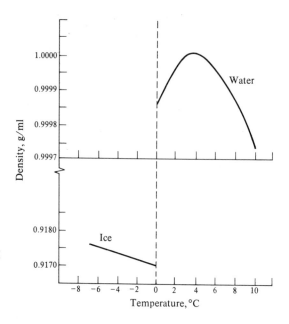

Fig. 8–17 The density of water increases to its maximum value at 4°C, and then decreases as the temperature of the water rises.

Table 8–1   Heat Capacities of Some Common Substances

Substance	Heat capacity (cal/g°C)
Aluminum	0.216
Copper	0.092
Gold	0.031
Iron	0.108
Silver	0.057
Water (liquid)	1.000

Table 8–2   Properties of Some Compounds with Small Molecules

Compound	Molecular weight	Melting point (°C)	Boiling point (°C)	Heat of melting (cal/g)	Heat of vaporization (cal/g)
H_2O	18	0	100	80	540
H_2S	34	−88	−59	17	132
H_2Se	81	−64	−42	7	60
NH_3	17	−78	−33	79	327
CH_4	16	−182	−161		122

The large heat capacity of water has important climatic influences upon our watery planet. It means that a large body of water can absorb or lose a large amount of heat without greatly increasing or decreasing its temperature. For this reason, land areas downwind from large bodies of water generally suffer much smaller temperature fluctuations than far inland areas.

Molecules of a liquid escape into the vapor phase when the kinetic energies of the molecules are sufficient, that is, when the temperature is high enough, to overcome the attractive forces between the molecules in the liquid. If molecules of two different compounds have equally strong attractive forces, but different molecular weights, the lighter molecules of the low-molecular-weight compound enter the vapor state at much lower energies, and thus lower temperatures, than those of the heavier molecules. Compounds with low molecular weight have low melting and boiling points unless the attraction between molecules is quite strong.

In Table 8–2 we see that the melting point and the boiling point of water are much higher than those of other compounds of about the same molecular weight. Ammonia, NH_3,

is somewhat like water in that its molecules are polar and are attracted to each other by hydrogen bonds, however, these bonds are much weaker in ammonia than in water.

In the last two columns of Table 8–2 are listed the amounts of heat needed to melt and vaporize a gram of each substance. Water requires an unusually large amount of heat to melt a gram of ice or convert a gram of water into steam. The latter is called the *heat of vaporization of water* or the *latent heat of steam*. Because of the large amounts of energy involved, evaporation and condensation of water are widely used as mechanisms for the transfer of energy in the environment. For example, the evaporation of 1 g of perspiration water removes about 540 cal from one's body, cooling the skin. The sun heats the surface of the oceans in the tropics and causes evaporation of water. When the water vapor rises in the atmosphere and cools, it condenses into droplets of water, giving up its latent heat to the atmosphere. This is a mechanism for transferring heat from water bodies to the atmosphere. The latent heat of steam is the main source of the energy that drives hurricanes. These storms pick up large amounts of energy from water evaporated from the hot ocean

Hurricane Gladys as seen from the Apollo satellite. (Courtesy of NASA.)

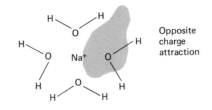

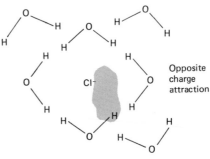

Fig. 8–18 Because of hydrogen bonding, a sheath of H_2O molecules is formed around the Na^+ and Cl^- ions of a sodium chloride solution.

water of the tropics. To put man in his proper perspective in nature, we should note that the kinetic energy of a large hurricane is about 10,000 times the amount of energy released in the Hiroshima nuclear explosion! Even a one-inch rainfall over a 10-mile square area (that is, 100 sq mi) releases about 9×10^{15} cal, equivalent to the energy released in 100 Hiroshima bombs.

8–4 WATER AS A CHEMICAL SYSTEM: ACIDS AND BASES

Water is probably the most important reagent in the chemical laboratory. Many chemical reactions are carried out in water, that is, in *aqueous solutions*. Water is an excellent solvent for a wide variety of substances, especially ionic substances which dissolve well in water, because of the polar nature of H_2O molecules. For example, if sodium chloride is dissolved in water, the salt completely breaks up into Na^+ and Cl^- ions. The slightly negative oxygen atoms of water molecules are electrostatically attracted by the positive charge of the Na^+ ions and the hydrogen atoms of the water molecules are

attracted by the negative charge of the Cl^- ions. The ions are surrounded by sheaths of water molecules, as shown in Fig. 8–18. The bonds with the water molecules are not as strong as covalent bonds, but the sheaths of water around the ions keep the ions in solution separated from each other.

Although there are more complicated and elegant definitions, it will be adequate for our purposes to define an acid *as a compound that yields hydrogen ions (H^+) in water solution*, e.g., hydrochloric acid:

$$HCl \text{ (aq)} \longrightarrow H^+ \text{ (aq)} + Cl^- \text{ (aq).} \tag{8–2}$$

Since an H^+ ion is just a proton (or deuteron), it probably is not present in solution as a bare nucleus. Instead, most of the hydrogen nuclei probably attach themselves to unused electron pairs of water molecules:

$$H^+ \;+\; :\!\overset{..}{\underset{..}{O}}\!:\!H \;\longrightarrow\; H\!:\!\overset{H}{\underset{..}{\overset{..}{O}}}\!:\!H \tag{8–3}$$

The hydrated proton H_3O^+ is called the *hydronium ion*. In other chemical writing, you may find that H_3O^+ is used where we use H^+. For simplicity, we write H^+, but one should keep in mind that it exists as H_3O^+. The hydronium ion probably also has more loosely attached water molecules nearby, just like any charged species in aqueous solution.

A base is a *compound that yields hydroxide ions* (OH^-) *in solution*, e.g., sodium hydroxide:

$$NaOH\,(aq) \longrightarrow Na^+\,(aq) + OH^-\,(aq). \qquad (8–4)$$

If a solution of HCl is mixed with an NaOH solution, the H^+ ions of the former react with the OH^- ions of the latter to form H_2O molecules and a salt:

$$H^+\,(aq) + Cl^-\,(aq) + Na^+\,(aq) + OH^-\,(aq) \rightleftharpoons$$
$$H_2O\,(l) + Na^+\,(aq) + Cl^-\,(aq) \qquad (8–5)$$

In Eq. (8–5) we have cancelled out the Na^+ and Cl^- ions on both sides, as they do not participate in the reaction. They are present in both the initial and final solutions as free ions. After removing these "bystander" ions, we see that the net *reaction* is simply the combination of H^+ and OH^- ions to form water:

$$H^+\,(aq) + OH^-\,(aq) \rightleftharpoons H_2O\,(l). \qquad (8–6)$$

If equal numbers of moles (M) of HCl and NaOH are mixed together, the final solution consists of equal numbers of Na^+ and Cl^- ions dissolved in water. For example, if equal volumes of $1M$ NaOH and $1M$ HCl solutions are mixed together, the final solution is an $0.5M$ solution of NaCl. Then if we boil off the water, the residue is solid sodium chloride, i.e., common table salt.

Hydrochloric acid and sodium hydroxide are examples of a *strong acid* and a *strong base*, respectively. When they are dissolved in water, their molecules break up completely into ions. There are also weak acids and bases, for example, acetic acid and ammonia, which largely remain as molecules in solution, with only a small percent breaking into ions:

Acetic acid: $HC_2H_3O_2\,(aq) \rightleftharpoons$
$$H^+\,(aq) + C_2H_3O_2^-\,(aq). \qquad (8–7)$$

Ammonia: $NH_3\,(aq) + H_2O\,(l) \rightleftharpoons$
$$NH_4^+\,(aq) + OH^-\,(aq). \qquad (8–8)$$

In other chemical writing or on reagent bottles in the laboratory, you may find the name "ammonium hydroxide" and the formula NH_4OH. It was previously thought that when ammonia gas, NH_3, dissolves in water it forms molecules of ammonium hydroxide.

$$NH_3\,(g) + H_2O\,(l) \rightleftharpoons NH_4OH\,(aq). \qquad (8–9)$$

However, there is no evidence that actual molecules of NH_4OH are formed. Thus we regard the solution as being simply that of ammonia dissolved in water, NH_3 (aq).

These reactions are written with arrows in both directions to indicate that they are *reversible equilibria*. At all times in an acetic acid solution, some $HC_2H_3O_2$ molecules are dissociating into ions and, at the same time, H^+ and $C_2H_3O_2^-$ ions are coming together at the same rate to form $HC_2H_3O_2$ molecules.

Water itself can be considered as a weak acid or weak base, as it dissociates slightly into H^+ and OH^- ions:

$$H_2O\,(l) \rightleftharpoons H^+\,(aq) + OH^-\,(aq). \qquad (8–10)$$

It is for this reason that we have shown arrows going in both directions in Eqs. (8–5) and (8–6). Even when the H^+ ions of HCl are exactly neutralized by OH^- ions of NaOH, a small and equal number of H^+ and OH^- ions remain in solution.

A reversible, equilibrium reaction can be forced to go farther to the left or right by adding or removing some of the reactants or products involved in the reaction. As enunciated qualitatively by Le Chatelier in 1884, "if a stress is applied to a system at equilibrium, the system readjusts, if possible, to reduce the stress." In practical terms, Le Chatelier's principle means that if a reaction,

$$A + B \rightleftharpoons C + D, \qquad (8–11)$$

is at equilibrium and we "stress" the system by adding more of species A, the system will readjust: some of the added A reacts with B to form more C and D. In general, the addition of reactants appearing on the left forces the reaction to proceed farther to the right, and addition of products appearing on the right forces the reaction toward the left. Conversely, the removal of reactants or products also forces the system towards the left or right, respectively.

A simple example of Le Chatelier's principle is shown by the reaction of carbon dioxide gas with water to form carbonic acid, H_2CO_3:

$$CO_2\,(g) + H_2O\,(l) \rightleftharpoons H_2CO_3\,(aq). \qquad (8–12)$$

If water is placed in a closed container along with CO_2 gas, some of the gas dissolves to form carbonic acid. If the pressure of CO_2 over the solution is increased, the system removes

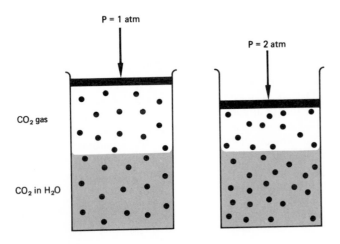

Fig. 8–19 The higher the pressure of carbon dioxide over water, the greater the amount of CO_2 that dissolves in the water.

the "stress"—more CO_2 goes into solution, yielding a higher concentration of H_2CO_3 (see Fig. 8–19). If, on the other hand, the pressure of CO_2 gas is reduced, some of the CO_2 comes out of solution and reenters the gas phase. This reaction occurs every time one opens a bottle of carbonated beverage: Cokes and similar drinks are bottled under a high pressure of CO_2. When the bottle cap is removed, the pressure is released, so that bubbles ("fizz") of CO_2 comes out of solution. If a beverage stands open in the room and warms up, more bubbles form as more CO_2 escapes. As with most gases, the solubility of CO_2 decreases at higher temperatures.

Carbonic acid is a weak acid that partially dissociates into H^+ ions and bicarbonate ions, HCO_3^-:

$$H_2CO_3(aq) \rightleftharpoons H^+(aq) + HCO_3^-(aq) \qquad (8-13)$$

This equilibrium is predominantly toward the left: only about one H_2CO_3 in ten thousand is dissociated into H^+ and HCO_3^- ions. If the concentration of H_2CO_3 is raised by increasing the CO_2 pressure over the solution, the system readjusts: some of the added H_2CO_3 dissociates to form more H^+ and HCO_3^- ions. Conversely, if one adds H^+ ions by adding HCl to the solution, some of the H^+ ions combine with bicarbonate ions to form more carbonic acid. One can "pull" the equilibrium toward the right by removing one of the products. For example, if NaOH is added to the solution, some of the OH^- ions

combine with H^+ ions to form H_2O, as in Eq. (8–6). When H^+ ions are thus removed, the equilibrium of Eq. (8–13) readjusts—additional H_2CO_3 dissociates to form more HCO_3^- and H^+ ions.

Carbonic acid is a *diprotic acid*, that is, one with two hydrogens that come off as H^+ ions. In this case, the bicarbonate formed in reaction (8–13) further dissociates very weakly, releasing a second H^+ ion and a carbonate ion, CO_3^{2-}:

$$HCO_3^-(aq) \rightleftharpoons H^+(aq) + CO_3^{2-}(aq). \qquad (8-14)$$

This equilibrium is even further favored on the left than that of Eq. (8–13): in a carbonic acid solution, only about one carbonate ion in 10^{11} breaks up to form H^+ and CO_3^{2-} ions. This reaction is strongly inhibited by the H^+ ions produced by reaction (8–13). Just as in Eq. (8–13), the equilibrium can be shifted to the right by adding NaOH, which reacts with the H^+ ions and reduces their concentration. If we start with a solution of pure carbonic acid and add two moles of NaOH for every mole of H_2CO_3 originally present, we obtain a solution containing mainly CO_3^{2-} and Na^+ ions:

$$2Na^+(aq) + 2OH^-(aq) + H_2CO_3(aq) \rightleftharpoons$$
$$2Na^+(aq) + 2H_2O(l) + CO_3^{2-}(aq). \qquad (7-15)$$

If we evaporate the water, we obtain solid, crystalline salt, sodium carbonate, Na_2CO_3.

It is also interesting to consider what happens if we dissolve pure sodium carbonate in water. First of all, Na_2CO_3 is strongly ionic, so that upon dissolution the salt immediately forms Na^+ and CO_3^{2-} ions:

$$Na_2CO_3(s) \xrightarrow{H_2O} 2Na^+(aq) + CO_3^{2-}(aq). \qquad (8-16)$$

We have seen above that HCO_3^- is very weakly dissociated; if there were a source of H^+ ions in the solution, the CO_3^{2-} ions would react with them to form HCO_3^- ions, i.e., we would have the reverse of Eq. (8–14). In this solution there is a source of H^+ ions: according to Eq. (8–10), water breaks up slightly to form H^+ and OH^- ions. Thus in the Na_2CO_3 solution, there is a competition between the H_2O molecules and CO_3^{2-} ions to see "who gets the H^+ ions." Water is a much weaker acid than HCO_3^- ions, but a small fraction of the CO_3^{2-} ions pick up H^+ from H_2O molecules:

$$CO_3^{2-}(aq) + H_2O(l) \rightleftharpoons HCO_3^-(aq) + OH^-(aq). \qquad (8-17)$$

By this process, called *hydrolysis*, the salt of a very weak acid reacts with water molecules to release OH^- ions, thus making the solution somewhat basic. We should note that Na_2CO_3 has been used as a substitute for phosphate in some household detergents (see Ch. 11). Although the use of carbonate eliminates some of the problems caused by phosphates, it is not a satisfactory substitute because of the strongly basic nature of its water solutions. The carbonate is quite harmful if ingested and the basic solution is irritating to the skin and corrosive to many surfaces.

The salts of other weak acids also hydrolyze. Consider the following examples.

Acetic acid:

$$C_2H_3O_2^-(aq) + H_2O(l) \rightleftharpoons HC_2H_3O_2(aq) + OH^-(aq).$$

acetate ion   acetic acid   (8–18)

Hydrogen sulfide:

$$S^{2-}(aq) + H_2O(l) \rightleftharpoons HS^-(aq) + OH^-(aq). \qquad (8-19)$$

sulfide   bisulfide
ion   ion

The salts of weak bases also hydrolyze. For example, we saw in Eq. (8–8) that ammonia is weakly dissociated. Thus if the salt ammonium chloride, NH_4Cl, is dissolved in water, it immediately dissociates into NH_4^+ and Cl^- ions. Some of the ammonium ions then react with water to release H^+ ions and make the solution acidic:*

$$NH_4^+(aq) \xrightarrow{H_2O} NH_3(aq) + H^+(aq). \qquad (8-20)$$

Let us return to the carbonate ion reactions. If we supply the carbonate ions with an abundance of H^+ ions by adding two moles of HCl for every mole of CO_3^{2-}, then each carbonate ion takes up two H^+ ions to form H_2CO_3:

$$2H^+(aq) + CO_3^{2-}(aq) \rightleftharpoons H_2CO_3(aq). \qquad (8-21)$$

Unless the CO_3^{2-} concentration was originally very small, the final solution contains so much carbonic acid that CO_2 gas is evolved,

$$H_2CO_3(aq) \rightleftharpoons CO_2(g) + H_2O(l), \qquad (8-22)$$

causing the solution to bubble like a carbonated beverage.

***

* In this case the involvement of water molecules is more obvious if we use the hydronium ion:

$$NH_4^+(aq) + H_2O(l) \rightleftharpoons NH_3(aq) + H_3O^+(aq)$$

This reaction is used as a standard geological test for rocks containing carbonate (e.g., limestone or marble, which are mostly $CaCO_3$). A few drops of HCl solution are dripped onto the rock. If it reacts vigorously with bubbling, the rock most likely contains carbonate.

The acidity of a solution is usually denoted by its pH value, defined by:

$$pH \equiv -\log M_{H^+}, \qquad (8-23)$$

where M_{H^+} is the molarity of the H^+ ions in the solution, that is, the number of moles of H^+ ions per liter of solution (see Section 2.5).†

Recall that if a number x is expressed as ten raised to some power y, then y is the logarithm of x:

$$x = 10^y, \qquad y = \log x. \qquad (8-24)$$

Thus if the H^+ ion concentration in a solution is exactly 10^{-4} M, then the pH of the solution is 4. A $1M$ solution of a strong acid such as HCl would dissociate completely to yield $M_{H^+} = 1 = 10^0$, giving pH = 0. Similarly, $0.1M$ HCl would yield $M_{H^+} = 0.1 = 10^{-1}$, or pH = 1. Since acetic acid is only partially dissociated, a $1M$ solution of $HC_2H_3O_2$ produces an H^+ concentration of only 4.2×10^{-3} $M = 10^{-2.35}$, giving pH = 2.35.

We have seen that water dissociates slightly to produce small, equal concentrations of H^+ and OH^- ions. At 25°C, about one molecule of H_2O per billion is dissociated to produce a H^+ ion concentration of $10^{-7}M$, so pure water has a pH of 7. If we neutralize a solution of HCl with an equal number of OH^- ions from an NaOH solution, the pH of the final solution is 7, just as in pure water.

If we add a base such as NaOH to pure water, according to Le Chatelier's principle, the dissociation of H_2O is repressed so that the concentration of H^+ ions is reduced. In $1M$ NaOH solution, M_{H^+} is reduced to 10^{-14}, giving pH = 14. Since ammonia is weak base, the OH^- concentration in a $1M$ NH$_3$ solution is only 4.25×10^{-3} M, and so the H^+ concentration is greater, $M_{H^+} = 2.4 \times 10^{-12}$, and the pH is 11.6. The pH range for various acid and base solutions is summarized in Fig. 8–20.

Water can be decomposed into its component elements by adding a little salt or other ionized substance to it (to carry

***

† For a review of logarithms, see Appendix A.

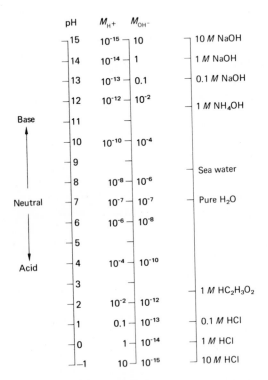

Fig. 8–20 The pH values of solutions at various concentrations of H⁺ and OH⁻ ions.

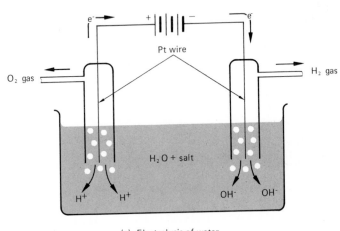

(a). Electrolysis of water

an electric current), dipping electrodes into it, and passing an electric current through. As shown in Fig. 8–21a, water is decomposed at the negative electrode to form hydrogen gas and hydroxide ions:

$$2H_2O\,(l) + 2e^- \rightleftharpoons H_2\,(g) + 2OH^-\,(aq). \qquad (8-25)$$

At the positive electrode, water is decomposed into oxygen gas and hydrogen ions:

$$2H_2O\,(l) \rightleftharpoons O_2\,(g) + 2H^+\,(aq) + 2e^-. \qquad (8-26)$$

The OH⁻ and H⁺ ions formed at the electrodes eventually combine with each other to form H_2O molecules.

These reactions are also reversible. If a battery is connected as shown in Fig. 8–21a, electrical energy from the

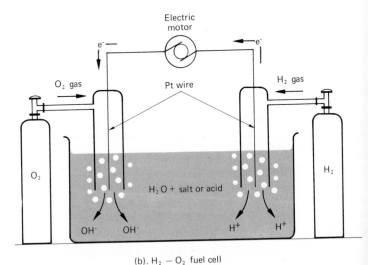

(b). $H_2 - O_2$ fuel cell

Fig. 8–21 (a) Water can be electrolyzed into H_2 and O_2 gases by passing an electric current through the water. (b) This fuel cell produces electrical energy when H_2 and O_2 gases combine to form water.

battery decomposes water into H_2 and O_2 gases. On the other hand, if we initially have H_2 and O_2 gases (as in Fig. 8–21b), we can do just the opposite: combine the gases to form water and obtain the energy in the form of electricity. This is the major energy source used in the space capsules that have gone to the moon. Hydrogen and oxygen gases are combined in a *fuel cell* to produce electricity. These fuel cells also have the advantage of producing water which can be consumed by the astronauts. Many research laboratories are trying to develop fuel cells which would serve as sources of energy here on Earth, particularly as a nonpolluting energy source for automobiles (see Chapter 14).

8–5 QUANTITATIVE TREATMENT OF ACID-BASE EQUILIBRIA*

In this section we demonstrate how Le Chatelier's principle can be applied quantitatively to some of the equilibrium reactions discussed in Section 8-4. As noted above, when the pressure of carbon dioxide gas over an aqueous solution of carbonic acid is increased, additional CO_2 dissolves in the water to form H_2CO_3, shifting the equilibrium

$$CO_2(g) + H_2O(l) \rightleftharpoons H_2CO_3(aq). \qquad (8-27)$$

toward the right. The relationship between the pressure of CO_2, P_{CO_2}, and the concentration of H_2CO_3 may be expressed quantitatively as

$$[H_2CO_3] = K \times P_{CO_2}, \qquad (8-28)$$

where for simplicity of notation we use brackets, [ ], to denote the concentration in moles per liter. The quantity K is the *equilibrium constant* for the reaction. The value of K is constant at a given temperature, being 0.03 M/atm at 25°C but decreasing at higher temperatures. Equation (8–28) shows, as one might suppose, that the concentration of H_2CO_3 in solution is proportional to the pressure of CO_2 gas over the solution.

The dissociation of water is expressed quantitatively by the expression

$$[H^+] \cdot [OH^-] = K_w = 10^{-14} \text{ (moles/liter)}^2 \qquad \text{(at 25°C),} \qquad (8-29)$$

---

* In courses employing a minimum of mathematics this section may be omitted as it is not essential for understanding the following chapters.

The power pack for an Apollo fuel cell being lowered into its container. (Courtesy of NASA.)

where K_w is another equilibrium constant, in this case specifically referred to as the *ionization constant of water*. In pure water, the dissociation produces equal numbers of H^+ and OH^- ions, so $[H^+] = [OH^-]$. Substituting in Eq. (8–29), we find

$$[H^+]^2 = 10^{-14} \text{ (moles/liter)}^2$$

and

$$[H^+] = 10^{-7} \text{ moles/liter,}$$

accounting for the pH of 7 for neutral water noted above.

If we add H^+ or OH^- ions to water, its dissociation is suppressed. For example, in a 1M HCl solution, the 1M H^+ ion concentration is far greater than that of neutral water.

Substituting this value in Eq. (8–29), we find the OH^- concentration to be

$$[OH^-] = \frac{10^{-14}}{[H^+]} = 10^{-14} \text{ moles/liter},$$

a factor of 10^7 less than in pure water. Likewise, a strong base, say $1M$ NaOH, produces $1M$ OH^- ions, also suppressing the dissociation of water, thus reducing the H^+ ion concentration. When $[OH^-] = 1M$, the H^+ concentration is given by:

$$[H^+] = \frac{10^{-14}}{[OH^-]} = \frac{10^{-14}}{1} = 10^{-14} \text{ moles/liter},$$

giving a pH of 14. This method is used to calculate the pH for the basic solutions given in Section 8–4 and Fig. 8–20.

The equilibrium expression for the dissociation of acetic acid is:

$$\frac{[H^+][C_2H_3O_2^-]}{[HC_2H_3O_2]} = 1.8 \times 10^{-5} \text{ moles/liter} \qquad \text{(at 25°C)}.$$
$$\text{(8–30)}$$

If we have a $1M$ solution of $HC_2H_3O_2$, what fraction of the acetic acid molecules is dissociated into H^+ ions and acetate ions? Let x represent the concentrations of H^+ and $C_2H_3O_2^-$ ions at equilibrium and let $1 - x$ be the concentration of the remaining $HC_2H_3O_2$; that is, let

$$HC_2H_3O_2(aq) \rightleftharpoons H^+(aq) + C_2H_3O_2^-(aq).$$

Equilibrium
concentration: $1 - x$ x x

Substituting these values into Eq. (8–30), we obtain

$$\frac{x \times x}{1 - x} = \frac{x^2}{1 - x} = 1.8 \times 10^{-5} \text{ moles/liter}.$$

If we assume that x is much smaller than one, we can use the approximation that $1 - x \approx 1$ in the denominator, yielding

$$x^2 = 1.8 \times 10^{-5} \text{ (moles/liter)}^2,$$

$$x = 4.2 \times 10^{-3} M.$$

Thus the fraction of acetic acid molecules dissociated is 0.0042 $M/1M$, or 0.42%. The H^+ ion concentration is $[H^+] = 4.2 \times 10^{-3} = 10^{-2.35}$, giving a pH of 2.35. If we dilute the solution by adding more water, the concentrations of all three

species are reduced, but the fraction of $HC_2H_3O_2$ molecules dissociated increases, as we can demonstrate by solving Eq. (8–30) for initial $HC_2H_3O_2$ concentrations of $0.1M$, $0.01M$, and so on.

The dissociation of the weak base, ammonium hydroxide, can be treated in the same way as that of acetic acid. The equilibrium expression is

$$\frac{[NH_4^+][OH^-]}{[NH_3]} = 1.8 \times 10^{-5} \text{ moles/liter} \qquad \text{(at 25°C)}.$$
$$\text{(8–31)}$$

If the initial concentration of NH_4OH is $0.1M$, let x equal the concentrations of NH_4^+ and OH^- ions at equilibrium and let the concentration of the remaining NH_4OH equal $0.1 - x$. Substituting into Eq. 8–31 we have

$$\frac{x \times x}{0.1 - x} = \frac{x^2}{0.1 - x} = 1.8 \times 10^{-5} \text{ moles/liter}.$$

Again we approximate $0.1 - x \approx 0.1$ and solve as follows:

$$x^2 = 1.8 \times 10^{-5} \times 0.1 = 1.8 \times 10^{-6} \text{ (moles/liter)}^2,$$

$$x = 1.3 \times 10^{-3} M.$$

The fraction dissociated is $0.0013/0.1$ or 1.3%. We can calculate the H^+ ion concentration by substituting for $[OH^-]$ in Eq. (8–29):

$$[H^+] = \frac{10^{-14}}{[OH^-]} = \frac{10^{-14}}{1.3 \times 10^{-3}} = 7.7 \times 10^{-12} M,$$

giving pH = 11.2.

Some acids and bases contain more than one H^+ or OH^- unit per molecule. Examples of such *polyprotic* acids are sulfuric acid (H_2SO_4), carbonic acid (H_2CO_3), and phosphoric acid (H_3PO_4). In all these acids, the hydrogen atoms can be dissociated to form H^+ ions. The first H^+ always comes off much more readily than succeeding ones. For example, sulfuric acid is considered a strong acid because the first dissociation is essentially complete (unless the H_2SO_4 solution is extremely concentrated) i.e., the equilibrium reaction

$$\text{Step 1:} \quad H_2SO_4(aq) \rightleftharpoons H^+(aq) + HSO_4^-(aq) \qquad \text{(8–32)}$$

goes almost completely to the right. However, in the second dissociation,

$$\text{Step 2:} \quad HSO_4^-(aq) \rightleftharpoons H^+(aq) + SO_4^{2-}(aq), \qquad \text{(8–33)}$$

Table 8–3   Equilibrium Constants for Dissociation of Acids

Acid	Reaction	K (25°C)
Acetic acid	$HC_2H_3O_2 \rightleftharpoons H^+ + C_2H_3O_2^-$	1.8×10^{-5}
Hydrocyanic acid	$HCN \rightleftharpoons H^+ + CN^-$	5×10^{-10}
Hydrofluoric acid	$HF \rightleftharpoons H^+ + F^-$	6.8×10^{-4}
Chloroacetic acid	$HC_2H_2ClO_2 \rightleftharpoons H^+ + C_2H_2ClO_2^-$	1.4×10^{-3}
Nitrous acid	$HNO_2 \rightleftharpoons H^+ + NO_2^-$	4.5×10^{-4}
Carbonic acid	$H_2CO_3 \rightleftharpoons H^+ + HCO_3^-$	$K_1 = 4.2 \times 10^{-7}$
	$HCO_3^- \rightleftharpoons H^+ + CO_3^{2-}$	$K_2 = 4.8 \times 10^{-11}$
Sulfuric acid	$H_2SO_4 \rightleftharpoons H^+ + HSO_4^-$	$K_1 > 10$
	$HSO_4^- \rightleftharpoons H^+ + SO_4^{2-}$	$K_2 = 1.2 \times 10^{-2}$
Sulfurous acid	$H_2SO_3 \rightleftharpoons H^+ + HSO_3^-$	$K_1 = 1.5 \times 10^{-2}$
	$HSO_3^- \rightleftharpoons H^+ + SO_3^{2-}$	$K_2 = 1.0 \times 10^{-7}$
Hydrogen sulfide	$H_2S \rightleftharpoons H^+ + HS^-$	$K_1 = 1.1 \times 10^{-7}$
	$HS^- \rightleftharpoons H^+ + S^{2-}$	$K_2 = 1 \times 10^{-14}$
Phosphoric acid	$H_3PO_4 \rightleftharpoons H^+ + H_2PO_4^-$	$K_1 = 7.5 \times 10^{-3}$
	$H_2PO_4^- \rightleftharpoons H^+ + HPO_4^{2-}$	$K_2 = 6.2 \times 10^{-8}$
	$HPO_4^{2-} \rightleftharpoons H^+ + PO_4^{3-}$	$K_3 = 2 \times 10^{-13}$

only about 1% of the HSO_4^- is dissociated. Note that according to Le Chatelier's principle, the H^+ ions produced in Step 1 suppress the second step.

The equilibria of polyprotic acids are treated like that of acetic acid discussed above, except that there is a separate equilibrium expression and equilibrium constant for each dissociation step. For sulfuric acid, the equilibrium expressions for reactions (8–32) and (8–33) would be

$$\frac{[H][HSO_4^-]}{[H_2SO_4]} = K_1 > 10 \tag{8–34}$$

and

$$\frac{[H^+][SO_3^{2-}]}{[HSO_4^-]} = K_2 = 1.2 \times 10^{-2} \quad \text{(at 25°C).} \tag{8–35}$$

The large K_1 of H_2SO_4 reflects the fact that the first dissociation is virtually complete. The smaller K_2 shows that the second dissociation is small. In a $1M$ H_2SO_4 solution, Step 1 would produce approximately $1M$ concentrations of H^+ and HSO_4^- ions. Substituting these values into Eq. (8–37), we see that

$$[SO_4^{2-}] = K_2\frac{[HSO_4^-]}{[H^+]} = K_2 = 1.2 \times 10^{-2} \tag{8–36}$$

and thus only about 1.2% of the HSO_4^- is dissociated.

The equilibrium expressions for the dissociation of carbonic acid, H_2CO_3, via reactions (8–13) and (8–14) are

$$\frac{[H^+][HCO_3^-]}{[H_2CO_3]} = K_1 = 4.2 \times 10^{-7} \tag{8–37}$$

and

$$\frac{[H^+][CO_3^{2-}]}{[HCO_3^-]} = K_2 = 4.8 \times 10^{-11}. \qquad (8-38)$$

If there is a pressure of 1 atm of CO_2 above a carbonic acid solution, then according to Eq. (8–28), $[H_2CO_3] = 0.03M$. Substituting this value in Eq. (8–37) and letting $x = [H^+] = [HCO_3^-]$, we obtain

$$x^2 = 4.2 \times 10^{-7} \times 3 \times 10^{-2} = 1.26 \times 10^{-8},$$
$$x = 1.1 \times 10^{-4}M,$$

yielding a pH of about 4. (This is approximately the condition in a bottle of carbonated beverage such as Coke or soda water.) Just as in sulfuric acid, the second dissociation of carbonic acid is much smaller than the first. For the reasons noted above, $[CO_3^{2-}] = K_2 = 4.8 \times 10^{-11}$.

For the triprotic phosphoric acid, H_3PO_4, there are three dissociations with progressively smaller ionization constants. All these reactions take place at 25°C.

$$H_3PO_4(aq) \rightleftharpoons H^+(aq) + H_2PO_4^-(aq), \quad K_1 = 7.5 \times 10^{-3}$$

$$H_2PO_4^-(aq) \rightleftharpoons H^+(aq) + HPO_4^{2-}(aq), \quad K_2 = 6.2 \times 10^{-8}$$

$$HPO_4^{2-}(aq) \rightleftharpoons H^+(aq) + PO_4^{3-}(aq), \quad K_3 = 2 \times 10^{-13}$$

Table 8–3 lists the equilibrium constants for the dissociation of a number of acids, most of them weak.

SUGGESTED READING

B. H. Mahan, *University Chemistry*, 2nd ed. (Addison-Wesley, Reading, Mass., 1969).

G. C. Pimental and R. D. Spratly, *Chemical Bonding Clarified through Quantum Mechanics* (Holden-Day, San Francisco, 1969).

J. H. Secrist and W. M. Powers, *General Chemistry* (Van Nostrand, Princeton, N.J., 1966).

R. M. Silverstein and G. C. Bassler, *Spectrometric Identification of Organic Compounds* (John Wiley, New York, 1963).

P. F. Weller and J. H. Supple, *Chemistry: Elementary Principles* (Addison-Wesley, Reading, Mass., 1971).

PROBLEMS AND QUESTIONS

1. Below we show the position of atoms in several kinds of molecules. On a separate sheet of paper, draw in the electrons that would make reasonable structures and satisfy the octet rule. Then depict the structure by bond notation.

   a) I_2 I   I

   b) C_2F_2 F   C   C   F

   c) C_2Cl_4 Cl   Cl
                C   C
            Cl   Cl

   d) H_2S H   S
               H

   e) H_2O_2 H   O   O   H

2. Describe or define each of the following terms.
   a) specific heat   b) heat of vaporization
   c) hydrogen bond   d) polar bond
   e) resonance

3. Write net, balanced reactions that describe what occurs in the following experiments.
   a) KOH is added to HBr (aq).
   b) Concentrated HCl solution is dropped on the mineral dolomite, $MgCO_3$.
   c) NaOH solution is mixed with $HC_2H_3O_2$ solution.
   d) K_2S is dissolved in water.
   e) NH_4Cl is dissolved in water.

4. Draw electron dot structures for the compounds listed below. If the structure exhibits resonance, indicate resonance and draw bond structures for the various resonance forms.
   a) GeH_4 b) SO_2
   c) Na_2SO_4 d) HNO_3
   e) H_2N-NH_2 f) KOH
   g) ClO^- h) CS_2

5. a) The longest wavelength in visible light is about 7600 Å. What is the frequency v and energy (in eV) of a photon of that wavelength?

b) As shown in Fig. 8–9, one of the principal vibrations of CO_2 is excited by infrared light of 15μ wavelength.
 i) What is the wavelength in cm?
 ii) What is the frequency?
 iii) What is the energy?

c) What is the wavelength in cm of the 2450 Hz radiations used in microwave ovens? What is the energy of one such photon (in eV)?

6. Draw electron dot structures for the three possible forms of the carbonate ion (CO_3^{2-}), indicating the electrons from each different type of atom.

7. a) Why is ice less dense than water?
   b) What part does hydrogen bonding play in ice formation?

8. a) Define an acid and a base.
   b) What is the difference between a weak acid and a strong acid?

9. a) What is the meaning of pH?
   b) What is the pH of $1M$ HCl, a strong acid?
   c) What is the pH of $1M$ NaOH, a strong base?

10. Ammonia gas, NH_3, dissolves in water to form ammonium hydroxide, NH_3(aq), a weak base. Suppose we have a flask containing NH_3 gas in equilibrium with an aqueous solution of NH_3 (aq), which is also in equilibrium with NH_4^+ and OH^- ions.

$$NH_3(g) \xrightarrow{H_2O} NH_3(aq)$$

$$NH_3(aq) + H_2O(l) \rightleftharpoons NH_4^+(aq) + OH^-(aq)$$

Using Le Chatelier's principle, answer the following questions.
a) Does the pressure of NH_3 increase or decrease if NaOH is added to the solution?
b) Does the pressure of NH_3 increase or decrease if HCl is added to the solution?
c) Does the pressure of NH_3 increase or decrease if NH_4Cl is added to the solution?
d) Does the OH^- concentration in solution increase or decrease if the NH_3 pressure is increased?

11. a) What is the pH of a $0.01M$ solution of HCl?
    b) What is the OH^- concentration of a solution made by dissolving 4 g of KOH in 100 ml of water?
    c) What is the pH of the solution in part (b)?

12. For the equilibrium $NH_3(aq) + H_2O(l) \rightleftharpoons NH_4^+(aq) + OH^-(aq)$, the equilibrium expression is

$$\frac{[NH_4^+][OH^-]}{[NH_3]} = 1.8 \times 10^{-5} \text{ moles/liter.}$$

a) What is the concentration of OH^- ions in a $1M$ solution of NH_3(aq)?
b) What is the pH of the solution?

13. Calculate the equilibrium concentrations of H_2S, HS^-, H^+, S^{2-}, and OH^- in a $0.01M$ solution of H_2S.

9 MAN'S EFFECT ON THE CLIMATE

Sunlight reflected from cloud cover. (Courtesy of The National Oceanic and Atmospheric Administration.)

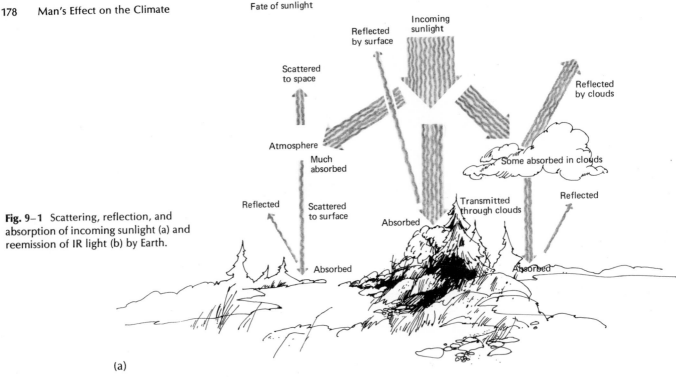

Fate of sunlight

Incoming sunlight

Reflected by surface

Scattered to space

Reflected by clouds

Atmosphere

Much absorbed

Some absorbed in clouds

Reflected

Scattered to surface

Transmitted through clouds

Reflected

Absorbed

Absorbed

Absorbed

Fig. 9–1 Scattering, reflection, and absorption of incoming sunlight (a) and reemission of IR light (b) by Earth.

(a)

9–1 INTRODUCTION

In some of the following chapters we examine environmental problems that may have serious consequences today or within the next few years. In general, these problems involve large, localized concentrations of people, and technological activities—such as power and heat production, automobile traffic and manufacturing—that produce high levels of air and water pollution in the area. Although these conditions are often uncomfortable to people and bad for their health, they do not, for the most part, threaten massive changes in the nature of the planet. In this chapter we consider some problems that *may have* global consequences—changes in global climate and perhaps the bombardment of the earth's surface with dangerous ultraviolet (UV) radiation. We emphasize "may have" because predictions of global problems are very difficult to make and the results of such calculations are fraught with large uncertainties.

9–2 THE WARMING OF EARTH

Earth is a "spaceship" that receives most of its surface warmth from the sun's radiation (the remainder is produced by natural radioactive decay; see Section 5.3).

The temperature of Earth's surface is governed by the amount of light received from the sun, the fraction that is reflected, the fraction absorbed, and the amount of light that Earth radiates into space. If Earth were initially very cold, it would be warmed by the absorbed sunlight. Just like any object having a temperature above 0°K, Earth would also emit radiation into space. As the temperature increased, the rate of energy radiation (as low-energy infrared, IR, radiation) would increase until the rate of radiation was equal to the rate of energy absorption. When such a point of *thermal equilibrium* is attained, the temperature remains constant with time thereafter until some condition affecting one or both rates is changed. The temperature at which equilibrium is achieved depends on the rates of energy absorption and emission: the greater the absorption rate, the greater is the equilibrium temperature. Earth is now approximately in thermal equilibrium, but man may now have the ability to alter the rates, possibly causing the average surface temperature to change.

When sunlight strikes the earth, some is reflected back into space by the tops of clouds, the surface of the oceans, snow on the ground, etc. Some light is scattered back into space by particles in the atmosphere; some is absorbed by

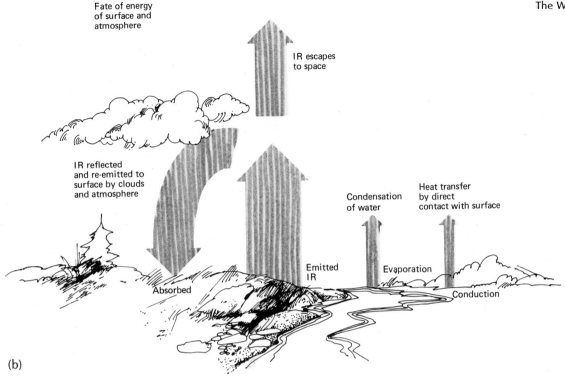

Fate of energy
of surface and
atmosphere

IR escapes
to space

IR reflected
and re-emitted to
surface by clouds
and atmosphere

Condensation
of water

Heat transfer
by direct
contact with surface

Emitted
IR

Evaporation

Conduction

Absorbed

(b)

molecules of the air and some by the materials of the Earth's surface, warming the atmosphere and the surface (see Fig. 9–1). Earth's *albedo*, the fraction of incoming sunlight that is reflected into space, is about 34%. About 19% of the incoming sunlight is absorbed in the atmosphere and 47% at Earth's surface.

Since the sun's surface temperature is about 6000 K, over half of its energy is radiated in the high-energy UV and visible portions of the spectrum (see Chapter 7, especially Fig. 7–10). The cooler Earth (about 280 K) emits only low-energy IR light. Sunlight warms the ground and the air only during the daytime, but the surface emits radiation during the night, causing the cooling that usually produces a minimum temperature at dawn. When skies are overcast at night, Earth's radiation is trapped below the cloud layer, so that the temperature remains higher during cloudy nights than when the sky is clear.

Earth's surface absorbs much more of the sun's radiation than the atmosphere. Thus the ground or water surface generally becomes warmer than the air. Air is heated by contact with the ground or water and also by the evaporation of water from the surface followed by its condensation in the

atmosphere. Some of the heat transferred to the air near ground level works its way higher into the atmosphere mainly by the movement of molecules of the air (convection). However, under normal atmospheric conditions, the temperature of the atmosphere decreases with increasing altitude (a drop of about 7°C for each kilometer) up to the *tropopause* at 10 to 12 km. In the *stratosphere* above the tropopause, the temperature remains in the vicinity of −55°C up to about 25 km, where it rises again with increasing altitude, reaching a maximum near 10°C at around 50 km. Figure 9–2 shows a profile of these temperature changes. The upper stratosphere is warmed because the ozone (O_3) present in the stratosphere absorbs UV radiation from incoming sunlight.

Over a year's time more sunlight falls on Earth's surface near the equator than on polar regions. If the temperature of an area were determined solely by the amount of radiation falling on it, the tropics would be unbearably hot and, in their winter seasons, polar regions would be unimaginably cold. Fortunately, the atmosphere and the *hydrosphere*, i.e., the oceans and seas, distribute heat from warm regions to colder areas. When a large temperature difference exists within a gas or liquid, movements of the molecules distribute

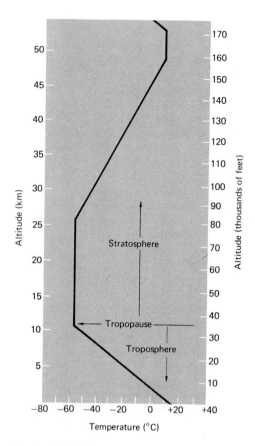

Fig. 9–2 Typical temperature profile of the atmosphere at latitude 30° to 40°.

heat from the warmer to the cooler regions, a process called *convection*. For example, if there is a stove in one corner of a room, convection carries heat to cooler parts of the room, although the air remains warmest near the stove. Convection via the atmosphere and oceans transfers heat from the tropics to the polar regions. It is the large temperature difference between the tropics and the poles that provides much of the driving force for the ocean currents and winds of the air that transport the heat.

Because of the large heat capacity of water (see Section 8–3), large bodies of water have important moderating effects on the climate of nearby land masses. For example, the Canadian cities of Victoria, Winnipeg, and St. John's are at about the same latitude (see Fig. 9–3); however, the average July maximum temperatures are 68°, 80°, and 69°F, respectively,

Fig. 9–3 Coastal cities such as Victoria and St. John's have more moderate climates than Winnipeg because of their proximity to oceans.

and the average January minima are 36°, −8°, and 19°F at the three cities.* The temperature range is quite small at Victoria because of the proximity of the Pacific Ocean. Winnipeg, near the center of the continent, has an enormous temperature range. St. John's is colder in winter than Victoria because the wind on the East Coast generally blows from the continent and the Labrador Current brings cold water to the St. John's area. Perhaps surprisingly, the coldest winter temperatures in the Northern Hemisphere occur not at the North Pole, but in Siberia at about latitude 65°N in the western part of the Soviet Union.

Without the heat transport of the atmosphere and oceans and the large heat capacity of the oceans, temperature fluctuations, both seasonal and diurnal (day-night) at many points on Earth would be quite large. On the moon, which has no atmosphere or oceans, the surface rocks have temperatures of about 240°F during the middle of the lunar day and drop to −300°F at night just before dawn.

---

*   R. W. Stewart, "The Atmosphere and the Ocean," *Scientific American* **221** (3), 76–86 (Sept. 1969).

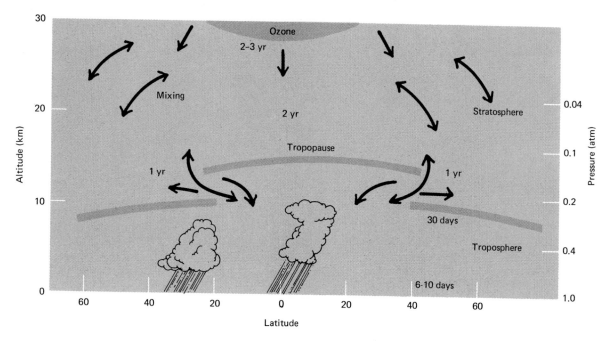

Fig. 9-4 Residence times of gases and particles in various regions of the atmosphere. [After R. E. Newell, "The Global Circulation of Atmospheric Pollutants," *Scientific American* **224** (1), 32–42 (Jan. 1971).]

Temperature differences also influence the vertical movement of air. In Fig. 9-2, we saw that temperature normally decreases with increasing altitude in the troposphere. Because of the temperature changes, there is a lot of vertical circulation of the air in the troposphere except during periods of temperature inversion (see Section 9-3). But in the lower stratosphere, which has an almost constant temperature from about 11 km up to 26 km, there is little vertical movement of air. For that reason, gas molecules and small particles are transported very slowly through the tropopause.

In fact, the tropopause prevents most particles and pollutant gases of the troposphere from rising into the stratosphere. However, once injected into the stratosphere, as by volcanic eruptions or nuclear explosions, particles and gas molecules remain in the stratosphere for average periods, called *residence times*, ranging from about four months in the polar stratosphere to two years above the equator (see Fig. 9-4). These long stratospheric residence times account for the slow removal of radioactive fallout from the stratosphere following large nuclear explosions, as discussed in Chapter 5.

Most cloud formation, rain, and snow are confined to the troposphere. These processes remove particles and gases from the air, bringing them down to the surface of the earth. As a result, residence times in the troposphere are much shorter than in the stratosphere, ranging from about one month near the tropopause to about a week in the lower troposphere.

9-3 THE GREENHOUSE EFFECT

The *greenhouse effect* is probably not the most pressing environmental problem on a global scale, but it illustrates the complexity of global-scale effects. Also, it shows that although man's activities may alter a global parameter only slightly, the results can be devastating.

The greenhouse effect is caused by carbon dioxide, CO_2, added to the atmosphere by man's combustion of large quantities of fossil fuels. Carbon dioxide absorbs certain frequencies of Earth's IR light quite strongly, but has little effect on the incoming visible and UV light from the sun. Thus it reduces the rate of Earth's radiation of energy into space, causing the surface and atmosphere to heat up. This trapping of radiation energy is analogous to the warming of a greenhouse. Visible photons of sunlight pass through the glass of the greenhouse, warming the plants inside. The warmed

Table 9–1   Average Composition of the Atmosphere

Gas	Composition by volume (ppm unless % indicated)
N_2	78.09%
O_2	20.95%
Ar	0.93%
CO_2	320
H_2O	$(0.4 \text{ to } 400) \times 10^2$
Ne	18
He	5.2
CH_4	1.5
Kr	1
N_2O	0.5
H_2	0.5
O_3	$(0 \text{ to } 5) \times 10^{-2}$ Depends on altitude
Xe	0.08
CO	$(1 \text{ to } 20) \times 10^{-2}$

Data taken from B. Mason, *Principles of Geochemistry*, 3rd ed. (John Wiley, New York, 1966), p. 208, and from C. E. Junge, *Air Chemistry and Radioactivity* (Academic Press, New York, 1963) p. 3.

plants emit IR radiation that is trapped inside, as glass is not transparent to IR, and warms the interior of the greenhouse. (In addition, the walls of the greenhouse shield the interior from cooling winds.)

The major gases of the atmosphere, O_2, N_2, and Ar (see Table 9–1) absorb little IR radiation. The most important absorbing species are the minor components O_3, H_2O, and CO_2, whose triatomic molecules absorb low-energy IR photons and are excited to the vibrational and rotational states discussed in Chapter 8. Man's effects on O_3 and H_2O are noted in following sections, but the major concern in our discussion of the greenhouse effect is with CO_2.

Carbon dioxide is not an air pollutant in the usual sense. It is produced by biological processes and is an essential nutrient for plant life. Men and animals breathe oxygen from the air and exhale CO_2. If this were the only important reaction of the *biosphere* (the mass of living materials at Earth's surface) with the atmosphere, the O_2 would gradually be replaced by CO_2. But during sunlight hours, living plants take in CO_2 from the atmosphere and return O_2, in the process called *photosynthesis*. Respiration of plants also returns some of the CO_2 to the atmosphere. When plants die, their decomposition releases additional CO_2 to the atmosphere. Over a year's time, the amount of CO_2 released by respiration of plants and animals and by decomposition of plants is approximately the same as the amount used up by photo-

synthesis. Thus, in nature, the amount of CO_2 in the atmosphere remains about constant.

The concept of residence time is useful in our discussion of the carbon cycle. To understand this, let's use the analogy of a water reservoir, initially assumed to be empty. If water flows into the reservoir at a constant rate R_{in} (tons/year), water W contained in the reservoir increases until it starts flowing over the spillway on the other side at a rate R_{out}. Eventually R_{out} becomes equal to R_{in} and W remains constant with time—a situation called *steady-state equilibrium*. Although water steadily flows in and out of the reservoir, the amount present, W, remains constant unless R_{in} or R_{out} is changed.

What is the average time, i.e., the *residence time* τ that a drop of water remains in the reservoir? Each year a fraction R_{out}/W of the water flows out of the reservoir; therefore, a drop of water, on the average, remains in the reservoir for W/R_{out} years. Thus the residence time for a steady-state system is given by

$$\tau = \frac{W}{R_{out}} \quad \text{or} \quad \frac{W}{R_{in}} \quad (\text{since } R_{in} = R_{out}). \quad (9-1)$$

We can apply this same concept to the carbon in any of the "boxes" of Fig. 9–5, which we consider carbon reservoirs. For example, the atmosphere has $W = 700$ and the amounts of carbon flowing out of the atmosphere each year are $R_{out} = 100 + 35 = 135$. Thus the residence time for carbon atoms in the atmosphere is

$$\tau = 700/135 = 5.2 \text{ years};$$

that is, carbon atoms remain in the atmosphere for 5.2 years on the average. Residence times are used in analyzing many environmental systems. In Chapter 5, the residence times of particles in the troposphere and stratosphere were used to estimate the times at which fallout from nuclear explosions reach ground level. These residence times are also shown in Fig. 9–4. The residence time of a chemical species dissolved in water of the oceans can be determined from R_{in}, the rate at which it is carried into the oceans by rivers, or from R_{out}, the rate at which it is deposited in sediments on the ocean floor. In Chapter 11, the residence times of water droplets in the various Great Lakes are used to estimate the rate at which the lakes are becoming polluted or the rate at which they could be cleaned up.

In the water reservoir analogy, if R_{in} of a system in steady state were suddenly increased, the amount of water in the

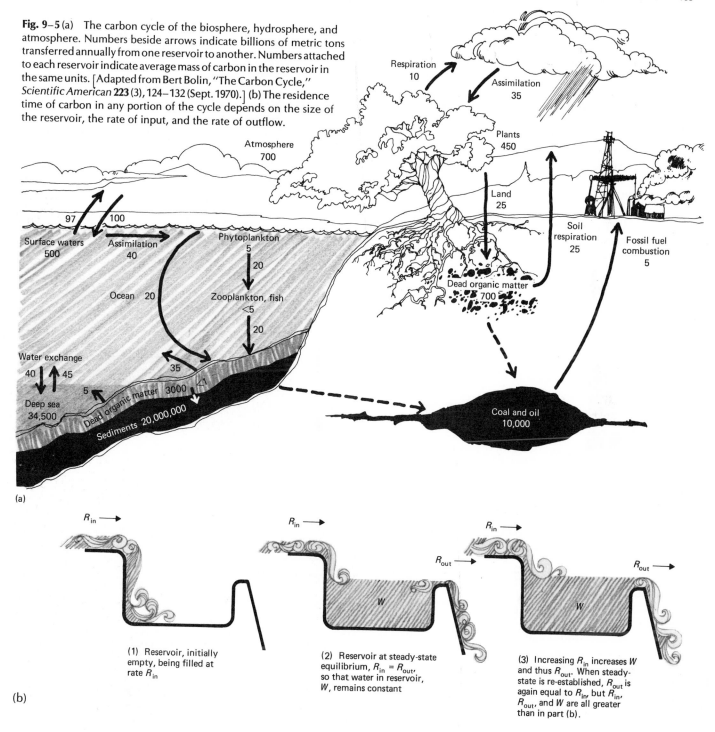

Fig. 9–5 (a) The carbon cycle of the biosphere, hydrosphere, and atmosphere. Numbers beside arrows indicate billions of metric tons transferred annually from one reservoir to another. Numbers attached to each reservoir indicate average mass of carbon in the reservoir in the same units. [Adapted from Bert Bolin, "The Carbon Cycle," *Scientific American* **223** (3), 124–132 (Sept. 1970).] (b) The residence time of carbon in any portion of the cycle depends on the size of the reservoir, the rate of input, and the rate of outflow.

Respiration
10

Assimilation
35

Plants
450

Atmosphere
700

Land
25

Soil
respiration
25

Fossil fuel
combustion
5

97   100

Surface waters
500

Assimilation
40

Phytoplankton
5

20

Ocean   20

Zooplankton, fish
<5

20

Water exchange

40   45

Deep sea
34,500

5

Dead organic matter   3000   <1

Sediments 20,000,000

Dead organic matter
700

Coal and oil
10,000

(a)

R_{in} →

R_{in} →

R_{in} →

R_{out} →

R_{out} →

W

W

(1) Reservoir, initially empty, being filled at rate R_{in}

(2) Reservoir at steady-state equilibrium, $R_{in} = R_{out}$, so that water in reservoir, W, remains constant

(3) Increasing R_{in} increases W and thus R_{out}. When steady-state is re-established, R_{out} is again equal to R_{in}, but R_{in}, R_{out}, and W are all greater than in part (b).

(b)

reservoir, W, would increase. Because of the higher water level, R_{out} would also increase until it was equal to the new value of R_{in}, thus reestablishing the steady-state system, but at higher values of R_{in}, R_{out}, and W. Concern about the greenhouse effect has arisen because man has increased the rate (R_{in}) at which carbon dioxide is added to the atmosphere.

As shown in Fig. 9–5, a small but unknown amount of dead organic material is buried under sediments, rocks, and soil each year. Instead of decomposing in the normal way, these materials are slowly converted to fossil fuels: coal (which is mainly carbon) and petroleum and natural gas, which consist of carbon and hydrogen compounds. In recent years, man has been extracting fossil fuels from Earth at an unprecedented rate and burning them to obtain energy. As shown in Table 4–1, the overwhelming majority of the energy that is used in the United States is derived from fossil fuels.

The combustion of fossil fuels releases CO_2.

Coal:

$$C(s) + O_2(g) \longrightarrow CO_2(g) \tag{9–2}$$

Methane (major constituent of natural gas):

$$CH_4(g) + 2O_2(g) \longrightarrow CO_2(g) + 2H_2O(g) \tag{9–3}$$

Octane (typical gasoline molecule):

$$2C_8H_{18}(l) + 25O_2(g) \longrightarrow 16CO_2(g) + 18H_2O(g) \tag{9–4}$$

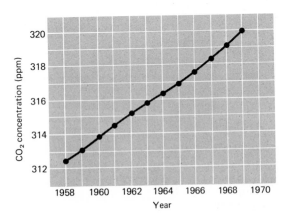

Fig. 9–6 Atmospheric concentration of carbon dioxide, 1958 to 1970. [Average of several sets of data from *Man's Impact on the Global Environment: Study of Critical Environmental Problems*, SCEP, (MIT Press, Cambridge, Mass., 1970).]

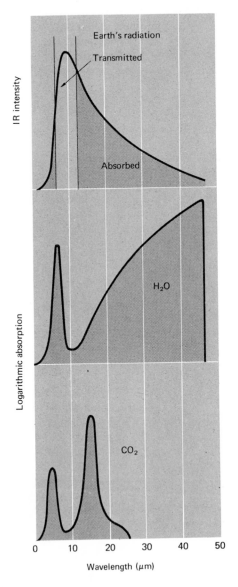

Fig. 9–7 Absorption spectra of carbon dioxide and water vapor.

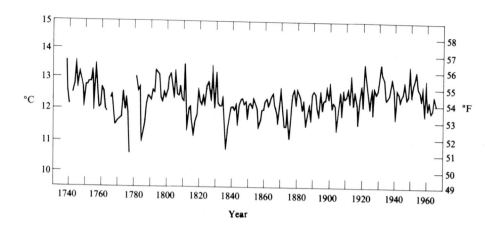

Fig. 9–8 Average annual temperature record for Philadelphia area. [From H. E. Landsberg, "Man-Made Climatic Change," *Science* **170**, 1265–1274 (1970), Fig. 1. Copyright © 1970 by the American Association for the Advancement of Science.]

Man's contribution of CO_2 is now a significant fraction of the amount that occurs naturally in the atmosphere. Between 1850 and 1950, CO_2 added to the atmosphere by worldwide fossil fuel combustion was about 10% of the total present in the atmosphere in 1950.* During the 1970s alone, man will add an amount of CO_2 estimated to be equal to about 10% of that already present.

Because of the burning of fossil fuels, more CO_2 is being added to the atmosphere each year than is being removed. In Fig. 9–5 we see that $R_{in} = 97 + 10 + 25 + 5 = 137$ units, whereas $R_{out} = 100 + 35 = 135$; that is, $R_{in} - R_{out} = 2$. If R_{in} is maintained at its present value for many centuries, R_{out} will eventually increase to the same value, but W, the amount of CO_2 in the atmosphere, will have increased above its natural amount. Thus, although the CO_2 added to the atmosphere does not remain there for many years, the atmospheric concentration of CO_2 is definitely increasing. Accurate measurements between 1958 and 1970 showed an increase of CO_2 concentration of about 3%, from 312 to 320 parts per million, by volume (see Fig. 9–6).

Since carbon dioxide molecules absorb IR radiation rather strongly (see Fig. 9–7), especially near wavelength $15\,\mu$, there is concern that the increased CO_2 will trap enough of Earth's radiation to warm the atmosphere and surface by several degrees centigrade.

Why should a few-degree increase in temperature alarm anyone? Most places in the temperate zone are cold in the winter and would become more comfortable. Unfortunately, a few-degree increase in the average temperature would have more serious consequences than one might suppose. It would expand the hot, arid desert regions several degrees of latitude toward the North and South Poles. In the United States, areas of Kansas, Nebraska, and perhaps the Dakotas, which are now fertile, might become desolate expanses of sand. Perhaps the most dramatic effect would be the melting of glaciers in polar regions. If the warming trend continued, even the polar ice caps would melt! If they melted completely, sea level would rise by about 200 feet.

Has Earth's temperature already risen as a result of the increased CO_2? Figure 9–8 shows a 230-year temperature record from Philadelphia, Pennsylvania. We see that fluctuations of temperature over periods of a few years are so great that one must have a very long record in order to observe long-term trends. Careful analyses of temperature records such as this one suggest that the average atmospheric temperature did rise by about 0.8°C between 1860 and 1940.† However, between about 1940 and 1960, there appears to have been a slight cooling, approximately 0.2°C, which seems to be continuing into the 1970s. This apparent trend suggests that the greenhouse prediction is too simple and that other effects are canceling that of the rising CO_2 concentration.

Many other factors influence the temperature of the earth, some of which could lower the temperature, as observed since about 1940.

---

* Report of the Study of Critical Environmental Problems (SCEP), *Man's Impact on the Global Environment* (MIT Press, Cambridge, Mass., 1970).

† E. K. Peterson, "Carbon Dioxide Affects Global Ecology," *Environmental Science and Technology* **3**, 1162–1169 (1969). See also Fig. 9–8.

1. Perhaps there are century-long cycles in the amount of sunlight striking Earth's surface, produced by variations either in the sun's light output or in Earth's orientation relative to its orbit.

2. Much of the CO_2 added to the atmosphere dissolves in the ocean. The equilibrium [Eq. (8–22)] between CO_2 in the atmosphere and H_2CO_3 (and other dissolved carbonate forms) must be maintained: when the pressure of CO_2 gas in the atmosphere is increased, the concentration of H_2CO_3 in the oceans must increase. Thus the oceans, which contain about 60 times as much CO_2 as the atmosphere (mostly as HCO_3^-), reduce fluctuations of atmospheric CO_2. The oceans are such large bodies of water that a long time is required for equilibrium to be established with the atmosphere. The surface waters, down to about 500 meters' depth, probably adjust to changes of atmospheric CO_2 within five to ten years, but deep waters mix with the surface so slowly that times of the order of 1000 years are required.

3. Both man-made and natural activities release particles into the atmosphere. As discussed below, particles can scatter sunlight back into space, raising Earth's albedo, thus tending to cool the lower atmosphere and Earth's surface. The observed cooling trend may be the result of increased man-made or natural (e.g., volcanic) injection of particles into the air.

We see that many factors complicate our simple prediction of the warming of Earth by fossil-fuel combustion. Detailed calculations predict that even if the CO_2 concentration were increased eightfold (which seems unlikely to occur within several thousand years), the increase of the earth's surface temperature would be less than 2°C.* The reason that the effect is so small is that CO_2 absorption of 15 μm (0.0015 cm) IR radiation "saturates"; i.e., after there is a certain amount of CO_2 in the atmosphere, it absorbs essentially all of Earth's radiation of wavelength near 15 μm, so that further increases of CO_2 concentration have little effect. Thus it appears that the greenhouse effect of CO_2 is not a serious global problem. However, as in most environmental situations, evaluating the various factors involved is a complicated process. The global effects of particles produced by man's activities are probably more serious than the effects of CO_2.

_____

* S. I. Rasool and S. H. Schneider, "Atmospheric Carbon Dioxide and Aerosols: Effects of Large Increases on Global Climate," *Science* **173**, 138–141 (1971).

Both man and nature inject particles into the atmosphere. Shown here are the volcanic island of Surtsey, an airplane trailing smoke, and a forest fire in Idaho. [Courtesy, respectively, of S. Thorarinsson, the National Oceanic and Atmospheric Administration, and the U.S. Dept. of Agriculture (photo by Bluford W. Muir).]

9–4 GLOBAL EFFECTS OF ATMOSPHERIC PARTICLES

Both man and nature inject particles into the atmosphere. These *aerosols*† range from agglomerations of a few molecules up to particles having radii of about 20 μm (0.002 cm). Most of the mass of aerosols is contained in particles with radii between 0.01 and 10 μm.‡ The smaller particles gradually coalesce to form larger ones. Particles with radii greater than about 10 μm tend to settle to the ground within a few hours or days. These very large particles are present in the dense, black smoke that one often sees coming from factory or power-plant smokestacks. However, the visible plumes and the hazy appearance of polluted air are caused by particles with radii between 0.1 and 1 μm. Their radii span the wavelength range of visible light (0.4 to 0.75 μm), and so the particles are just the right size to scatter or absorb light.

Natural sources of particles include volcanoes, forest fires, the bursting of bubbles at the sea surface, wind erosion of rocks and soil, and the evaporation of organic compounds called terpenes from pine forests (which cause the hazy appearance of the Great Smoky Mountains). Man-made sources include automobile exhaust, smoke from power

_____

† Aerosols are liquid droplets or solid particles suspended in a gas.
‡ *Inadvertent Climate Modification, SMIC* (MIT Press, Cambridge, Mass., 1971). See also Table 9–1, especially the Mason reference.

Fig. 9–9 Atmospheric particulate material comes from many sources, both natural and man-made.

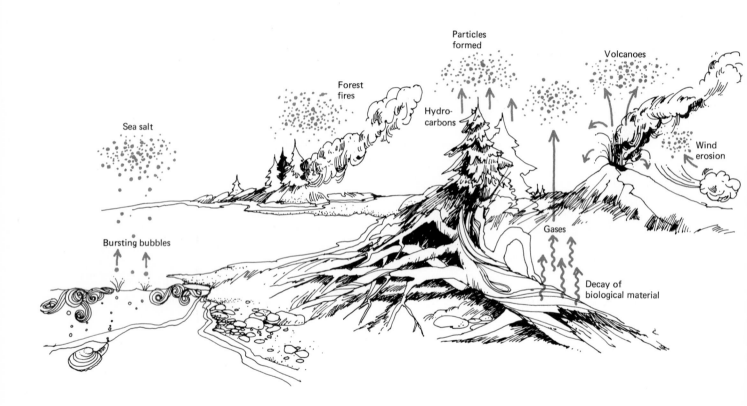

plants and factories, and ground-clearing operations (see Fig. 9–9). An indirect source of aerosols, both natural and man-made, is gases, especially those containing sulfur and nitrogen, which react in the atmosphere to form particles containing sulfates, nitrates, and ammonium ions.

Our present knowledge of the magnitudes of various natural and man-made sources is not very accurate, but in Table 9–2 we see that man's contribution is between about 10% and 50% of that from natural sources. However, it is clear that man's contribution has been increasing steadily over recent decades and there are several indications that the atmospheric concentration of particles is increasing. Professor Vincent Schaefer of the State University of New York at Albany has found that the concentration of particles with radii less than 0.1 μm at Flagstaff, Arizona, and Yellowstone National Park increased tenfold over a five-year period during the middle 1960s.* Particulate concentrations rose by

---

* V. J. Schaefer, "The Inadvertent Modification of the Atmosphere by Air Pollution," *Bulletin. American Meterological Society* **50**, 199–206 (1969).

an average of 12% between 1962 and 1966 at twenty nonurban sites in the United States.*

Although it is clear that man's activities have significantly raised particle concentrations in the troposphere, the situation in the stratosphere is much less certain. The greatest fluctuating sources are large volcanoes, which spew out copious quantities of particles, some of which penetrate the

tropopause. Furthermore, they release large quantities of gases (predominantly sulfurous gases) that can form particles in the stratosphere. The spectacular Krakatoa eruption in 1883 injected so many particles into the stratosphere that sunlight reaching ground level was measurably reduced for some years afterwards.† The particles remained in the stratosphere for about five years, causing dramatic red sunsets

*   J. H. Ludwig, G. B. Morgan, and T. B. McMullen, "Trends in Urban Air Quality," *EOS Transactions. American Geophysical Union* **51**, 468–475 (1970).

†   See H. E. Landsberg, "Man-Made Climatic Changes," *Science* **170**, 1265–1274 (1970).

Table 9-2   Estimated Sources of Global Atmospheric Aerosols

Sources	Annual total emissions* (millions of metric tons)
Man-made	
Dust and smoke	10 to 90
Particles formed from gases	130 to 200
Sulfates	130 to 200
Nitrates, ammonium	30 to 35
Hydrocarbons	15 to 90
Total man-made sources	185 to 415
Natural	
Sea salt, soil, volcanic ash, forest fires	428 to 1100
Particles formed from gases	
Sulfates	130 to 200
Nitrates, ammonium	140 to 700
Hydrocarbons	75 to 200
Total natural sources	773 to 2200

From *Inadvertent Climate Modification*, SMIC (MIT Press, Cambridge, Mass., 1971).

*   Authors' note: We favor the higher values of the ranges given for the man-made sources. See estimates given for U.S. sources alone in Chapter 10.

throughout the world (because of light scattering from the particles).

Depending on the size and nature of aerosols, they can either raise or lower temperatures near Earth's surface. If their main effect is to scatter incoming sunlight back into space, they raise Earth's albedo and lower its temperature. On the other hand, if they mainly absorb sunlight, they can lower the albedo and cause a warming trend. For example, following the eruption of the Mt. Agung volcano on the island of Bali in 1963, which injected particles into the stratosphere, the temperature of the air at 19.5 km altitude near the equator was raised by as much as 8°C for more than a year afterwards.* Aerosols affect the amount of visible and UV light that reaches ground level, but they have little effect on IR light, whose wavelengths are much greater than the diameters of the particles.

The most detailed predictions of the effects of aerosols on Earth's temperature are those by Rasool and Schneider.†

---

*   Newell, "The Global Circulation of Atmospheric Pollutants."
†   Rasool and Schneider, "Atmospheric Carbon Dioxide and Aerosols."

They conclude that a fourfold increase in the dust concentration of the atmosphere, which could occur within the next century, could lower Earth's surface temperature by up to 3.5°C. If such a decreased temperature continued for several years, a new ice age could begin. Although this is a rather alarming prediction, we must realize that it is very uncertain. Landsberg notes, for example, that the settling of man-made particles on ice and snow may have more important effects on worldwide temperature than suspended aerosols.‡ The particles could reduce the albedo of these surfaces, thus increasing the absorption of sunlight and causing the frozen water to melt. This in turn could lead to a rise of sea level.

In summary, the greenhouse effect *may* cause an increase of worldwide temperature and aerosols *may* cause an increase or decrease. A large change in either direction would have devastating effects on world climate and sea level. However, the predictions are so uncertain that we cannot even know whether man's effects will cause a temperature increase or decrease. (If we're lucky, the various effects will cancel each other so as to cause no change.) Although there is probably no danger of a massive change over the next few years, the magnitude of possible changes is so great that these global problems must be studied carefully. The United States has set up stations at remote locations such as Hawaii and the south pole in order to observe trends in such quantities as temperature, sunlight intensity, and concentrations of atmospheric gases and particles. Satellites will also be used to measure quantities such as cloud cover and light scattering from the atmosphere, clouds, and the earth's surface. By monitoring these global parameters, it is hoped that it will be possible to detect any strong trends in time to take corrective measures.

9-5   THE SUPERSONIC TRANSPORT AND THE STRATOSPHERE

After commercial jet transport planes were introduced in the late 1950s, the next logical step in air transportation was the development of supersonic transports (SST's), jet liners that would travel faster than the velocity of sound (about 770 mph at sea level and 660 mph at the reduced atmospheric density of the lower stratosphere).

In 1972, the Soviet Union and a French-British combine each built and flight-tested prototype SST's. The U.S. government sponsored design work on an SST, but by a close vote in

---

‡   Landsberg, "Man-Made Climatic Changes."

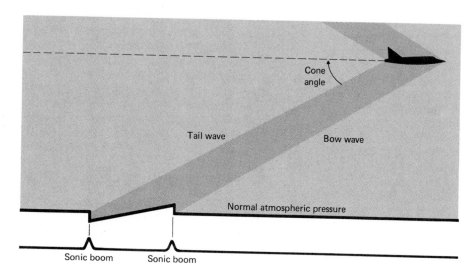

Cone angle

Tail wave

Bow wave

Normal atmospheric pressure

Sonic boom   Sonic boom

Fig. 9–10   The shock wave from an object moving at supersonic speed. [From Joseph Priest, *Problems of Our Environment* (Addison-Wesley, Reading Mass., 1973) p. 297.]

the Senate early in 1971, funds for further development were cut off. This vote followed an intense debate on the merits of such a plane. Most arguments against the SST were economic, but much of the opposition stemmed from concern about the environmental effects of the SST.

The most obvious environmental effect of an SST is its associated *sonic boom*. Any object traveling faster than sound sends out a shock wave of high air pressure similar to the waves produced in water by a high-speed motorboat. The shock wave from an SST, traveling away from the plane with the speed of sound, produces a burst of sound much like an explosion that is very annoying to people (see Fig. 9–10). The high air pressures of the sonic booms from military aircraft have caused considerable breakage of windows.

Our concern here is with atmospheric chemical effects rather than sonic booms, which are described in more detail elsewhere.* The SST's have cruising altitudes of about 20 km (65,000 ft) in the stratosphere, well above the tropopause. At 20 km the atmosphere is quite rarefied, having a pressure of only about one-sixteenth that at sea level.

Emissions from jet planes that would only slightly increase existing air pollution problems at low altitudes may have serious consequences in the more fragile stratosphere, where concentrations of gases are twelvefold smaller than at sea level, and residence times of pollutants are long.

The stratosphere is very cold and dry, having a water vapor concentration of about 3 ppm by weight, or 30 μg/m^3 versus 10 g/m^3 at sea level. Most water vapor rising from the ground is condensed as water droplets or ice crystals before reaching the stratosphere. Jet fuels are hydrocarbons, and their combustion produces CO_2 and water vapor by reactions similar to that in Eq. (9–4). A four-engine SST would emit about 80 tons of water vapor per hour of flight,[†] leaving a water vapor pressure in its wake that would exceed the sublimation pressure of ice (see Fig. 3–4). The water would form a myriad of fine ice crystals that make up the familiar contrails you have seen behind high-flying jet planes. On calm days the contrails sometimes linger long after the plane has passed and eventually become thin, high-altitude (cirrus) clouds. There is fear that a large fleet of SST's could produce ap-

---

*   See, for example, W. A. Shurcliff, *S/S/T and Sonic Boom Handbook* (Ballantine Books, New York, 1970).

---

†   *Inadvertent Climate Modification,* SMIC.

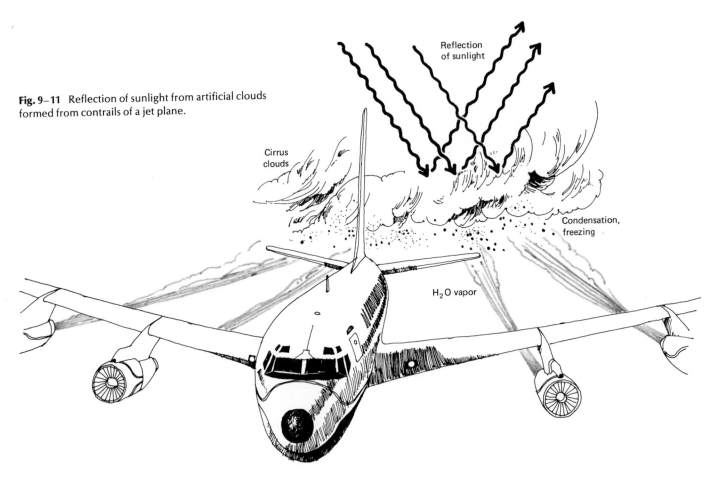

Fig. 9–11 Reflection of sunlight from artificial clouds formed from contrails of a jet plane.

preciable artificial cloudiness. Cloud tops are quite effective in reflecting sunlight into space (see Fig. 9–11). If the artificial cloudiness is an appreciable fraction of normal cloud cover, it could have a significant cooling effect on the lower atmosphere and on Earth's surface.

Most experts do not feel that the artificial clouds pose a serious problem. One calculation, for example, predicts that jet emissions would raise the global stratospheric water concentration by about 7% (from 3.0 to 3.2 ppm), although along heavily traveled air routes, the increase might be 70% (from 3 to 5 ppm).* However, one can hardly be complacent when he realizes that in one year's time, a 500-plane fleet would inject into the stratosphere, in the vicinity of heavily traveled

*   *Man's Impact on the Global Environment, SCEP.*

routes, an amount of water vapor equal to about 30% of the total amount that occurs naturally there!

Potentially, the most serious global problem of SST'S may be their effect on the ozone layer of the stratosphere. Figure 9–12 shows a number of parameters of the atmosphere. Well above the tropopause, at about 25 km, the atmospheric temperature rises, reaching a maximum of about +10°C at 50 km. The reason for this increase is the high concentration of ozone, O_3, in the stratosphere.

The important feature of ozone is that it strongly absorbs UV light. Figure 9–13 shows portions of the spectrum of solar radiation as received at the outside of Earth's atmosphere and at ground level after passage through the atmosphere. The inset shows the UV absorption spectrum of ozone. Ozone absorbs nearly all radiations of wavelength less than 2900 Å

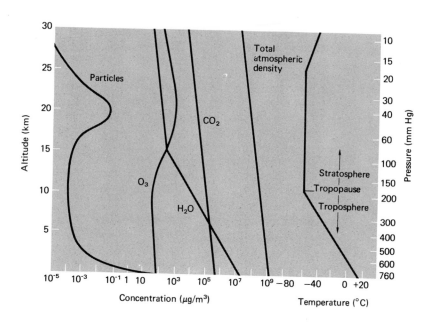

Fig. 9–12 The pressure, temperature, and typical composition of the atmosphere changes with altitude.

and some up to about 3200 Å. Much of the absorbed energy is converted to kinetic energy by re-radiation and collision of molecules, raising the temperature of the stratosphere. The UV absorption by ozone is extremely important for life at the earth's surface. Although the two spectra of Fig. 9–13 do not appear to be drastically different, the solar radiations of wavelengths less than 3000 Å that are absorbed by O_3 would do great harm if they reached the earth's surface: among other things, the incidence of skin cancer would soar and we would suffer eye damage. Although we could wear protective clothes and sunglasses, as astronauts do on the moon, which has no atmosphere, many forms of animal life could probably not survive.

Ozone is both created and destroyed by UV light. When UV photons of wavelength 1700 to 1900 Å strike O_2 molecules, the molecules can break into two free oxygen atoms:

$$O_2(g) + h\nu \ (1700\text{–}1900 \ \overset{\circ}{A}) \longrightarrow O(g) + O(g). \qquad (9\text{–}5)$$

(Although no photons of those energies appear in the solar spectrum as shown in Fig. 9–13, an expanded plot of the spectrum would reveal a small intensity at those wavelengths.) One of the oxygen atoms from reaction (9–5) can combine with an O_2 molecule to form ozone:

$$O(g) + O_2(g) \longrightarrow O_3(g). \qquad (9\text{–}6)$$

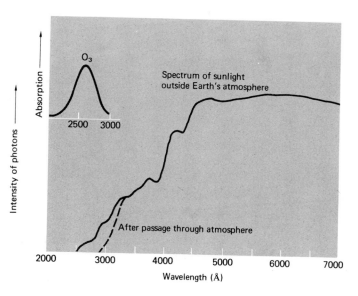

Fig. 9–13 Partial spectrum of sunlight before and after passage through the earth's atmosphere. Inset: Absorption spectrum of ozone.

If nothing destroyed ozone, its buildup would continue until all the O_2 was used up. Fortunately, there are several reactions that destroy O_3, including the reaction with O atoms,

$$O(g) + O_3(g) \longrightarrow 2O_2(g), \qquad (9-7)$$

and photochemical decomposition following absorption of UV light,

$$O_3(g) + h\nu \ (2000-3000 \ \overset{\circ}{A}) \longrightarrow O_2(g) + O(g). \qquad (9-8)$$

(This is the absorption responsible for the removal of UV from sunlight, as discussed above.) Several other reactions also destroy ozone. The total amount of ozone in the stratosphere remains approximately constant, as it is produced at about the same rate that it is destroyed. More is produced over the equator and during summertime in the temperate zones because of increased sunlight. Thus the concentration of ozone depends both on season and latitude. The concentration profile shown in Fig. 9–12 is a typical distribution at about latitude 30° or 40°.

There is much concern that nitrogen oxides from SST engines might destroy some of the ozone. Whenever air is drawn through a hot combustion chamber, reactions between the nitrogen and the oxygen of the air produce some nitric oxide:

$$N_2(g) + O_2(g) \longrightarrow 2NO(g). \qquad (9-9)$$

This reaction occurs in all high-temperature combustion processes utilizing air. The NO produced by automobiles and power plants is a vital ingredient of photochemical smog (see Section 14–4).

In 1971, Professor Harold Johnston of the University of California at Berkeley shook up the aviation world by suggesting that nitrogen oxides from SST's might destroy up to half of the stratospheric ozone![*] The important reactions in Johnston's scheme are

$$\text{NO}(g) + O_3(g) \longrightarrow NO_2(g) + O_2(g) \qquad (9-10)$$

and

$$NO_2(g) + O(g) \longrightarrow NO(g) + O_2(g) \qquad (9-11)$$

Net reaction:

$$O(g) + O_3(g) \longrightarrow O_2(g) + O_2(g) \qquad (9-12)$$

These reactions are very interesting: first, the net reaction is the combination of O and O_3 to form O_2, the same as reaction (9–7). Second, the NO destroyed in the first reaction is regenerated by the second one, so it is not consumed in the overall process. The NO is thus a *catalyst*, a substance that participates in and speeds up the reaction but is not itself used up. If each NO molecule destroyed only one O_3 molecule, the problem would not be too serious. But according to Johnston's catalytic scheme, each NO molecule can go on destroying more ozone until some other kind of reaction removes NO or NO_2 from the cycle (e.g., by converting it to HNO_3 and subsequent particle formation) or until the gases diffuse to the troposphere.

Johnston estimated the extent of catalytic destruction of O_3 by NO. His results, shown in Fig. 9–14, are based on the assumption that the accumulation of NO_x (meaning both NO and NO_2) from two years of operation of a 500-plane fleet of SST's is uniformly distributed around the globe in a layer between 19 and 23 km altitude. We see that the ozone is strongly depleted in that layer. The total ozone in a column from the ground to the top of the atmosphere is depleted to 88% of its pre-SST level! Other assumptions reduce the ozone to as low as 50% of its natural levels beneath heavily traveled air routes.

How serious are these ozone depletions? Since the ozone layer now absorbs virtually all the high-energy UV radiation coming from the sun, a very small decrease of ozone leads to a much larger increase of UV reaching the earth's surface. It is estimated that a 1% decrease of ozone would increase the UV flux at sea level by 8%, a serious increase as far as human safety is concerned.[†]

Johnston points out that his calculations do not prove that the SST would have damaging consequences, but they show what *might* happen. There are many uncertainties in his calculations: for example, the rates of many of the pertinent reactions are well-known only at 25°C, if at all, and not at the low temperatures of the stratosphere. Furthermore, the natural concentrations of many species, particularly the nitrogen oxides, are poorly known, as are the rates of global and vertical movement of gases. It has been assumed, for example, that the residence time of the nitrogen oxides in the stratosphere is of the order of two years, but reactions that destroy them may be fast enough to remove them in a few weeks or months, thus reducing the severity of the NO_x

---

*   H. Johnston, "Reduction of Stratospheric Ozone by Nitrogen Oxide Catalysts from Supersonic Transport Exhaust," *Science* **173**, 517–522 (1971).

†   *Inadvertent Climate Modification*, SMIC.

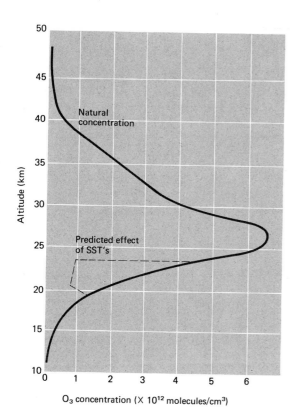

Fig. 9–14 Concentration profiles of ozone. Solid line: natural concentration. Dashed line: concentration after SST operations. (From H. Johnston, "Reduction of Stratospheric Ozone by Nitrogen Oxide Catalysts from Supersonic Transport Exhaust," *Science* **173**, 517–522 (6 August 1971). Copyright © 1971 by the American Association for the Advancement of Science.)

9–6 INTENTIONAL AND INADVERTENT WEATHER MODIFICATION

Before discussing man-made effects on the weather, let us consider some of the natural phenomena involved in cloud formation and subsequent precipitation of rain or snow. Suppose air at ground level has a temperature of 20°C and a relative humidity of 60%. Referring to Fig. 3–4, we see that the pressure of water vapor is then about 10 mm and the air contains 10 g of H_2O per m^3. If a parcel of this air rises to about 1 km altitude, the air will expand and cool, reaching a temperature in the neighborhood of 10°C. The air will be saturated with water vapor, if its vapor pressure remains 10 mm, since that is the equilibrium vapor pressure of liquid water at 10°C.* At ground level, we would say that the air had reached the *dew point* and any further cooling would cause the water vapor to start condensing as dew on cold surfaces. If the parcel of air rises further and cools, it becomes *supersaturated* with water vapor, i.e., the relative humidity is more than 100%. The air contains more water than it would if in equilibrium with liquid water.

If air is very clean (i.e., contains almost no particles), it can be cooled considerably below the dew point before formation of liquid droplets occurs, as condensation is difficult to initiate unless there is some suitable surface for it. However, as noted in Section 9–4, the air contains many aerosols of marine and continental origin that can serve as *condensation nuclei*. Some of them, especially sea salt and other *hygroscopic* particles, contain water-soluble materials that take up water quite readily. There are typically 200 to 1000 nuclei per cm^3 in unpolluted air. The growth of the particles is rather slow until the air becomes slightly supersaturated, at which point they quickly grow from radii of $1 \mu m$ or less to cloud droplets having radii of $10 \mu m$ or more.†

These droplets can form a cloud, but not rain. Droplets of this size have a fall rate of 1 cm/s, but updrafts in the cloud keep them aloft. In order for rain to occur, the droplets must grow to radii of $150 \mu m$ or more before they are heavy enough to fall out of the cloud. If the cloud droplets stay in a supersaturated atmosphere long enough, they mix and collide with each other in the cloud and many of the smaller droplets

problem. Since many aerosols are of just the right size to scatter UV radiation efficiently, they may prevent excess UV from reaching ground level.

Johnston's calculations do indicate clearly that we must investigate all these questions much more carefully *before* decisions to build and operate large SST fleets become irrevocable. Fortunately, many of the problems of the SST are unique among global environmental problems in that we have the capability of obtaining answers to many of the open questions within two or three years. The U.S. Department of Transportation has funded an intensive two-year study of all aspects of the problem that should answer many of the questions before SST fleets come into being.

---

*   Actually the situation is somewhat more complicated, because expansion of the air will slightly decrease the pressure of the water contained in it.

†   A more detailed, yet simple account of precipitation phenomena is given by J. A. Day and G. L. Sternes, *Climate and Weather* (Addison-Wesley, Reading, Mass., 1970).

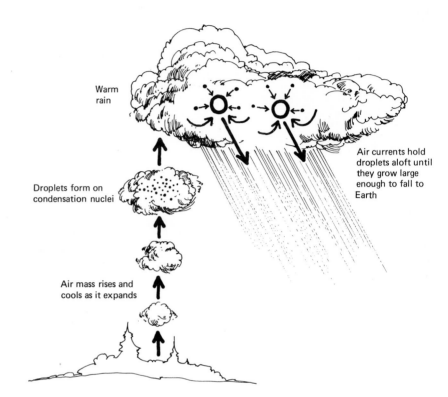

Fig. 9–15 Warm rain may occur following growth of liquid water droplets.

Warm rain

Droplets form on condensation nuclei

Air mass rises and cools as it expands

Air currents hold droplets aloft until they grow large enough to fall to Earth

coalesce into larger drops. Some grow large enough to fall to earth before evaporating in the warmer, unsaturated air beneath the cloud. This sort of precipitation is called "warm rain" (see Fig. 9–15).

The more interesting cloud physics occurs when the clouds reach altitudes at which the temperature drops below 0°C. Ice always melts when the temperature is raised above 0°C, but droplets of extremely pure water can be cooled to as low as −40°C before freezing occurs. Such *supercooled* water needs nucleation centers to provide surfaces for the crystallization of ice.

If some appropriate *freezing nuclei*, e.g., silicate particles from wind erosion of rocks and soil, are introduced into the supercooled cloud, ice forms around the particles. Once freezing is initiated, the ice crystals grow quickly at the ex-

pense of vapor from the liquid droplets, so that the cloud becomes a suspension of ice crystals. If the ice crystals grow large enough, they fall to Earth as snow or, if they melt on the way down, as rain. If there are extremely violent updrafts in the cloud (e.g., winds above 100 mph), the ice crystals may be kept aloft until they grow to such size that hail falls from the cloud.

Most schemes for intentional weather modification involve attempts to induce freezing in clouds of supercooled droplets. One of the simplest methods is that of dropping fine particles of dry ice (frozen CO_2) into the supercooled cloud. Dry ice sublimes at −79°C at atmospheric pressure. A small plane can be flown through the cloud to sprinkle the dry ice particles. As the particles fall through the cloud, the air around them is cooled to as low as −79°C. Since water

The hole in these supercooled clouds was caused by dry-ice seeding. (Courtesy of U.S. Air Force.)

Three silver iodide generators in Colorado. (U.S. Dept. of the Interior, Bureau of Reclamation photo by W. L. Rusho.)

droplets cannot remain liquid below $-40°C$, droplets close enough to the path to reach that temperature are frozen, providing an abundance of freezing nuclei. If conditions are right, precipitation will then occur (see Fig 9–16).

The other principal method for weather modification is that of seeding supercooled clouds with particles that serve as good freezing nuclei. The best known material is crystalline silver iodide, as the size of the cells of its crystal lattice is almost identical to that of ice. Fine crystals of AgI can induce freezing of water droplets at all temperatures below $-5.2°C$.* Another important substance for this purpose is lead iodide, PbI_2, which induces freezing at $-6.5°C$ and below. Silver iodide is used extensively in weather modification experi-

* Day and Sternes, *Climate and Weather.*

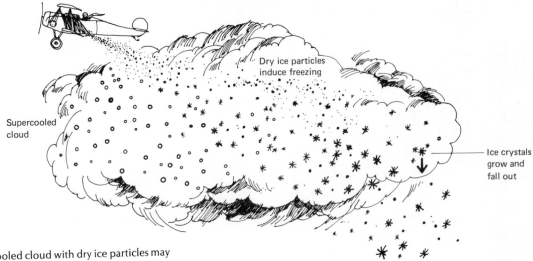

Dry ice particles
induce freezing

Supercooled
cloud

Ice crystals
grow and
fall out

Fig. 9–16 Seeding of supercooled cloud with dry ice particles may initiate freezing, growth of ice crystals, and precipitation.

ments in the United States. It can be vaporized from a torch at ground level upwind from the cloud or released from an airplane in order to cause rain or snow. Lead iodide is used a lot in the Soviet Union in attempts to suppress destructive hail. When a hail-forming cloud appears, shells loaded with PbI_2 crystals are fired into the cloud with a cannon. The intent is to provide so many freezing nuclei that few will acquire enough water to grow to hail size.

What have been the results of weather modification experiments? The question is rather difficult to answer. One case in which weather modification has clearly been successful is in removal of fog from airports. When supercooled fog prevents the landing of planes, the fog can be seeded with dry ice, causing a brief snow storm in the seeded area. The area around the landing strip is cleared of fog for about an hour. The success rate for this operation is about 80%.[*]

The success of other cloud-seeding operations is still in question. There have been numerous cases in which individual clouds have been seeded and a rain or snow shower has resulted. The real question is whether seeding conducted over a several-month period significantly alters the amount of precipitation over a sizable region or whether the

water deposited by the seeding would have fallen nearby anyway. There are such large variations in precipitation from one year to another that one must conduct seeding for several years to determine the overall effect of the seeding.

Many large-scale weather modification experiments are planned or in progress.[†] One of the largest involves an attempt to enhance snowfall in the mountains near Durango, Colorado, to increase the runoff of water into the Colorado River. As in all weather modification experiments, it is necessary to proceed carefully. If successful, the project will be beneficial to the people who need water from the river. But, residents around Ouray, Colorado, fear that the added snowfall will increase the frequency of avalanches. Furthermore, when seeding is done to increase precipitation in one area, some fear that precipitation will be decreased downwind from the seeded area. At present this would not appear to be a serious problem: only about 10% of the water vapor passing across the United States is precipitated naturally and seeding typically removes only an additional 1 or 2%.[‡]

Project Stormfury is a large-scale experiment designed

_____

[*]   Ibid.

_____

[†]   A. L. Hammond, "Weather Modification: A Technology Coming of Age," *Science* **172**, 548–549 (1971).

[‡]   Ibid.

Heavy snowfall obscures visibility on mountain pass between Durango and Ouray, Colorado. (Bureau of Reclamation photo by W. L. Rusho.)

Fig. 9–17 Buffalo, New York, receives excess snowfall from the "lake effect" of winds and clouds from Lake Erie.

to reduce the destructive forces of hurricanes by seeding them.* This is a difficult project because the processes involved in hurricanes are not well understood. Moreover, seeding can be performed on only a few hurricanes, as they must be far enough from land, so that effects of the seeding can be observed before the hurricane hits a land area and breaks up. Seeding of two hurricanes, Beulah in 1963 and Debbie in 1969, produced some encouraging results, as wind speeds dropped significantly after the seeding. Although hurricane modification may be desirable for residents of coastal areas in the path of the hurricane, this large-scale modification must be approached with caution. Inland areas throughout the eastern United States receive large amounts of rainfall via the less destructive storms produced after the hurricanes come ashore. The possible loss of this source of water might cause serious drought conditions.†

An interesting project near Buffalo, New York, was initiated in the winter of 1971–1972.‡ Buffalo and nearby areas at the eastern end of Lake Erie receive as much as 100 inches of snow annually because of the "lake effect" (see Fig. 9–17). Cold winds from the west pick up heat and water vapor from the somewhat warmer surface of the unfrozen lake and dump the water as snow on the eastern shore areas. In the weather modification project, some clouds will be seeded heavily in order that an enormous number of very fine ice crystals, rather than a small number of large ones, will be formed. The hope is that the finer crystals will pass over Buffalo and be deposited over a much larger area to the east. In another scheme, seeding will be done in such a way that the snow will fall heavily right at the shore line before the clouds reach Buffalo.

---

* Louise Purrett, "How to Subdue a Hurricane," *Science News* **100**, 128–129 (1971).

† Landsberg, "Man-Made Climatic Changes."

‡ "Spreading Out Buffalo's Burden of Snow," *Science News* **100**, 341 (1971).

Man has long been inadvertently modifying the climate and weather. Simply by bringing together large masses of people and their activities in metropolitan areas, man affects the local climate. One major effect is that of the *heat island* around a city.* You've probably noticed in weather reports that downtown temperatures are frequently 5° to 10°F higher than those of outlying suburbs. There are several reasons for the increased temperatures in central cities: (1) the use of energy for heating, cooling, transportation, etc., releases significant amounts of heat; (2) buildings reduce wind velocities in the city; (3) cities are covered by many surfaces, especially asphalt, that are good absorbers of sunlight; and (4) following rains, water drains off quickly from paved surfaces into sewers, eliminating the cooling effect of water evaporation.

Precipitation patterns are also altered near cities. The warmer air of a city rises in the atmosphere, carrying with it many particles produced by man's activities. When the rising air cools to the dew point or the freezing point, some of the particles of the polluted air masses serve as condensation or freezing nuclei. Condensation causes significantly more cloudiness over cities than over upwind rural areas. Freezing apparently causes an increase in precipitation in the region of cities (see Fig. 9–18). Precipitation in some cities is markedly greater on weekdays than on weekend days.† The only obvious difference that could cause this result is that on weekend days there are fewer particle-forming activities (e.g. factory operations, commuter traffic) than on weekdays.

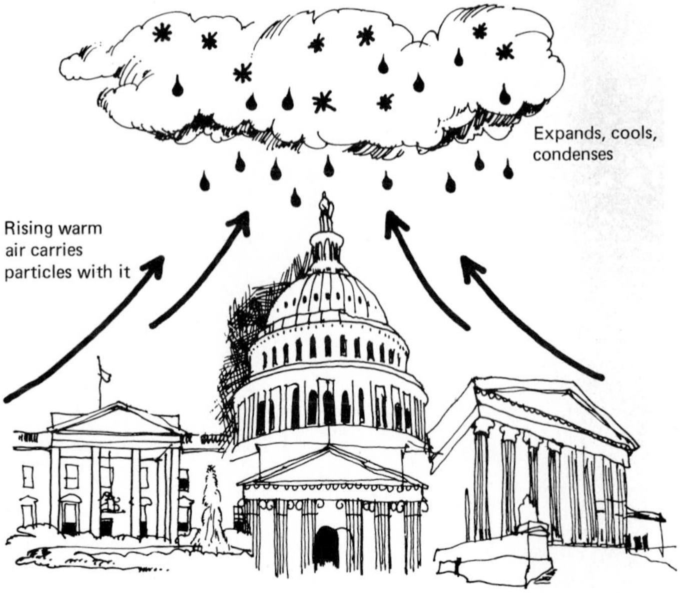

Aside from the nuisance of warmer summers and excess precipitation in and around cities, are there serious regional climatic effects caused by the particles injected into the atmosphere by urban activities? We cannot be certain, but some observations by Professor Schaefer must arouse concern. He has noted that hazy plumes are often seen by observers in airplanes that are hundreds of miles downwind from major cities, even when the most offending visible emissions of smoke stacks have been cleaned up.‡

What causes the haze? One particular pollution source that may be important is the combustion of leaded gasoline in automobiles. Lead is emitted as extremely small particles

Rising warm air carries particles with it

Expands, cools, condenses

Fig. 9–18   The heat-island effect in a city may cause inadvertent cloud seeding and excess precipitation.

of mixed chloride and bromide salts of lead ($PbCl_2$, $PbBr_2$, and $PbBrCl$). Schaefer has demonstrated that if these lead-bearing particles come in contact with iodine vapor, many tiny freezing nuclei consisting of PbI_2 are formed. As noted above, PbI_2 is used in deliberate attempts to suppress hail. The lead-bearing particles from auto exhaust probably can find enough iodine vapor (much of it vaporized from oceans) in the air to react with and form freezing nuclei in great abundance. When these freezing nuclei encounter supercooled water vapor, they can form fine ice crystals. This mechanism may be responsible for the abundance of ice crystals observed in plumes downwind from cities.

When a mass of polluted air containing freezing nuclei and ice crystals encounters supercooled clouds, the latter are seeded and precipitation may result. Schaefer is concerned about the possible overseeding of clouds by pollutants. If there are too many nuclei in a cloud, few of them can attract enough water vapor to grow large enough to fall to the ground. Thus, depending on the concentration of seed

---

*   Landsberg, "Man-Made Climatic Changes." See also W. P. Lowry, "The Climate of Cities," *Scientific American* **217** (2) 15–23 (Aug. 1967).

†   Landsberg, "Man-Made Climatic Changes."

‡   Schaefer, "Inadvertent Modification of the Atmosphere." See also V. J. Schaefer, "Auto Exhaust, Pollution and Weather Patterns," *Science and Public Affairs* **26** (8), 31–33 (1970).

nuclei, precipitation can be either enhanced or decreased. Schaefer has noted an increased incidence of storms characterized by snowflakes or raindrops of very small size. Raindrops seem to drift to Earth rather than fall. If urban air pollution continues to increase, he fears that these unusual storms will occur with increasing frequency. The overseeding of clouds by pollutants could deprive large areas downwind from metropolitan areas of badly needed water. Serious droughts in regions that were once fertile may occur and municipal water supplies may be depleted.

Smog in New York City at high noon. (Courtesy of the Environmental Protection Agency.)

Probably the first air pollution disease in history was rickets, a deformation or weakening of bones, particularly in children, caused by vitamin D deficiency. As discussed in Chapter 15, vitamin D is essential for deposition of calcium in bones. We take in some vitamin D in food, but much of it is made in the body by the irradiation of chemicals just under the skin by UV photons. Suspended pollutant particles in urban atmospheres absorb and scatter incoming sunlight. In winter, when large amounts of fuel are burned for heat and power production, the intensity of visible light is often reduced by 30% or more and UV light, with wavelengths less than 4000 Å, is virtually eliminated. As we have seen above, most UV light with wavelength less than 3000 Å is removed by stratospheric ozone. Although such short-wavelength UV of high intensity would be harmful to biological life, small amounts of longer wavelength UV are needed to form vitamin D.

Since about 1650, when soft-coal burning was introduced in England, rickets has been a common affliction among children of industrialized cities in northern Europe.* In winter, areas at high latitudes receive barely enough light to form the necessary amount of vitamin D even when the air is clean. But in the heavily polluted cities, so little UV reached ground level that rickets took a heavy toll, with childhood deaths and severe crippling. Today rickets has largely been eradicated from advanced nations by supplementing diets with foods rich in vitamin D, such as cod-liver oil. Also, such staples as milk and bread are often fortified with vitamin D. However, in less advanced cities with heavy air pollution, rickets may still affect sizable numbers of children.

9–7 WHERE DO WE STAND?

In this chapter we have discussed a number of problems that could affect the earth's climate on a global or regional scale. As we have seen, in most cases there is a great deal of uncertainty about the predicted magnitudes and sometimes even about the direction of the effects. If some of the effects discussed turn out to be large, the changes in climate would necessitate significant movements of people to different areas of Earth. Even greater changes of the surroundings could cause such large alterations of temperature, rainfall,

---

* W. F. Loomis, "Rickets," *Scientific American* **223** (6), 76–91 (Dec. 1970).

or UV radiation that life would become impossible for many species.

How likely is it that some of the predicted climatic effects will present serious problems? It is difficult to say. It now appears that the greenhouse effect of increased atmospheric CO_2 is unlikely to cause large temperature changes over the next several centuries. Predictions of the effects of particles released by man on global temperature and regional precipitation are much less certain. Present dangers do not now appear great enough to necessitate drastic cutbacks in man's release of particles. Although, as discussed in Chapter 10, there are other reasons for wanting to reduce particle emissions in areas of high population density. Perhaps the best approach is that which is now being undertaken: continue to study the problems and carefully monitor atmospheric concentrations of CO_2, particles and other species, rainfall, temperature, and influx of radiation from the sun. From these records, it should be possible to observe trends of these quantities soon enough to take some corrective action.

The potential problems of the SST are somewhat different. The SST is not essential to man's well-being (unlike the combustion of fuels for heat, which releases CO_2 and particles). The possible hazards of UV radiation increases caused by depletion of the ozone layer are so great that we can surely hold up deployment of large numbers of SST's until studies assess the magnitude of the problem. The necessary experiments are now in progress and, within two years, we should know a great deal more about reactions in the stratosphere and the effects of SST emissions.

SUGGESTED READING

H. R. Byers, *General Meteorology*, 3rd ed. (McGraw-Hill, New York, 1959).

J. A. Day and G. L. Sternes, *Climate and Weather* (Addison-Wesley, Reading, Mass., 1970).

C. L. Hamilton, ed., *Chemistry in the Environment* (W. H. Freeman, San Francisco, 1973). Excellent collection of reprints from *Scientific American*, including several articles referred to in this chapter. Paperback.

Inadvertent Climate Modification: Study of Man's Impact on Climate (SMIC) (MIT Press, Cambridge, Mass., 1971). Paperback.

H. E. Landsberg, "Man-Made Climatic Changes," *Science* **170**, 1265–1274 (1970).

Man's Impact on the Global Environment: Study of Critical Environmental Problems (SCEP) (MIT Press, Cambridge, Mass., 1970). Paperback.

R. E. Newell, "The Global Circulation of Atmospheric Pollutants," *Scientific American* **224** (1), 32–42 (Jan. 1971).

A. H. Oort, "The Energy Cycle of the Earth," *Scientific American* **223** (3), 54–63 (Sept. 1970).

W. O. Roberts, "We're Doing Something about the Weather," *National Geographic* **141**, 518–555 (1972).

V. J. Schaefer, "Auto Exhaust, Pollution, and Weather Patterns," *Science and Public Affairs* **26** (8), 31–34 (1970).

Scientific American **221** (3) (Sept. 1969). Entire issue on the oceans.

S. K. Singer, ed., *Global Effects of Environmental Pollution* (D. Reidel, Amsterdam, 1970).

QUESTIONS AND PROBLEMS

1. What and where are the major layers in the atmosphere, beginning at the earth's surface?

2. Why are the temperature differences between day and night so small on the earth compared with the moon?

3. What is the difference between condensation nuclei and freezing nuclei? Which are found in greater quantities in the urban atmosphere?

4. The greenhouse effect of carbon dioxide involves absorption of IR radiation of about 15 μm wavelength.

   a) Sketch the kind of excitation (e.g., bending, stretching, rotation) caused in CO_2 by the absorption (see Chapter 8).

   b) What is the energy, in eV, of a 15 μm photon?

5. If the earth's atmosphere became undesirably hot, which of the following actions would tend to cool it down? Explain how each works.

   a) Cover a large area of the earth with asphalt.

   b) Cover a large area of the earth with aluminum foil.

   c) Cut down on our use of fossil fuels (coal, gas, oil).

   d) Run more jet planes high in the atmosphere to create artificial clouds.

   e) Shoot off some large nuclear weapons to blow some additional particulate matter into the stratosphere.

   f) Dissolve a lot of limestone with acid to put more carbon dioxide into the atmosphere:

$$CaCO_3(s) + 2HCl(aq) \longrightarrow Ca^{2+}(aq) + H_2O(l) + CO_2(g) + 2Cl^-(aq)$$

6. Describe or define each of the following terms.
   a) relative humidity   b) dew point
   c) cloud seeding   d) freezing nucleus
   e) condensation nucleus   f) supersaturated (water vapor)
   g) supercooled (water)

7. a) What are the two substances used predominantly in United States weather modification experiments?
   b) Very briefly describe the effect of each substance on a supercooled cloud.

8. What are some of the reasons for the fact that urban areas are "heat islands"?

9. Explain how the heat island effect and an excessive number of particles can cause cities to have excessive amounts of rainfall.

10. a) Suppose the temperature at ground level is 25°C and the water vapor pressure is 8 mm. What is the relative humidity? (See Fig. 3–6).
   b) If the air temperature drops by about 7°C/km with altitude, at what altitude will the air from part (a) become saturated? (Ignore the decrease of water vapor pressure because of expansion of the air as it rises.)

11. Using the following calculation, you can probably demonstrate the need for a humidifier attached to home heating systems. Suppose the outside air temperature is 0°C and the relative humidity 50%. Outside air leaks into the house, where it is warmed to 22°C (72°F).
   a) What is the water vapor pressure in the outside air?
   b) Ignoring expansion of the air when it is heated (an 8% increase), what is the relative humidity of the air at 22°C if no water is added during heating? (See Fig. 3–6).

12. Most calculations of SST effects assume a fleet of 500 planes, each operating an average of seven hours per day in the stratosphere. Of these, 334 would have four engines and 166 would have the equivalent of two engines.
   a) The gases of the stratosphere weigh about 8×10^{20} g, and water vapor makes up about 3 ppm by mass. If each jet engine puts out 41,400 lb of H_2O per hour of operation, how many grams of water would be deposited in the stratosphere in two years by the SST fleet? To what fraction of the total amount of stratospheric water vapor does this amount correspond?

   b) An SST engine is expected to emit about 1400 lb of NO per hour. What weight of NO is emitted by the fleet over a two-year period? How many moles of NO? The weight of O_3 in the stratosphere is about 3×10^{15} g. If each molecule of NO destroys just one molecule of O_3, what fraction of the O_3 would be destroyed over the two-year period? (Note that much of the destroyed O_3 would be replaced by generation of additional O_3.) What fraction of the O_3 would be destroyed if each NO destroyed 1000 molecules of O_3? [Data mostly from the *SCEP* Report, *Man's Impact on the Global Environment* (MIT Press, Cambridge, Mass., 1970.]

13. a) Calculate the residence time of carbon in plants on land.
   b) Calculate the residence time of carbon in the deep sea. (This long residence time explains the prediction that many centuries will be required for CO_2 dissolved in the oceans to come into equilibrium with the increased CO_2 concentration in the atmosphere.)

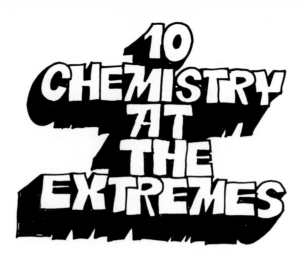

10 CHEMISTRY AT THE EXTREMES

Sulfur compounds play a major role in urban air pollution. Shown here is a London bus during the sulfurous smog of December 8, 1952. (Courtesy of the Radio Times Hulton Picture Library.)

10-1 INTRODUCTION

In Chapters 7 and 8 we noted that the chemical behavior of an element depends mainly on the number of electrons in the valence shell of its neutral atoms. Even before the electronic structures of atoms were known, Mendeleev and others constructed the periodic table by placing elements of similar chemical behavior in the same column of the table. We now know that members of a column, called a *chemical group*, have chemical similarity because they have the same numbers of valence electrons.

In this chapter, we discuss the groups at the far left and far right of the periodic table (ignoring the noble gases, which have little chemistry). These groups include the alkali metals of Group IA, the alkaline earths of Group IIA, and the halogens of Group VIIA. In these groups the chemical similarities among the elements are quite striking. Then we discuss oxygen and sulfur of Group VIA. Although these two elements have considerable similarity to each other, elements further down the column (Se, Te, and Po) bear increasingly little chemical relationship with oxygen. In Group VIA, as in Groups IIIA, IVA, and VA (discussed in later chapters), the groups are headed by nonmetals or semimetals, with metallic character increasing as we move down the column.

Finally, in connection with the chemistry of sulfur, we discuss some problems of urban air pollution, in which sulfur compounds play a major role.

10-2 ALKALI METALS

The alkali metals of Group IA (Li, Na, I, Rb, Cs, Fr) are soft metals.   Each of their neutral atoms has but one loosely bound electron in its valence shell. Since this electron can easily come off, the alkali metals react strongly with many other compounds and elements to form strongly ionic compounds. For example, sodium reacts vigorously with chlorine gas to form sodium chloride (ordinary table salt):

$$2Na(s) + Cl_2(g) \longrightarrow 2Na^+ + 2Cl^- \text{ (solid salt).} \qquad (10-1)$$

The alkali metals also react with water, e.g.,

$$2Na(s) + 2H_2O(l) \longrightarrow H_2(g) + 2Na^+(aq) + 2OH^-(aq).$$
$$(10-2)$$

The latter reaction occurs with increasing vigor as we move down the group. The reaction of water with potassium and heavier alkali metals releases so much heat that the hydrogen gas released invariably catches fire. With sodium, the hydrogen ignites only if one uses a large piece of the metal and, with lithium, there is usually no ignition. At the completion of the reaction, the solution is a base; in the example above it is sodium hydroxide. Alkali metals react so readily with oxygen and water vapor in the air that they must be stored under kerosene (an organic liquid).

The compounds of sodium are among the most extensively used chemicals of American industry, sodium hydroxide or "lye," NaOH, being the fourth largest volume item of the chemical industry. It is used in the conversion of fats into soap and is the major ingredient in commercial drain cleaners. Sodium metal is made by electrolysis of molten sodium chloride, the sodium being drawn off at the negative electrode and chlorine gas at the positive electrode.

Sodium hydroxide is made by the *chloro-alkali process*, which is the electrolysis of aqueous solutions of sodium chloride ("brine"). As shown in Fig. 10-1, the solution is electrolyzed in the two end tanks, the Cl_2 gas being liberated from the positive graphite (carbon) electrodes in the end sections of the tank. The liquid metal mercury at the bottom of the tank serves as the negative electrode at which Na^+ ions are converted to Na atoms:

$$Na^+(aq) + e^- \longrightarrow Na \text{ (amalgam).} \qquad (10-3)$$

Sodium dissolves in the mercury to form an *amalgam*, a solution of a metal in mercury. The electrolysis tank is rocked back and forth to carry the amalgam to the center section, where sodium atoms react with water to form NaOH, as in Eq. (10-2). The tank is divided into sections to prevent the reaction of Cl_2 gas with the sodium hydroxide solution, which would form Cl^- and ClO^- (hypochlorite ion):

$$Cl_2(g) + 2OH^-(g) \longrightarrow Cl^-(aq) + ClO^-(aq) + H_2O(l).$$
$$(10-4)$$

The chlorine gas produced by the chloro-alkali process is also quite useful in the chemical industry, ranking just after sodium hydroxide on the list of most-used compounds.

Unless precautions are taken, small amounts of mercury from the tanks escape in the waste water from chloro-alkali plants. Before strict clean-up of mercury effluents was initiated in 1971, chloro-alkali plants were major sources of mercury pollution (see Chapter 12).

Nearly all salts of alkali metals with various negative ions are soluble in water. Since the alkali metals are strongly reactive, one never finds deposits of the pure metals in nature.

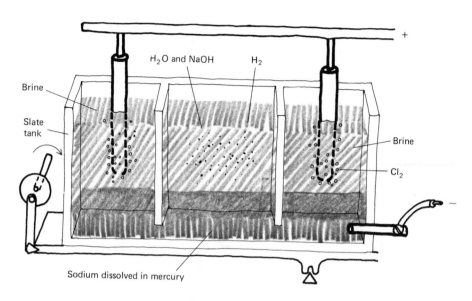

Brine

Slate
tank

H₂O and NaOH

H₂

Brine

Cl₂

Sodium dissolved in mercury

Fig. 10–1 The chloro-alkali process for making sodium hydroxide and chlorine gas.

Salt beds and processing plant near the Great Salt Lake. (Courtesy of Morton-Norwich Products, Inc.)

Because of their solubility, alkali metal salts are carried in the runoff water from rocks and soil in land areas. In any water body that has no outlet, e.g., the oceans, the Dead Sea, and the Great Salt Lake, water escapes only by evaporation, leaving the salts behind. As a result, the oceans are on the average about 3.5% salt by weight, that is, 35 grams of salt per liter of ocean water. In Table 10–1 are listed the concentrations of major ions in the ocean. The alkali metal ions (mainly Na^+) and alkaline earth ions are the most prominent *cations* (positive ions) in ocean water (nearly 36% of dissolved ions). The most prominent *anions* (negative ions) are chloride ions.

In Table 10–1 are also listed the concentrations of several ions in blood plasma. The ions that are prominent in ocean water are also important constituents of blood plasma. This fact is not a coincidence, but a result of the fact that we have evolved from animals that once lived in the sea.

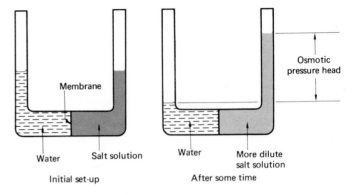

Fig. 10–2 The osmotic pressure of pure water causes it to pass through the membrane and dilute the salt solution.

Table 10–1 Concentrations of a Few Important Species in Sea Water and Blood Plasma

Species	Concentration (grams/liter)	
	Sea water	Blood plasma
Cations		
Na^+	10.6	3.3
K^+	0.38	0.195
Mg^{2+}	1.27	0.036
Ca^{2+}	0.40	0.10
	12.70	
Anions		
Cl^-	18.98	3.66
SO_4^{2-}	2.65	0.048
HCO_3^-	0.14	1.65
	21.8*	

* Sum of major cations and anions = 34.5 g/l, or about 98% of the total dissolved salts.

The dissolved salts in various body fluids are important in controlling the flow of liquids through membranes. Consider Fig. 10–2, in which two liquids, water and a salt solution, are separated by a "semipermeable membrane"—a barrier that allows the water molecules to pass through, but not the dissolved species. If pure water is placed on one side and a salt solution on the other, the pure water has a strong tendency to move through the barrier to dilute the salt solution.

The pure water exerts great pressure, called *osmotic pressure*, on the membrane. Water molecules move through the barrier and dilute the salt solution, raising the level of the solution. Osmotic pressures are quite large—about twelve times atmospheric pressure (enough to raise the column on the right 360 feet high) if the salt concentration is 35g/l as in sea water. Osmosis regulates the passage of liquids through cell walls and other body membranes. Beverages such as Gatorade contain salt concentrations similar to those of body fluids and thus cause much less disruption of the body's fluid balance than pure water when one consumes a large quantity of it.

Alkali metal ions are important in the transmission of messages through the nervous system. Perhaps related to this fact is the discovery that lithium compounds are helpful in the treatment of the mental condition known as "manic depression." Administration of Li^+ has a calming influence during the manic phase of the cycle. Rubidium, also an alkali metal, has just the opposite effect of lithium: rats injected with rubidium become irritable and aggressive.* A University of Texas biochemist, Dr. Earl B. Dawson, has reported a significantly higher admission rate at mental hospitals in Dallas, Texas, than in El Paso. He suggests that there may be a correlation between the better mental health in El Paso and the high concentration of lithium in the city's water supply, which is

* B. J. Carroll and P. T. Sharp, "Rubidium and Lithium: Opposite Effects on Amine-Mediated Excitement," *Science* **172**, 1355–1357 (1971).

obtained from deep wells.* Further study of this hypothesis is needed, as there may be variables other than the lithium concentration of the water supplies.

10–3 ALKALINE EARTHS

The alkaline earths of Group IIA of the periodic table, $_4$Be, $_{12}$Mg, $_{20}$Ca, $_{38}$Sr, $_{56}$Ba, and $_{88}$Ra, are similar in behavior to the alkali metals, except that alkaline earth atoms have two electrons in their outermost shells, and thus lose two electrons to form 2^+ ions instead of the 1^+ ions of the alkali metals. Calcium and heavier alkaline earths react with water to release hydrogen:

$$Ca(s) + 2H_2O(l) \longrightarrow Ca^{2+}(aq) + 2OH^-(aq) + H_2(g).$$
$$(10–5)$$

Alkaline earths react rather strongly in air to form oxides. Magnesium ribbon, for example, burns in the atmosphere to form the oxide

$$2Mg(s) + O_2(g) \longrightarrow 2MgO(s). \qquad (10–6)$$

The oxides react with water to form hydroxides (bases), for example,

$$CaO(s) + H_2O(l) \longrightarrow Ca(OH)_2(s). \qquad (10–7)$$

Because reaction (10–7) uses up water, the oxides are good "drying agents." Calcium oxide can react with the water in an alcohol-water mixture, using up the water and thus "drying" the alcohol. Alkaline earth hydroxides are not so soluble as those of the alkali metals. Magnesium hydroxide is sufficiently insoluble to be used as a neutralizer of stomach acid, as in Milk of Magnesia. In the stomach it dissolves slowly as the hydroxide present in solution reacts with and uses up excess stomach acid.

Many compounds of the alkaline earths are soluble, but several important ones are only slightly soluble. Notable among the latter are the carbonates (for example, $CaCO_3$), the hydroxides as noted above, and the sulfates (for example, $BaSO_4$). When an ionic solid such as $BaSO_4$ is placed in contact with water, ions go into solution until the concentration equals the solubility of the compound at the given temperature, thus forming a *saturated solution*. If there is excess solid in contact with the saturated solution, there is an equili-

---

* "Tranquilizer in El Paso Water," Washington *Post*, Sept. 2, 1971.

brium between ions of the solid and those in solution, e.g., in the case of $BaSO_4$:

$$BaSO_4(s) \rightleftarrows Ba^{2+}(aq) + SO_4^{2-}(aq). \qquad (10–8)$$

This type of equilibrium, just like all others, can be understood according to LeChatelier's principle. In pure water, the solubility of $BaSO_4$ is about $10^{-5}M$ (0.0023 g/l). If we add one of the type of ions that appear on the right-hand side of Eq. (10–8), the equilibrium is pushed toward the left. For example, if we add some Na_2SO_4 (which is highly soluble) to the saturated solution, some of the sulfate and barium ions precipitate out of the solution. The solubility of $BaSO_4$ in $1M$ Na_2SO_4 is reduced to 10^{-10} M.

Calcium carbonate is another slightly soluble alkaline earth compound. It occurs in many rocks in the form of limestone or marble. When ground water trickles over rocks and soil it dissolves small amounts of calcium carbonate and other minerals. Natural water containing small amounts of Ca^{2+} ions, as well as Mg^{2+} and Fe^{3+} ions, is called "hard water." Hard water is undesirable for washing because the dissolved ions form insoluble precipitates with soap molecules (see Chapter 12). These precipitates form bathtub rings and make deposits on hands and clothes washed with soap in hard water.

Calcium carbonate dissolves in water only to the extent of about 10^{-4} moles/liter to form Ca^{2+}, CO_3^{2-} and HCO_3^- ions. Because of the rather high concentration of CO_3^{2-} ions present in the ocean, only limited amounts of Ca^{2+} ions can be present without causing $CaCO_3$ to be precipitated:

$$Ca^{2+}(aq) + CO_3^{2-}(aq) \rightleftarrows CaCO_3(s). \qquad (10–9)$$

The oceans are saturated with calcium carbonate. If a large amount of Ca^{2+} ions were added to the ocean, calcium carbonate would be precipitated. Large amounts of Ca^{2+} are carried into the oceans by rivers, and at the same time about equal amounts are deposited on the ocean floor. The major portion is deposited in the form of skeletons and shells of organisms, although under certain conditions, chemical precipitation of $CaCO_3$ occurs. The carbonate of the oceans is an important part of the carbon cycle discussed in Chapter 9. The dissolved carbonate species (CO_3^{2-}, HCO_3^-, and H_2CO_3) are approximately in equilibrium with each other and with carbon dioxide in the atmosphere and carbonates at the ocean bottom.

The pH of the oceans is possibly regulated by the reaction of the carbonate system with calcium, since the observed pH, about 8.2, is what one would predict from considering only the reactions of calcium, carbon dioxide, and the carbonates. However, this may be a coincidence, as the pH is perhaps regulated by interaction between dissolved forms of silicates in the sea and the silicates of the rocks of the sea floor.*

Calcium carbonate in the form of limestone is important in agriculture. The ability of some plants to grow depends sensitively upon the acidity of the soil. If the soil is too acidic, important legumes (e.g., peas, beans, clover, and alfalfa) show poor growth. Farmers spread limestone on the soil to neutralize the acidity by the reaction

$$CaCO_3(s) + H^+(aq) \longrightarrow Ca^{2+}(aq) + HCO_3^-(aq). \quad (10-10)$$

When carbonates are heated to very high temperatures (several hundred degrees), they decompose to form oxides and CO_2 gas, for example,

$$CaCO_3(s) \longrightarrow CaO(s) + CO_2(g). \quad (10-11)$$

Sodium bicarbonate, $NaHCO_3$, is added to the batter of bakery products to cause them to rise during baking. The CO_2 gas released by the soda forms bubbles in the cake, giving it a light texture. Sodium bicarbonate is also used as a neutralizer or "buffer" of stomach acid. The "acid indigestion" caused by excess stomach acid, HCl, can be relieved by taking some baking soda, which reacts with the stomach acid,

$$Na^+(aq) + HCO_3^-(aq) + H^+(aq) \longrightarrow$$
$$Na^+(aq) + H_2CO_3(aq), \quad (10-12)$$

thus using up the excess acid. Magnesium hydroxide, as noted above, is also used as a neutralizer of stomach acid.

10-4 HALOGENS AND HALIDES

Now let us jump to the opposite side of the periodic table and consider the elements of Group VIIA, the *halogens* $_9F$, $_{17}Cl$, $_{35}Br$, $_{53}I$, and $_{85}At$. Neutral atoms of the halogens contain seven electrons in their valence shells; thus in many of their reactions they pick up an additional electron to become 1− *halide* ions with eight electrons in the valence shell, for example, Cl^- in NaCl.

---

* For an excellent discussion of the carbonate system and other features of the chemistry of the oceans, see L. G. Sillen, "The Physical Chemistry of Sea Water," in *Oceanography*, edited by Mary Sears (American Association for the Advancement of Science, Washington, D.C., 1961).

Halogens are highly reactive because of their strong affinity for additional electrons (strongly electronegative). Fluorine, which has the highest electronegativity of all the elements, forms compounds with more other single elements than any other element. Most compounds of the noble gases contain fluorine or oxygen because of their high electronegativity. Fluorine gas is so reactive that for many years chemists could carry out measurements of it only in pure nickel or copper vessels. Now it is possible to contain fluorine in Teflon (see below).

When a pure element which is quite reactive has reacted and satisfied the requirements of its valence shell, it often becomes so "satisfied" as to be quite unreactive. Sodium metal, for instance, is extremely reactive, as we saw in Section 10-2; however, after achieving their 1+ charges, Na^+ ions are quite inert, generally being an "innocent bystander" in reactions. Likewise, fluorine becomes quite inert after reacting. One of the most chemically inert materials found in a household is Teflon. Teflon is a compound in which the hydrogen atoms of a hydrocarbon are replaced by fluorine atoms:

hydrocarbon   Teflon

(Here we show only a segment of a long molecule that contains hundreds of carbon atoms; see Chapter 13). Teflon, which is used to coat cooking and baking surfaces of pots and pans, is almost completely unreactive with any of the materials with which it comes in contact. However, if Teflon-coated pans are heated to a very high temperature (above those normally used in cooking), the Teflon can decompose to form poisonous gases. Thus, some caution must be observed in using them.

In pure form, halogens exist as diatomic molecules. Since there is little attraction between molecules, the low-molecular-weight halogens, F_2 and Cl_2, are gases at room temperature. Bromine is a volatile liquid, and I_2 occurs as solid crystals that have high sublimation pressures. Not much is known about astatine, as all of its isotopes are radioactive.

With the alkali metals and alkaline earths, the halogens form ionic compounds such as NaCl that are generally quite soluble in water. As shown in Table 10-1, chloride ions are common in both sea water and body fluids. The only im-

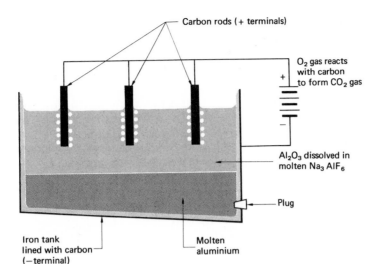

Carbon rods (+ terminals)

O$_2$ gas reacts
with carbon
to form CO$_2$ gas

Al$_2$O$_3$ dissolved in
molten Na$_3$ AlF$_6$

Plug

Fig. 10-3 Electrolysis of a fluoride melt
with dissolved Al^{3+} produces Al metal
and O$_2$ gas.

Iron tank
lined with carbon
(−terminal)

Molten
aluminium

portant function of chloride ions in the body seems to be that of neutralizing the charge of the positive ions.

In combination with hydrogen atoms, the halogens, except fluorine, form strong acids, the most prominent being hydrochloric acid, HCl. Hydrofluoric acid, HF, has the unique property of dissolving (or "etching") glass. The F$^-$ ions react with the SiO$_2$ (silica, the major constituent of sand and glass) to form hexafluorosilicate ions, SiF$_6^{2-}$:

$$SiO_2(s) + 6HF(aq) \longrightarrow$$
$$SiF_6^{2-}(aq) + 2H_2O(l) + 2H^+(aq). \quad (10-13)$$

Thus reactions involving HF must be performed in containers other than glass, e.g., polyethylene.

All the nonradioactive halogens are important either in environment or for practical uses. Various compounds of carbon, chlorine, and fluorine, such as CCl$_2$F$_2$ under the trade name Freon, are used as the circulating cooling gas in the coils of refrigerators. They are also used in aerosol cans of shaving cream, deodorant, hair spray, etc., as the propellant gas that pushes the substance out of the can.

Fluorine is a minor but important constituent of bones and teeth in the form of minerals such as apatite, Ca$_5$(PO$_4$)$_3$F. People who drink water containing about 1 ppm (part per million by weight) of fluoride generally have excellent teeth, free of cavities. Most natural water supplies contain less than that amount, so many communities add small amounts of fluoride (about one pound per 100,000 gallons) to their water systems at the water purification plants. There is little doubt that the addition of fluoride has had a beneficial effect upon the teeth of growing children. However, fluoride has become

a subject of heated political controversy in many communities. Opponents feel that local governments should not be forcing all citizens to take such "medicine." Some feel that fluoride may have harmful side effects, since concentrated sodium fluoride is used as a poison for rodents and ants.

The best evidence against the possible harmful effects of fluorides is that in some communities the water supply comes from regions where the rocks and soil have unusually large concentrations of fluorides. About one million people in the U.S. drink water containing more than 3 ppm fluoride. People have been drinking such water for generations with no apparent adverse effect except, ironically, upon their teeth. About one ppm of fluoride is the optimum. At higher concentrations the teeth become speckled with dark spots. At several ppm there may be some erosion of the teeth over the years and fluorosis of bones.

Perhaps the strongest argument against fluoridation is that it is wasteful. Only a minor fraction of a city's water is drunk by children. Fluoridated water and toothpaste do little apparent good in adults whose teeth are fully formed. It would be more efficient for children to take fluoride tablets or vitamins with added fluoride. This would provide an adequate solution for children who see a pediatrician regularly, but unfortunately many do not.

Fluoride is used extensively in the refining of aluminum. Aluminum oxide (Al$_2$O$_3$) is obtained from the aluminum ore *bauxite* and dissolved in a molten mineral containing fluorides of sodium aluminum, and sometimes calcium (see Fig. 10-3). The molten mass is electrolyzed and at the negative terminal (essentially, the walls of the container), electrons

A bank of three silver iodide generators is used to seed clouds passing over the Rocky Mountains in Colorado. The generators burn silver iodide mixed in acetone and propane and burn with the sound of a blowtorch. (U.S. Dept. of the Interior Bureau of Reclamation photo by W. L. Rusho.)

are added to the Al^{3+} ions to form molten metallic aluminum:

$$Al^{3+} + 3e^- \longrightarrow Al(s). \qquad (10-14)$$

Oxygen gas is the major product at the positive terminal, but some fluorides can be carried along with it. Most of the fluoride (about 99%) is trapped in the plant's stack, but the amount that escapes can be significant. Atmospheric concentrations of fluorides in the vicinity of aluminum plants are sometimes quite high and may damage local vegetation.

Pure chlorine, Cl_2, and many of its compounds are used extensively as household bleaches and municipal water disinfectants. Both of these uses result from chlorine's action as an oxidizing agent, as discussed below. Chlorine is a poisonous gas and has been used in warfare. However, the more common chlorine-containing poison is phosgene, $COCl_2$. Most persistent pesticides such as DDT contain chlorine (see Section 13–12).

Compounds of halogens, usually bromine with silver, are used in photography. Black-and-white film consists of many small grains of silver halides suspended in a gelatin. When light strikes a grain of silver bromide, a chemical reaction is initiated—probably the liberation of free metallic silver and bromine atoms.

$$AgBr(s) \xrightarrow{\text{light}} Ag(s) + Br(g). \qquad (10-15)$$

The bromine atom probably escapes, leaving a free Ag atom in the grain. The remarkable, and as yet poorly understood feature of these silver halide crystals is the fact that only about four or five of the billion or so Ag^+ ions of a grain need to react in order to "activate" the entire grain (see Fig. 10–4).

In order to develop the latent image (convert it to visible form), the film is treated with a developer, generally hydroquinone. Hydroquinone converts activated grains into pure silver metal, thus forming the negative image on the film.

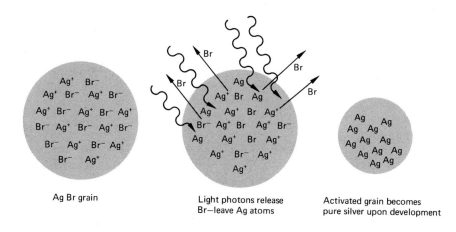

Ag Br grain

Light photons release
Br—leave Ag atoms

Activated grain becomes
pure silver upon development

Fig. 10–4 When several photons of light strike a grain of AgBr on a photographic plate, a few atoms of metallic silver are formed, thus "activating" the grain, that is, making it possible for the entire grain (containing about a billion AgBr molecules) to be converted to silver when the film is developed.

(The negative has dark deposits of silver wherever light has struck the film. In order to convert the picture to the positive form, one must expose photographic paper to the negative.) The fact that only a few atoms of silver must be formed to activate the grain containing a billion Ag^+ ions means that we obtain an amplification factor of one billion in the light effect upon the film. Only because of this amplification are we able to have "fast" films that can record a picture in a fraction of a second.

Although the concentration of bromine in sea water is only 0.065 g/l, it is one of the four elements (along with sodium, magnesium, and chlorine) for which sea water is a major commercial source. The principal use of bromine is in leaded gasolines. Organic compounds containing chlorine and bromine are added to gasoline along with tetraethyl lead in order to prevent the formation of lead deposits in automobile engines (see Section 14–3).

Iodine (an alcohol solution) is occasionally used as a disinfectant for simple wounds. Iodine also is among the elements that are essential for human life. It is needed mainly by the thyroid gland for synthesis of hormones such as thyroxine, $C_{25}H_{11}O_4NI_4$. Deficiencies of iodine in the diet can cause an enlargement of the thyroid gland, a condition called "goiter." Goiter is an *endemic* disease, meaning a disease that has high incidence among people in certain populations. Iodine deficiencies occur particularly in mountainous areas and in

regions that have been subjected to extensive glaciation (e.g., the states around the Great Lakes), where the soil is deficient in iodine.* Goiters are usually rare near sea coasts, as seaweed, fish and marine aerosols (which fall onto the land) are high in iodine. Goiters also occur among people whose diets include large amounts of such foods as brussels sprouts, cabbage, turnips, and soybeans, all of which contain compounds that tie up iodine in the body, preventing its absorption into the bloodstream. Since the 1920s, most table salt sold in the United States has contained trace amounts of sodium iodide to make up for any dietary iodine deficiencies.

10–5 OXIDATION AND REDUCTION REACTIONS

Before discussing some more complicated halogen compounds, let us consider the concepts of *oxidation* and *reduction* in more detail (see Section 7–6). There are two general types of chemical reactions: (1) those in which there is no transfer of electrons, and (2) oxidation-reduction reactions, "redox" for short, in which there is a transfer of sharing of electrons. The former category includes those reactions in which dissolved ions come together to from undissociated molecules or insoluble solids, or the reverse occurs.

* R. B. Gillie, "Endemic Goiter," *Scientific American* **224** (6) 92–101 (June 1971).

Originally the term *oxidation* meant the combination of a substance with oxygen as, for example, occurs when coal is burned in air:

$$C(s) + O_2(g) \longrightarrow CO_2(g) \tag{10-16}$$

Another example is that of the oxidation (rusting) of iron:*

$$4Fe(s) + 3O_2(g) \longrightarrow 2Fe_2O_3(s). \tag{10-17}$$

In modern chemistry, the term *oxidation* has come to denote a more general class of reactions, in many cases not involving oxygen at all. In oxidation-reduction reactions there is a change in the electrons, either a complete transfer as when a neutral atom is ionized or a change of electrons shared with other atoms. An example of the former is the reaction of sodium metal with chlorine gas in which one electron is transferred from each sodium atom to a chlorine atom. An example of a redox reaction involving shared electrons is that between hydrogen and oxygen gas to form water:

$$2H_2(g) + O_2(g) \longrightarrow 2H_2O(g). \tag{10-18}$$

Each hydrogen atom, which formerly shared a pair of electrons with another hydrogen, now shares a pair of electrons with an oxygen atom.

In order to better understand redox reactions, let us extend the concept of oxidation number (or valence) introduced in Section 7-6. We should realize that although the concept is a useful scheme for keeping track of electrons, we should not interpret the oxidation numbers too literally. For simple ions such as Li^+, Sr^{2+}, and I^-, the oxidation number is just the charge of the ion. For compounds in which there is covalent bonding, oxidation numbers can be assigned on the basis of the following rules.

1. Pure elements are assigned oxidation number zero, regardless of the molecular form of the element. For example, oxygen atoms, whether O, O_2, or O_3 (ozone), have oxidation number zero.

2. Hydrogen in compounds is assigned oxidation number 1+ unless it is in combination with a more electropositive element, as in CaH_2 (calcium hydride), in which we would consider the hydrogen to be 1−.

3. Oxygen atoms in compounds are assigned oxidation number 2−, unless the oxygen atoms are attached to each

other, as in peroxides, e.g., hydrogen peroxide, $H-O-O-H$.

4. The algebraic sum of the oxidation numbers of the atoms in the formula for a compound or complex ion (a charged group of atoms such as SO_4^{2-}) must equal the observed charge of the species.

Now let us apply these rules in a few examples. Calcium carbonate, $CaCO_3$, breaks up in solution to form Ca^{2+} and CO_3^{2-} ions. Using the rules given above, Ca^{2+} has oxidation number 2+, and so the oxidation numbers of the carbon and oxygens must sum to 2−. We assign each oxygen a 2− oxidation number, for a total of 6−, so the carbon must be 4+ to yield the net value of 2− for the carbonate ion:

```
1C @ 4+    = 4+
3O @ 2−    = 6−
Net charge = 2−
```

The rules may seem to give a result that is not consistent with the true condition of the atom, i.e., the carbon is *not* a C^{4+} ion nor are the oxygens O^{2-} ions. Nevertheless, the concept of oxidation number is useful in balancing chemical reactions.

In Na_2SO_4, the sodiums are assigned 1+ values. Since the sulfate ion, SO_4^{2-}, has a 2− charge, the sulfur must be assigned a 6+ oxidation number, that is,

```
1S @ 6+    = 6+
4O @ 2−    = 8−
Net charge = 2−
```

but this does *not* imply the existence of S^{6+} ions. If we replace one of the oxygen atoms of SO_4^{2-} with a sulfur atom, we obtain the *thiosulfate* ion, $S_2O_3^{2-}$, in which the prefix *thio* indicates the replacement of sulfur for oxygen:

sulfate ion   thiosulfate ion

In thiosulfates, the oxidation number of the sulfur atoms is 2+:

```
2S @ 2+    = 4+
3O @ 2−    = 6−
Net charge = 2−
```

---

*   This is a highly simplified representation for the rusting of iron, a process that is so complex that it is not yet fully understood.

We saw in Table 7–2 that some of the transition metals can have oxidation numbers of 6+ or 7+, although they do not form simple ions with such high charges. Examples of such high oxidation numbers are the chromium in potassium chromate, K_2CrO_4, and manganese in potassium permanganate, $KMnO_4$:

Chromate ion, CrO_4^{2-}

1Cr @ 6+ = 6+
4O @ 2−   = 8−
Net charge = 2−

Permanganate ion, MnO_4^-

1Mn @ 7+ = 7+
4O @ 2− = 8−
Net charge = 1−

Now that we have established the concept of oxidation numbers more precisely, we can define redox reactions as those in which the oxidation number of at least one element is changed during the reaction. Atoms whose oxidation numbers are raised are said to be oxidized and those which are decreased are reduced. Thus in the reaction

$$2Rb(s) + Br_2(g) \longrightarrow 2Rb^+ + 2Br^-, \tag{10-19}$$

rubidium is oxidized from the zero state to the 1+ state by bromine. We would say that Br_2 is an oxidizing agent. In general, an oxidizing agent takes away electrons and, in the process, is itself reduced.

Likewise, in reaction (10–19), the bromine is reduced from oxidation number zero to 1−. We would say that rubidium is the reducing agent. A reducing agent gives up electrons to the other reactant and is itself oxidized. Although it is not the custom, one could also refer to reaction (9–19) as a reduction reaction, as the bromine is reduced.

We can see that the modern definition of oxidation also encompasses the older, more limited definition. For example, in the burning of coal, reaction (10–16), carbon is oxidized from oxidation number 0 to 4+. Similarly, the rusting of iron, reaction (10–17), raises the oxidation number of iron from 0 to 3+. Oxidation and reduction reactions are an important feature of the chemical behavior of many elements, including the halogens. Strong oxidizing agents often react vigorously with other substances and thus are frequently used in explosives and rocket fuels. Less vigorous oxidizing agents are used in the household as disinfectants and bleaches.

10–6 OXYGENATED HALOGEN COMPOUNDS

In addition to the 1− halide ions, the halogens also form a wide variety of ions in combination with oxygen. Chlorine, for example, forms the series of ions listed in Table 10–2.

Table 10–2 Oxygenated Ions of Chlorine

Name of ion	Formula	Electron dot structure	Chlorine oxidation state
Hypochlorite	ClO^-		1+
Chlorite	ClO_2^-		3+
Chlorate	ClO_3^-		5+
Perchlorate	ClO_4^-		7+

Thus chlorine can have oxidation numbers all the way from the 1− of chloride up to the 7+ of perchlorate. All these species with oxidation numbers greater than 1−, including Cl_2 with oxidation number zero, can act as oxidizing agents, either by actually releasing oxygen or by taking electrons away from another species and dropping down to the Cl^- state. It is the oxidizing power of Cl_2 and higher oxidation states that make them good disinfectants and bleaches. Chlorine is also used to kill bacteria in municipal water systems and in swimming pool water. Oxidation of the molecules of bacteria apparently renders them harmless.

The standard laundry bleach is an aqueous solution of sodium hypochlorite, NaClO, marketed under various trade names such as Clorox. Most colored organic compounds have molecules involving resonant structure with excess electrons (see Chapter 13). Oxidizing agents such as ClO^- oxidize them by removing some of the electrons, leaving them decolorized.

The oxyhalogen ions form acids with H^+ ions. Perchloric acid, $HClO_4$, is both a strong acid and a strong oxidizing agent. It is a dangerous chemical as it reacts vigorously, even explosively, with wood or other organic compounds that can be oxidized.

The potassium salt of chloric acid, $KClO_3$ (potassium chlorate), is a standard source of oxygen gas in the laboratory. By heating $KClO_3$, oxygen is given off:

$$2KClO_3(s) \xrightarrow{\text{heat}} 2KCl(s) + 3O_2(g). \tag{10-20}$$

The reaction goes much faster if we add some manganese dioxide, MnO_2, which acts as a catalyst.

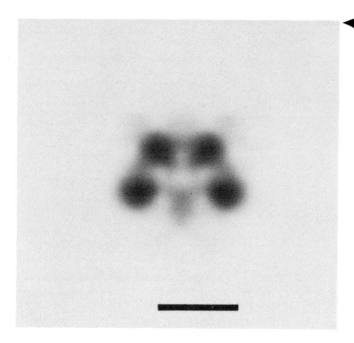

Sulfur atom, in a remarkable electron micrograph obtained by scientists at the Ontario Cancer Institute, Toronto. As shown here, the image is magnified about 22 million times! The scale marker represents 10 angstroms. (Courtesy of F. P. Ottensmeyer.)

$$SO_2(g) + H_2O(l) \longrightarrow H_2SO_3(aq) \text{ (sulfurous acid)},$$
$$(10-24)$$
$$SO_3(g) + H_2O(l) \longrightarrow H_2SO_4(aq) \text{ (sulfuric acid)}.$$

Sulfuric acid is an important strong acid both in the laboratory and in the chemical industry.

Oxygen and sulfur combine with a wide variety of other elements. Many elements such as iron combine with oxygen to form oxides, in this case rust, Fe_2O_3. The reaction of oxygen with some metals forms thin oxide layers that protect the metal from further attack by the air. For example, a clean surface of aluminum metal reacts quickly with the oxygen of the atmosphere to form a thin oxide layer that protects the bulk of the metal from further reaction.

$$6Al(s) + 3O_2(g) \longrightarrow 2Al_2O_3(s). \tag{10-25}$$

Deposits of some metals occur naturally as sulfides, e.g., copper (Cu_2S, chalcocite), silver (Ag_2S, argentite), zinc (ZnS, zinc blende), mercury (HgS, cinnabar), and lead (PbS, galena), to name but a few. In order to obtain the metals in pure form, one must remove the sulfur. First, the ore is separated from rocky minerals by grinding to small particle size and "floating" with wetting agents in a stream of air bubbles. Then the sulfide is "roasted" by heating it in a stream of air. In some cases, roasting yields the metal directly, for example,

$$Cu_2S(s) + O_2(g) \longrightarrow SO_2(g) + 2Cu. \tag{10-26}$$

Copper produced in this way contains impurities from other metallic sulfides present in the chalcocite. They are removed by dissolving the impure copper in acid and then passing a current through the solution to deposit the copper electrolytically. Roasting of some other sulfide ores leaves the oxide of the metal, for example,

$$2ZnS(s) + 3O_2(g) \longrightarrow 2ZnO(s) + 2SO_2(g). \tag{10-27}$$

The oxide must then be reduced to obtain the pure metal. Often this is done by using carbon (in the form of coke) as the reducing agent:

$$ZnO(s) + C(s) \longrightarrow Zn(s) + CO(g). \tag{10-28}$$

Reactions such as (10–26) and (10–27) are major sources of

10–7 OXYGEN AND SULFUR

Oxygen and sulfur have six electrons in their outermost shells; thus one of their important valence states is 2−. With hydrogen, for example, they form water, H_2O, and hydrogen sulfide, H_2S. The latter is a gas that smells like rotten eggs. In water it is a weak acid, dissociating just as water does:

$$H_2S(aq) \rightleftarrows H^+(aq) + HS^-(aq). \tag{10-21}$$

Sulfur and oxygen form a number of important compounds and ions in combination with each other. When sulfur is burned in air, as in the head of a match, sulfur and oxygen combine to form sulfur dioxide:

$$S(s) + O_2(g) \longrightarrow SO_2(g). \tag{10-22}$$

In the presence of more oxygen and catalysts such as vanadium pentoxide, V_2O_5, sulfur dioxide can be oxidized to sulfur trioxide:

$$2SO_2(g) + O_2(g) \xrightarrow{V_2O_5} 2SO_3(g). \tag{10-23}$$

Sulfur dioxide and SO_3 gases are called acid anhydrides because they combine with water to form acids.

Acid mine water flowing from an old anthracite mine tunnel in Shamokin, Pennsylvania. The bottom of the stream is coated with iron and sulfur. (Courtesy of the U.S. Dept. of Agriculture.)

atmospheric sulfur dioxide in areas in which there is extensive metal refining.

Sulfur compounds are water pollutants, particularly in the form of "acid mine drainage." Many types of mines contain metallic sulfides; for example, most coal contains several percent of sulfur in the form of sulfides such as pyrite, FeS_2. Certain microorganisms catalyze the oxidation of pyrite by oxygen from the air, which raises the sulfur to the 4+ and 6+ oxidation states, that is, SO_3^{2-} and SO_4^{2-}, forming sulfurous and sulfuric acids. Water draining from sulfur-containing mines is often highly acidic (pH as low as 2 or 3) and it pollutes the streams and rivers into which it drains. Acid mine drainage is a serious water pollution problem in mining areas, particularly in the Appalachian regions of the eastern United States. Acidic water is detrimental to fish and other life forms of a river; few species can exist when the water is more acidic than pH 5.

No satisfactory solution to the acid mine drainage problem, has been devised. It is not generally possible to seal all outlets from the mine into the general ground water system. When excessive acid drainage poses a large threat to aquatic life downstream, the acid is sometimes neutralized with lime, CaO. Not only is this an expensive process, but if the mine contains much iron, neutralization of the water causes formation of "yellow boy," a yellow-brown precipitate of ferric hydroxide that destroys aquatic life by coating it.*

---

* T. Aaronsen, "Problems Underfoot," *Environment* **12** (9), 17 (1970).

Dusting sulfur on grapevines to retard mildew. (EPA—Documerica photo by Gene Daniels.)

10-8 AIR POLLUTION BY SULFUR COMPOUNDS

In Chapter 9 we discussed global air pollution problems and possible effects of cities on their own climates. We now consider some more localized air pollution effects, mostly of urban areas, in which sulfur oxides play an important role.

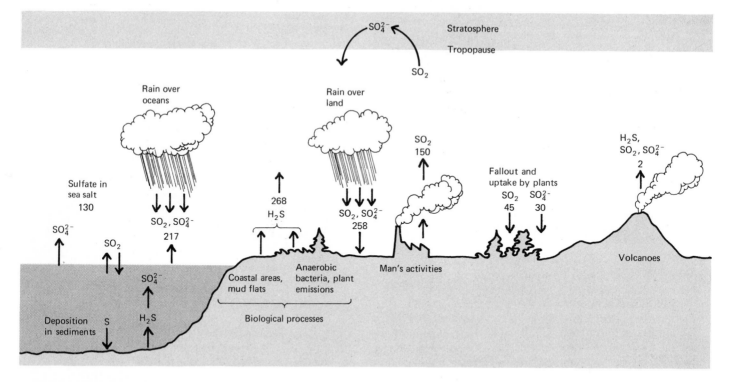

Fig. 10–5 Estimated amounts of sulfur (in millions of metric tons/yr) added to and removed from the atmosphere by various natural and man-made processes. [Adapted from W. W. Kellogg, R. D. Cadle, E. R. Allen. A. L. Lazarus, and E. A. Martell, "The Sulfur Cycle," *Science* **175**, 587–596 (1972).]

The major problem of urban air pollution is the heavy concentration both of sources of air pollution and of people breathing the pollutants. If there were always strong winds blowing through our cities, they would bring in fresh air and simultaneously flush pollutants out, dispersing and diluting them in the air of the countryside. Students are often surprised to learn that, in the case of most air pollutants, natural processes introduce as much of the "pollutant" into the atmosphere as man's activities. Figure 10–5 shows estimates of the amounts and chemical forms of sulfur added to and removed from the atmosphere each year by various processes. We see that man's activities contribute only about 30% of the total sulfur injected into the atmosphere. Natural sources of atmospheric sulfur are generally spread over large areas far from population centers, e.g., sulfates from sea spray, H_2S from biological decay. Because they are so dispersed, the natural forms of these pollutants usually do little harm to people, animals, and vegetation. Likewise, urban air pollution would not be very serious if the pollutants could be rapidly dispersed over the countryside. However, this would not be a complete solution. In many areas of Europe, the rainfall has become acidic in recent years, probably because of man-made SO_2 in the atmosphere.

Air pollution problems could be cleaned up quickly if strenuous measures were taken. Most pollutants are returned to the surface by settling, rain, and snow after residence times of only 6 to 10 days. The lower atmosphere would return to an almost unpolluted state within a few weeks if all air pollution activities were suddenly halted. By contrast, it would take many years, in some cases centuries, to return large polluted water bodies to their natural state.

The most serious air pollution problems arise in cities when the air remains almost motionless for several days. Stagnation of the atmosphere is often caused by a *temperature inversion.* Whereas the atmosphere normally gets cooler at higher altitudes, in an inversion, it does just the opposite—it gets warmer (see Fig. 10–6). This is a very stable situation—warm air, which is less dense, above cold dense air. Pollutants are trapped below the layer of warm air. If wind speeds are low, there is also little horizontal movement of the pollutants. If stagnation continues for several days, concentrations of pollutants in the city air may build up to alarming levels, with serious consequences to the inhabitants.

The classic example of SO_2 and particulate pollution is the city of London.* Until recently, most buildings were heated by burning soft coal, which produced large quantities of sulfur oxides, particulates, and other pollutants. As long as a strong wind was blowing, the pollutants did little obvious damage. However, London occasionally has thick fogs lasting for a week or more which trap the pollutants, building up their concentrations to the disaster level.

In 1952, a thick fog enveloped London from the fifth to the eighth of December. Analysis of vital statistics showed that there were about 4000 more deaths in London than would normally have occurred during that period. Another pollution episode, resulting in about 1000 excess deaths, occurred in 1956. These incidents shocked the people of London into action. The government has since required the cleanup of smokestack emissions and has subsidized the switch from soft coal to less polluting fuels throughout London. The program is beginning to pay off. The severity of air pollution episodes has greatly decreased and even the weather has improved. Apparently the 80% reduction in particulate material in the atmosphere has increased the amount of sunlight reaching the ground by about 50%, and the incidence of fog and precipitation has been reduced.†

Similar air pollution incidents have occurred in the United States. In 1948 a four-day fog trapped zinc sulfate and sulfur oxides in the small mill town of Donora, Pennsylvania. About half of the 14,000 residents became sick, and 20 died. Often in recent years the entire Eastern Seaboard has been covered by stagnant air for several days at a time. One of the worst smogs, during Thanksgiving week 1966, caused 100 or more excess deaths in New York City alone.‡

---

*   E. Iglauer, "A Reporter at Large—The Ambient Air," *The New Yorker*, April 13, 1968.
†   Gordon Young, "Threat to Man's Only Home," *National Geographic* **138**, 739 (1970); see also "Out of the Fog," *Time Magazine*, Dec. 12, 1969, p. 57.
‡   Iglauer, "The Ambient Air."

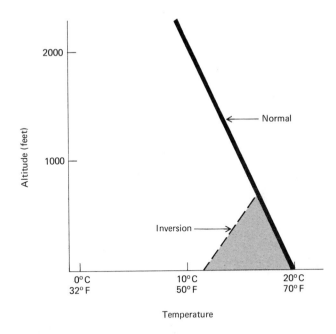

Fig. 10–6 Normally, air gets colder at higher altitudes, but in an inversion, air temperature increases with altitude.

Some of the most serious pollutants are the oxides of sulfur. Table 10–3 lists the total weight of the major air pollutants produced in the United States in the year 1970. By weight, carbon monoxide is the most prominent air pollutant, but sulfur oxides probably present a more serious problem. Large quantities of sulfur dioxide, SO_2, are produced in cities of the North and East by burning coal and residual fuel for heat and electricity. Residual fuel makes up much of the "left-overs" from crude oil after all the more volatile fractions (gasoline, kerosene, etc.) have been boiled off (see Chapter 14). Residual fuel (often called "Bunker-C" or "#6 oil") is almost like tar, being so viscous at room temperature that it must be heated to get it to flow through pipes. Both coal and residual fuel normally contain several percent of sulfur, which is oxidized to SO_2 during combustion. We see in Table 10–4 that combustion of fossil fuels accounts for about three-fourths of the sulfur oxides released in the United States. Most of the remainder comes from industrial operations, especially the smelting of sulfide ores and the refining of petroleum.

Table 10–3   U.S. Air Pollutant Emissions in 1970

Source	Emission (millions of tons/yr)				
	Carbon monoxide	Particulates	Sulfur oxides	Hydrocarbons	Nitrogen oxides
Transportation	111.0	0.7	1.0	19.5	11.7
Fuel combustion for heat and electric power generation	0.8	6.8	26.5	0.6	10.0
Industry	11.4	13.1	6.0	5.5	0.2
Solid waste disposal	7.2	1.4	0.1	2.0	0.4
Miscellaneous	16.8	3.4	0.3	7.1	0.4
Total	147.2	25.4	33.9	34.7	22.7

Council on Environmental Quality, *Third Annual Report*, Aug. 1972 (U.S. Government Printing Office, Washington, D.C.).

Table 10–4   U.S. Emissions of Sulfur Oxides in 1968

Source	Emissions	
	Millions of tons/yr	Percent of total
Transportation	0.8	2.4
Heat and electric power generation	24.4	73.6
Coal	20.1	60.6
Fuel oil	0.4	1.2
Residual oil	3.9	11.8
Industry	7.2	21.9
Petroleum refineries	2.1	6.4
Smelters	3.9	11.9
Sulfuric acid plants	0.6	1.8
Coking operations	0.6	1.8
Other	0.7	2.1

Natural Air Pollution Control Administration, *Nationwide Inventory of Air Pollutant Emissions 1968*, NAPCA Publication AP–73 (U.S. Government Printing Office, Washington, D.C.).

Sulfur dioxide injury to a white birch leaf is shown on the left. The leaf on the right is normal. (Courtesy of the U.S. Dept. of Agriculture.)

What are the effects of sulfur dioxides in the atmosphere? Sulfur dioxide, SO_2, at concentrations of 700 $\mu g/m^3$ or more is very harmful to the leaves or needles of many plants and trees. Throughout the United States there are areas where all vegetation on entire fields or mountainsides downwind from large smelters or power plants has been killed by the pollutant gas.

The more serious problems caused by sulfur are probably those that arise after the SO_2 is oxidized to SO_3:

$$SO_2(g) + [O] \longrightarrow SO_3(g). \qquad (10\text{--}29)$$

We have written the reaction with oxygen in brackets because the mechanism by which SO_2 is oxidized in the atmosphere is complicated and poorly understood. Some SO_2 may be oxidized by gases such as ozone (O_3), but some undoubtedly reacts with oxygen dissolved in water on moist particles in the atmosphere.

Most of the SO_3 reacts with water droplets or moist particles to form sulfuric acid:

$$SO_3(g) + H_2O(l) \longrightarrow H_2SO_4(aq). \qquad (10-24)$$

The sulfuric acid may react further with ammonia (NH_3) to form $(NH_4)_2SO_4$ or with metal oxides to form sulfates. Sulfuric acid droplets are very damaging to material surfaces. Many building facades are made of limestone or marble, which are forms of $CaCO_3$. When an acidic droplet falls on such a surface, the reaction

$$CaCO_3(s) + H_2SO_4(aq) \longrightarrow CaSO_4(s) + CO_2(g) + H_2O(l)$$
$$(10-30)$$

can cause a pitting or flaking of the surface. Famous landmarks such as the Lincoln Memorial in Washington, D.C., are suffering severe damage from air pollution effects of this sort. The Greek ruins on the Acropolis in Athens have suffered more deterioration during the past 40 years than during the previous 20 centuries! Direct costs of air pollution effects on materials and vegetation in the United States are estimated to be $5 billion annually, a large fraction of which is attributed to sulfur oxide problems.* The acidic droplets are eventually brought to ground by rain, where they contribute to the acidity of soil and ground water. As the release of sulfur oxides has increased in recent years, the pH of rainfall over northern Europe has gradually decreased, occasionally reaching pH 4 or less.

Our ultimate concern with air pollutants is their effect on human beings. How do sulfur oxides affect breathing? Apparently the oxides themselves or the acids cause irritation of the small passages in the lungs where the air is separated from the bloodstream only by thin membranes. Irritation closes some of the passages and reduces the capacity of the lungs to exchange O_2 and CO_2. These effects, especially on people who already have severe, chronic respiratory ailments, probably account for a majority of the excess deaths observed during air pollution episodes. The effects of long-term exposure to sulfur oxides and acid mists in healthy persons is not well established, but they are surely adverse.

To reduce the damages of air pollution to health and materials, we must lower the concentrations of sulfur oxides in the atmosphere. Following passage of the Air Quality Acts of 1967 and 1970, about 250 federal air quality control regions were established in the United States. During 1972, the individual states established air quality standards for a number of air pollutants and devised plans for keeping pollutant concentrations below the limits specified.

In the case of SO_2, the U.S. Environmental Protection Agency (EPA) has recommended that the annual average concentration of SO_2 be kept below 80 $\mu g/m^3$ (0.03 ppm) and that the 24-hr average concentration not exceed 365 $\mu g/m^3$ (0.14 ppm) more than once a year.† (Appendix D discusses the relationship between ppm and $\mu g/m^3$).

Plans for strict implementation will have to be devised in some areas to bring the SO_2 levels below these values, particularly in the highly industrialized states of the Northeast, which produce nearly half of the total SO_2 in this country.‡ Average SO_2 concentrations during 1968 (before much clean-up of sources was started) in several U.S. cities were: Chicago, 307 $\mu g/m^3$; Cincinnati, 44; Denver, 34; Philadelphia, 212; St. Louis, 73; Washington, D.C., 97; and New York, about 350. During periods of stagnant air, SO_2 concentrations rise far higher. During the Thanksgiving 1966 episode in New York City, the SO_2 concentration rose as high as 2600 $\mu g/m^3$ for two brief periods.§

A Supreme Court decision in 1973 may have more far-reaching effects on air pollution than the regulations noted above.‖ The court upheld a lower court decision that the Environmental Protection Agency must not permit deterioration of air quality even in areas where pollutant levels are far below the guideline values noted above. The effects of this decision will probably be felt most strongly in the sparsely

---

*   Council on Environmental Quality, *Third Annual Report*.

---

†   S. S. Miller, "National Air Quality Standards Finalized," *Environmental Science Technology* **5**, 503 (1971).

‡   NAPCA, *Nationwide Inventory of Air Pollution Emissions, 1968*.

§   U.S. Dept. of Health, Education and Welfare, Public Health Service, National Air Pollution Control Administration, "Thanksgiving 1966 Air Pollution Episode in the Eastern United States" (U.S. Government Printing Office, Washington, D.C., 1968).

‖   J. P. Mackenzie, "Court Bars Pollution of Cleanest Air," Washington *Post*, June 12, 1973.

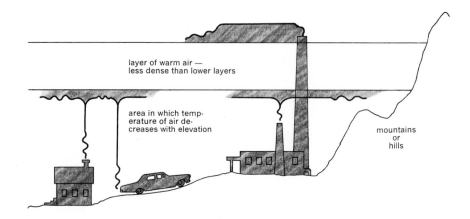

Fig. 10–7 Tall stacks may release pollutants above inversion layers, thus preventing the pollutants from reaching ground level in the vicinity.

populated areas of Western states that now have very clean air. Industrial developers in these areas fear that the ruling will prevent industrial growth there. Especially hard hit will be the huge power plants being built near large deposits of low-grade coal in order to supply electricity mainly for the more populated areas near the West Coast.

What can we do to reduce the amount of sulfur oxides in the atmosphere of our cities? Since the oxides come mainly from the burning of coal and residual oil for heat and electricity, we could shift to nuclear reactors as our major energy source. This would be expensive, however, and would take many years. Furthermore, although reactors are chemically very "clean," we have seen that they have a number of adverse side effects. Natural gas produces very little SO_2, particulates, or other pollutants, but our gas reserves are much too small for gas to be considered a major alternative.

A partial solution consists of building very tall smokestacks that release effluents at such a high altitude that most gases and particles leave the urban area before coming to the ground. Some of today's fossil-fuel power plants have stacks several hundred or even 1000 feet tall, so that their emissions are usually released above inversion layers (see Fig. 10–7). Tall stacks must be viewed as a provisional solution, but they could alleviate urban air pollution crises. Air pollution far downwind from the stack would be increased, but to less serious levels than in cities, as the pollutants would disperse gradually as they move away from the stack.

Many areas have established limits on the sulfur content of the fuels that may be burned there. For example, during 1970 the New York–New Jersey Metropolitan Area restricted fuels for existing power plants to 1% sulfur, and in 1971 the limit for soft coal dropped to 0.3% sulfur. In regions such as New York–New Jersey, where these measures were first taken, air quality has begun to show some improvement.*

Before restrictions on the sulfur content of fuels can become an effective solution throughout the land, means for removing sulfur from coal or trapping SO_2 in smokestacks will be needed, because supplies of natural low-sulfur fuels are not sufficient to meet the need. Sulfur can be removed from residual oil at modest cost, but not enough oil is available to replace all the fossil fuels being used.

As shown in Tables 4–1 and 4–2, projected needs for coal are so great that we must develop methods for using coal with a minimum release of SO_2. The sulfur content of most coals ranges as high as 6 to 8%, about half as pyrite (FeS_2) and half as complex organic materials. The former can be removed rather readily, but the latter are difficult to remove.

The search is now being pressed for methods of trapping SO_2 from coal combustion before it leaves the plant stack. About 20 different schemes for the removal of SO_2 from effluent gases are under investigation. One of the simplest involves the injection of limestone, $CaCO_3$, into the combustion chamber along with the coal. The idea is that SO_2 will react with the limestone to produce $CaSO_4$ and harmless CO_2:

$$2CaCO_3(s) + 2SO_2(g) + O_2(g) \longrightarrow 2CaSO_4(s) + 2CO_2(g).$$
$$(10\text{–}31)$$

---

* Council on Environmental Quality, *Third Annual Report*

Unfortunately, it is found that the limestone particles form a coating of $CaSO_4$ that prevents reaction of the $CaCO_3$ inside the particles.*

Many of the proposed systems involve "scrubbers" which spray water or a water solution into the effluent gases to trap the sulfur oxides, eventually converting them to sulfuric acid. One would suppose that this would serve as an excellent opportunity to recycle a "resource-out-of-place." However, the concentration of the sulfuric acid produced is too low for most industrial uses. Many large sources of SO_2 are located so far from the markets for sulfuric acid that the cost of transportation is greater than the value of the product. Nevertheless, if we were able to convert all SO_2 from combustion and smelter sources into sulfuric acid, the amount produced would exceed that now consumed by United States industry.

Perhaps ultimately the solution of the coal emissions problem (as well as that of the shortage of natural gas) will involve the conversion of coal to gas. Many possible processes are now under investigation.† From the standpoint of both the energy crisis and of the need to reduce air pollution, research on gasification of coal must be considered one of our nation's highest priorities.

10–9 AIR POLLUTION BY PARTICULATE MATERIAL

Another major air pollutant of cities and other industrial areas is particulate material. In Chapter 9 we discussed the possible effects of particles from human activities on global and regional climate. Here we consider the effects of particles on humans and materials.

Atmospheric particulates result from many natural and man-made processes. Typically, the concentration of particles from natural processes in nonurban, continental areas is about 5 to 50 $\mu g/m^3$. Man's activities produce much higher concentrations of particulate matter around power plants, cement factories, metal refining plants, and in cities generally. Most sizable cities have particulate concentrations of about 80 $\mu g/m^3$ in normal times, which may rise to 1000 $\mu g/m^3$ or more during periods of stagnant air.

The size of pollutant aerosols depends on their mode of formation. Automobile exhaust particles are quite small— 90% are of less than 0.5 μm diameter. At the other extreme, the burning of coal releases very large particles called "fly ash," largely because of the presence of rocky fragments of silicate minerals that do not burn.

The importance of the different sources of particulates varies greatly from city to city. In Table 10–5, we see that heat and power generation account for more than half of the particulate matter in the atmosphere of New York and Washington, whereas in Los Angeles, about half results from transportation, most of it from automobiles.

Man-made particulates have many effects on the environment. Particulate matter is dirty—it falls on buildings, automobiles, laundry on the line, and executives' white shirts, greatly increasing the nation's cleaning bill. In many cities, dustfall amounts to about fifty tons/mi² every month, or about 600 tons/yr.

What happens when we breathe air containing large concentrations of particulates? Fortunately, the body has several lines of defense against aerosols. The hairs of the nostrils, the wetness of membranes, and the sizes and shapes of passages from the nose to the lungs are all designed to trap many particles that would otherwise be deposited in the lungs. These defenses remove nearly all particles of diameters greater than 1 μm but smaller particles do get into the lungs.

The air passages in the lung divide into a multitude of smaller and smaller branches, terminating in the *alveolar* (air) *sacs*. There are about 300 million of these chambers, 150 to 400 μm in diameter, which provide a huge surface area (about 50 m², nearly the size of a tennis court) for exchange of oxygen and carbon dioxide between the air and the blood (see Fig. 10–8). In healthy individuals, most of the particles with a diameter less than 1 μm that reach the lungs are removed by a secondary defense system. Lung surfaces are covered with hairlike structures called *cilia*. These wave back and forth moving along the current of *mucus*, the sticky substance that traps most particles, and flushes them out of the respiratory system and into the gastrointestinal tract for transport out of the body.

These defense mechanisms efficiently rid the lung of foreign particles unless it has been damaged or the particles have an unusual ability to "dig into" it. Years of "insult," such as cigarette smoking or exposure to SO_2 or sulfuric acid mist, break down the lungs' defense mechanisms. The passages may become constricted, the flow of mucus and the "beat" of the cilia may be reduced. With defenses weakened, the lungs

---

*   A. M. Squires, "Capturing Sulfur during Combustion," *Technology Review* **74** (2), 52–59 (1971).

†   Ibid, See also J. P. Henry, Jr., and B. M. Louks, "An Economic Study of Pipeline Gas Production from Coal," *Chemical Technology*, Apr. 1971, pp. 238–247.

Table 10–5   Important Sources of Particulate Matter in Three Major U.S. Metropolitan Areas

Source	Percentage contribution to particulates of indicated city		
	New York–New Jersey	Washington, D.C.	Los Angeles
Power generation	17.3	28.5	10.5
Coal	13.7	28.4	—
Oil	3.3	0.1	—
Industrial energy	14.5	1.0	1.6
Home heating	17.8	9.1	6.6
Coal	7.7	2.1	—
Oil	9.3	5.3	—
Gas	0.7	1.7	—
Commercial and govt.	8.5	16.8	Incl. with homes
Refuse disposal	18.0	23.4	0.8
Transportation	15.2	18.0	47.0
Industrial processes	8.6	3.2	33.5
Total tons/yr	231,000	34,800	44,300

National Air Pollution Control Administration, "Air Quality Criteria for Particulate Matter," NAPCA Publication No. AP–49, Jan. 1969.

are more susceptible to damage from inhaled particles and gases. Under these conditions, lungs are subject to attack by particles containing *carcinogenic* compounds, i.e., those which are thought to induce cancer.* Most of the suspected carcinogens in the atmosphere are organic compounds containing "aromatic" rings of carbon atoms bonded together (see Chapter 13). Such compounds arise for the most part in the incomplete combustion of organic compounds, e.g., materials such as wood, gasoline, a well-done steak, burned in a limited amount of oxygen that is not sufficient to convert the carbon and hydrogen to carbon dioxide and water vapor.

Particulates from asbestos are especially dangerous. Asbestos is a silicate mineral that can easily be broken into long, thin threadlike fibers. The fibers can be woven into a clothlike material that is fireproof and an insulator of heat and electricity. It is used in roofing and siding, wall and ceiling insula-

tion, fireproof drapes, ironing-board covers, automobile brake linings, clutch facings, spark plugs, and hundreds of other applications. For years it has been known that workers in the asbestos industry who have inhaled vast amounts of asbestos dust are often afflicted by a fibrosis of the lungs called "asbestosis." About 10% of the deaths of asbestos insulation workers in New York over the past 25 years resulted from this disease.†

Reduction of the amount of airborne asbestos dust has reduced the incidence of asbestosis. Now it appears that asbestos workers have an alarming rate of lung cancer. The malignancies are apparently caused by the inability of the lungs to cleanse themselves of asbestos fibers (because of their needlelike shapes) or dissolve the silicate minerals. Fibers embed themselves in the lung tissue, causing an irritation that may, over many years, lead to cancer. Although problems caused by asbestos are most serious among workers in the industry, asbestos fibers are present, according to Selikoff, in 40 to 50% of the general population *who have had*

---

*   "Carcinogenic" compounds are those which have been found to cause malignant tumor growths when applied to laboratory animal skins. However, there is rarely direct proof of their carcinogenic effect on humans.

---

†   I. J. Selikoff, "Asbestos," *Environment* **11** (2), 2 (1969).

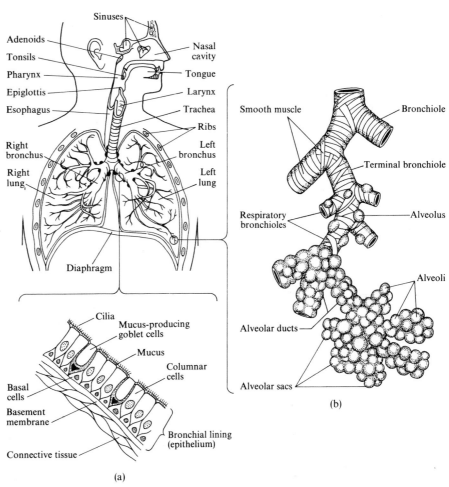

Sinuses

Adenoids

Tonsils

Pharynx

Epiglottis

Esophagus

Right bronchus

Right lung

Nasal cavity

Tongue

Larynx

Trachea

Ribs

Left bronchus

Left lung

Diaphragm

Smooth muscle

Bronchiole

Terminal bronchiole

Respiratory bronchioles

Alveolus

Alveoli

Alveolar ducts

Alveolar sacs

(b)

Cilia

Mucus-producing goblet cells

Mucus

Columnar cells

Basal cells

Basement membrane

Connective tissue

Bronchial lining (epithelium)

(a)

Fig. 10–8 Diagram of the human lung. [From Samuel J. Williamson, *Fundamentals of Air Pollution* (Addison-Wesley, Reading, Mass., 1973), p. 37.]

One example of lung disease caused by inhaling particulates is "black lung," found among coal miners. Above, normal lung; below, diseased lung. (EPA.—Documerica photos by LeRoy Woodson.)

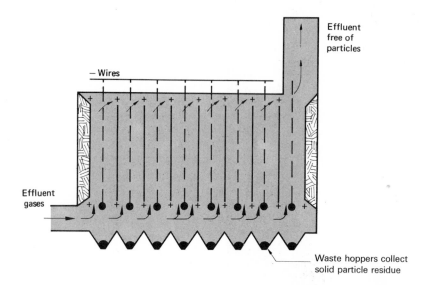

Fig. 10-9 An electrostatic precipitator.

no occupational exposure to asbestos, apparently because of the extensive use of asbestos around us.

Many pollutant particles in the atmosphere contain toxic metals that are deposited in the lungs. Lead from automobile exhaust contributes to lead poisoning. Beryllium, although a very rare element in the air, can cause a degenerative lung disease called "berylliosis" at very low levels of exposure. Nickel carbonyl, $Ni(CO)_4$, is carcinogenic. Airborne cadmium, mercury, and several other metals are thought to be quite harmful to life, as discussed in more detail in Chapter 12.

Partial solutions to the problems of particulate emissions exist. One of the best ways of collecting particulates is to pass effluent gases from a plant through an electrostatic precipitator (see Fig. 10-9). The precipitator may be a series of parallel metal plates with large electric potentials that alternate from positive to negative. Particles passing through the large electric field between the plates become electrically charged and are attracted to a plate of the opposite charge. Electrostatic precipitators effectively remove particles from the effluent gases from "dry" processes such as coal fires. But they are not effective on "wet" gases, as produced by oil combustion, since the burning of hydrocarbons yields both carbon dioxide and water vapor. There are many devices such as settling chambers, scrubbers, and filters for removal of particles, but at present none works very well with oil fires.

In recent years, methods such as these have been used to clean up the dense black smoke from many obvious, large air pollution sources. However, most methods do not remove particles with radii well below 1 μm, and yet these tiny particles may be quite harmful to people. Effects of these particles and methods for their removal are now under intense investigation.

SUGGESTED READING

American Chemical Society, *Cleaning Our Environment—The Chemical Basis for Action* (ACS, Washington, D.C., 1969).

J. C. Esposito, *Vanishing Air* (Grossman, New York, 1970).

N. de Nevers, "Enforcing the Clean Air Act of 1970," *Scientific American* **228** (6), 14–21 (June 1973).

National Air Pollution Control Administration, "Air Quality Criteria for Particulate Matter," NAPCA Publication No. AP–49 Jan. 1969; see also "Air Quality Criteria for Sulfur Oxides," NAPCA Publication No. AP–50 Jan. 1969.

Linus Pauling, *General Chemistry*, 3rd. ed. (W. H. Freeman, San Francisco, 1970).

L. G. Sillen, "The Physical Chemistry of Sea Water," in *Oceanography*, edited by Mary Sears (American Association for the Advancement of Science, Washington, D.C., 1961).

[handwritten top-left margin:]

$-\delta$ $-\delta\delta$
$H^+ \overset{\cup}{\underset{+\,\delta^+}{S}} H$ $H \overset{\cup}{\underset{\delta\delta}{S}} H$

✓ no more electrons than S

[handwritten top-right margin:]

C with a triple bond is -1 charge

QUESTIONS AND PROBLEMS

[handwritten left margin: "ng" "bond"]

1. a) What are the chemical species in hard water that cause trouble with soap?

   b) What chemical system involving alkaline earth elements is believed to be responsible for controlling the pH of the world's oceans?

2. What is Teflon, and why is it so inert to chemical attack?

3. What chemical changes occur during the exposure, developing, and printing of a black and white photograph?

4. Iodine is necessary for the synthesis of hormones such as thyroxine in the thyroid gland. What sources of iodine do we have in our diets in addition to iodized salt?

5. What is the source of sulfur usually found in acid mine drainage?

6. If elements 117, 118, and 119 were discovered, what would you expect to be the valence states of these elements when and if they formed ions? Would they tend to lose or gain electrons in forming compounds?

7. On a separate sheet of paper, complete and balance the following equations. If no chemical change occurs, write "no reaction."

   a) $Li(s) + H_2O(l) \longrightarrow$

   ✓ b) $HBr(aq) + KOH(aq) \longrightarrow$ *[handwritten:]* $KBr + H_2O$
   $K^+ \quad OH^-$
   $H^+ \quad Br^-$

   ✓ c) $Cl_2(g) + Na(s) \longrightarrow$

   d) $BaO(s) + H_2O(l) \longrightarrow$

   ? ✓ e) $Ar(g) + HCl(aq) \longrightarrow$

   ✓ f) $Ca^{2+}(aq) + CO_3^{2-}(aq) \longrightarrow$

   ✓ g) $Na_2CO_3(s) + 2HCl(aq) \longrightarrow$ *[handwritten:]* $(2H + Cl^-) - NaCl + H_2CO_3$ $(2Na\,CO_3)$ ↓ $H_2O + CO_2$

   ✓ h) $SO_2(g) + O_2(g) \longrightarrow$

   ✓ i) $Al(s) + O_2(g) \longrightarrow$ *[handwritten:]* Al_2O_3

8. Draw Lewis (electron dot) structures for the following chemical species and note those bonds that would be ionic.

   a) $HBrO_3$ b) Br_2

   c) CCl_4 d) C_2F_2

   e) Na_2O f) HI

   g) C_3H_8 h) H_2SO_4

9. What is the oxidation number of each of the following?

   a) Li in Li metal   b) Li in Li_2O

   c) S in SO_3 d) Br in $HBrO_2$

   e) Ge in GeO_2 f) C in C_2H_2

   g) N in NH_3

   [handwritten near f:] H w/ metal (-1) C $(-4$ to $+4)$

10. Explain the following observations. (a) At room temperature F_2 and Cl_2 are gases and Br_2 is a liquid, but (b) at room temperature H_2O is liquid and H_2S is a gas.

    [handwritten right margin: "lon greater ms have higher boiling pt"]

11. Classify each of the following as an *oxidizing* or a *reducing* agent.

    a) K   *[handwritten: give electron reducing agent]*   b) C in reaction with ZnO

    c) C in reactions with H_2 d) H_2O in reactions with Na

    [handwritten: graphite are reducing agents]

    e) H_2 in reactions with Ca   f) $HClO_4$

12. In an attempt to improve water quality, some states have passed laws requiring boats to grind up their on-board sewage and add amounts of Cl_2 to it before dumping it into the rivers or lakes on which the boats are located.

    a) Can you see any environmental dangers of such an operation?

    b) Can you suggest a better way of handling the problem of sewage from boats?

13. Briefly list the possible ways in which we can reduce the concentration of sulfur oxides in the atmosphere of American cities.

14. Briefly list the possible ways in which we can reduce the amount of particulate matter in the atmosphere of American cities.

15. Let us see how SO_2 concentrations can become large in a city during periods of stagnant air. Consider a city of 10 million inhabitants occupying a square area 40 km on a side. Let's assume that an inversion layer prevents the rise of pollutants above 300 m altitude. To estimate the amount of SO_2 emitted into the city's air, assume that each of the 200 million persons in the United States accounts for emission of his proportionate share of the total SO_2 emitted annually in the United States (see Table 10–3).

    a) If the air is so stagnant for 24 hours that there is no horizontal movement of pollutants away from the city and their vertical movement is limited by the inversion at 300 m, estimate the average SO_2 concentra-

[handwritten bottom:] Carbonates are basic   predict   ~~acid base~~ ✓ metal, nonmetal reactions

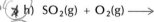

tion (in $\mu g/m^3$) in the city at the end of the 24-hr period, assuming no removal by chemical reactions or washout by rain.

b) Now let's make the problem a little more realistic— let's allow a small wind of 5 km/hr to blow across the city at right angles to one of its boundaries. Assuming that all the SO_2 in the city immediately mixes thoroughly in the air over the city, calculate the equilibrium concentration of SO_2 ($\mu g/m^3$) in the air. (Hint: At equilibrium, the rate of injection of SO_2 into the atmosphere will be equal to the rate of removal by wind at the edge of the city.)

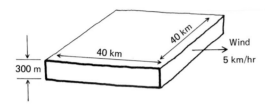

c) How does your result in part (b) compare with the maximum 24-hr average for SO_2 recommended by the EPA?

d) Does this model suggest to you the importance of building stacks as high as 1000 ft?

e) If our model city represents New York City, what kinds of errors did we make by assuming that each person's activities emit into the city air an amount of SO_2 equal to the average amount released per person in the United States?

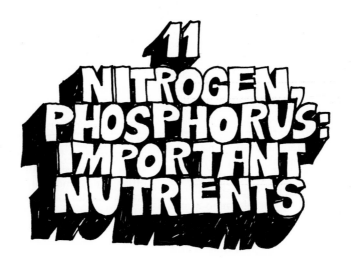

11 NITROGEN, PHOSPHORUS: IMPORTANT NUTRIENTS

Algae scum on Lake Tahoe. LEPA–Documerica photo by Belinda Rain.)

11-1 ELEMENTS OF GROUP VA

Nitrogen, phosphorus, arsenic, antimony, and bismuth make up Group VA of the periodic table. The first two, which are decidedly nonmetals, are quite important elements in the chemistry of living systems and the chemical industry. As discussed in Chapter 12, the metallic character of elements of a group increases down the column. Thus arsenic is a semimetal and antimony and bismuth exhibit metallic properties.

Nitrogen is a very important element in man's environment. Nitrogen gas (N_2) makes up about 78% of the atmosphere. It is one of the five major elements required by living systems, in part because of its presence in all amino acids, the "building blocks" from which proteins are constructed (Chapters 13 and 15). Nitrogen is one of the three major elements present in commercial fertilizers (the others being phosphorus and potassium). Animal wastes contain large amounts of nitrogen compounds that are beneficial when used as natural fertilizer, but troublesome as water pollutants.

Nitrogen compounds are important products of the chemical industry: in addition to fertilizers, products such as ammoniated household cleansers, fabric dyes, and nylon contain nitrogen. Nearly all chemical explosives contain nitrogen.

Phosphorus is also a major element essential to living systems. As discussed in Chapter 15, phosphate linkages (containing one phosphorus atom surrounded by four oxygens) hold together the skeletons of the enormous molecules of deoxyribose nucleic acid, DNA, that contain the genetic "code" in animal cells. A series of adenosine phosphate compounds store and later release the energy obtained when high-energy compounds such as glucose are oxidized in the body.

Arsenic, antimony, and bismuth are much less abundant in Earth's crust and much less important than nitrogen and phosphorus. Although As is known to be toxic in large doses, small amounts of it raise the resistance of biological life against the toxic effects of selenium. Antimony is used in alloys with lead, especially in type metal. These elements are discussed in more detail in Chapter 12.

11-2 COMPOUNDS OF NITROGEN

The chemical behavior of nitrogen is complicated because of the many oxidation states it exhibits: with five electrons in its valence shell it occurs in compounds with every possible oxidation state from 3− to 5+. Pure nitrogen, in oxidation state zero, exists as a diatomic molecule, N_2, in which the two atoms are held together by a triple bond:

$$:N\!:\,\overset{\times}{\underset{\times}{\times}}N\overset{\times}{\underset{\times}{}}$$

The N_2 molecule is very stable (i.e., not very reactive) because of the great strength of this triple bond: nearly 10 eV of energy is required to break the bond. This is one of the highest known bond strengths among diatomic molecules. On the other hand, since the N_2 molecule is nonpolar, there is very little attraction between one molecule and another. The small attraction and the low molecular weight cause solid N_2 to melt at the very low temperature of −210°C (63 K) and the liquid to boil at −196°C (77 K). Liquid nitrogen slowly boiling in a Dewar vessel (similar to a thermos bottle) at −196°C is used in laboratories when extremely low temperatures are needed for experiments.

The lowest oxidation state of nitrogen, 3−, occurs in ammonia, NH_3, a covalent compound with the structure shown in Fig. 11–1. The NH_3 molecule has a pyramidal shape, with hydrogen atoms at the three corners of the base of the pyramid. An orbital containing a pair of electrons is directed away from the hydrogen atoms. With the electropositive hydrogen atoms on one side of the nitrogen and a pair of electrons on the other, the ammonia molecule is polar and, thus, attracted to other polar molecules, such as H_2O or another NH_3. Because of hydrogen bonding, pure ammonia has a much higher boiling point (−33°C) than normal for a compound of such low molecular weight (see Table 8–2). As described in Chapter 8, NH_3 dissolves in water to form a weakly basic solution.

An interesting feature of ammonia molecules is their ability to form *complex ions*: the unshared pair of electrons in the orbital directed away from the hydrogen atoms is strongly attracted to positive ions (see Fig. 11–2). Many metal ions form such strong bonds with NH_3 molecules that ions of definite chemical composition and structure are formed, for example, $Ag(NH_3)_2^+$, $Zn(NH_3)_4^{2+}$, $Cu(NH_3)_4^{2+}$, and $Cr(NH_3)_6^{3+}$. The formation of strongly bonded complex ions is often a way of getting "insoluble" materials to dissolve. For example, the very slightly soluble salt, silver chloride, can be dissolved by adding ammonia:

$$AgCl(s) + 2NH_3(aq) \longrightarrow Ag(NH_3)_2^+(aq) + Cl^-(aq). \quad (11-1)$$

We can think of the ammonia molecule as a fundamental building block: removal of one hydrogen atom leaves an

Fig. 11–1 Structure of the ammonia molecule, depicted in three different ways. [Parts (b) and (c) from Gregory R. Choppin and Russell H. Johnson, *Introductory Chemistry* (Addison-Wesley, Reading, Mass., 1972) p. 247.]

(a)

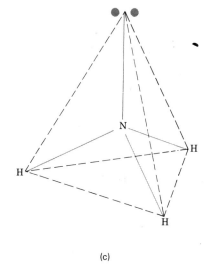

(b)

(c)

An ammonia injector fertilizes winter wheat in Lovelock, Nevada. This farmer buys NH_3 in railroad tank car lots. (Courtesy of the U.S. Dept. of Agriculture.)

Fig. 11–2 Complex ions formed from ammonia and metal ions.

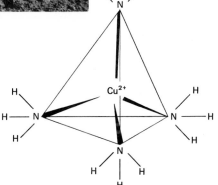

Diamminesilver ion

Tetraamminecopper ion

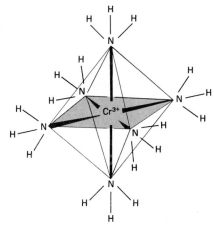

Hexaamminechromium ion

Table 11–1   Oxidation States of Nitrogen

Oxidation number	Example of nitrogen compound	Structure	Description
3−	Ammonia	NH_3	Weak base
	Urea	$H_2N-\overset{\overset{\textstyle O}{\|}}{C}-NH_2$	Decomposes to release NH_3; present in urine; high-quality nitrogen fertilizer
2−	Hydrazine	H_2N-NH_2	Weak base; reducing agent; used in rocket fuel
1−	Hydroxylamine	H_2N-OH	Weak base
0	Nitrogen gas	N_2	Fairly inert gas; major gas of the atmosphere
1+	Nitrous oxide	N_2O	"Laughing gas"; a general anesthetic
2+	Nitric oxide	NO	Air pollutant from high-temperature combustion; produced naturally by lightning
3+	Nitrogen trioxide	N_2O_3	
	Nitrous acid	HNO_2	Weak acid, decomposes to nitric acid and NO
	Nitrite salts such as	$NaNO_2$	Rather stable; used as a meat additive to retain red color
4+	Nitrogen dioxide	NO_2	Reddish-brown gas; formed in smog from NO
	Dinitrogen tetroxide	N_2O_4	Colorless gas formed by combination of two NO_2 molecules
5+	Dinitrogen pentoxide	N_2O_5	Unstable gas
	Nitric acid	HNO_3	Strong acid; good oxidizing agent
	Nitrate salts such as	$NaNO_3$	Most nitrates quite water soluble

amine group, $-NH_2$, that can be attached to other structures. For example, two amine groups can be joined together to form hydrazine, N_2H_4:

Hydrazine is a weak base somewhat like ammonia. It is dangerous to handle as it can react explosively with strong oxidizing agents; its reaction with oxidants is sometimes used to propel liquid-fueled rockets.

According to the rules given in Section 10–5, the oxidation number of nitrogen in hydrazine is 2−. The oxidation numbers of nitrogen in various compounds are shown in Table 11–1.

Some organic compounds containing amine groups form even more tightly bound complex ions with metals than ammonia does. One example of such a complex-former is ethylene diamine, $H_2N-CH_2-CH_2-NH_2$, which has an amine group at each end of the molecule, both of which can be bonded to a cation as in the example of Ni(ethylene diamine)$_3^{2+}$ depicted in Fig. 11–3. One of the strongest known complexing agents is ethylene diamine tetraacetate ion (ab-

Fig. 11–3 Ethylene diamine forms a complex ion by attaching itself at both ends to a cation such as Ni^{2+}. ▶

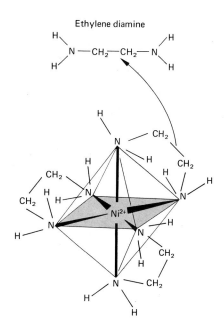

Ethylene diamine

breviated EDTA, also called "versene"). The EDTA ion has six points of attachment (two nitrogen and four oxygen atoms) to the cation. The EDTA ion literally wraps itself around the central ion to form a stable, soluble complex ion (see Fig. 11–4). Complexes in which the surrounding groups, the *ligands*, each have more than one point of attachment are called *chelates*, a descriptive term from the Greek root for "crab's claw." Complexing agents such as EDTA are sometimes injected into persons suffering toxic metal poisoning. They form water-soluble chelates with the metal atoms, causing them to be excreted in the urine.

As shown in Table 11–1, there are several nitrogen oxides. The one with oxidation number $1+$ is nitrous oxide, N_2O, a colorless gas called "laughing gas" because of its strange effect on persons who breathe small amounts of it: some claim that it produces a mildly hallucinatory experience! Larger amounts, mixed with oxygen in the proper proportion, cause unconsciousness, and the mixture is used by dentists to remove teeth "under gas."

Nitric oxide, NO, is an important product of reactions between N_2 and O_2 at high temperature:

$$N_2(g) + O_2(g) \xrightarrow{\text{heat}} 2NO(g). \qquad (11-2)$$

In part because the bond between nitrogen atoms in N_2 is considerably stronger than that between nitrogen and oxygen in NO (10 eV versus 7 eV), the reaction is *endothermic*; i.e., heat must be supplied for the reaction to occur. The reaction proceeds to a very small extent at normal atmospheric temperatures, so we don't have to worry about its using up the oxygen of the atmosphere. However, as we would expect from LeChatelier's principle, the reaction proceeds to a much greater extent at high temperature. Reaction (11–2) produces large amounts of NO when air is drawn through hot combus-

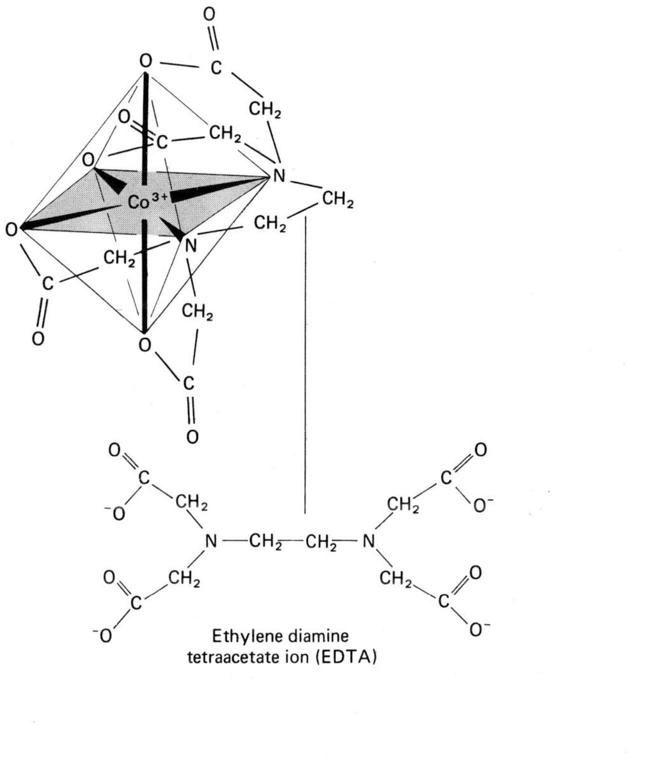

Fig. 11–4 Ethylene diamine tetraacetate literally wraps itself around the central metal ion, attaching itself in six places to form a chelate. ▶

Ethylene diamine
tetraacetate ion (EDTA)

tion chambers of power plants or engines. In Chapter 9, we noted the possible importance of NO from jet engines of SST's in the stratosphere and in Chapter 14, we discuss the role of NO from engines and power plants in the production of smog. Some NO is also produced naturally via reaction (11–2) by lightning discharges in the atmosphere.

Nitrogen trioxide, N_2O_3, is not a very stable compound. It exists as a solid at low temperatures, but in the liquid and gaseous states, it is largely dissociated into NO and NO_2. However, the reaction of such a gas mixture with a basic water solution produces nitrous acid, HNO_2, or the nitrite ion, NO_2^-, an important and reasonably stable species. Nitrous acid is a weak acid that decomposes to form nitric oxide gas, leaving behind a nitric acid solution:

$$3HNO_2(aq) \; \rightleftharpoons \; H^+(aq) + NO_3^-(aq) + H_2O(l) + 2NO(g) \tag{11–3}$$

Despite the instability of nitrous acid, many of its salts are known, such as sodium nitrite, $NaNO_2$, a common additive to fresh meats to prevent natural air oxidation of the meat, which turns the meat brown.* The nitrite is added mainly to make the meat look more attractive to the consumer, although it does have the beneficial effect of preventing the growth of spores that produce the lethal botulinus toxin. However, the nitrites can have serious side effects. In combination with certain amine compounds in the body, the nitrite can form nitrosamines, a class of compounds known to be carcinogenic in animals. It has been suggested that the permitted level of nitrites in meat be dropped from its present value for most meat products, 200 ppm, to the FDA limit of 10 ppm for smoked tuna.† This would reduce the excessive "cosmetic" use of nitrites in meat, but would leave enough of the additive to prevent growth of botulinal spores.

Nitrogen dioxide, NO_2, a reddish-brown gas, is produced commercially or in the laboratory by the reaction of nitric oxide with oxygen:

$$2NO(g) + O_2(g) \longrightarrow 2NO_2(g). \tag{11–4}$$

The oxidation of NO to NO_2 is an important reaction in the formation of smog. At the low concentration of NO in pol-

luted atmospheres, reaction (11–4) is too slow to account for the observed rate of oxidation, but the exact mechanism by which NO is oxidized is not clear (see Chapter 14).

At high concentrations, pairs of NO_2 molecules come together to form the "dimer" molecules of the colorless gas dinitrogen tetroxide, N_2O_4:

$$2NO_2(g) \; \rightleftharpoons \; N_2O_4(g). \tag{11–5}$$

The highest oxidation state of nitrogen normally observed is the $5+$ state, as in the gas dinitrogen pentoxide (N_2O_5), nitric acid (HNO_3), and the nitrate ion (NO_3^-). Dinitrogen pentoxide decomposes readily at room temperature to form NO_2 and O_2:

$$2N_2O_5(g) \longrightarrow 4NO_2(g) + O_2(g). \tag{11–6}$$

Nitric acid is a strong acid, being nearly completely dissociated in water solutions. The nitrate ion, NO_3^-, is quite stable in neutral or basic solutions, but under acidic conditions it gives up its oxygen atoms quite readily. Thus nitric acid is both a strong acid and a powerful oxidizing agent, unlike some other strong acids such as HCl and H_2SO_4. Because of its oxidizing strength, HNO_3 dissolves many metals that do not readily dissolve in other strong acids. As shown by the examples below, the final state of the nitrogen depends on conditions, particularly on the nature of the metal with which it reacts:

$$Ag(s) + 2H^+(aq) + NO_3^-(aq) \longrightarrow$$
$$Ag^+(aq) + H_2O(l) + NO_2(g), \tag{11–7}$$

$$3Cu(s) + 8H^+(aq) + 2NO_3^-(aq) \longrightarrow$$
$$3Cu^{2+}(aq) + 2NO(g) + 4H_2O(l), \tag{11–8}$$

$$4Zn(s) + 10H^+(aq) + NO_3^-(aq) \longrightarrow$$
$$4Zn^{2+}(aq) + NH_4^+(aq) + 3H_2O(l). \tag{11–9}$$

Nitric acid is a very important industrial chemical (e.g., for the manufacture of explosives and dyes). It is made by two principal methods, one of which is the reaction of sulfuric acid with sodium nitrate, a mineral obtained from large deposits in Chile. The other method starts with nitrogen from the atmosphere, which reacts with hydrogen gas at high

---

* H. J. Sanders, "Food Additives Makers Face Intensified Attack," *Chemical and Engineering News*, July 12, 1971, pp. 16–23.
† "Meat Color Additives Linked to Cancer," Washington *Post*, March 17, 1971.

temperature and pressure in the presence of iron as a catalyst to form ammonia (this is known as the Haber process):

$$N_2(g) + 3H_2(g) \xrightarrow{\text{heat, catalyst}} 2NH_3(g). \qquad (11-10)$$

Much of the ammonia is used for fertilizers, but some is used in the synthesis of nitric acid. The next step involves the reaction of ammonia with oxygen. At 1000°C in the presence of a platinum catalyst, the main reaction is

$$4NH_3(g) + 5O_2(g) \xrightarrow{\text{heat, catalyst}} 4NO(g) + 6H_2O(g). \qquad (11-11)$$

The NO produced then reacts with O_2 gas, as in Eq. (11–4), to produce NO_2 gas. Finally, the NO_2 gas is allowed to react with water to form the nitric acid:

$$3NO_2(g) + H_2O(l) \longrightarrow 2H^+(aq) + 2NO_3^-(aq) + NO(g). \qquad (11-12)$$

The NO gas that is released further reacts with O_2 to form NO_2, which is recycled as in reaction (11–12).

Many organic materials can be made into explosives by treating them with a combination of nitric and sulfuric acids, as described in Sections 13–8 and 13–10. The explosives so formed have nitro- ($-NO_2$) or nitrate ($-O-NO_2$) groups attached to the organic molecules. Explosives are highly unstable compounds; i.e., they have high potential energies. The products formed in the explosive reaction are much more stable compounds of lower potential energy. The difference in energy is given off as heat which causes a sudden large expansion of the gases formed in the reaction, thus producing the shock wave of the explosion. Consider nitroglycerin, for which the main explosion reaction is:

$$4C_3H_5(NO_3)_3(l) \longrightarrow 12CO_2(g) + 10H_2O(g) + 6N_2(g) + O_2(g). \qquad (11-13)$$

This reaction releases about 1500 calories per gram of nitroglycerin.* Note that at the high temperature of the explosion, all the products are gases. One gram of nitroglycerin occupying less than 1 cm^3 suddenly produces about 700 cm^3 of hot expanding gases, creating the shock wave that causes the destruction.

---

* In all explosive reactions, the amount of the products obtained depends on the temperature of the explosive material and the composition of the device. Under highly oxidizing conditions, one might obtain some oxides of nitrogen instead of N_2 or, under reducing conditions, some CO instead of CO_2.

Although most nitrogen-containing explosives are organic compounds, there are also inorganic nitrogen explosives, for example, ammonium nitrate, NH_4NO_3, a common laboratory reagent which is the salt formed in the neutralization of ammonia with nitric acid. The major use of NH_4NO_3 is as an agricultural fertilizer. It is stable under normal handling conditions, but if subjected to a severe shock from a primary explosion, the ammonium nitrate will explode just as TNT or nitroglycerin. Perhaps the largest nonnuclear explosion ever caused by man was that of an entire shipload of ammonium nitrate fertilizer which blew up in the Texas City harbor near Galveston, Texas, in April 1947, destroying much of the city, killing 561 people, and doing about $70 million damage.

11–3 NITROGEN IN LIFE SYSTEMS

There is a nitrogen cycle involving plants and animals, soil, water, and the atmosphere as sketched in Fig. 11–5. The cycle includes several important chemical states of nitrogen, ranging from ammonia and amino acids to the nitrates of the soil. Several types of biological life are able to "fix" elemental nitrogen, N_2, from the atmosphere. That is, they convert it to more useful states needed for plant and animal nutrition, in the form of amino acids which are then linked together as proteins. The *legumes*, such as peas, beans, clover, and alfalfa, have nodules on their roots that contain symbiotic bacteria that effect this transformation. After the plants die, their decomposition releases ammonia and ammonium salts into the soil and water where they serve as nitrogen sources for plants that cannot fix atmospheric nitrogen. In aquatic ecosystems, some types of algae are able to fix atmospheric nitrogen. Their decay releases nitrogen into the water in forms usable by other aquatic species.

Plants can utilize nitrogen in several chemical forms, especially NH_3, NH_4^+ and NO_3^-. Two groups of bacteria in soil convert NH_3 and NH_4^+ into nitrates: the first type yield nitrites, NO_2^- and the second type convert nitrites to nitrates, a very soluble, mobile form for use by plants.

A further, smaller source of "fixed" nitrogen in soil and water is the atmosphere itself. Lightning forms nitrogen oxide gases, some of which react in the atmosphere to produce nitrates that are brought down by rain.

Animals obtain nitrogen as proteins from plants or other animals lower in the food chain. The proteins are broken down into their constituent amino acids and reassembled into

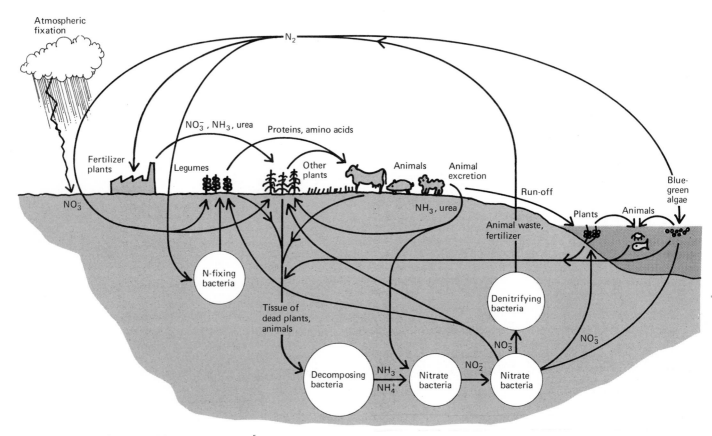

Fig. 11–5 Major features of the nitrogen cycle. [Adapted from C. C. Delwiche, "The Nitrogen Cycle," *Scientific American* **223** (3), 136 (Sept. 1970).]

new protein structures. Animals (including humans) excrete nitrogen compounds both in feces and urine. A major constituent of urine is urea, which, when applied to soil either as animal wastes or as commercial urea fertilizer, reacts with water to release ammonia:

$$H_2N-\overset{\overset{\displaystyle O}{\|}}{\underset{\text{urea}}{C}}-NH_2(s) + H_2O(l) \longrightarrow 2NH_3(aq) + CO_2(g).$$

$$(11-14)$$

To complete the cycle and maintain a stable balance of nitrogen between the atmosphere, hydrosphere, and biosphere, nitrogen must ultimately be returned to the atmosphere. The "denitrifying" bacteria of the soil take in nitrates and release nitrogen or nitrous oxide gases that escape to the atmosphere.

Ever since man became agricultural, he had caused imbalances of the ecological system. Although we think of environmental problems as being a recent development, some effects caused by man many centuries ago were large even by today's standards. Vast areas of fertile land in Africa, the Middle East, and western India were transformed into deserts by overgrazing animals. In more recent times man has cleared large areas of land of their natural ground cover in order to grow crops of his own choosing. Lying bare of vegetation for much of the year, cleared land loses

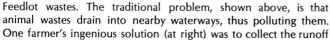

Feedlot wastes. The traditional problem, shown above, is that animal wastes drain into nearby waterways, thus polluting them. One farmer's ingenious solution (at right) was to collect the runoff in a special pond and pump it back to the fields as liquid fertilizer. (Photos courtesy of the U.S. Dept of Agriculture.)

moisture and is subject to losses of topsoil by wind and water erosion, leading to such massive disasters as the Dust Bowl of the central plains of the United States during the middle 1930s.

Until recent times however, man generally maintained a sort of ecological balance of key nutrient cycles, including that of nitrogen. If the land was grazed, the animal wastes were dropped rather uniformly on the land. The only nut-rients that actually left the farm or ranch were in the form of the meat or milk sold to others. In other types of farms, especially in the midwestern United States, animals were fed corn, hay, and other products grown on the farm. Nutrients in the animal wastes were removed periodically from the barns and feedlots and returned to the soil with manure spreaders. Again, the only nutrients to leave the farm were those in the commodities sold, that is, the meat, eggs, milk, etc.

Feedlot wastes. The modern problem, shown above, is huge feedlots like this one near Greeley, Colorado, where thousands of beef cattle fill the pens. Feedlots like this one contribute heavily to the pollution of nearby rivers and/or creeks. (Courtesy of Environmental Protection Agency.)

In recent years changes in farm operation have caused some severe environmental problems. It has become more economical to specialize in production of a smaller number of commodities per farm and to increase the size of each operating unit. Instead of each farmer raising a few hundred chickens, perhaps only one farmer in a hundred today raises chickens and he might typically have a flock of hundreds of thousands—enough to support the cost of the automatic feeding and egg-collecting machines that reduce labor costs.

The specialization and increased scale of agricultural operations has broken up some of the artificial ecological balances noted above. Today, much of the livestock is fattened in huge feedlots, often located near urban centers, in which thousands of cattle or hogs are fed. The feedlots are not parts of large farms that can supply enough feed for the animals or that can absorb all the animal wastes as fertilizer for the land. This method of livestock production causes ecological imbalances at both ends of the line.* On one hand, farms of the Midwest and West that grow the feed no longer have animal wastes to use as fertilizer to replenish the nutrients removed from the land in the feed. These nutrients must now be supplied in the form of chemical fertilizers, especially in the case of nitrogen, which is applied in vast quantities as liquid ammonia, urea, and ammonium nitrate.

The economics of farming also encourage the use of excessive amounts of fertilizer. While the cost of land, equip-

* An excellent discussion of man's "breaking" of the nitrogen cycle and the consequences is given by Barry Commoner in *The Closing Circle* (Knopf, New York, 1971).

ment, and labor has increased enormously over the past three decades, the price of fertilizer has remained fairly stable. In order to make a profit, the farmer must apply a lot of cheap fertilizer to enhance crop yields. Only a portion of the nitrogen of the fertilizers is actually consumed by the plants; the rest of it runs off the fields in ground water, adding to the problems discussed below.

The problems at the other end of the line are more serious. The animal wastes from large feedlots are so overwhelming that they cannot be used effectively as fertilizer and are thus just material to be disposed of. Generally this is done by locating the feedlot on a hillside so that the wastes drain into a creek. Animal wastes can create very serious water pollution problems. The U.S. total of about 120 million cattle, hogs, and sheep and the three billion poultry create wastes equal to that of a human population of about two billion, about ten times the amount of waste from the actual U.S. human population.

Animal wastes serve as nutrients (especially the nitrogen compounds) that enhance the growth of algae in bodies of water. Bacteria and viruses in untreated human and animal wastes may cause disease among the people that drink or swim in the water, but at present this does not seem to be a major problem. If the wastes are allowed to percolate through soil before becoming available for additional use, most pathogens are removed or decomposed. However, nitrates from both animal wastes and chemical fertilizers, are not removed effectively in soil because of their high solubility in water.

Nitrates in small amounts are readily taken up by aquatic plants as nutrients. But large amounts cannot be removed and thus become a water pollutant. Nitrate itself is not particularly toxic to humans, but certain intestinal bacteria that are prevalent in infants convert some of the nitrate to nitrite. The latter converts hemoglobin of the blood to methemoglobin, so that the hemoglobin can no longer transport oxygen from the lungs to other parts of the body. This condition, called methemoglobinemia or "blue baby," is similar to that caused by breathing excessive carbon monoxide. The affected infant has a shortage of oxygen in his body and may die in extreme cases. This problem has not received much public attention, but it is widespread: infant deaths from excess nitrate have occurred in France, Germany, Czechoslovakia, and Israel.* Between 1947 and 1950, there were 139 reported cases

of infant methemoglobinemia in Minnesota alone, 14 of them resulting in death.†

Recently, excessive nitrate levels have been observed in the public water supply of Decatur, Illinois. Professor Barry Commoner and his coworkers from Washington University in St. Louis have investigated the problem and concluded that at least 60% of the nitrate in the water supply originates from fertilizer runoff.‡

Man has intervened in a large-scale way in the nitrogen cycle: in 1968, industrial fixation of nitrogen was about 30 million tons annually, a sizable fraction of the amount fixed by all natural processes, an estimated 52 million tons.§ By the year 2000, industrial fixation may exceed 100 million tons! As we have seen, the addition of man-made nitrogen fertilizers has some serious ecological consequences. Therefore, we are faced with a dilemma: how can we feed the expanding world population without polluting our water supply with modern fertilizers?

11–4 NUTRITIONAL ASPECTS OF NITROGEN

In spite of all efforts to limit the birthrate, we will almost certainly have to provide for a world population of five to seven billion people by the year 2000. Whether or not we will be able to feed such a vast population is a matter of considerable debate. Paul Ehrlich,‖ for example, feels that "the battle to feed humanity is already lost" and that we cannot prevent "large-scale famines in the next decade or so." By contrast, Margaret Mead# has stated that "for the first time in history, we have the technical competence to feed all the hungry." Some optimists feel that the world could feed as many as 100 billion people!

We feel it will be possible, *in principle*, to feed the world population of A.D. 2000 with well-balanced diets. *However, we doubt that the effort and resources required to feed the population will be expended!* A major requirement for the task would be abundant, cheap energy to desalt water for irrigation and to produce fertilizers (e.g., convert N_2 of the

---

*   Ibid.

---

†   American Chemical Society, *Cleaning Our Environment—The Chemical Basis for Action* (ACS, Washington, D.C., 1969), pp. 142–144.
‡   Commoner, *The Closing Circle.*
§   Delwiche, "The Nitrogen Cycle."
‖   P. R. Ehrlich, *The Population Bomb* (Ballantine Books, New York, 1968).
Margaret Mead, ed., *Hunger*, A Scientists Institute for Public Information Workbook, New York, 1970, p. 3.

Ammonium nitrate is separated from waste water in these ion exchange units at the Farmers Chemical Association plant in Harrison, Tennessee. The ammonium nitrate is later used as commercial fertilizer. (Courtesy of the Farmers Chemical Association, Inc.)

Nobel prizewinner Dr. Norman E. Borlang recording the vigor and stage of growth of wheat in a breeding plot. (Courtesy of U.S. Dept. of Agriculture.)

atmosphere to usable chemical forms). Once the food is produced, it must be transported throughout the world and distributed to the people—a formidable job in an underdeveloped nation. Perhaps most difficult, the people may need to change some of their customs to improve their diets. For example, in order to obtain adequate protein, people may need to learn how to use fish protein concentrate in their food. New, improved hybrids of cereal grains have somewhat different characteristics from the grains that people have been using for centuries and they require changes in cooking and eating habits.

The number of deaths from inadequate or improper nutrition in the world today is difficult to ascertain since "starvation" or "malnutrition" is seldom listed as the cause of death. Poorly nourished persons are very susceptible to disease. Illnesses such as measles, chicken pox, and whooping cough, which are usually not too serious in well-nourished persons, are often fatal to those existing on the brink of starvation. Thus famine is not a problem that will suddenly occur when the population outgrows the food supply at some date

in the future—it is here today and affects a sizable portion of the world's people. One estimate suggests that over half the people in the less industrialized nations, amounting to one-third of the world population, are undernourished.*

The problem of nutrition is twofold: first, sufficient energy content in food, i.e., an adequate number of calories (about 2400 to 2700 calories per day per person) and, second, a properly balanced diet containing adequate amounts of various amino acids, vitamins, minerals, etc. The first condition is easier to satisfy: Borgstrom estimates that 10 to 15% of the world's population suffers from caloric shortages, whereas more than 1.5 billion are malnourished because of deficiencies in proteins, minerals, or vitamins.† The most widespread problem is that of providing the approximately

---

* President's Science Advisory Panel on the World Food Supply, *The World Food Problem* (U.S. Government Printing Office, Washington, D.C., 1967).

† G. A. Borgstrom, "The Dual Challenge of Health and Hunger: A Global Crisis," *Science and Public Affairs* **26** (8), 42–46 (1970).

70 g of protein per day that each person needs. There must be an adequate amount of protein, and the protein sources must include sufficient quantities of each of the ten essential amino acids that cannot be synthesized within human bodies (see Section 15–6).

A great deal of progress has been made on the problem of satisfying human needs for caloric content. We are in the midst of the "Green Revolution," a term referring to the development of new high-yield hybrids of the important cereal crops such as wheat, rice, and corn ("maize").* The new hybrids have yields two or three times as great as those of older types of crops. Increased yields of dwarf wheat in India and Pakistan, new strains of rice in Southeast Asia, and improved corn varieties throughout the tropics and subtropics have provided enormous gains in the food supply in those regions.

Improved seeds, however, are only one component of the Green Revolution. Take the case of dwarf wheat: the main reason for its success is its ability to utilize greater quantities of fertilizer than older varieties. Heavy fertilization causes ordinary wheat to grow so tall that it is easily damaged by wind and rain; dwarf wheat increases its yield when heavily fertilized. However, the dwarf wheat is of value only if it is heavily fertilized. Other requirements for increasing yields of crops in some areas are heavy application of insecticides and herbicides, provision of irrigation water, education of farmers in new techniques, and a supply of new tools and equipment.

These resources are not only expensive but, unless handled carefully, they exact enormous environmental and social costs: pollution from energy sources, runoff of fertilizers, entry of pesticides into the environment, displacement of farm laborers, and the conglomeration of small family farms into large state or corporate farms for efficient use of heavy machinery. Furthermore, the Green Revolution, while providing calories in the form of carbohydrates, has done little to alleviate the almost universal shortage of good protein sources.

If proteins are inadequate in the diet, serious problems of malnutrition can develop, particularly in young children. The affected children often have retarded physical growth and subnormal development of mental capacity.† They fall behind their well-nourished age-mates in school and they seldom catch up. Malnutrition is not confined to underdeveloped nations; there is a great deal of it in the United States. In part, the malnutrition is caused by poverty—low-protein starchy foods cost less than high-protein beefsteak. However, some of the problem is doubtless caused by ignorance of nutritional requirements and of food composition. For instance, 20 g of protein from peanut butter costs only one-third as much as 20 g of protein from beefsteak. Fortunately, if protein deficiency is caught in time, addition of high-protein foods to a child's diet can partially correct the situation. Thus supplemental breakfast and lunch programs in schools may be of considerable value in helping children to learn.

Humans require that about 17% of their food (by weight) be protein. We can obtain an estimate of the protein content of a food from its carbon to nitrogen (C/N) ratio, since carbon is the major element of all food, and nitrogen is mainly present just in proteins. The average C/N ratio of proteins is about 3.2/1; thus to obtain the necessary protein, one's diet must have a C/N ratio of 17/1 or less.‡ Unfortunately, the high-yield crops of the Green Revolution have very low protein content, as shown by C/N ratios in the range 26/1 to 48/1 for modern hybrids of wheat, corn, sugar beets, etc. The older, low-yield types of corn and wheat had C/N ratios in the range 19/1 to 16/1. Thus most of the increased weight of modern crops has been in the form of starch and little as protein.

In the developed nations we can balance our diets by eating high-protein meat and fish. The C/N ratio of beef, for example, is about 4.3/1. But in many developing nations, there are not enough high-protein foods to supplement the starchy cereal grains that form the major dietary staples. Despite the high quality and quantity of protein in cattle and other animals, increased livestock production in the developing nations is probably not a solution to their need for more protein. The transfer of energy and nutrients from one level of a food chain to the next higher one is inefficient: for each added step in a food chain, both energy and nutrients are cut by a factor of about ten. Thus if we feed grain to cattle and eat the beef instead of eating the grain itself, we throw away about 90% of the value of the grain. For example, legume crops, which are much higher in protein than most cereal crops, produce about 13 pounds of protein per acre,

---

* N. E. Borlaug, "The Green Revolution: For Bread and Peace," *Science and Public Affairs* **27** (6), 6–48 (1971).

† N. S. Scrimshaw, "Infant Malnutrition and Adult Learning," *Saturday Review*, March 16, 1968.

‡ W. D. Russell-Hunter, *Aquatic Productivity* (Macmillan, New York, 1970).

whereas only about 1.6 pounds are obtained by raising chickens, cows, sheep, or hogs on the same land.*

The solution of the protein shortage must come from a combination of three sources: improved cereal grains with higher protein content, new high-yield varieties of legume crops, and more high-protein food from the world's oceans. On the first point, the development of high-lysine corn is an important first step (lysine is one of the essential amino acids).† Corn, which is a major food in the tropics and subtropics, is a poor source of protein for humans and single-stomach animals which digest only about half the protein of the corn. Furthermore, the protein of corn contains inadequate amounts of two essential amino acids, lysine and tryptophan.

In areas where corn is the principal food, many children suffer the protein-deficiency syndrome called "kwashiorkor." During the 1950s and early 1960s, scientists at Purdue University cross-fertilized ordinary corn with two mutant strains of corn (one of them having soft, floury kernels) to produce hybrids containing 60% more lysine and increased tryptophan and other amino acids. The new, softer corn contains a higher fraction of easily digestible protein than older varieties. Children suffering from protein-deficiency conditions show impressive improvements when high-lysine corn is used in their food. Despite this success, the job is not done: yields of high-lysine corn are less than those of ordinary corn. Also, the new hybrid is less resistant to insect attack and, because of its softer kernels, must be handled differently in milling and cooking. There is optimism that these deficiencies can be alleviated by further crossbreeding.‡

Little has been done to increase yields of legume crops. These crops, which are good protein sources, generally have much lower yields than cereal grains. Although meats and fish are the best sources of proteins it is possible to achieve a balanced diet containing the required amino acids, vitamins, and minerals solely from plant sources.§ Such legumes as peanuts and soybeans are very rich sources of proteins. Thus research is needed to develop new, high-yield hybrids of legumes. Since the nodules on the roots of the legumes fix nitrogen from the atmosphere, high-yield legumes would probably not require the heavy application of nitrogenous fertilizers required by the "miracle" cereal grains. Another approach is the attempt to breed high-yield cereal grains that have similar nodules on their roots, thus reducing their need for artificially applied nitrogen.

Finally, what about the oceans? Can't they supply almost infinite numbers of fish, another high-protein source? One might suppose so, but most experts disagree. Today fish contribute only an estimated 10% of the world's supply of animal protein. The best projections suggest that the annual world fish catch can be raised from the present 60 million metric tons to only 140 to 180 million tons.‖ Over the short range, therefore, fish products could be used to alleviate protein shortages. One promising development is that of fish-protein concentrate (FPC), an almost tasteless, odorless powder made from fish by extraction of the fish with isopropyl alcohol, which dissolves the fat and water, leaving behind mostly protein.# The FPC needs no regrigeration and is so high in protein that only 15 g per day are needed to supplement most persons' protein-deficient diets (at a cost of about $2.50/year for each person). It can be made from fish such as hake that are not customarily considered edible.**

Much controversy surrounds the question of FPC because it can be most economically produced from the whole fish, including the head and guts that are not usually eaten. The United States Food and Drug Administration (FDA) has been slow to approve its use in the United States because of the use of entire fish bodies.†† Millions of people throughout the world are desperately short of protein, but the United States can ill afford to export FPC or the technology for its manufacture if it is considered unfit for our own residents. Fortunately some progress is being made: the FDA has approved the use of FPC; however, it is categorized as a "food additive" rather than a food, subjecting it to some restrictions. The most difficult (and stupid) restriction is that it cannot be sold within the United States in packages of greater than one pound each—a great inconvenience for commercial users.

Over the longer range, we cannot simply rely upon fish that can be caught to supply major new amounts of protein. Natural variations of climate and water currents or man's

*   S. K. Majumder, "Vegetarianism: Fad, Faith or Fact," *American Scientist* **60**, 175–179 (1972).
†   D. D. Harpstead, "High-Lysine Corn," *Scientific American* **225** (2), 34–42 (Aug. 1971).
‡   Ibid.
§   Majumder, "Vegetarianism."

‖   S. J. Holt, "The Food Resources of the Ocean," *Scientific American* **221** (3), 178–914 (Sept. 1969).
C. Holden, "Fish Flour: Protein Supplement Has Yet to Fulfil Expectations," *Science* **173**, 410–412 (1971).
**   Russell-Hunter, *Aquatic Productivity.*
††   Holden, "Fish Flour."

overfishing in certain areas sometimes substantially reduces the population of certain species. Both factors probably contributed to the disastrous fall-off of the once enormous anchovy catch off the coast of Peru in 1972.* Instead of relying on fish that can be caught, we must learn how to do ocean farming, or "aquaculture." That is, we must grow fish of our choice in appropriate areas of the sea, protect them from predators, make sure they have the necessary nutrients, and then harvest the crop, just as in ordinary agriculture. Pilot studies suggest that certain areas of the sea could be highly productive, but a great deal of research is needed before we can learn to do as well as farmers of the land.†

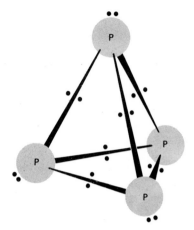

Fig. 11–6 Tetrahedral P_4 molecules are the same both in the gas phase and in solid white phosphorus.

11–5 CHEMISTRY OF PHOSPHORUS

Phosphorus, like nitrogen, exhibits oxidation states ranging from 3− to 5+. However, the 3− state of phosphorus, as in the gas phosphine, PH_3 (analogous to ammonia, NH_3), is much less important than that of nitrogen. On the other hand, the 5+ oxidation state in orthophosphoric acid, H_3PO_4, is more stable than the 5+ state of nitrogen, as in HNO_3. This means that orthophosphoric acid and the phosphates (containing PO_4^{3-}) are not effective oxidizing agents like nitric acid.

Pure phosphorus occurs as P_4 tetrahedra in the gas phase, with each phosphorus atom sharing a pair of electrons with each of the other three atoms in the molecule, as in Fig. 11–6. At 280°C or below, P_4 vapor condenses to the liquid form, which freezes at 44°C to form solid *white phosphorus*. White phosphorus, a soft, waxy, slightly yellow solid, is just one of several forms of pure solid phosphorus that can exist. *Red phosphorus* is the more stable form at room temperature, but the conversion from white to red is very slow, taking several hours even at 250°C. There are several other types of phosphorus, one of which is *black phosphorus*, formed from white phosphorus at high pressures.

The phenomenon of many different types of a pure element, especially in the same phase, as in solid phosphorus,

is called *allotropy;* the different allotropic forms are the result of differing crystal structures. Solid white phosphorus consists of P_4 tetrahedra packed closely together. Black and red phosphorus are *polymeric* solids, i.e., they contain long chains of phosphorus atoms bonded together. The difference in crystal structures accounts for the difference in chemical behavior of the various allotropes. White phosphorus is very reactive igniting spontaneously in air at 40°C. Below that temperature, slow oxidation in the air causes the phosphorus to glow faintly in the dark, from which the word "phosphorescence" is derived. To prevent its reaction with the atmosphere, one must store white phosphorus under water, in which it is insoluble. It is both poisonous and corrosive, causing painful skin burns upon contact. Red phosphorus does not react with O_2 of the atmosphere nor is it poisonous. It can be heated to about 400°C, at which point it sublimes (passes directly to the vapor state).

Phosphorus shells exploding during the Korean War. (U.S. Army photograph.)

---

* C. P. Idyll, "The Anchovy Crisis," *Scientific American* **228** (6), 22–29 (June, 1973). Much of the anchovy is converted to fish meal, a high-protein animal feed supplement. The fall-off of the anchovy catch and a poor yield of soybeans (the other principal protein supplement) in the United States during the same year contributed to the meat shortages and price rises in the United States.
† C. O. Hodge, "Farming the Oceans: Lagging Technology," *Technology Review* **75** (7), 72–73 (June 1973).

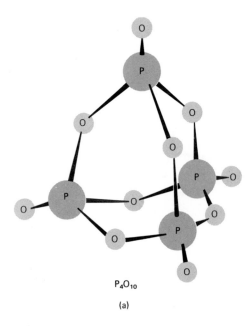

P_4O_{10}

(a)

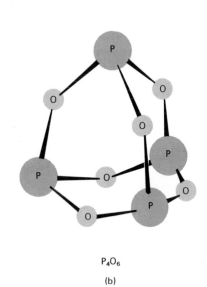

P_4O_6

(b)

Fig. 11–7 Structure of phosphorus oxides: (a) P_4O_{10} (b) P_4O_6.

A major use of phosphorus is in matches. Years ago, matchheads contained white phosphorus, which would ignite at low temperature. However, because of the hazards of spontaneous combustion and the poisonous character of white phosphorus, it is no longer used. Instead, "strike anywhere" matches contain P_4S_3, which ignites easily but is much safer than white phosphorus. The matchhead contains pumice or ground glass to create friction and ignite the P_4S_3, as well as $KClO_3$, a good oxidizing agent, which causes the burning to be more vigorous.

The most important compounds of phosphorus contain oxygen. Phosphorus pentoxide, whose formula is usually given as P_2O_5, actually consists of P_4O_{10} molecular units, as shown in Fig. 11–7a. The four phosphorus atoms still form a tetrahedron as in P_4 but the distances between phosphorus atoms are greater to allow insertion of oxygen atoms. Six of the ten oxygen atoms are in the tetrahedron, and each of the remaining four is attached to a phosphorus atom at an apex of the tetrahedron. Phosphorus trioxide, P_2O_3 or P_4O_6, has a similar structure (Fig. 11–7b), but without the four oxygen atoms at the apexes.

Phosphorus pentoxide is formed when phosphorus burns in air. It is a good drying agent, as it reacts strongly with water to form orthophosphoric acid:

$$P_2O_5(s) + 3H_2O(l) \longrightarrow 2H_3PO_4(aq). \qquad (11–15)$$

Orthophosphoric acid is only a moderately strong acid, but all three hydrogens are acidic and can react with strong bases such as NaOH to form salts like NaH_2PO_4 (sodium dihydrogen phosphate), Na_2HPO_4 (disodium hydrogen phosphate), and Na_3PO_4 (trisodium phosphate).

Mixtures of phosphate salts are good *buffers* i.e., they keep the pH of a solution almost constant despite additions of acid or base. For example, if a solution contains mainly HPO_4^{2-}, addition of H^+ ions would simply cause some formation of $H_2PO_4^-$:

$$HPO_4^{2-}(aq) + H^+(aq) \rightleftharpoons H_2PO_4^-(aq), \qquad (11–16)$$

and addition of OH^- would be counteracted by the reaction

$$HPO_4^{2-}(aq) + OH^-(aq) \rightleftharpoons PO_4^{3-}(aq) + H_2O(l), \qquad (11–17)$$

with the pH remaining about the same.

Orthophosphoric acid and the phosphates contain PO_4 tetrahedra, as shown in Fig. 11–8a. Other compounds of phosphorus and oxygen contain PO_4 tetrahedra linked to-

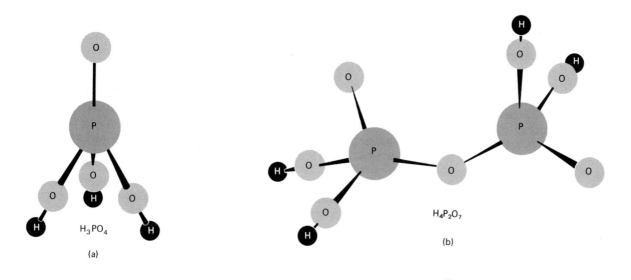

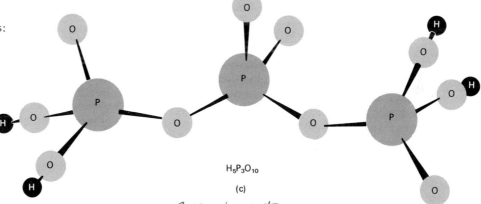

Fig. 11–8 Structures of phosphoric acids:
(a) H_3PO_4, orthophosphoric acid; (b)
$H_4P_2O_7$, pyrophosphoric acid; and (c) a
meta-phosphoric acid, $H_5P_3O_{10}$.

gether in various ways. Several types of *condensed* phosphoric acids can be formed by combining H_3PO_4 units and driving off water molecules. For example, *pyrophosphoric acid* can be made by condensation of two orthophosphoric acid molecules, driving off the water with heat:

$$2H_3PO_4\,(s) \underset{\text{heat}}{\overset{}{\rightleftharpoons}} H_4P_2O_7\,(s) + H_2O\,(g). \qquad (11\text{–}18)$$

Pyrophosphoric acid consists of two PO_4 tetrahedra sharing an oxygen atom (Fig. 11–8b). The next member of the series, $H_5P_3O_{10}$, contains three linked PO_4 tetrahedra.

↗ Hard water ions

The various phosphates are important components of modern laundry detergents. Their buffering action controls the pH during the washing, keeping the wash water slightly basic. The phosphates also form soluble complexes with the hard water ions Ca^{2+}, Mg^{2+}, and so on, that would otherwise form insoluble precipitates with detergent molecules. The salt of the acid $H_5P_3O_{10}$, sodium tripolyphosphate, $Na_5P_3O_{10}$ (often denoted "STPP"), is the most commonly used laundry phosphate. Oxygen atoms of the phosphate tetrahedra are strongly attracted to the positive metal ions, so that the poly-

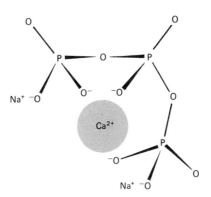

Fig. 11–9 Chelates such as this one formed by STPP and a Ca^{2+} ion are often found in laundry detergents.

phosphate group "wraps" itself around the central ion, as in Fig. 11–9, forming a stable, soluble chelate similar to those of the amines and EDTA discussed in Section 11–2. Many longer chain molecules of HPO_4 units can be made. As discussed in the following section, great controversy has arisen about the effects of detergent phosphates on water pollution.

The major naturally occurring forms of phosphorus are the minerals *apatite*, $Ca_5(PO_4)_3F$, *hydroxyapatite*, $Ca_5(PO_4)_3(OH)$, and *phosphate rock*, which is a mixture of tricalcium phosphate, $Ca_3(PO_4)_2$, and some hydroxyapatite. Hydroxyapatite is an important constituent of human and animal bones and teeth. The beneficial effect of fluorides on teeth probably involves some conversion of hydroxyapatite to apatite.

Since phosphorus is one of the major essential nutrients for plant and animal life, phosphates are included in commercial fertilizers. Naturally occurring phosphate rock $[Ca_3(PO_4)_2$ and $Ca_5(PO_4)_3(OH)]$ is not soluble enough to serve as an effective source of phosphates for plants. Thus it is treated with sulfuric acid and water to form calcium dihydrogen phosphate and gypsum:

$$Ca_3(PO_4)_2(s) + 2H_2SO_4(l) + 4H_2O(l) \longrightarrow$$
$$2CaSO_4 \cdot 2H_2O(s) + Ca(H_2PO_4)_2(s). \quad (11-19)$$

The mixture, called "superphosphate," is used as a fertilizer and soil conditioner.

Phosphorus is important in living systems not only in bones and teeth, but also in many organic compounds of bio-chemical significance. Among the most important of these are *adenosine triphosphate* (ATP) and *adenosine diphosphate* (ADP), which are involved in the transport and storage of chemical energy in the body. Also, phosphate groups serve as the linkages that hold together the enormous molecules of DNA (*deoxyribose nucleic acid*) that are found in the nuclei of cells and carry genetic information through the division of cells (see Chapter 15).

11–6 EUTROPHICATION OF WATER BODIES

Lake Erie is one of America's most seriously polluted bodies of water. It has been said that Lake Erie is dying or dead. Although it is dying under present conditions, it is not yet dead. It could probably be salvaged if steps are taken now to reverse the trend of its demise. One reason that pollution has shown up so alarmingly in Lake Erie is that it contains the smallest volume of water of all the Great Lakes (see Fig. 11–10), yet all the water flowing out of Lakes Michigan, Superior, and Huron flow through Erie on their way to Ontario. In addition, enormous quantities of pollutants are added to Lake Erie from industrialized areas such as Detroit, Toledo, Cleveland, and Erie, Pennsylvania. The small volume of Lake Erie and the large flow of water through it mean that an average drop of water stays in the lake for an average residence time of only about three years.[*] (See discussion of residence time in Section 9–3.) The short residence time causes pollution problems to show up quickly; on the other hand, if we start to clean up the lake, we can expect to see results much sooner than in the other Great Lakes.

There are many possible reasons for the deterioration of Lake Erie and the quality of its aquatic life over the past two centuries. Some of it may be due to natural causes (e.g., the gradual warming of the atmosphere and water in that area), but these effects are probably quite small compared to those of man such as:[†] the removal of natural ground cover followed by cultivation of land in the lake's watershed, over-fishing of certain desirable fish, damming of streams feeding the lake, the introduction of unfamiliar predators by the opening of the Welland Canal and the Erie Canal, and water pollution from the surrounding metropolitan, industrial communities.

[*]   J. W. Winchester, "Pollution Pathways in the Great Lakes," *Limnos* **2** (1), 20–24 (1969).

[†]   H. A. Regier and W. L. Hartman, "Lake Erie's Fish Community: 150 Years of Cultural Stresses," *Science* **180**, 1248–1255 (1973).

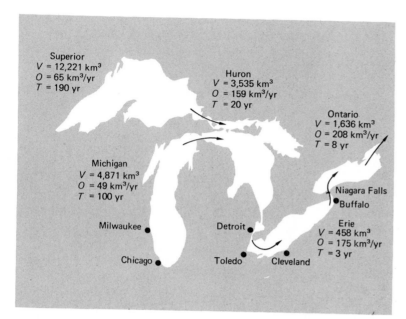

Fig. 11–10 The Great Lakes. The following key is used: volume (*V*), outflow rate (*O*), and residence time (*T*). (After Winchester, "Pollution Pathways in the Great Lakes.")

Although toxic chemicals certainly have an adverse effect upon species in a water body, some of the most serious water pollution problems involve *eutrophication*—literally the "over-nutrition" of the lake or stream. Lake Erie is a particularly bad example—water flowing into the lake carries nutrients from natural and man-made sources that are essential to aquatic life. Nutrients are, of course, needed by aquatic life. However, man's activities release such vast quantities of nutrients into some water bodies, such as Lake Erie, that they may become overfed and thus "die." That is, they become filled with silt and choked with algae and aquatic weeds to the point that they no longer exist as bodies of water. Lakes undergo natural aging processes and have lifetimes of 1000 to 100,000 years,* but by dumping excessive quantities of nutrients into the water, man is greatly accelerating the aging of lakes.

Excessive nutrients cause the growth of vast quantities of algae—the green algal "bloom" that often occurs in the late spring or early summer. In Fig. 11–11, we see that the con-

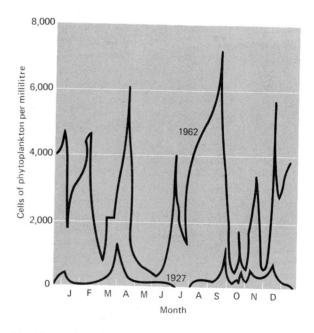

Fig. 11–11 Phytoplankton in Lake Erie found at the Cleveland water intake, 1927 and 1962. (After Federal Water Pollution Control Administration, Great Lakes Region, *Lake Erie Report—A Plan for Water Pollution Control*, August 1968.)

---

*   The natural lifetime of a lake depends on its size, width, depth, and the climate and terrain in its vicinity. A shallow lake of large surface area in a region of flat, productive land (e.g., Lake Erie) has a shorter natural life than a deep, narrow lake with steep, rocky sides in a mountainous area.

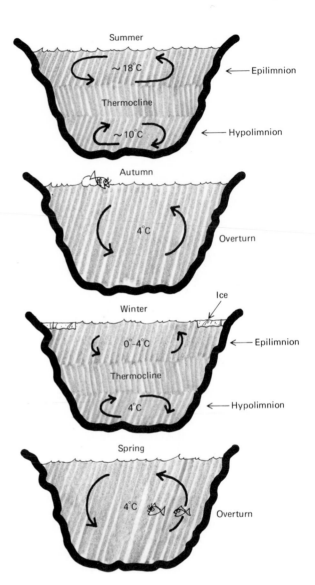

Fig. 11–12 Layering of lakes during (a) summer and (b) winter.

centration of phytoplankton (photosynthetic plant life, including algae) in Lake Erie has gone up dramatically between 1927 and 1962. The green slime is unpleasant, but while the algae are alive, they have a beneficial effect on the life of the lake. The algae are photosynthetic, i.e., they absorb sunlight and use the energy to convert carbon dioxide (CO_2), water, nitrogen and phosphorus compounds, and minor nutrients into high-energy compounds and proteins that supply the energy and amino acids for aquatic animals. Algae also produce oxygen (O_2), some of which remains dissolved in the water and is consumed by fish.

When algae die, however, their decomposition is the reverse of photosynthesis—large amounts of oxygen are needed to decompose the algae. This process uses up oxygen needed by fish and other desirable animal life of the lake. A major reason for passing municipal wastes through sewage treatment plants is that of oxidizing them to minimize the *biological oxygen demand* (BOD), the amount of oxygen that sewage would consume from the water if allowed to decompose there. Unfortunately, by allowing nutrients for the algae to pass into the water, we indirectly raise the BOD.

The depletion of oxygen by decaying algae wouldn't be too serious if the lake were kept stirred up enough so that contact with the air maintained the oxygen concentration at near saturation level (about 8 mg/l) throughout the lake. However, unless the lake is quite shallow, it becomes thermally stratified (layered) at certain times (see Fig. 11–12). For example, between about June and September, the central basin of Lake Erie forms two layers: an upper layer, or *epilimnion*, of 75°F water floating on top of cooler (55°F), denser water in the lower layer, the *hypolimnion*. A depth profile of Lake Erie is shown in Fig. 11–13. There is little mixing of water and dissolved material across the *thermocline*, about 50 to 60 ft below the surface, that separates the two layers. Although the oxygen content in the epilimnion usually remains near saturation in the central basin, the oxygen is quickly depleted

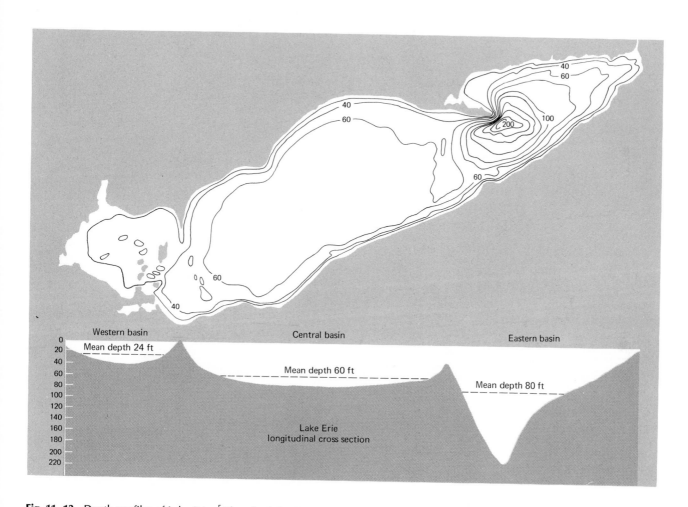

Fig. 11–13 Depth profiles of Lake Erie. [After the *Lake Erie Report—A Plan for Water Pollution Control*, see Fig. 11–11.]

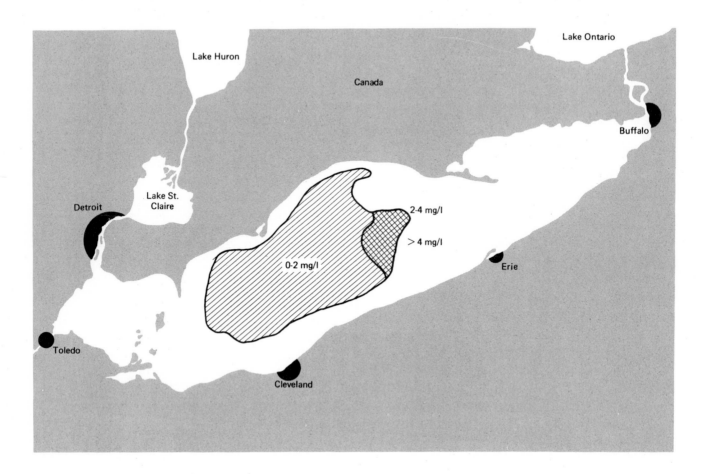

Fig. 11–14 Oxygen concentration of bottom waters in Lake Erie, August, 1964. [After the *Lake Erie Report—A Plan for Water Pollution Control.*]

in the hypolimnion and is only slowly replenished from above. As we see in Fig. 11–14, the bottom layer of the central basin becomes depleted of oxygen in late summer. A thermocline is only occasionally established in the shallow western basin. It exists throughout the summer in the deep eastern basin, but the oxygen depletion is not so serious there because of low biological activity. As the water cools in the autumn, the thermal stratification breaks up. There is an overturn of water that establishes a new layering before the onset of winter (see Fig. 11–12). As noted in Section 8–3, the

water has its highest density at 4°C (39°F); thus, during winter, there is typically a layer of 4°C water at the bottom of the lake with water near or at the 0°C freezing point on top. There is another overturn of the water in the spring.

What are the effects of the decreased level of oxygen caused by eutrophication? One of the major effects is the disappearance of certain species of fish which are very sensitive to the oxygen level of the water. Invariably, desirable game fish are the most sensitive to changes such as this (as well as the heating caused by thermal pollution) and they

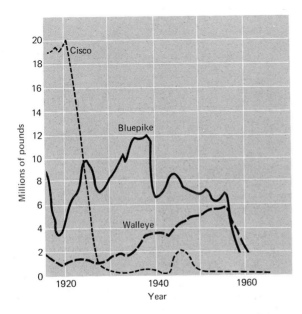

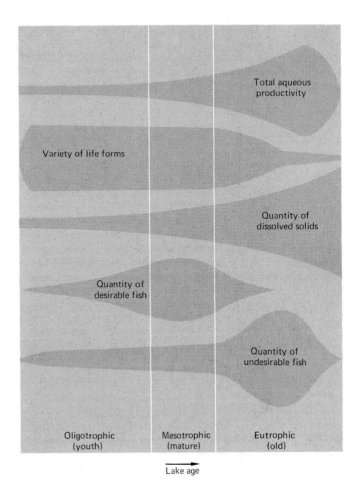

Fig. 11–15 Decline of desirable fish in Lake Erie. [After the *Lake Erie Report—A Plan for Water Pollution Control.*]

Fig. 11–16 Various changes that occur during the aging of a lake. [After the *Lake Erie Report—A Plan for Water Pollution Control.*]

leave, allowing less desirable "rough" fish such as carp to take over. In Lake Erie, for example (see Fig. 11–15), catches of cisco, blue pike, and walleyes have dropped off markedly over the last few years. Other symptoms of the aging of a lake are shown in Fig. 11–16.

What can we do to halt the demise of a lake? Most suggestions for halting eutrophication of Lake Erie are based on the idea of a "balanced diet" for the algae. For optimum growth, algae need a certain ratio of carbon to nitrogen to phosphate to minor nutrients. Very crudely we can represent the photo-

synthetic production of algae from the major nutrients by the average chemical equation

$$106CO_2(g) + 90H_2O(l) + 16NO_3(aq) + PO_4(aq) + \text{sunlight}$$
$$\longrightarrow C_{106}H_{180}O_{45}N_{16}P(s) + 154.5O_2(g).* \qquad (11–20)$$

This equation is only an approximation that represents the

---

* W. Stumm and J. J. Morgan, "Stream Pollution by Algal Nutrients," in *Transactions of the 12th Annual Conference on Sanitary Engineering*, University of Kansas, 1962, pp. 16–26.

overall process of algal growth, as we have not placed charges on the ions nor have we included species such as S, K, Na, and other minor and trace elements. The formula $C_{106}H_{180}O_{45}N_{16}P$ is not that of a single compound but represents the overall average composition of algae. In terms of numbers of atoms, the balanced diet for algae would have a C/N/P ratio of 106/16/1, giving a weight ratio of 41/7/1.

If algae are presented with certain quantities of the various nutrients (plus sunlight and appropriate temperature), they grow until they exhaust one of the essential nutrients. No matter how great the excess of the other nutrients, their growth is limited by the nutrient exhausted first. Thus if the concentration of some essential element in Lake Erie could be reduced, the growth of algae would be limited despite the presence of excesses of the remaining nutrients.

Although there is general agreement with our statements to this point, when it comes to choosing an element to serve as the growth-limiting factor, great controversy arises. Algae need more carbon than anything else, but they can usually obtain it from carbonate ions (CO_3^{2-}), bicarbonate ions (HCO_3^-), or carbonic acid (H_2CO_3) in the water in equilibrium with CO_2 in the atmosphere. Furthermore, bacterial decomposition of organic materials in the water releases additional carbonaceous compounds. Although carbon is generally not a limiting element, studies have suggested that under some circumstances, carbon can become limiting.[*]

Occasionally, nitrogen can become the growth-limiting nutrient; however, it is next to impossible to control the nitrogen input to water bodies.[†] Nitrogen compounds originate in municipal sewage plants, but in most areas, runoff of fertilizer and feedlot wastes contributes more than municipal sewage. Also, the prevalent blue-green algae are able to utilize atmospheric nitrogen.

Although a shortage of carbon, nitrogen, or occasionally a trace element, may limit the growth of algae, it is generally thought that phosphorus, can be controlled well enough to become the limiting factor. For example, an estimated 70% of the phosphates in Lake Erie enter at specific points that can be controlled, such as municipal sewage plants.[‡]

About 70% of all phosphates used annually in the United States are in the form of fertilizers, 13% are in detergents, and

---

[*]   L. E. Kuentzel, "Bacteria, Carbon Dioxide and Algal Blooms," *Journal of the Water Pollution Control Association*, Oct. 1969.

[†]   A. L. Hammond, "Phosphate Replacements: Problems with the Washday Miracle," *Science* **172**, 361–363 (1971).

[‡]   Ibid.

Incompletely processed sewage being discharged from sewage treatment plant in Denver. (EPA–Documerica photo by Bruce McAllister.)

17% are in all other uses combined.* However, the phosphates used as fertilizer are partly taken up by plants, and the remainder, being not very soluble, are held rather tightly by the soil. On the other hand, phosphates in detergents are dumped directly into waste water systems. Most sewage-treatment facilities remove little of the phosphate before releasing the treated water into a river or lake. Although in rural areas, most of the phosphates may come from agricultural sources, in waters near urban centers, most of the phosphates come from detergents. Detergents contribute about 60% of the phosphates to such important bodies of water as Lake Erie and the Potomac Estuary.†

These, then, are the arguments that have created the pressure on detergent manufacturers to remove phosphates from their products: (1) the phosphate concentration in water bodies can be reduced to a level at which the phosphorus shortage limits the growth of algae and (2) the major source of phosphates is detergents. If we accept these assumptions, what should we do to control eutrophication? Obviously, we should remove the phosphates either from detergents or from waste water at sewage plants.

What can be done about phosphates in laundry and dishwashing detergents? Recall that they are added to control pH and to prevent the formation of precipitates of hard-water ions with the detergent molecules. An extensive search is under way for a nonphosphorus compound that will perform these functions. The leading candidate for a phosphate substitute was sodium nitrilotriacetate, NTA, whose structure is

NTA has properties similar to those of the polyphosphates. For example, NTA is the sodium salt of the weak triprotic nitrilotriacetic acid, in which the hydrogen atoms circled in our figure can be replaced by Na if the acid is neutralized with NaOH. In water, NTA dissociates into Na^+ ions and NTA^{3-} ions, whose hydrolysis keeps the wash water basic (about pH 9 to 10) as desired, just as the phosphate ions do.

Furthermore, as shown in Fig. 11–17, the NTA ions form stable, soluble chelates with the hard water ions to prevent their precipitation with detergent molecules. For a while, NTA appeared to be an ideal substitute for phosphates and was beginning to replace them in detergents during 1970. However, during December 1970, the detergent industry halted the use of NTA‡ after it was found that NTA in combination with heavy metals such as mercury and cadmium causes increased infant mortality and teratogenic effects in rats and mice.§ Apparently the chelate formed by NTA and the metal ion can more easily pass through the placental barrier that protects the fetus than the metal ions themselves and can thus cause death or deformity of the fetus.

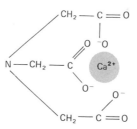

Fig. 11–17 Chelate formed by a Ca^{2+} ion and NTA.

That's where things stand today. Phosphates are under attack by many governmental units and environmental groups. Some localities have banned the sale of detergents that contain phosphates, but NTA may cause more serious problems than the phosphates. Other materials, such as

NTA

Nitrilotriacetic acid

*   "Phosphates in Detergents and the Eutrophication of America's Waters," Hearings by the Conservation and Natural Resources Subcommittee of the House Committee on Government Operations, 91st Congress of the United States, Dec. 15 and 16, 1969.

†   Ibid.

‡   Hammond, "Phosphate Replacements."

§   Teratogenic effects, literally "monster formations," are modifications, generally deformities, caused in fetuses. An example would be the flipperlike deformities caused by Thalidomide.

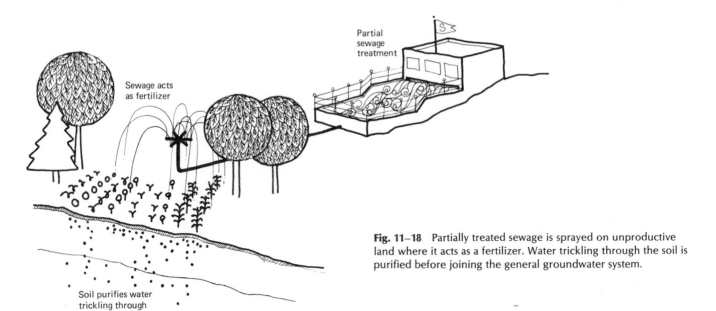

Partial sewage treatment

Sewage acts as fertilizer

Soil purifies water trickling through

Fig. 11–18   Partially treated sewage is sprayed on unproductive land where it acts as a fertilizer. Water trickling through the soil is purified before joining the general groundwater system.

carbonates and silicates, have been used as phosphate replacements. Instead of forming chelates with the hard water ions, these anions remove the cations from solution by forming granular precipitates that settle to the bottom of the tub. Although detergents containing carbonates and silicates do an adequate job of washing, they are rather dangerous materials to use in the household. They are so strongly hydrolyzed that they are quite basic, with pH values in the 10 to 11 range, with some metasilicates even above 11.* These very caustic materials cause serious illness and even death if swallowed by a child. Concentrated solutions of them cause irritation to the skin and damage to the eyes if splattered on a person.

At this time there is no obvious short-range solution to the problem. Both the detergent manufacturers and the housewives are left in a quandary as to the choice of a washing product that is both safe in the household and unlikely to cause environmental degradation. Perhaps the best solution is the oldest one—the use of plain, old soap made from animal fat or vegetable oils (see Section 13–8). Although soap does form precipitates with the hard-water ions, the deposits can be rinsed away with additional water. Communities which

have banned phosphates have gotten along fine with soap, except in automatic dishwashers, which have rather exacting specifications for detergents.

Over the longer range, detergent makers feel that we should add phosphate-removal steps to the sewage treatment process. Several methods for phosphate removal are under development. Eventually, phosphate-removal steps will probably be added as part of generally improved sewage-treatment facilities.

But there are even more imaginative solutions to the eutrophication problem. These approaches consider the nutrients to be a "resource-out-of-place," i.e., they are troublesome in water bodies because they support the growth of algae. But nutrients are highly desirable in other circumstances—farmers pay good money for nutrients such as nitrates, phosphates and potash in fertilizers. Thus it should be possible to put these nutrients to beneficial use. Various demonstration projects have been set up to try this approach. One of the first was set up in Nittany Valley on Spring Creek near Pennsylvania State University.† Sewage from the university and the town of State College, the neigh-

---

*   Hammond, "Phosphate Replacements."

†   L. T. Kardos, "A new prospect," *Environment* **12** (2), 11 (1970).

boring community, was partially treated, although it was still far below the quality of drinking water, and then sprayed onto plots of agricultural crops or groves of trees, in both cases on ground of very low fertility (see Fig. 11–18). The nutrients and the water sprayed on the ground greatly increased the yield of alfalfa, corn, oats, and white spruce trees. Furthermore, the soil on which the waste water was sprayed served as a "living filter." Excess nutrients and bacteria were removed as the water trickled through the soil before joining the natural groundwater system. The bacterial quality of water is normally determined by the "coliform count"—the number of coliform bacteria per unit volume. Coliform bacteria exist in intestines, and their presence in water indicates contamination by human fecal matter. A coliform count of 2400 per 100 ml is normally accepted as the upper limit for safe swimming water. By contrast, partially treated waste water, after trickling through two to four feet of soil, shows an average coliform count of only *one* per 100 ml. Similar but larger-scale projects of this type are being tested in other communities.

Some communities are also using sewage sludge—the solid material that settles out of waste water during treatment—as an excellent fertilizer and soil conditioner. Despite optimism about these uses of resources-out-of-place, a word of caution must be added. Since the sewage contains wastes from commercial and industrial users as well as individual households, the sewage and sludge is often contaminated with toxic metals. Tests are now in progress to determine if food crops grown with these fertilizers contain levels of toxic metals above those deemed safe for human consumption. If it turns out that the toxic metal content of food grown under these conditions is too high, it should still be possible to use the sewage products as fertilizers for trees grown for lumber or for other products that will not enter the food chain.

SUGGESTED READING

"The Biosphere," *Scientific American* **223** (3), entire issue, (Sept. 1970).

A. H. Boerma, "A World Agricultural Plan," *Scientific American* **223** (2), 54–69 (Aug. 1970).

Barry Commoner, *The Closing Circle* (Knopf, New York, 1971).

H. D. Embree and H. J. DeBey, *Introduction to the Chemistry of Life* (Addison-Wesley, Reading, Mass., 1968).

Federal Water Pollution Control Administration, Great Lakes Region, *Lake Erie Report—A Plan for Water Pollution Control*, Aug. 1968.

Margaret Mead, ed., *Hunger*, A Scientists' Institute for Public Information Workbook (1970). Available from SIPI, 30 East 68th St., New York, N.Y. 10021. Price $1.

Linus Pauling, *General Chemistry*, 3rd ed. (W. H. Freeman, San Francisco, 1970).

"Phosphates in Detergents and the Eutrophication of America's Waters," Hearings of the Conservation and Natural Resources Subcommittee of the House Committee on Government Operations, 91st Congress, Dec. 15–16, 1969. Available from the Supt. of Documents, U.S. Government Printing Office, Washington, D.C. 20402. Price $1.25.

W. D. Russell-Hunter, *Aquatic Productivity* (Macmillan, New York, 1970).

QUESTIONS AND PROBLEMS

1. What is the oxidation state of nitrogen or phosphorus in each of the following compounds?

   a) N_2O, nitrous oxide

   b) PH_3, phosphine

   c) NO_2, nitrogen dioxide

   d) $NaNO_2$, sodium nitrite

   e) $Na_5P_3O_{10}$, pentasodium tripolyphosphate

   f) N_2O_5, nitrogen pentoxide

   g) $NaNO_3$, sodium nitrate

   h) NH_4Cl, ammonium chloride

   i) HCN, hydrogen cyanide

2. Some compounds are able to "oxidize themselves." For example, when orange ammonium dichromate is heated or ignited, it is converted to green chromic oxide, with the evolution of some gas. The reaction is probably

$$(NH_4)_2Cr_2O_7(s) \longrightarrow N_2(g) + 4H_2O(g) + Cr_2O_3(s).$$

Check the oxidation states of the various elements before and after reaction. Indicate which element has been oxidized, which reduced, and the initial and final oxidation states of both.

3. Nitrogen is an important element in life systems because it is contained in

   a) amino acids   b) carbohydrates

   c) fats   d) proteins

4. Modern feedlots, located near urban centers, supply tens of thousands of animals with feed shipped in from specialized grain farms. Show how these methods have led to serious imbalances of the ecological system and serious pollution problems.

5. Why are large amounts of nitrogen and phosphorus nutrients generally considered undesirable in bodies of water?

6. For what reasons have phosphates been singled out as the essential nutrient of algae that can most easily be limited to control the growth of the algae?

7. It has been suggested by some scientists that small amounts of arsenic (As) are added to waste water as an impurity in laundry detergents. Although this question is not settled at this time, could you indicate chemical reasons for the possible presence of small amounts of arsenic in detergents?

8. One of the major functions of phosphates in detergents is that of buffering wash-water solution. If salts of ortho-phosphoric acid, H_3PO_4, can perform this function, why don't the salts of nitric acid, HNO_3, perform a similar function? Note that P and N are in the same oxidation states in the two compounds.

9. Suppose that during the spring or summer we collected several large tanks of Lake Blotto water (along with its algae, fish, and so on). To the various tanks we add: (a) NH_4NO_3, (b) Na_3PO_4, (c) NaCl, (d) bubbles of CO_2 gas, and (e) $Fe(NO_3)_3$. Nothing very exciting happens except in tank (e), in which a tremendous algal bloom develops. What can we conclude about limiting nutrients? Explain.

10. What is the connection between phosphate and fluoride in the protection of teeth?

11. Calculate the weight percent of nitrogen in ammonium nitrate (NH_4NO_3), which is used as fertilizer.

12. If a household detergent contains 35% of STPP ($Na_5P_3O_{10}$) by weight, what is the weight percent of phosphorus in the detergent?

12 METALS

Basic oxygen furnace in operation. (Courtesy of the American Iron and Steel Institute.)

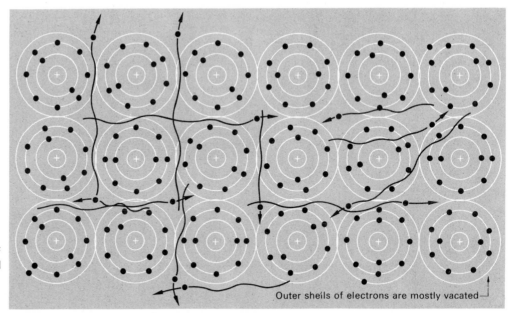

Outer shells of electrons are mostly vacated

Fig. 12–1 The valence electrons of metal atoms are so loosely attached that they become free to wander about when a metallic crystal is formed.

12–1 GENERAL PROPERTIES OF METALS

About 80 elements, a considerable majority of the known 105 elements, are classified as *metals*. Although there are great variations in the properties of the many different metallic elements, all of them have several properties in common by which we can distinguish them from nonmetals or semimetals. As you can see from the periodic table inside the front cover, nearly all the metals have their valence shells only half filled or less. Because of this the metals are all *electropositive*: their atoms tend to either give up electrons and become positive ions (for example, K^+, Mg^{2+}, Sc^{3+}) or acquire high oxidation states in combination with nonmetals, like the $7+$ oxidation number of Mn in MnO_4^-.

When metals form compounds with nonmetals, there are basically two types of bonds formed: ionic bonds such as those between Na^+ and Cl^- ions in salt crystals, and covalent bonds like those between chromium and oxygen in CrO_4^{2-} and $Cr_2O_7^{2-}$ ions. (Most metal-nonmetal bonds, of course, fall between the extremes of purely ionic and covalent character.) However, when metal atoms combine with other metal atoms it's generally not possible to form either type of bond. Since both atoms are electropositive, it would be difficult for elec-

trons to be transferred from one atom to the other to make a positive ion and a negative ion. On the other hand, it is also difficult to form strong covalent bonds between two metal ions because, with their valence shells only half full or less, even by sharing electrons they cannot attain outer shells of eight electrons. Covalent bonds in diatomic metal molecules are observed in the gas phase, but they are very weak compared to covalent bonds involving nonmetals.

Metals are bound together in a fundamentally different way called *metallic bonding*. In forming metallic bonds, large numbers of atoms come together and arrange themselves in a highly ordered three-dimensional crystal lattice, as in most solids, so that each atom is surrounded by many neighboring atoms with which it can share electrons. However, since the valence electrons of electropositive atoms are loosely attached, the valence electrons of metal atoms in a crystal lose their identities; i.e., they do not remain bound to their original atoms, but are so completely shared with other atoms that they become free to move throughout the crystal (see Fig. 12–1).

The presence of free electrons accounts for the unique properties of metals, especially their high *electrical* and *thermal conductivity*. When a metal is connected across the poles of a battery, the free electrons move through the crystal, carrying electrical current. In most nonmetals, most of the electrons are bound to the individual atoms or molecules, making them very poor conductors of electricity. The low conductivity of nonmetals increases with temperature, as the greater thermal agitation shakes additional electrons loose from their atoms. By contrast, the conductivity of metals decreases at higher temperatures, because the strongly agitating atoms of the metal interfere with the movement of free electrons. A distinction between metals and nonmetals or semimetals can thus be made by observing the change of conductivity with temperature. In Table 12–1, we compare the electrical conductivities of several metals and nonmetals. We see that silver, copper, gold, and aluminum are the best conductors. Because of the expense of gold and silver, however, copper is far more widely used in electrical wiring.* Although aluminum is a poorer conductor than copper, aluminum is being used increasingly for transmission lines because of its lower density.

If you place one end of a nonmetallic rod in a flame, the heat slowly moves up the rod, warming the end that you're holding. The heat of the flame causes the atoms of the one end of the rod to vibrate faster about their positions. The vibration is transferred from one atom to its neighbor, and so forth, gradually warming the rod along its entire length. If you heat a metal rod, you find that the heat travels up the rod much faster, so that the cool end quickly becomes too hot to handle. As shown in Table 12–1, metals have much greater thermal conductivity than nonmetals. Free electrons transfer heat much faster than the lattice vibrations of a non-conductor.

The highly reflective, "silvery" surface of metals also results from presence of free electrons. When a photon of light strikes a free electron, the electron oscillates, giving off light with the same wavelength and frequency (color) as that of the incoming photon (see Section 7–1). Metals in their pure state have a shiny surface, although for many metals, the shiny surface disappears when exposed to air, due to reactions with oxygen that form oxides on the surface. Two examples are iron and aluminum, which oxidize to form rust (Fe_2O_3) and aluminum oxide (Al_2O_3), respectively.

---

* During World War II, copper was in such short supply that silver from the United States depository at Ft. Knox was temporarily borrowed and made into wire for the coils of the huge electromagnetic separators used at Oak Ridge, Tennessee, to produce enriched ^{235}U for weapons and reactors. After the war, the silver was returned to Fort Knox and replaced by copper wires.

Table 12–1 Electrical and Thermal Conductivities Relative to Fe = 1.0

Material	Electrical conductivity	Thermal conductivity
Ag	6.2	6.1
Cu	5.9	5.6
Au	4.4	5.2
Al	3.8	3.0
Zn	1.7	1.6
Fe	1.0	1.0
Pb	0.45	0.5
C (graphite)	0.0073	0.044
Si	0.00012	———
S	5×10^{-23}	0.0045

Metallic bonds are strong enough so that most metals are solids at room temperature. The major exception is mercury (Hg), which melts at −38°C; it is the densest known liquid (13.6 g/cm³) at room temperature. The melting points of metals cover a wide range: gallium (Ga) and cesium (Cs) melt at about 29°C, whereas the melting point of tungsten (W) is 3410°C. Even in the liquid state, the valence electrons remain free to move throughout the metal, and so the liquid also has high electrical and thermal conductivity and a shiny surface.

Many metals are *ductile*, meaning that they can be pulled through a series of dies of decreasing radius to draw them into fine wires. Some metals are quite *malleable*: they can be beaten or rolled into extremely thin foils. Gold (Au), for example, can be rolled into gold "leaf" about 10^{-5} cm thick or even less; it can be used as a decorative coating or inlay for a variety of objects. Other metals, such as tungsten, are quite brittle.

Because metals possess such a great variety of chemical and physical properties, they are used in many different applications that take advantage of specific properties of particular metals. Iron occurs in extensive ore deposits near the earth's surface, so that it can be extracted rather cheaply. By adding small amounts of other elements such as carbon, vanadium, nickel, etc., we can convert iron into steels that have great strength and sufficient chemical inertness to be used for automobile bodies, building frameworks, and so on. The development of aircraft and high-speed trains has brought about increased use of strong, lightweight metals, especially aluminum (Al) and magnesium (Mg). Space hardware and supersonic airplanes require a metal that is strong, lightweight, and capable of withstanding the high temperatures generated by friction of the atmosphere. As a result, we are beginning to learn how to "work" the very difficult element titanium (Ti), which can fill these needs.

Copper (Cu) is used extensively in electrical wiring because of its ductility and good electrical conductivity. Because of its high thermal conductivity it is also used in cookware, automobile radiators, and other applications where heat transfer is important. Gold is useful in jewelry, coins, and table utensils because it is malleable and almost totally inert chemically. Silver is used more extensively in these applications because it is cheaper; however, silver is often less desirable because it is chemically more active. Silverware tarnishes, for example, when it reacts with the sulfur in eggs to form black Ag_2S. Tungsten's high melting point allows it to be used as the filament in incandescent bulbs, where electrical current heats it to white hot but it doesn't melt. Mercury is used in barometers, "quiet" electric switches, etc., where a dense or conducting liquid is needed.

12-2 REACTIVITIES OF THE METALS

We can correlate the chemical properties of the metallic elements with their positions in the periodic table. These positions, in turn, are related to the electronic structure of the atoms. Several classes of metallic elements are listed in Table 12-2. The properties of the alkali metals and alkaline earth elements were discussed in Chapter 10. In this chapter we deal with the rest of the metals in a brief overview, concentrating on the ones of interest that are especially useful and the ones that create environmental problems.

Table 12-2 Summary of Metallic Elements

Class	Chemical group	Elements
Alkali metals	IA	Li, Na, K, Rb, Cs, Fr
Alkaline earths	IIA	Be, Mg, Ca, Sr, Ba, Ra
Transition metals	IIIB–	First series: Sc–Ni
	VIII	Second series: Y–Pd
		Third series: Hf–Pt
Lanthanides	—	La–Lu
Actinides	—	Ac–Lw
Coinage metals	IB	Cu, Ag, Au
Zinc group	IIB	Zn, Cd, Hg
Aluminum group	IIIA	Al, Ga, In, Tl
Tin-lead	IVA	Sn, Pb
Antimony-bismuth group	VA	Sb, Bi

There are great differences in the electropositive strength and reactivity of the various classes of metals. One measure of the reactivity of metals is the energy released in the "half-cell" reaction, by which atoms of a metal give up electrons to become positive ions, for example,

$$Zn(s) \longrightarrow Zn^{2+}(aq) + 2e^-. \qquad (12-1)$$

Half-cell reactions such as the above occur at each of the electrodes of a battery. If a reaction such as (12-1) releases electrons at one terminal, there must be a second half-cell reaction at the other terminal which accepts electrons, for example,

$$Cu^{2+}(aq) + 2e^- \longrightarrow Cu(s). \qquad (12-2)$$

A chemical battery employing these two reactions can be made as shown in Fig. 12-2. Of course, it is impossible to set up the battery without negative ions such as SO_4^{2-} in the solution to balance the positive charges. The battery would develop a potential ("voltage") of about 0.4 volts, which could be used to run an electric motor or do some other kind of electrical work. During operation, Zn atoms leave the Zn electrode and enter the solution as Zn^{2+} ions. Simultaneously, Cu^{2+} ions are neutralized by electrons and deposited on the Cu electrode. The net reaction is the sum of (12-1) and (12-2):

$$Zn(s) + Cu^{2+}(aq) \rightleftharpoons Zn^{2+}(aq) + Cu(s) + 0.4\,V \qquad (12-3)$$

Just like all equilibrium reactions, this reaction is *reversible*. According to LeChatelier's principle, if we add a quantity appearing on the right to the system, the reaction will be forced towards the left. Thus if another battery of voltage greater than 0.4 V is placed in the circuit opposing the current flow of the Cu-Zn battery (Fig. 12-3), electrons flow through the circuit in the opposite direction. Reaction (12-3) is forced toward the left—Cu comes into solution and Zn is plated out.

Another example of a reversible chemical battery is the familiar lead storage battery used in automobiles. When the engine is running reasonably fast, the generator (or "alternator") produces more current than is used by the spark plugs, lights, and accessories. The excess electrical energy is stored in the battery. When electrical energy is needed, as when starting the engine, it can be withdrawn from the battery. The storage battery consists of a lead terminal, a lead dioxide (PbO_2) terminal, and a solution of sulfuric acid, H_2SO_4 (see

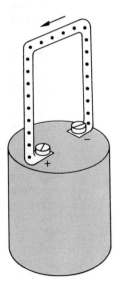

Fig. 12–2 When a metal is connected across the poles of a battery, the electrons move under the influence of the potential of the battery.

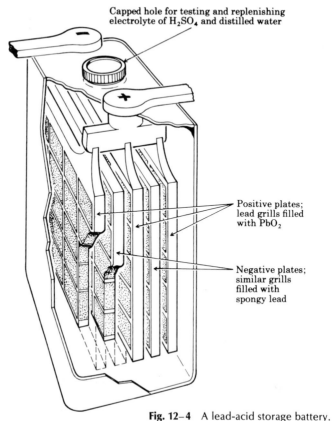

Capped hole for testing and replenishing electrolyte of H_2SO_4 and distilled water

Positive plates; lead grills filled with PbO_2

Negative plates; similar grills filled with spongy lead

Fig. 12–4 A lead-acid storage battery.

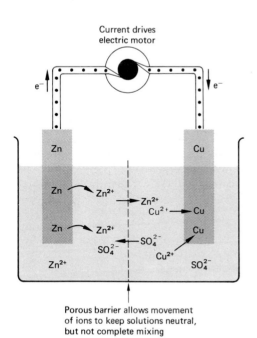

Current drives electric motor

Porous barrier allows movement of ions to keep solutions neutral, but not complete mixing

Fig. 12–3 A battery made with copper and zinc electrodes.

Fig. 12–4). When the battery gives off energy, the following reactions occur:

— terminal: (12—4)

$$Pb(s) + HSO_4^-(aq) \rightleftarrows PbSO_4(s) + H^+(aq) + 2e^-,$$

+ terminal:

$$PbO_2(s) + HSO_4^-(aq) + 3H^+(aq) + 2e^- \rightleftarrows$$
$$PbSO_4(s) + 2H_2O(l), \quad (12—5)$$

Combined overall reaction:

$$Pb(s) + PbO_2(s) + 2HSO_4^-(aq) + 2H^+(aq) \rightleftarrows$$
$$2PbSO_4(s) + 2H_2O(l). \quad (12—6)$$

Table 12–3   The Electromotive Series

Reduced form $\longrightarrow$ Oxidized form + electron(s)

Strong reducing agents ←

$$Li \longrightarrow Li^+ + e^-$$
$$K \longrightarrow K^+ + e^-$$
$$Cs \longrightarrow Cs^+ + e^-$$
$$Ba \longrightarrow Ba^{2+} + 2e^-$$
$$Ca \longrightarrow Ca^{2+} + 2e^-$$
$$Na \longrightarrow Na^+ + e^-$$
$$La \longrightarrow La^{3+} + 3e^-$$
$$Mg \longrightarrow Mg^{2+} + 2e^-$$
$$Al \longrightarrow Al^{3+} + 3e^-$$
$$Mn \longrightarrow Mn^{2+} + 2e^-$$
$$Zn \longrightarrow Zn^{2+} + 2e^-$$
$$Fe \longrightarrow Fe^{2+} + 2e^-$$
$$Co \longrightarrow Co^{2+} + 2e^-$$
$$Ni \longrightarrow Ni^{2+} + 2e^-$$
$$Sn \longrightarrow Sn^{2+} + 2e^-$$
$$Pb \longrightarrow Pb^{2+} + 2e^-$$
$$H_2(g) \longrightarrow 2H^+ + 2e^-$$
$$Cu \longrightarrow Cu^{2+} + 2e^-$$
$$2I^- \longrightarrow I_2(s) + 2e^-$$
$$Fe^{2+} \longrightarrow Fe^{3+} + e^-$$
$$Ag \longrightarrow Ag^+ + e^-$$
$$Hg \longrightarrow Hg^{2+} + 2e^-$$
$$2Br^- \longrightarrow Br_2(l) + 2e^-$$
$$2Cl^- \longrightarrow Cl_2(g) + 2e^-$$
$$Au \longrightarrow Au^{3+} + 3e^-$$
$$2F^- \longrightarrow F_2(g) + 2e^-$$

→ Strong oxidizing agents

When the battery is being charged, reaction (11–6) goes toward the left. The main species in solution is sulfuric acid. When the battery is discharged, the H_2SO_4 is used up, leaving nearly pure water. The state of "health" of a lead battery is determined by measuring the density of the fluid—the higher its charge, the greater the H_2SO_4 concentration and the higher the density of the fluid.

The reactivity of a metal is measured by the energy released by the half-cell oxidation reactions (see Eqs. 12–1 and 12–2). The strength of the tendency of pure metals and other reduced forms of elements to be oxidized is indicated by the position of the half-cell reaction in the electromotive series, a portion of which is shown in Table 12–3. The reduced species of any half-cell reaction will react with the oxidized species of any half-cell reaction below it in the series. For example, the Cu-Cu^{2+} half-cell lies higher in the series than

the Ag-Ag^+ half-cell. This means that Cu has a stronger tendency to go into solution than Ag; thus if we dip Cu metal into a solution containing Ag^+ ions, the Cu goes into solution, replacing the less active Ag^+ ions, which come out of solution as Ag metal:

$$Cu(s) + 2Ag^+(aq) \longrightarrow Cu^{2+}(aq) + 2Ag(s). \qquad (12-7)$$

We can make a beautiful demonstration of this reaction by fashioning a Christmas tree out of Cu wire and standing it in a solution of silver nitrate, $AgNO_3$ (see Fig. 12–5). The more active Cu atoms go into solution as Cu^{2+}, forming a blue solution, replacing the colorless Ag^+ ions, which precipitate out as pure silvery-gray Ag metal crystals. The latter form beautiful frosty silver desposits on the tree in the blue solution.

The most reactive metals are the alkali metals and alkaline earths, whose half-cell reactions are at the top of the electromotive series. The pure metals are such strong reducing agents that, as we saw in Chapter 10, they reduce the hydrogen of water to H_2 gas, for example:

$$2K(s) + 2H_2O(l) \longrightarrow 2K^+(aq) + 2OH^-(aq) + H_2(g). \qquad (12-8)$$

Metals in the vicinity of Zn in the series are not strong enough reducing agents to liberate H_2 gas from water, but they do react with strong acids such as sulfuric acid to release hydrogen:

$$Zn(s) + 2H^+(aq) + SO_4^{2-}(aq) \longrightarrow Zn^{2+}(aq) + H_2(g) + SO_4^{2-}(aq). \qquad (12-9)$$

Metals near or below the H_2-H^+ half-cell in the series are so unreactive that they don't react with strong acids. Gold, for example, is such a poor reducing agent that a mixture of nitric and hydrochloric acids is needed to dissolve it.

As indicated above in the discussion of the Cu-Zn battery, we can force the reactions in a chemical cell to go in the opposite direction by the use of another battery or other current source having a greater voltage than that of the cell. This technique is the basis of electroplating. The object to be electroplated is connected to the negative terminal of the external source and dipped into a solution of the metal to be plated. To complete the circuit a second electrode must be connected to the positive terminal and dipped into solution. When current flows through the system, the metal ions are plated out as the pure metal on the negative electrode.

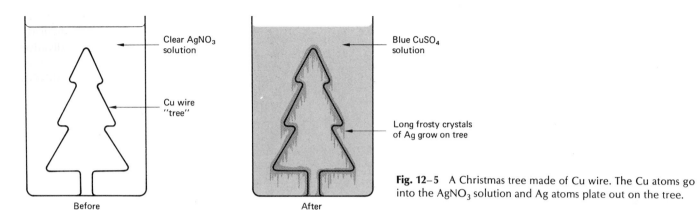

Before   After

Fig. 12–5 A Christmas tree made of Cu wire. The Cu atoms go into the AgNO₃ solution and Ag atoms plate out on the tree.

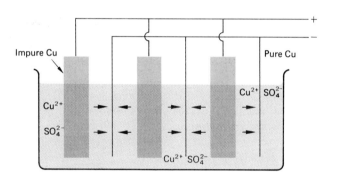

Fig. 12–6 Copper is purified by electrolysis through a CuSO₄ plating bath.

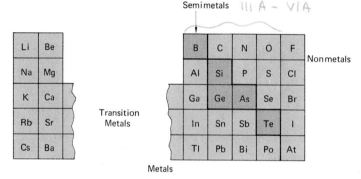

Fig. 12–7 Portion of the periodic table showing the division between metals, nonmetals, and semimetals.

Some elements are purified by electrolysis. The copper obtained from the smelting process discussed in Section 10–7 is quite impure. In order to clean it up, it is made into plates which serve as the positive terminal of an electroplating cell (see Fig. 12–6). Copper sulfate solution is placed in the cell and a thin sheet of pure copper serves as the negative terminal. When the current flows, copper from the impure electrode goes into solution and highly pure copper is plated on the negative electrode. Metals less active than copper settle to the bottom of the tank and more active metal impurities remain in solution.

Electroplating is frequently used to deposit thin layers of one metal on another. Shiny, protective layers of cadmium are electroplated on steel hand tools, bolts, and fancy engine parts. Thin deposits of chromium are electroplated on automobile trim.

12–3 ELEMENTS OF GROUPS IIIA THROUGH VIA

These groups of elements are headed by elements that are nonmetallic, but the elements at the bottoms of the columns are metallic. In Groups IIIA through VIA, the outer electrons of the small atoms of elements at the top of each column of the periodic table are more strongly attached to their atoms than those of the larger atoms farther down the columns. The electropositive character and thus the metallic tendency increase on moving down any one of the columns.

In some columns there are elements that can best be described as semimetals, for example, B, Si, Ge, As and Te. Since electropositive character generally decreases as the number of valence electrons increases, the line separating the metals from the nonmetals and semimetals cuts across these groups diagonally from the upper left to the lower right corner of that section of the table (see Fig. 12–7).

In Section 12–1, we discussed various physical properties such as electrical and thermal conductivity by which we can distinguish between metals and nonmetals. There are also chemical distinctions between them, for example, the behavior of the oxide of the element with water. Oxides of nonmetals are acid anhydrides; i.e., they react with water to form acids. For example, CO_2 forms carbonic acid (H_2CO_3), P_2O_5 yields phosphoric acid (H_3PO_4), and SO_3 forms sulfuric acid (H_2SO_4). The nonmetal atom is attached to the oxygen atoms so strongly that acidic H^+ ions are released when the compound is dissolved in water. On the other hand, metal oxides are mostly basic anhydrides or amphoteric (meaning that the oxide can act either as an acid or a base; see discussion of aluminum oxide below) except in very high oxidation states, such as the 6+ of chromium. For example, Na_2O reacts with water to form the strong base, NaOH, and CaO yields $Ca(OH)_2$. In these cases the bond between the metal and the oxygen is so highly polar (ionic), that the entire OH^- group is detached from the metal when the compound is dissolved in water.

Group IIIA

Atoms of Group IIIA have three electrons in their valence shells, and except for thallium (Tl), have 3+ oxidation states in most of their compounds. The electrons in the first member, boron, are so tightly bound that B^{3+} ions are not formed. Boron is a semimetal: its electrical conductivity is small and increases with temperature. The oxide B_2O_3 is the anhydride of the weak acid H_3BO_3, boric acid, solutions of which are used as an eyewash. The most important boron compound is borax, sodium tetraborate decahydrate ($NaB_4O_7 \cdot 10H_2O$), which is used in paints, water softeners, and household cleansers ("Boraxo"). Borax is also used in making glass, especially Pyrex glass, which contains about 12% B_2O_3. In compounds, boron atoms rarely attain outer shells of eight electrons, as the three electrons can form only three bond pairs for a total of six electrons, as in the compound boron trifluoride:

Aluminum is the most abundant metal of the earth's crust (see Table 6–2). The high conductivity of aluminum indicates that it is definitely a metal. One interesting feature of aluminum chemistry is the fact that it forms an amphoteric hydroxide, i.e., an insoluble hydroxide precipitate that will dissolve either in acid or in additional hydroxide.* Suppose we have precipitated $Al(OH)_3$ by adding NaOH solution to $Al(NO_3)_3$ solution. The precipitate will dissolve in acid,

$$Al(OH)_3(s) + 3H^+(aq) \rightleftharpoons Al^{3+}(aq) + 3H_2O(l), \quad (12–10)$$

or in additional hydroxide (base),

$$Al(OH)_3(s) + OH^-(aq) \rightleftharpoons Al(OH)_4^-(aq). \quad (12–11)$$

The soluble species $Al(OH)_4^-$ is a complex ion called the aluminate ion. Aluminum is just one example of a metal that forms an amphoteric hydroxide or oxide; others include gallium (Ga, also in Group IIIA), beryllium, zinc, tin, lead, and chromium.

As can be seen by its position in the electromotive series of Table 12–3, aluminum is an active metal and a better reducing agent than nearly all other metals except the alkali metals and the alkaline earths. Its reducing strength is further enhanced in reactions involving oxygen, because aluminum oxide has very strong bonds. One of the most important reducing reactions of aluminum is the thermite reaction, which occurs between iron oxide, Fe_2O_3, and aluminum:

$$2Al(s) + Fe_2O_3(s) \longrightarrow 2Fe + Al_2O_3(s) + energy \quad (12–12)$$

This reaction liberates such an enormous amount of heat that it produces red-hot molten iron that can melt or burn its way through many materials. This reaction has many important

---

*   Throughout this chapter the terms "hydroxide" and "oxide" can generally be used almost interchangably when discussing compounds in contact with water solutions, as they differ only by the addition or removal of some water molecules. For example, $Al(OH)_3(s)$ can also be considered aluminum oxide, Al_2O_3, by removal of water:

$$2Al(OH)_3(s) \longrightarrow Al_2O_3(s) + 3H_2O(l).$$

In many cases, large and often unknown numbers of water molecules remain attached to the precipitate. For example, when a chemical reaction shows the precipitation of ferric hydroxide, $Fe(OH)_3$, the actual gelatinous precipitate formed contains a large amount of water. One could probably refer to it most accurately as "hydrated ferric oxide," with the corresponding formula, $Fe_2O_3 \cdot nH_2O(s)$, where n is a large unknown number.

uses; for example, if the reaction mixture is placed at the seam between two pieces of iron, the heat and the molten iron welds the pieces together. Years ago, the transcontinental railroad was completed thanks to the ability of the thermite reaction to weld the tracks together. The reaction is also used occasionally in the refining of iron ore to reduce iron oxide to iron. The thermite reaction can also be used to reduce the oxides of other metals such as chromium.

Aluminum oxide (alumina) occurs in nature as corundum, the hardest natural substance except for diamond. Impurities in corundum cause colored variations that are sought as precious stones; for example, red (ruby) and blue (sapphire) are caused by chromic oxide and titanium oxide, respectively. The rocks of the interior of the earth are made up largely of SiO_4^{2-} tetrahedra packed together in various ways with Fe^{2+} and Mg^{2+} ions. At the earth's surface, many of the important rock types are alumino-silicates, in which the Si atoms of some tetrahedra have been replaced by Al atoms. Aluminum is present in such widespread substances as potassium feldspar ($KAlSi_3O_8$) and clay minerals.

Aluminum is used extensively, as it is a light-weight metal which is a good electrical conductor and rather strong when alloyed with small amounts of other metals. Dural, consisting of 95% Al, 4% Cu, 0.5% Mn and 0.5% Mg, is stronger than pure aluminum, but less resistant to corrosion. For this reason, Dural is often coated with pure aluminum which rapidly becomes coated with protective Al_2O_3, in reaction with the O_2 of the air. From the position of Al in Table 12-3, we would expect it to react with water to release H_2 gas. However, if the reaction occurs, it is too slow to be detected, perhaps because the Al sample is soon coated with Al_2O_3.

Aluminum is so reactive that the pure metal is never found in nature. Since it is so abundant, aluminum could be obtained from many sources such as clay. However, the only economically competitive source of aluminum is *bauxite*, a hydrated aluminum oxide. After impurities and water are removed from the bauxite, it is dissolved in cryolite, Na_3AlF_6, which melts at a low temperature (about 100°C). Aluminum metal is then separated from the mixture by electrolysis, as described in Section 10-4. This step consumes a great deal of electrical energy, about seven to eight kilowatt-hours per pound of aluminum. The enormous rate of growth of aluminum demand places an added drain upon our already scarce energy supplies. We could effect large savings of energy if aluminum products were more extensively recycled; how-

ever, most recycled aluminum is of low quality because of contamination by other undesirable metals that are difficult to remove from the aluminum.

Group IVA

The Group IVA elements have four valence electrons. The group is headed by carbon, definitely a nonmetal, whose chemical properties are discussed in Chapter 13.

Although many of the chemical properties of silicon (Si) are those of a nonmetal, highly pure silicon is a semiconductor and, in that respect, a semimetal. Silicon is the third most abundant element of the whole earth and the second most abundant of Earth's crust. It is a major constituent of nearly all types of rocks, whose minerals contain mostly tetratrahedral SiO_4^{2-} units linked together to form long chains, flat sheets, or three-dimensional lattices, as well as metals such as Fe, Mg, Ca, K, and Na. The rocks of Earth's mantle are thought to consist in large part of such minerals as olivine (Mg_2SiO_4 and Fe_2SiO_4) and pyroxenes (mainly $MgSiO_3$ and $FeSiO_3$). Many common minerals at the surface of the earth also have silicon-oxygen tetrahedra as their basic structures: "silica" or "quartz," SiO_2, present in large amounts in sand; potassium feldspar (mainly $KAlSi_3O_8$) and albite ($NaAlSi_3O_8$); montmorillonite $[AlSi_2O_5(OH)\cdot xH_2O]$ and kaolinite $[Mg_3Si_4O_{10}(OH)_2]$, to name but a few. In many of these minerals some aluminum atoms have taken the place of silicon atoms in the center of the tetrahedra. Sand, silica, and other silicon-bearing minerals are used in the manufacture of glass. As discussed in Section 10-9, one silicate-containing mineral, asbestos, is a serious air pollutant that causes a fibrosis disease or lung cancer.

The most important industrial uses of silicon and germanium take advantage of their properties as semiconductors. In a pure semiconductor material at 0 K, all the valence electrons are bound to their atoms, but not as strongly as in a nonconductor. As the temperature rises, a small fraction of the valence electrons become detached from their atoms. Both the free electrons and the positive "holes" left behind can move throughout the material and, thus, carry a small current.

The properties of semiconductor crystals can be greatly altered by adding impurities to the pure crystals. Each silicon or germanium atom has four valence electrons, which it shares with four neighboring atoms to form a complete valence shell of eight shared electrons. If an atom of a Group

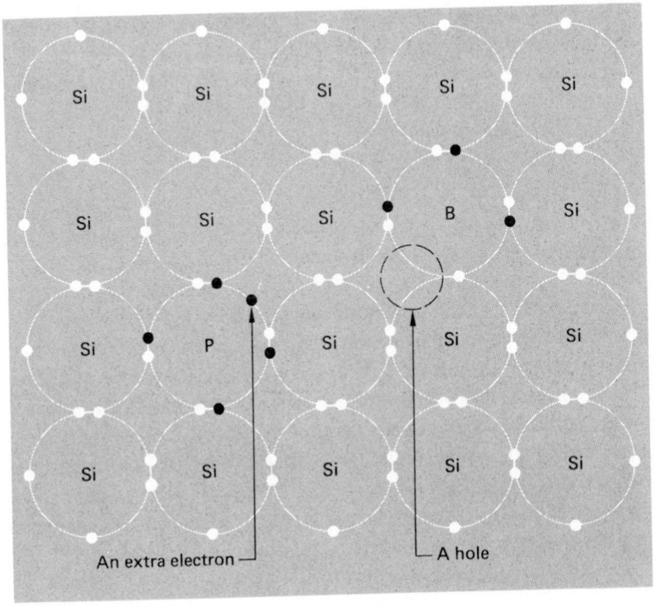

Fig. 12–8 In a crystal of pure silicon, each atom is surrounded by eight valence electrons, four pairs of electrons shared with four neighboring atoms. When phosphorus is substituted for a silicon atom, the ninth valence electron can easily become detached, becoming a free conduction electron. When boron is substituted for silicon, there are only seven valence electrons, and so an electron from one of the silicon atoms can move to the boron atom, leaving behind a positive electron "hole."

V element such as phosphorus is substituted for the silicon atom, the five electrons of phosphorus make a total of nine in the vicinity of the phosphorus atom (see Fig. 12–8). One of the nine electrons can easily be detached from its site in the crystal and become a conducting electron. Such "doped" silicon is called "n-type," because the conduction is by negative electrons. Conversely, if a silicon atom is replaced by a Group III atom such as boron, the latter has one fewer valence electron than silicon does. The resulting electron vacancy can act as a positive hole and promote conduction. Silicon or germanium doped with boron is called "p-type," because the conduction is by positive holes.

When a p-type semiconductor is in contact with an n-type, the resulting "p-n junction" acts as a diode in which the resistance to the flow of electricity depends on the direction of the applied voltage. Semiconductor diodes have replaced the diode vacuum tubes that were used some years ago in electronic circuits. Transistors are somewhat more complicated semiconductor devices that have replaced triodes and other types of vacuum tubes in modern electronic circuits. Transistors are much smaller and much more resistant to damage from mechanical shock, and they draw far less current than the vacuum tubes they have supplanted. Because of these advantages, transistors have had an enormous impact on the design of electronic equipment. They made possible the development of transistor radios, which are physically small and draw very low current, so that a single battery can supply the power for many weeks.

Transistorized circuits have made possible great reductions in the size of computers. Computers that would have occupied entire rooms in the early 1950s can now be fitted into a shoe box. The space program would have been impossible without transistors. Tube circuits for the necessary space communications would have been too bulky and unreliable and would have drawn too much current for space travel. Transistors in the late 1950s and early 1960s replaced individual vacuum tubes, but now, in the early 1970s, entire electronic amplifiers and other circuits can be etched onto a single chip of silicon smaller than the head of a pin.

Tin and lead are metallic in most ways, although one allotrope of pure tin, grey tin (stable below 13°C) is nonmetallic, whereas the white tin, stable at room temperature, is metallic. As one can see in Table 12–3, Sn and Pb are much less active than the other metals we have discussed. Tin is used a great deal as a protective coating, e.g., on steel in "tin" cans, because of its chemical inertness. For similar reasons, lead was previously used widely for pipes (and still is to some extent for drains in laboratories that use acids), and the word "plumbing" is derived from the Latin name of lead, *plumbum. White lead,* $Pb_3(OH)_2(CO_3)_2$, *red lead,* Pb_3O_4, and lead chromate, $PbCrO_4$ (also called "chrome yellow"), were previously used extensively in paints and pigments.

About one-third of the lead consumed in the United States goes into the manufacture of lead storage batteries, and one-sixth goes into lead compounds that are added to gasoline (Chapter 14). Other major uses of lead include ammunition, type metal, and solder. The lead in paint, gasoline and, occasionally, ceramics and plumbing can cause lead poisoning, as discussed in Section 12–8.

Group VA

In Group VA, nitrogen and phosphorus are clearly non-metals, arsenic (As) is a semimetal, and antimony (Sb) and bismuth (Bi) are considered metals on the basis of their electrical conductivity, although the oxides of both elements are acid anhydrides. Arsenic, Sb, and Bi are rare elements in nature. Arsenic is used mainly in pesticides. Antimony is used principally as an alloying additive to lead in storage batteries, type metal, solder, and ammunition. The melting point of the alloy is lower than that of lead, but the alloy is harder than the very soft pure lead.

12–4 THE TRANSITION METALS

Some of the most important metals are included in the several series (rows) of transition metals in the periodic table (see Fig. 12–9). The elements of the first transition series, particularly, are used extensively in industry and consumer goods. All the metals of the first transition series except Sc, Ti, and Ni are known to be essential elements of living systems. Because of chemical similarities, the coinage metals (see Table 12–2) are often discussed together with the transition metals, and we shall do the same.

Although there are chemical similarities among the transition metals, both among elements of a given horizontal row and among those of a column of the periodic table, there are also great differences, both vertically and horizontally. Common properties of the transition metals include the fact that most are rather good electrical conductors, most can exist in two or more oxidation states, and most form complex ions in which the metal ion is surrounded by two, four, or six electronegative groups. Many of the compounds and solutions of the transition metals are colored because they contain unpaired electrons.

The reactivity of the transition metals varies enormously. Although many of their half-cell reactions lie above that of hydrogen in the electromotive series (see Table 12–3), most

Transition Metals

	IIIB	IVB	VB	VIB	VIIB	VIII			IB	IIB	
Be 4											B 5
Mg 12											Al 13
Ca 20	Sc 21	Ti 22	V 23	Cr 24	Mn 25	Fe 26	Co 27	Ni 28	Cu 29	Zn 30	Ga 31
Sr 38	Y 39	Zr 40	Nb 41	Mo 42	Tc 43	Ru 44	Rh 45	Pd 46	Ag 47	Cd 48	In 49
Ba 56	La 57	Hf 72	Ta 73	W 74	Re 75	Os 76	Ir 77	Pt 78	Au 79	Hg 80	Tl 81

Fig. 12–9 Portion of the periodic table showing the transition metals.

react slowly, if at all, with acidic solutions. In many cases, the initial reaction apparently causes formation of a thin, protective layer at the surface of the metal that protects the bulk of the material from acid attack.

Some members of the second and third transition series, such as Au, Pt, Pd, and their neighboring elements, lie so low in the electromotive series that they can be dissolved only by very strong oxidizing agents. One of the few solvents that will dissolve gold, for example, is aqua regia, a mixture of hydrochloric and nitric acid. The nitric acid provides the oxidizing strength needed and the hydrochloric acid provides both additional H^+ ions and the Cl^- ions that help in the dissolution by forming complex ions with the gold:

$$Au\,(s) + 3NO_3^-(aq) + 4Cl^-(aq) + 6H^+(aq) \rightleftharpoons$$

$$AuCl_4^-(aq) + 2NO_2(g) + 3H_2O\,(l). \quad (12\text{–}13)$$

Because of their strong resistance to chemical attack, noble metals such as gold are commonly found as pure flecks or nuggets in nature. They are useful in jewelry, coins, or silverware because of their durability. Pure gold is so malleable

Ancient Mexican jewelry made of gold. (Courtesy of the American Museum of National History.)

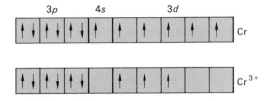

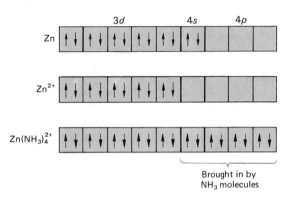

Fig. 12–10 Orbital-filling diagrams for Cr and Cr^{3+}. Each box stands for a Cr orbital and each arrow for an electron. Arrows pointing up and down denote different spin directions of the two electrons permitted to occupy each orbital.

Brought in by NH_3 molecules

Fig. 12–11 Orbital-filling diagram for Zn, Zn^{2+}, and $Zn(NH_3)_4^{2+}$.

that it does not keep its shape well, but when alloyed with copper and silver it retains its chemical inertness while becoming much harder.

Each transition series begins just after the s orbital of the valence shell has been filled. Electrons are then added to the d subshell of the next to outermost shell. Chemical similarities among members of a transition series stem from the fact that neutral atoms of a series have the same valence shell configuration with two electrons in the s orbital. Differences in chemical behavior and the many oxidation states of the transition metals result from the fact that electrons in the inner d subshell, as well as in the p orbitals of the valence shell, are rather close in energy to the s electrons. Therefore, some d-subshell electrons are occasionally removed when the atoms are ionized, as shown in Fig. 12–10.

The presence of the empty p orbitals and often incompletely filled d orbitals (that is, 4p and 3d in the first transition

series) with energies very close to that of the s orbital account for the ability of transition metals to form many kinds of soluble complex ions. Consider the formation of $Zn(NH_3)_4^{2+}$ ions. The nitrogen atom of NH_3 has an unshared pair of electrons in an orbital directed away from the hydrogens. In an oversimplified way, we can picture the formation of bonds in which the electrons of those pairs enter vacant 4s and 4p orbitals of the Zn^{2+} ion, as in Fig. 12–11.

In Zn^{2+}, the 4s and the three 4p orbitals are empty and thus available for occupancy by the incoming electron pairs of the NH_3 atoms to form a tetrahedral complex ion (Fig. 12–12).

Similarly, the orbitals of Co^{3+} have space available for the twelve additional electrons from six molecules of NH_3, thus forming the octahedral complex ion $Co(NH_3)_6^{3+}$, as shown in Fig. 12–13. Other more complicated complex ions of the transition metals are discussed in Section 11–2.

There are many ligands besides NH_3 that can form complex ions, such as CN^-, OH^-, Cl^-, and CO. Many of the transition metal ions have such a strong tendency to form complexes that it is virtually impossible to observe the "bare" ion in water solutions. If there are no other ligands present, the water molecules themselves can form complex ions by using the unshared electron pairs of the oxygen atoms, as in $Cr(H_2O)_6^{3+}$.

Fig. 12–12 The structure of the complex ion $Zn(NH_3)_4^{2+}$.

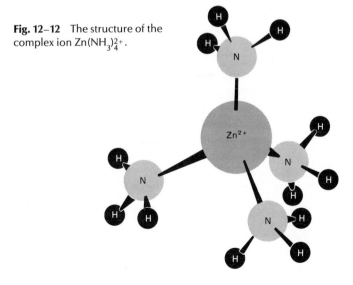

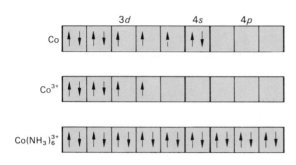

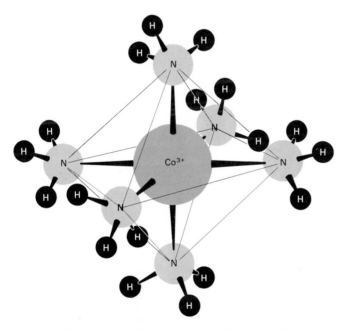

Fig. 12–13 Orbital-filling diagram for Co, Co^{3+} and $Co(NH_3)_6^{3+}$ and the structure of $Co(NH_3)_6^{3+}$.

12–5 THE TREATMENT OF METAL ORES

Iron is the most abundant element of the earth and the second most abundant metal of earth's crust (see Table 6–2). Because of the useful properties of iron and its great abundance, the volume of its use is far greater than that of any other metal.

Iron ores occur mainly as mixtures of various oxides: *hematite* (Fe_2O_3), *magnetite* (Fe_3O_4), and *limonite* ($Fe_2O_3 \cdot xH_2O$). Some iron occurs in the carbonate, *siderite* ($FeCO_3$). Before use, the ore is crushed into convenient chunks and washed to remove low-density impurities; then, if the ore is magnetic (like magnetite), it may be separated magnetically from rocky impurities. Unless the ore occurs close to coal deposits, it must be transported to a location where coal is available for the next treatment. In the United States, ores from the Lake Superior region are hauled in huge ships to the southern shores of Lake Erie and Lake Michigan for processing in the industrial complexes of Illinois, Indiana, Ohio, and Pennsylvania.

The iron ore, along with coke (a purer carbon form of coal) and limestone ($CaCO_3$), is placed in a *blast furnace*. The reactions that occur in a blast furnace are very complicated, but there are two main objects of the operation: (1) to reduce the iron oxide to metallic iron, and (2) to melt the mixture of iron and rocky materials in order to separate the two liquids formed (see Fig. 12–14).

Hot air blown into the bottom of the blast furnace reacts with coke to form mainly CO:

$$2C(s) + O_2(g) \longrightarrow 2CO(g). \tag{12–14}$$

The reaction releases heat needed to maintain a high temperature in the furnace. The CO gas rises into the ore, reacting with the iron oxide to reduce it to the metallic state, as in

$$Fe_2O_3(s) + 3CO(g) \longrightarrow 2Fe(s) + 3CO_2(g). \tag{12–15}$$

Also, in the presence of the iron, some of the CO forms CO_2 and carbon:

$$2CO(g) \longrightarrow C(s) + CO_2(g). \tag{12–16}$$

The carbon deposits itself on the iron and reduces its melting point from 835°C to 610°C, making it easier to melt. The $CaCO_3$ in the furnace serves mainly as a *flux*, combining with the silicate minerals to form a low-melting glassy material.

The blast furnace is continuously fed the "charge" of ore, limestone, and coke at the top. As the iron and silicates melt, they trickle down into a pool at the bottom. The "slag,"

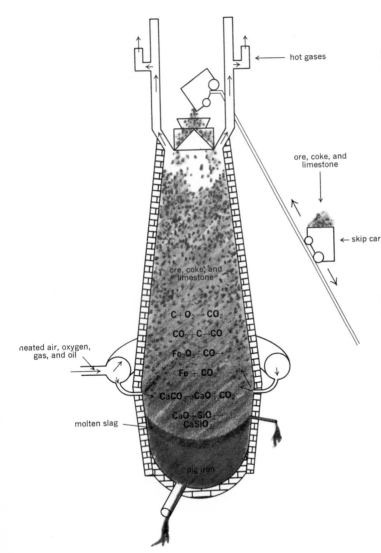

hot gases

ore, coke, and
limestone

← skip car

ore, coke, and
limestone

$C + O_2 \rightarrow CO_2$

$CO_2 + C \rightarrow CO$

$Fe_2O_3 + CO \rightarrow$

$Fe + CO_2$

$CaCO_3 \rightarrow CaO + CO_2$

$CaO + SiO_2 \rightarrow$
$CaSiO_3$

heated air, oxygen,
gas, and oil

molten slag

pig iron

Fig. 12–14 Blast furnace used to reduce
iron oxides to pig iron.

the molten silicates, float on top of the iron. They are periodically skimmed off, leaving the molten iron, which may be cast into "pig iron" ingots or transported directly to nearby machinery for further processing.

The pig iron obtained from the blast furnace is not a very useful product. It generally contains about 4% or more of carbon, small amounts of other impurities such as silicon, and some remaining iron oxide. Pig iron is further processed in one of several types of *converters*, such as an open-hearth furnace. In a converter, hot air or pure oxygen is blown through the mass of molten pig iron, literally burning out most of the carbon, releasing it as CO and CO_2 (see Fig. 12–15). Silicon and manganese impurities react with FeO to form SiO_2 and MnO, which separate from the iron to form slag. The converter is lined with silicate or carbonate materials which act as a flux and are gradually used up.

Most U.S. iron is processed in basic oxygen furnaces (37%) or open-hearth furnaces (50%). The basic oxygen process is fast, taking about 20 to 30 min per batch. The open-hearth process requires several hours per batch, allowing sufficient time for chemical analyses of the material to be made and additional materials, including scrap iron, to be

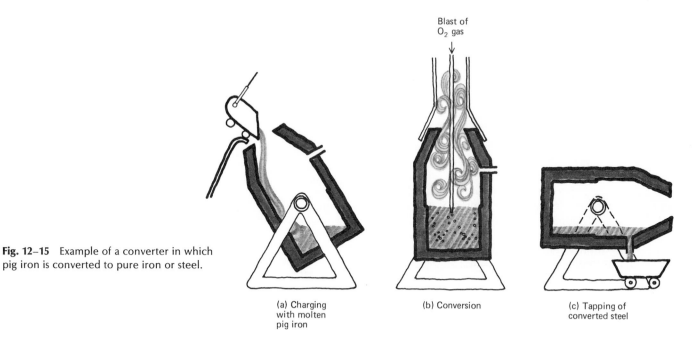

Fig. 12–15 Example of a converter in which pig iron is converted to pure iron or steel.

Blast of O₂ gas

(a) Charging
with molten
pig iron

(b) Conversion

(c) Tapping of
converted steel

added to obtain the desired final product. In both processes, the composition of the materials in the charge must be carefully controlled to ensure that the reactions generate enough heat to keep the material molten and that the reactions needed for removal of impurities occur. These requirements limit the amount of scrap iron that can be added to the "virgin" pig iron in the converter. The open-hearth can use up to 60% scrap, whereas the basic oxygen furnace can use only 30% or less. Electric furnaces, which now produce about 13% of the nation's steel can use about 100% scrap steel. These limitations are important in the recycling of iron.

During or after the conversion process, various elements are added to the iron to produce *steels*, i.e., iron containing impurities that alter its properties in a desired way. Only about one-sixth of the iron is used in fairly pure form, as it is not so hard and strong as steel (although pure iron is ductile and tough). The most common additive is carbon which, in amounts up to about 1.5%, makes the steel hard and strong. About three-fourths of the iron used in the United States is converted into some type of carbon steel. Many metals, especially transition metals, are added to give the steel special properties. For example, stainless steel contains 18% chromium and 8% nickel. These additives produce a steel that is

highly resistant to oxidation and many other chemical reactions. Manganese steel (up to 14% Mn), which is very hard, is used in machines that grind and crush other materials. Chromium-vanadium steels are used extensively in the manufacture of automobile parts, where a tough, elastic metal is required. High-speed cutting tools are made from steels containing 0.65% to 0.85% carbon and various amounts of tungsten (W), chromium, vanadium, molybdenum (Mo), and cobalt. Cobalt is also added to steels from which permanent magnets are made.

The properties of steel also depend on the way in which they are cooled and heat-treated (tempered). Fast cooling produces a hard, brittle steel, whereas slow cooling yields a softer, tougher metal. Knife blades are cooled faster on the cutting edge to produce a hard edge and more slowly on the back to give a tougher body to the blade.

Of the other important metals in use today, some such as manganese and tin occur as oxides and are treated by methods similar to those used for iron. Many others, such as copper, silver, and zinc, occur as sulfides. They are often found together in natural ores, so that it is economical to recover several metals from the same sulfide deposit. Nickel is found both as oxides and sulfides. As explained in Section 10–7, the sulfides are first "roasted" in air to drive off the sulfur as SO_2. In some cases this produces the metal itself (for example, Ag), whereas for other metals, roasting produces the oxide of the metal which must then be converted to the metal with a reducing agent such as carbon.

12–6 RESOURCES AND RECYCLING OF METALS

The world is running out of economically usable deposits of many metals that are critical to the maintenance of present standards in the technological nations and, especially, to the growth of technology in the developing nations. As we have noted throughout the book, the United States with about 6% of the world's population typically uses about 25% to 30% of the world's supply of energy and material resources. If every nation used as much of each resource *per capita* as we do, there wouldn't be enough of several important materials to satisfy the demand.

Professor Dennis Meadows of M.I.T. and his coworkers have examined the rates at which the world is using up valuable resources.* Often when this type of projection is made, one simply divides the known reserves of the resource by the amount presently used each year in order to calculate the time remaining before the resource is used up. This kind of result, called the "static index", is given for a number of resources in Table 12–4.

A major thesis of the Meadows model is that many quantities such as population and the rates of resource and energy usage increase exponentially with time. This is similar to the growth of a savings deposit invested at compound interest. If you invest $100 at 5% compounded annually, at the end of the first year $5 is added to your account, and at the end of the second year $5.25, the additional 25 cents being the interest on the first $5 added to the account. It would take about 14 years to double your money at this rate, whereas it would take 20 years if the interest were paid only on the original $100. It's the same with population: if the population of a country increased 5% per year, it would take only about 14 years for it to double.

Column 4 of Table 12–4 lists the annual increase in use of several resources, and column 5 lists the "exponential index," i.e., the number of years the resource will last when the rate of increase in use is taken into account. We see that unless

* D. H. Meadows, D. L. Meadows, J. Randers, and W. W. Behrens, III, *The Limits to Growth* (Universe Books, New York, 1972).

Table 12–4   Reserves, Usage Rates, and Remaining Years of Supply of Several Important Resources

Resource	Known world reserves (metric tons)	Static index (years)	Projected growth rate of usage (% per year)	Exponential index (years)
Aluminum	1.06×10^9	100	6.4	31
Chromium	7.1×10^8	420	2.6	95
Coal	4.5×10^{12}	2300	4.1	111
Copper	2.8×10^8	36	4.6	21
Gold	1.1×10^4	11	4.1	9
Iron	9.0×10^{10}	240	1.8	93
Lead	8.0×10^6	26	2.0	21
Mercury	1.2×10^5	13	2.6	13
Nickel	6.7×10^5	150	3.4	53
Petroleum	6.5×10^{10}	31	3.9	20
Silver	1.7×10^5	16	2.7	13
Tin	4.3×10^6	17	1.1	15
Zinc	1.1×10^8	23	2.9	18

After Meadows et al., *The Limits to Growth.*

enormous new reserves are discovered, the world will soon encounter severe shortages of many key resources, especially copper, gold, lead, mercury, silver, tin, and zinc. Even if new discoveries increase the known reserves fivefold, the exponential growth rate of usage causes the exponential index to only double or triple.[†]

When a resource runs low, we don't suddenly find some day that we've used up the last ton. Instead, the richest ores are used up first. Then lower-grade ores are used, driving up the price of the product until it becomes impractical to use the resource any more except for the most critical applications. Thus we will never completely exhaust the supply of any particular resource: it will just become too expensive to recover it from low-grade ores. Perhaps it will become economical to extract needed metals from very low-grade deposits, even granite rocks, at some time in the future when new sources of cheap, abundant energy are available.[‡]

[†]   Meadows et al., *The Limits to Growth.*
[‡]   A. M. Weinberg, *reflections on Big Science* (MIT Press, Cambridge, Mass., 1967); see also A. M. Weinberg and R. P. Hammond, "Limits to the Use of Energy," *American Scientist* **58**, 412–418 (1970).

Although energy can help make up for a deficiency of high-grade ores, large amounts of other materials such as acids may be needed to produce the pure metals. Furthermore, the mining of enormous quantities of rocks would scar the landscape and leave vast piles of solid wastes. Thus for the time being, we feel that much greater efforts to conserve Earth's resources should be made.

There are several ways in which we could reduce the rates at which we use metals. One way that is often forgotten is that of keeping the products in service longer. Many consumer products in the United States are constructed so that they are costly or even impossible to repair. In view of high labor costs, it is often cheaper to throw a product away and buy a new one, even when it has a minor malfunction. This practice is a wasteful use of resources.

Today there is considerable interest in *recycling* as a way of conserving resources. Recycling has many other advantages as well.

1. It often substantially reduces the amount of energy needed to manufacture products.

2. It eliminates the SO_2 released to the atmosphere by processing of sulfide ores.

3. It cuts the need for reducing agents such as coal or coke in the processing of oxide ores.

4. It reduces the volume of trash that must be hauled to sanitary landfills or otherwise disposed of.

5. In some cases, it prevents the release of toxic metals such as mercury and cadmium into the environment.

The most economical form of recycling is the reuse of the product itself as, for example, by returnable beverage containers that may be used a dozen or more times. Even if the product itself cannot be reused, many of the advantages noted above are achieved if the material in the product is recovered and reused in a new product. Many used products are a more concentrated "deposit" of the resource than the original ore. A junked car, for example, is a richer source of iron than the same weight of ore. Many metals used by man are more concentrated after he has finished with them than they were in their natural states.

There are, however, some notable exceptions. Much of the silver mined goes into photographic film as silver halide salts. After use, the silver is deposited in family photo albums

and in the negatives of instant-development photographs that litter scenic spots throughout America. Much of the lead produced is used as gasoline additives which become widely dispersed throughout the air, land and water.

Substantial amounts of some metals are recycled today. About half of the iron and steel used is made from scrap. Although some of this is obtained from discarded consumer products, about 60% of it is scrap material from the steel plants and foundries themselves.[*] Iron and steel plants prefer to use this "runaround" scrap because they know its composition better than that of other kinds of scrap. The percentage of scrap used in steel has been declining in recent years, but it is expected to increase in the future with greater use of the electric furnace, which can use entirely scrap material. About one-third of the lead used in the United States goes into the manufacture of lead-acid storage batteries, and nearly all of it is recycled.

Since 1970 many recycling centers have been organized by various civic groups. Although they have evoked much favorable response and cooperation from citizens, the volume of material recycled is still a small fraction of the total used. For example, during 1971 about one-tenth of the ten billion all-aluminum cans manufactured were returned for recycling.[†] More than 100,000 "tin" cans, i.e., steel cans coated with a thin layer of tin, were recovered for further use, accounting for about 2 to 3% of the total manufactured. Because of the tin layer, the cans are not useful as scrap steel, but instead are used in recovering copper from ore-processing solutions[‡] The copper is dissolved from the ore with sulfuric acid and then brought in contact with ground-up tin cans. Since, as shown in Table 12–3, iron is a more active metal than copper, the iron replaces copper in the solution and the metallic copper precipitates out on the cans.

In order to achieve close to 100% recovery of resources, it will be necessary to develop large-scale processes for separation and recycling of materials present in municipal trash. The U.S. Bureau of Mines laboratory at College Park, Maryland, has set up a pilot plant in which the residue from municipal incinerators is processed with machinery of the

*   H. T. Reno and F. E. Brantley, "Iron," in U.S. Bureau of Mines Bulletin 650, *Mineral Facts and Problems* (U.S. Government Printing Office, Washington, D.C., 1970), pp. 291–314.
†   R. R. Grinstead, "Bottlenecks," *Environment* **14** (3), 2–13 (1972).
‡   Grinstead, "Bottlenecks."

Junked cars ready for recycling. (Courtesy of the Institute of Scrap Iron and Steel.)

This mountain of steel fragments is the remains of 30,000 old automobiles that have been processed for remelting by steel mills and foundries. (Courtesy of the Institute of Scrap Iron and Steel.)

sort used in standard treatment of ores.* The refuse is subjected to screening, grinding, shredding, flotation, magnetic separation, and photosensitive separation of glass fragments into clear and colored fractions. For every ton of residue processed, they obtain about 610 lb of ferrous metal, 32 lb of aluminum, 24 lb of copper and zinc and 950 lb of glass.

---

* P. M. Sullivan and M. H. Stanczyk, "Economics of Recycling Metals and Minerals from Urban Refuse," Bureau of Mines Solid Waste Research Program Technical Progress Report 33, April 1971; see also C. B. Kenahan, "Solid Waste—Resources Out of Place," *Environmental Science and Technology* **5**, 594–600 (1971).

H 1																	He 2
Li 3	Be 4											B 5	C 6	N 7	O 8	F 9	Ne 10
Na 11	Mg 12											Al 13	Si 14	P 15	S 16	Cl 17	Ar 18
K 19	Ca 20	Sc 21	Ti 22	V 23	Cr 24	Mn 25	Fe 26	Co 27	Ni 28	Cu 29	Zn 30	Ga 31	Ge 32	As 33	Se 34	Br 35	Kr 36
Rb 37	Sr 38	Y 39	Zr 40	Nb 41	Mo 42	Tc 43	Ru 44	Rh 45	Pd 46	Ag 47	Cd 48	In 49	Sn 50	Sb 51	Te 52	I 53	Xe 54
Cs 55	Ba 56	La 57	Hf 72	Ta 73	W 74	Re 75	Os 76	Ir 77	Pt 78	Au 79	Hg 80	Tl 81	Pb 82	Bi 83	Po 84	At 85	Rn 86

Essential elements shaded

Fig. 12–16 Periodic table showing elements known to be essential to animal life (shaded). (From "The Chemical Elements of Life," by Earl Frieden. Copyright © 1972 by Scientific American, Inc. All rights reserved.)

They are able to make a number of useful products from the materials collected, e.g., high-quality bricks from the glass (with added clay), glass wool, high-purity zinc, a copper-aluminum alloy. The Bureau of Mines has estimated that refuse can be processed at a cost of about two to four dollars per ton and that the recovered materials have a market value of about sixteen dollars per ton. Thus it should be possible for municipal governments or private companies to operate such recycling plants at a profit.*

The Environmental Protection Agency and the Bureau of Mines are conducting studies on handling raw urban trash prior to incineration. This is a more difficult task of separation of materials, but it has greater potential benefits, as air pollution from the incinerator is avoided and some of the paper and other combustibles can be salvaged or used as fuels.

Recycling of materials will surely increase in the future. In order to hasten that day, most products (especially packaging materials) should be designed with their ultimate recycling in mind. For example, all-aluminum cans are much more valuable and easy to process than cans containing metals that are difficult to separate from aluminum. Used steel cans would be more valuable if not coated with tin. In fact, the fraction of steel cans without the tin coating is increasing rapidly today (some are now coated with chromium, which forms desirable alloys with iron). Screw-top glass bottles would be easier to recycle if they weren't left with a ring of metal around the neck.

* The cost of collecting the trash, the major expense of solid-waste handling, should not be charged against this process, since the trash has to be collected regardless of its final disposition. Furthermore, any savings on the cost of transportation and landfilling operations should be credited to the process.

12–7 ESSENTIAL AND TOXIC METALS

Although today we often hear of environmental contamination by metals such as mercury, lead, and cadmium, we must realize that many metals play essential roles in the biochemical systems of plants and animals (see Fig. 12–16). Six of these essential elements typically form 0.05 to 1% of the weight of living matter: Na, Mg, K, Ca, Fe, and Cu. Of the 24 elements that have now been established as essential to animal life, 13 are metals.†

Some of the important functions of the essential metals (as well as some nonmetals required in trace amounts) and typical body concentrations are listed in Table 12–5. Our knowledge of essential elements is obtained by observing the growth of animals whose diets are deficient in certain elements. It is difficult to determine whether some trace elements are essential or not, as it is almost impossible to remove all traces of the element from the animals' diets. Other trace elements, such as those listed in the last part of Table 12–5, may eventually be found to be essential. At present, the functions of several essential elements are not known. Possibly they are connected with enzymes, biochemical catalysts that have large effects on the body although present in very small quantities (see Chapter 15).

† Earl Frieden, "The Chemical Elements of Life," *Scientific American* **227** (1), 52–60 (July 1972).

Table 12–5   Functions of Essential Metals and Other Trace Elements in Man and Animals

Element	Approx. conc. in man and mammals (ppm)*	Principal functions
Essential abundant metals		
Sodium	2,600	Principal cation in body fluids outside of cells
Magnesium	400	Required for certain enzyme activities; important in plants as the central atom of chlorophyll, which absorbs sunlight to initiate photosynthesis
Potassium	2,200	Principal cation of fluids inside cells
Calcium	13,800	Present in minerals of bones, teeth, shells; needed by some enzymes
Iron	50	Central atom of the heme part of hemoglobin; present in many enzymes
Copper	4	Present in enzymes controlling oxidation
Essential trace elements		
Fluorine (a nonmetal)	—	Constituent of bones and teeth
Silicon (a semimetal)	40	Essential for structure of diatoms; found to be essential for growth of chickens but function not known
Vanadium	0.026	Established as essential, but functions not known
Chromium	—	Linked to the action of insulin
Manganese	1	Present in several enzymes
Cobalt	0.04	Present in several enzymes and in vitamin B_{12}
Zinc (a semimetal)	25	Present in several enzymes, especially those that are important in the healing of wounds
Selenium (a nonmetal)	—	Essential for function of the liver
Molybdenum	0.2	Present in several enzymes
Tin	2	Established as essential in rats, but function unknown
Iodine (a nonmetal)	1	Present in thyroid hormones
Other elements that may be essential		
Lithium	0.03	Some influence on nervous system (see Section 10–2)
Rubidium	9	
Boron (a semimetal)	0.2	Not known to be essential in animals, but is needed by some plants for unknown functions
Aluminum	0.5	Not established as essential, but now under study
Nickel	0.025	

Based in part on Frieden, "The Chemical Elements of Life."
*   Brian Mason, *Principles of Geochemistry* (John Wiley and Sons, Inc.,
New York, 1966) 3rd edition, p. 231.

Many metals and other trace elements are toxic to biological life at concentrations above certain levels. Under some conditions, even the essential elements can be harmful. For example, nickel may be an essential element, or at least nontoxic if ingested via food or water. But if nickel is breathed into the lungs in the form of nickel carbonyl, $Ni(CO)_4$, it is a powerful carcinogen.

A great deal has been learned about the toxicity of metals by studies with animals that are fed high levels of various metals and then compared with control groups on the same diet excluding the metal. These studies must be done very carefully in order to obtain reliable results. One complication often encountered is that the toxicity of one element is dependent on the amounts of other elements present. For example, the toxicity of cadmium is reduced if plenty of zinc is available in the diet. High levels of arsenic offer protection against excess selenium. Cattle suffer the "blind staggers," if they eat vegetation which has the ability to concentrate selenium quite strongly. By adding arsenic to the cattle feed, a rancher can protect his cattle against the selenium effects.

Another interesting way to observe relationships between diseases and trace elements is by looking for areas of the world where there is an abnormal incidence of certain diseases. By determining the concentration of various elements of the soil and water in that area, one can see whether excesses or deficiencies of certain elements correlate strongly with the incidence of the disease. As noted in Section 10–4, for example, the incidence of goiter (thyroid gland enlargement) is high in areas of glacially deposited soil that is deficient in iodine. In a small area of Taiwan, an unusually high incidence of "blackfoot disease" has been noted.[*] This condition of dry gangrene, which affected up to 0.5% of the people and was fatal to about 15% of those affected, was noticed after 1910 when some deep wells were dug for a water supply. Analysis of the water showed that it contained 1.2 to 2 ppm of arsenic, which caused about 60% of the affected persons to have arsenical dermatitis and 5% to develop skin cancer.

The soil and water in some areas of the western United States are deficient in zinc, thus stunting the growth of cattle and hogs. Their growth rates can be raised to normal by adding zinc to their feed. Zinc deficiencies in humans are observed in some areas of the world. So far, zinc deficiencies have not been a problem in the United States because our food and water pick up traces of zinc from contact with galvanized (zinc-coated) surfaces. However, as galvanized containers and water pipes are gradually replaced with other materials such as plastic, we may have to start supplementing our diets with zinc.

Many other conditions are known to occur with high incidence in certain areas, but in many cases the causative factors have not been identified. For example, Japanese men have an abnormally high incidence of stomach cancer (seven times the rate for American men). It has been suggested that the condition is caused by the Japanese preference for talc-coated rice, since talc frequently contains small asbestos particle impurities,[†] which are known to cause cancer when embedded in lung tissue (see Section 10–9). However, the condition could also be related to some excessive trace element.

It is becoming more difficult to correlate diseases with trace element concentrations in soil and water in the United States because great quantities of our food and beverages are shipped in from other parts of the world. Nevertheless, distinct differences in death rates for similar population groups living in different areas of the United States have been found.[‡] Figure 12–17 shows the death rate from all causes among white males between the ages of 45 and 64.[§] Death rates in the most heavily shaded areas (e.g., the South Atlantic States) are about twice as high as those in much of the central United States. When the death rate is broken down separately into various causes, one finds a pattern very similar to that of Fig. 12–17 for deaths involving the heart and circulatory system, e.g., hypertension (high blood pressure). Not all diseases exhibit the same geographic pattern: the death rate for rheumatic heart diseases is high in Montana, Idaho, Utah and Colorado, and low in the South Atlantic States. The death rates for females often exhibit quite different patterns from those for men.

* H. L. Cannon and H. C. Hopps, eds., Environmental Geochemistry in Health and Disease (Geological Society of America, Boulder, Colo., 1971).

† R. R. Merliss, "Talc-Treated Rice and Japanese Stomach Cancer," Science 173, 1141–1142 (1971).
‡ H. I. Sauer and F. R. Brand, "Geographic Patterns in the Risk of Dying" in Cannon and Hopps, Environmental Geochemistry in Health and Disease.
§ These various specifications are made to eliminate any death rate differences due to race or sex. One important remaining variable is economic status, which may be quite different in the various areas, leading to differences in medical care, nutrition, and so on.

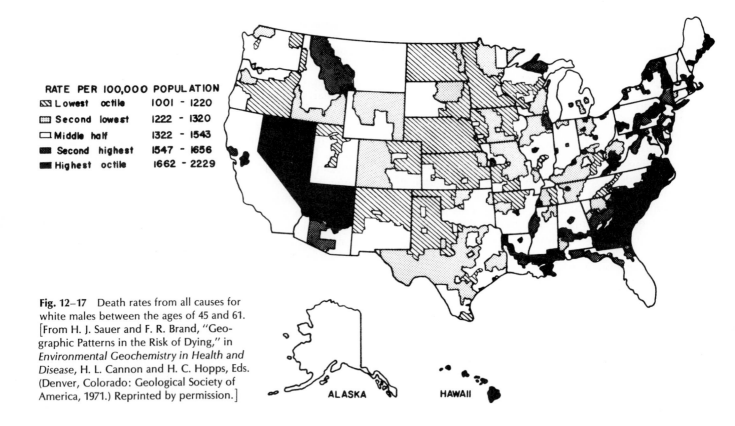

RATE PER 100,000 POPULATION	
Lowest octile	1001 - 1220
Second lowest	1222 - 1320
Middle half	1322 - 1543
Second highest	1547 - 1656
Highest octile	1662 - 2229

ALASKA   **HAWAII**

Fig. 12–17 Death rates from all causes for white males between the ages of 45 and 61. [From H. J. Sauer and F. R. Brand, "Geographic Patterns in the Risk of Dying," in *Environmental Geochemistry in Health and Disease*, H. L. Cannon and H. C. Hopps, Eds. (Denver, Colorado: Geological Society of America, 1971.) Reprinted by permission.]

The high death rate among men in the South Atlantic states is not yet definitely correlated with an environmental effect. One possible explanation is the fact that ground water in the area is much "softer" than in other areas of the country; i.e., it contains smaller concentrations of Ca^{2+}, Mg^{2+}, and HCO_3^- ions (see Section 10–3). Soft water dissolves trace quantities of various metals more readily than hard water; thus it may be that the soft water picks up trace quantities of toxic metals from soil, water mains, or pipes in the house (see also the discussion of cadmium in Section 12–8).

From studies with animals, accidental exposure of certain population groups, and geographical occurrence of some diseases, we know the symptoms caused by high levels of certain elements. Table 12–6 lists the symptoms caused by several known toxic elements, major sources of the elements, and an indication of the elemental concentrations at which effects may become apparent. There are large differences in

the toxicities of the various elements. Although arsenic is a classical poison from murder mysteries, it is not as toxic as some other elements. Beryllium is far more toxic, as just a few micrograms deposited in the lungs can cause berylliosis, a disease that is often fatal.

12–8 METALS IN THE ENVIRONMENT

Among the elements listed in Table 12–6, those that probably present the most severe threat to human health are lead, mercury, cadmium, and possibly beryllium. They are discussed in detail in this section.

Lead

Lead is a toxic metal whose natural abundance in the environment is quite small. As shown in Fig. 6–8, the cosmic abun-

Table 12–6   Major Sources and Biological Effects of Several Toxic Trace Elements

Element	Sources	Symptoms	Levels at which symptoms may occur
Beryllium	Coal burning, rocket fuel; industrial exposure in nuclear energy industry	Degenerative lung disease called "berylliosis." Lung cancer	Only about 10 μg in lungs of chronic sufferers. Could accumulate in 250 days breathing air at 2 μg/m^3
Nickel	Diesel oil, residual oil, additive in some gasolines, metal industries	Lung cancer. Nickel carbonyl, $Ni(CO)_4$ is a known carcinogen	
Arsenic	Fossil fuel burning; impurity in phosphates in detergents and fertilizers; processing of sulfide ores	Variety of gastrointestinal disorders; skin disorders, probably carcinogenic. Requires very large doses for toxic effects. Moderate amounts protect against selenium toxicity	Lethal dose over 24-hr period about 40 mg per kg of body weight in mammals
Selenium	Combustion of any high-sulfur fuel; roasting of sulfide ores	May cause dental cavities; carcinogenic in rats, but an essential element. Causes "blind staggers" in animals	Lethal dose over 24-hr period about 4 μg per kg of body weight in mammals
Cadmium	Electroplating industry; associated with zinc in galvanized products; some gasolines; tobacco smoke	Hypertension and other cardiovascular diseases; in very high doses causes "ouch-ouch" disease, painful affliction of bones, joints; possible carcinogen	Ingestion of tens of mg over few days may cause severe problems
Mercury	Chloro-alkali process for mfg. of Na, Cl_2; fungicides and mildewcides in paints, seeds, paper and pulp industry	Organic Hg compounds most toxic—cause damage to brain and nervous system	Daily intake of >0.5 μg leading to 5 ppm in brain
Lead	Leaded gasolines; lead-based paint; some insecticides	Damage to brain and nervous system; at low levels nausea, irritability.	Blood level of lead >0.8 ppm

See also "Trace Metals: Unknown, Unseen Pollution Threat," *Chemical and Engineering News*, July 17, 1971, pp. 29–33. In addition, see Anthony Tucker, *The Toxic Metals* (Ballantine Books, New York, 1972), and H. A. Schroeder, "Metals in the Air," *Environment* **13** (8), 18–32 (1971).

dance of lead is about ten thousand times smaller than that of iron. However, so much lead is used in modern society that its concentration in our surroundings is far greater than its natural occurrence. Despite the small cosmic abundance of lead, its concentration in urban atmospheres (typically 1 to 3 μg/m^3) is about the same as that of the most common metals (Na, Fe, Al, and so on). Many of the environmental problems we have discussed throughout the book are potential problems—the greenhouse effect, the destruction of stratospheric ozone, and others that may cause serious environ-

mental degradation in the future. Unfortunately, lead is a clear and present danger. Hundreds, perhaps even thousands of persons, mostly children, have died of lead poisoning in this country over the past several decades. Thousands more have survived, but many have been left with permanent disabilities such as mental retardation.

There are two major sources of environmental lead, lead-based paint and leaded gasoline, plus many smaller sources. Until about twenty years ago, most paint contained large amounts of lead compounds. Today the lead has largely been

Old paint, which is likely to be high in lead content, is a hazard to young children. (Courtesy of the Food and Drug Administration.)

replaced by compounds of titanium and zinc, but many older houses still have interior surfaces containing lead-based paints. This problem is especially prevalent in older, inner-city houses and apartments. In a survey of housing in Washington, D.C., for example, about 40% of the rooms tested were found to contain lead paint.* Small children often ingest toxic amounts of lead by eating paint chips that flake from the walls and woodwork or by chewing on chunks of painted putty that fall from the windows. In many large cities, 25 to 50% of the inner-city children are found to have an abnormally high lead concentration in their blood.†

The combustion of leaded gasolines by automobiles accounts for most of the lead on particles suspended in the urban atmosphere as well as the high lead concentration (typically 5000 to 15,000 ppm, or 0.5 to 1.5% by weight) in dust along city streets. Although most lead poisoning of urban children has been blamed on lead paint, automotive lead is being increasingly recognized as a contributor to the prob-

lem. If a child gets street dust on his lollipop, pacifier, or hands, he can easily transfer a significant amount to his mouth. It is estimated that a child could suffer lead poisoning after ingestion of only 1/24th of a teaspoon of street dust a day for eight months.‡ Under normal circumstances, an urban resident takes more lead into his body each day from food (400 μg) and water (10 μg) than from the atmosphere (27 μg). However, only about 5% of the lead in food and water is absorbed by the body, while about 40% of the inhaled lead is retained.§ Thus the daily retention of lead is about 20 μg from food and water and 10 μg from the air. An unknown fraction of the lead in food and water originates in the atmosphere. Airbone particles are deposited effectively on leaves of plants and trees. Rain and snow eventually bring all lead-bearing particles down to the ground, where the lead contaminates soil or water.

Other sources of lead affect certain segments of the population. A $1\frac{1}{2}$ pack-a-day cigarette smoker inhales about 24 μg of lead per day in the smoke, retaining about 10 μg. Some people receive small amounts of lead from water passed through lead pipes that are still in use in some older houses and city water mains. Canned foods and beverages may dissolve some lead from solder joints on the cans. The glaze on most ceramics contains lead. If the glaze is fired at a high enough temperature, the lead presents no hazard. But if the glaze is improperly fired, some of the lead from the glaze is taken up by food or beverages that come in contact with it, especially if they are acidic (e.g., vinegar). One unusual but often potent source of lead is "moonshine" liquor, which is often made in stills put together from junked car radiators and other parts containing solder.

An average United States urban resident has a lead concentration in his blood of from 0.05 to 0.2 ppm, about 100 times that estimated for an average pretechnological man.§ Present-day lead levels are so high that there is little margin for safety. Any unusual exposure to lead from the sources noted above can easily raise the lead concentration in the blood to the range of 0.5 to 0.8 ppm, at which symptoms of lead poisoning start to appear. Excessive lead interferes with the action of enzymes that are needed for the manufacture of

*   J. Y. Smith, "City Sees Greater Lead Risk," Washington *Post*, Oct. 19, 1971.

†   R. J. Bazell, "Lead Poisoning: Combating the Threat from the Air," *Science* **174**, 574–576 (1971); see also B. Barnes, "Car Fumes Seen Child Threat," Washington *Post*, Feb. 13, 1972.

‡   "A Stringent Goal on Lead in Urban Air," *Science News* **101**, 133 (1972).

§   C. C. Patterson, "Contaminated and Natural Lead Environments of Man," *Archives of Environmental Health* **11**, 344–360 (1965).

Scientists at the Food and Drug Administration test pottery samples for lead content in the glaze (above). (Courtesy of the Food and Drug Administration.)

The glaze on this dish has a lead content of 160 ppm, compared to a safe level of 7 ppm (below). (Courtesy of the Food and Drug Administration.)

red blood cells, causing a form of anemia. Lead also causes nausea, vomiting, constipation, and stomach cramps. The most serious effects of lead are those on the nervous system, starting with irritability and depression at low levels and progressing through convulsions, coma, and death at higher levels.

There are typically 100 or more cases of lead poisoning each year in each of several large United States cities, mostly involving children. A few years ago, before quick methods of diagnosis and treatment were widespread, many of the victims died. Today, doctors are more aware of the problem and the poisoning is more accurately diagnosed. Complexing agents such as EDTA (see Section 11–2) are given to the patient in order to form soluble complexes with the lead and flush it out of the body. However, in severe cases, even if the victim survives, he is often left with permanent mental disability or retardation.

What can be done about the problems of lead in the environment? As a start, lead-based paints have been prohibited for interior use. Several cities have initiated massive surveys to detect lead paint in houses. When some is found it is removed or covered over with plasterboard (whichever can be done more cheaply). Possibly in the future, lead paints could be rendered harmless by covering them with paint that contains a complexing agent such as EDTA. The complexing agent would be ingested along with the paint, thus carrying the lead out of the body before it could harm the person. Several petroleum companies now market gasolines containing little or no lead. The Environmental Protection Agency will probably have a timetable for the elimination of lead from gasoline over the next few years (see Section 14–6). Clearly, stern measures must be taken to reduce man's exposure to lead. The known cases of lead poisoning may be just the tip of the iceberg. Various massive screening programs have shown that 30% or more urban children have elevated

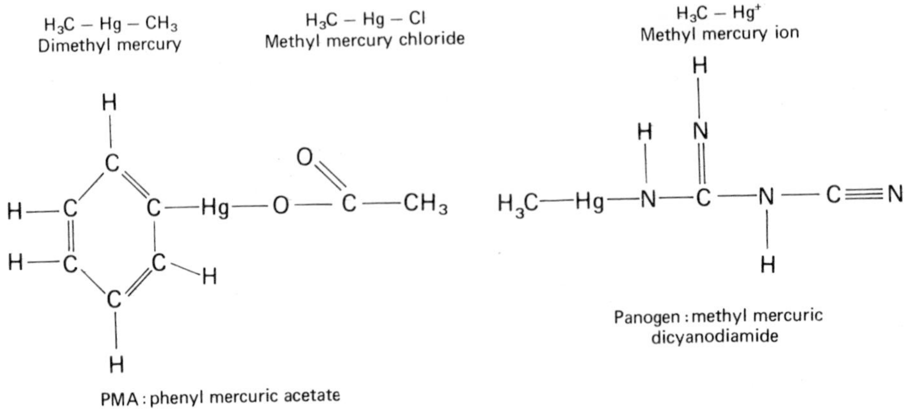

H₃C — Hg — CH₃
Dimethyl mercury

H₃C — Hg — Cl
Methyl mercury chloride

H₃C — Hg⁺
Methyl mercury ion

Fig. 12–18 Examples of organic mercury compounds.

PMA: phenyl mercuric acetate

Panogen: methyl mercuric dicyanodiamide

body burdens of lead. What are the effects of lead on children who do not have obvious symptoms? We don't know, but one wonders how much mental retardation has resulted from excessive lead levels.

The Mercury Scare

Mercury, which is a dense, conducting liquid at room temperature, has many uses in modern society. It is used extensively in electrical equipment such as the "silent" switch, in which the circuit is completed by dipping a wire in a pool of mercury. The bright lights along streets and highways as well as the fluorescent lights in homes and offices contain mercury vapor. As noted in Section 10–2, mercury is used as a solvent for metallic sodium in the chloro-alkali process in which Cl_2 gas, NaOCl bleach, and sodium hydroxide are made by electrolysis of NaCl. Amalgams of silver and tin in mercury are frequently used for dental fillings.*

Compounds of mercury have long been used as medicines, for instance, calomel (Hg_2Cl_2), which was administered for various internal disorders. As early as the sixteenth century, mercury compounds were used in the treatment of syphilis, leading to the old maxim, "One night with Venus and six months with Mercury." As shown in Table 12–7, mercury is still used in phamaceuticals such as the disinfectant mercurochrome. Mercury is also used as a catalyst in the manufacture of consumer products such as plastics (although little mercury is present in the final product).

In recent years, organic mercury compounds have come into wide use. In these compounds, mercury atoms are bonded to carbon atoms, as shown in Fig. 12–18. Many of the organic mercury compounds, especially Panogen and PMA, are used extensively as fungicides and mildewcides to prevent fungal attack on seeds in moist ground, on wood pulp used in the paper industry, and on latex paints while still in the can, which would release gases that could explode the can.†

The toxicity of metallic mercury and inorganic mercury compounds has been known for many years. The inhalation

---

* The concentration of mercury in a person's blood remains unusually high for a couple of weeks after teeth are filled with silver amalgams. Since this mercury is in the elemental, or at least inorganic form, it probably causes no great hazard. The greater hazard is for dentists and their assistants who constantly breathe air containing about a thousand times as much mercury vapor (1 to 5 $\mu g/m^3$) as is present in normal outside air (about 0.003 $\mu g/m^3$). See R. S. Foote, "Mercury Vapor Concentrations Inside Buildings," *Science* 177, 513–514 (1972).

---

† Foote, in "Mercury Vapor Concentrations," finds that rooms painted with latex paints containing mercury mildewcides have high atmospheric concentrations of mercury vapor, typically 0.1 to 0.2 $\mu g/m^3$, for many months after application. The mercury compounds could be, and probably will be, replaced with nonmercuric mildewcides.

Table 12–7   U.S. Uses of Mercury, 1969*

Use	Amount (metric tons/yr)
Chloro-alkali plants	714
Electrical apparatus	634
Paints (mainly as mildewcide in latex paint)	335
Industrial controls in instruments	240
Dental preparations	105
Catalysts	102
Agriculture (including fungicides and bactericides in industry)	93
General laboratory use	70
Phamaceuticals	25
Pulp and paper manufacture	19
Other	340
Total	2729

*   "Mercury in the Environment," U.S. Geological Survey Professional Paper 713 (U.S. Government Printing Office, Washington, D.C., 1970).

of sizable amounts of mercury vapor, which has frequently occurred in industrial plants and mining operations, causes irritation and destruction of lung tissue, chills, fever, coughing, pains in the chest, and tremors. Ingestion of inorganic compounds such as mercuric nitrate, $Hg(NO_3)_2$, causes such symptoms as inflammation of the gums, general irritability, and tremors.

The effects of overexposure to mercury can often be cured by "sweating" the mercury out of the body with heat lamps and ultraviolet light.* But if the exposure is too great, the victim is left with severe tremors and mental disability.

Years ago, felt-hat workers placed furs into vats of mercuric nitrate for preparation of the felt. Many workers suffered mercury poisoning because of absorption of $Hg(NO_3)_2$ through the skin or inhalation of mercury vapor. Many suffered tremors, a slurring of speech, loss of teeth, walking difficulty, and mental disorders. It's quite possible that the Mad Hatter of Lewis Carroll's *Alice in Wonderland* was based on typical victims of that occupational disease.†

For years industries have been releasing pure mercury and its compounds into water and the atmosphere. Until recently there was little concern about the mercury dumped into water. Being a very dense liquid, metallic mercury would presumably settle quickly. Inorganic mercury compounds would also present little hazard, because mercury forms such insoluble compounds with many anions (such as sulfide, S^{2-}) that it is precipitated in sediments close to the point of release. It is found that twenty miles downstream from large sources of mercury, precipitation of mercury is so effective that mercury concentrations drop to normal background levels.

During the 1950s, some disturbing occurrences suggested that the hazards of mercury could not be so easily dismissed. Between 1953 and 1960, 43 Japanese fishermen or members of their families who consumed fish from Minamata Bay died and 68 became severely ill because of mercury poisoning.‡ The mercury pollution was traced to a plastics plant that was releasing mercury into the bay. Although most of the mercury released was in inorganic forms, the problem was caused by the approximately 1% of mercury in organic forms.

During the late 1950s and early 1960s in Sweden, various species of seed-eating birds—pigeons, partridges, pheasants—were found to be dying and having trouble reproducing. Populations of various birds of prey—eagles, owls, peregrines, and so on—were also found to be declining. These problems were traced to the use of mercury fungicides on seeds which were then eaten by birds or by small animals which were consumed by birds.§ Because of these problems and the possible ultimate danger to humans, the use of mercury fungicides was discontinued in Sweden in 1966. Since that time the bird population has been recovering and the elimination of mercury fungicides has had no apparent effect on crop yields.

Seeds treated with mercury fungicides are colored pink, so that they will not be eaten by people or fed to animals. Unfortunately, this coding is not always effective as some farmers are not aware of the hazard or do not know what the pink color means. In 1969, several children in Almagordo, New Mexico, became quite sick after eating the meat of a hog that had been fed treated seed.‖ They suffered fever, loss of

---

*   J. J. Putman and K. W. Madden, "Quicksilver and Slow Death," *National Geographic* **142**, 507–527 (1972).
†   Putnam and Madden, "Quicksilver and Slow Death."

‡   Tucker, *The Toxic Metals.*
§   G. Lofroth and M. E. Duffy, "Birds Give Warning," *Environment* **11** (4), 10–17 (1969).
‖   Berton Roueche, "Annals of Medicine—Insufficient Evidence," *The New Yorker*, Aug. 22, 1970, pp. 64–81.

stability, impaired vision, mental disability, and coma. The children survived the poisoning, but were left with severe, permanent damage to the brain, nervous system, and vision.

During 1972, massive mercury poisoning was suffered by people in Iraq. An estimated 1000 persons died and 60,000 became severely ill after eating foods made from mercury-treated wheat and barley that was intended for use only as seed.*

The symptoms suffered by the family in New Mexico, the Iraqis, and the victims of the Minamata Bay disaster are different from and more serious than those of classical mercury poisoning. When the mercury is attached to a methyl group $(-CH_3)$ similar organic constituent, it behaves differently from inorganic mercury in the body. Most inorganic mercury passes through the body, but 90 to 95% of methyl mercury compounds are absorbed through the intestinal walls into the blood stream, which carries it throughout the body. Large amounts of the circulating mercury become concentrated in the brain, leading to the symptoms noted above.

Although inorganic mercury dumped into rivers and lakes was thought to be harmless, in 1967 two Norwegian scientists discovered that anaerobic bacteria living at the bottoms of lakes and rivers convert inorganic mercury into the more toxic methyl mercury compounds. Despite this finding, the appropriate officials of our government were seemingly taken by surprise when a graduate student at the University of Western Ontario, Norvald Fimriete, in March 1970 found that fish from Lake St. Clair (on the U.S.–Canadian border near Detroit) contained several ppm of mercury, well above the limit set by the Food and Drug Administration of 0.5 ppm on food. Following this discovery, various agencies went around measuring mercury concentrations in rivers, lakes, and fish. Some fish were pulled off the market because of excessive mercury, consumption of fish from contaminated water bodies was prohibited, and the major mercury pollution sources (mostly chloro-alkali plants that had been dumping ten pounds per day into water) were forced to greatly reduce their discharges of mercury.

One might suppose that we are now protected from the threat of mercury poisoning. That comforting thought is, unfortunately, not true.† As of January 1972, methyl mercury

compounds were still in use as fungicides on seeds. Although some major sources of mercury discharge have been cleaned up, large amounts of mercury are still being dumped into the environment. All the mercury listed in Table 12–7 that is not recovered and recycled eventually returns to the environment one way or another. For example, consumer products such as batteries and switches are thrown in the trash, hauled to the incinerator, and burned, allowing the mercury to escape to the air and eventually settle to the ground. An estimated 20 to 50% of the mercury used annually in the United States is returned to the environment—about 1000 metric tons per year. Furthermore, fossil fuels such as coal and oil contain small concentrations of mercury that is released to the atmosphere when the fuel is burned, contributing another 500 to 1000 metric tons. Thus shutting off dumping by chloro-alkali plants hardly solves the problem.

The thousands of tons of mercury that have been deposited in the bottoms of rivers and lakes over the past several decades pose a continuing threat. Anaerobic bacteria in bottom sediments can slowly convert inorganic mercury into the more toxic methyl mercury compounds. They are taken up by aquatic plants and animals and, eventually, by the birds.

Methyl mercury behaves like DDT and other synthetic, unnatural substances released to the environment. It is stored and concentrated more strongly as it moves up through a food chain which might involve several levels, such as: aquatic plants are eaten by microscopic aquatic animals; they are eaten by small fish; small fish are eaten by larger fish; and they, in turn, are eaten by predator birds or man. As discussed in more detail in Section 13–12, a substance such as methyl mercury that is rather insoluble in water and more strongly attracted to body tissue, often increases in concentration in the higher members of a food chain. Inorganic mercury is also concentrated in the food chain, but not as strongly as methyl mercury. David Klein has noted that in the Minamata Bay disaster, only about 1% of the mercury released by the chemical plant was in the form of methyl mercury.‡ However, most of the 10 ppm mercury in the contaminated fish was of the methyl form. The methyl mercury concentration had increased by at least a millionfold between the water (0.01 to 0.001 parts per billion, ppb) and the fish, 10 ppm.

* M. Mintz, "Iraqi Mercury Deaths May Hit 1,000," Washington *Post*, June 15, 1972.
† J. M. Wood, "A Progress Report on Mercury," *Environment* **14** (1), 33–39 (1972).

‡ D. H. Klein, Statement in Hearings on "The Effects of Mercury and Other Toxic Metals on Man and the Environment," Part 1, p. 89; see Suggested Reading.

Inorganic mercury is also concentrated by fish, but only by a factor of about 1000 relative to water. Methyl mercury is concentrated 1000 times as much as inorganic mercury and is at least ten times as toxic. Thus at the same concentration in water, methyl mercury is 10,000 times as dangerous as the same amount of inorganic mercury.

Even if all release of mercury to water bodies were completely halted, the mercury released over the past decades mostly remains in the sediments at the bottom of lakes and rivers. This mercury is slowly being converted to the methyl form that can enter the food chain, and thus poses a constant threat of poisoning to people who consume fish taken from the waters. As yet no suitable way of eliminating this threat has been devised.

The FDA limit of 0.5 ppm mercury in edible fish has created an economic disaster for fishermen around contaminated rivers and lakes and even for some ocean fishermen. Is the FDA limit too stringent, or should the limit be placed even lower? Dr. Neville Grant has argued that the present limit provides an acceptable margin of safety *only* if we assume that *all* Americans consume far less fish per capita than the Japanese who were involved in the Minamata disaster.[*] The fish involved had mercury concentrations in the range of 1 to 10 ppm, not so far above the FDA limits. If all of a person's food contained 0.5 ppm methyl mercury, his daily intake would be 0.25 mg/day, producing a concentration in the brain of about 2.5 ppm. This is not very far below the brain concentrations at which symptoms begin to appear: damage to fetuses of pregnant women at 3 ppm, some overt symptoms in sensitive persons at 5 ppm, and death at about 12 ppm and above. Thus the margin of safety for persons who consume large amounts of fish works out to be less than a factor of ten.

What about ocean fish such as tuna and swordfish that have been found to have high mercury concentrations—has man contaminated them, too, or are they simply concentrating natural mercury? We can't really say. The total amount of mercury dumped into the oceans as a result of man's activities is quite small compared to the total amount naturally present. Thus it would seem unlikely that man has contaminated ocean fish unless there is some mechanism by which man's contribution of mercury is somehow concentrated more effectively in the food chain than we now realize.

Beryllium and Cadmium

Although they have not received as much attention as mercury and lead, beryllium and cadmium may prove to be just as hazardous to health.

Beryllium is one of the most toxic elements. The danger of beryllium is almost entirely one of airborne material. The inhalation of minute amounts of fine particles of beryllium and its compounds can cause quite serious, even fatal lung conditions.[†] Short-term exposure to high levels of atmospheric beryllium cause an acute beryllium disease involving inflammation of membranes throughout the respiratory tract; the victim usually recovers within a few weeks. Of greater concern from a public health point of view is the chronic condition "berylliosis," whose symptoms don't show up until several years after exposure to airborne beryllium. The symptoms are somewhat similar to those of tuberculosis—formation of nodules in the lungs, a thickening of lung tissue, and a loss of respiratory function. Often berylliosis is fatal, in large part because the decreased lung capacity puts an added strain on the heart. Some of the major exposures to beryllium occurred during the 1940s, when fluorescent lighting came into widespread use. Early fluorescent tubes were coated on the inside with beryllium zinc sulfate. Both workers in the industry and consumers handling broken tubes were affected. Since 1949 other compounds have been used instead of the beryllium salt.

Beryllium is so toxic that only about 10 μg is found in the lungs of most chronic sufferers.[‡] Professor Henry Schroeder has suggested that atmospheric concentrations as low as 2 ng/m^3 (1 nanogram $= 10^{-3}$ μg) of Be may present a serious problem. Present measuring techniques for Be are inadequate, but concentrations in some cities are in that range.

Beryllium is not among the major metals used in the United States, but the demand for it is growing faster than for almost any other resource: an increase of 500% between 1948 and 1968, to a total consumption of about 300 metric tons.[§] Much of the increase has occurred because of the importance of beryllium in the nuclear energy industry and as a fuel additive for rockets.

* Neville Grant, "Mercury in Man," *Environment* **13** (4), 2–15 (1971).

† C. E. Knapp, "Beryllium—Hazardous Air Pollutant," *Environmental Science and Technology* **5**, 584–585 (1971).
‡ H. A. Schroeder, "Metals in the Air," *Environment* **13** (8), 18–32 (1971).
§ U.S. Bureau of Mines, *Mineral Facts and Problems* (U.S. Government Printing Office, Washington, D.C., 1970).

Some experts feel that cadmium has more "lethal possibilities than any of the other metals."[*] Since the early 1960s, there have been over 200 cases in Japan, including 50 deaths, of the painful "ouch-ouch" disease, a degenerative bone disease caused by ingestion of excessive quantities of cadmium. The victims lived along the Jintsu River, which was contaminated with cadmium because of drainage from the area of an old mine and zinc smelter. Most victims were women over 50 who had had three or more children. Apparently the women had a preexisting deficiency of calcium that made them more susceptible to the effects of cadmium.

At lower levels of exposure, cadmium concentrates strongly in the liver and kidneys, producing a number of effects by interfering in some enzyme-controlled reactions. A major effect of low-level cadmium ingestion is hypertension (high blood pressure) and associated cardiovascular ailments. Hypertension is expected to occur at an intake rate of 175 μg Cd/day.[†] Various experts have indicated that the average American ingests about 60 to 80 μg/day, and so our margin of safety is only a factor of two or three. Any appreciable increase of exposure to cadmium may well bring on disorders such as hypertension or worse.

Aside from the Japanese incidents, there has been little evidence of mass poisoning by cadmium. However, there are various indications that vast segments of the population of the United States may have had their lives considerably shortened by the presence of elevated cadmium levels. In Sauer and Brand's study (cited above) we noted the high death rates for white males in the South Atlantic states, and we added that a similar pattern of death rates was obtained when deaths from cardiovascular illnesses were singled out. The only obvious difference between this area and others is the fact that the water is generally quite soft in the South Atlantic region. Distinct correlations have been observed between the softness of water and the death rate from heart disease.[‡] Soft water dissolves trace metals more effectively than hard water, so it is possible that the soft water dissolves cadmium from water pipes. Cadmium is an impurity in the zinc of galvanized pipes and an additive in black polyethylene pipes. Professor Schroeder and others found than when soft water stands overnight in such pipes, it dissolves cadmium up to concentrations of 77 μg/liter, nearly eight times the

Public Health Service guideline for cadmium in drinkable water.[§] Other investigators have found a direct correlation between the atmospheric concentration of cadmium in 28 American cities and the death rates from hypertension and arteriosclerotic heart disease,[||] although these results have been challenged by other investigators. None of these observations proves that our lives are being shortened by the effects of cadmium; however, the indications strongly suggest that we make much more careful studies of cadmium concentrations in the environment and its effects and that we begin to control the release of cadmium.

The United States consumes only about 6000 metric tons of cadmium per year, mainly for electroplating of shiny surfaces on tools, machine and automotive parts, and in batteries, paints, and plastics. However, the uses of cadmium per se tell us only part of the story about the release of cadmium into the environment. Since cadmium is chemically similar to zinc, it is always present as an impurity in zinc. (In fact there are no cadmium ores, as Cd is always obtained as a by-product of Zn.) Thus cadmium is released into the environment not only via the uses of cadmium itself, but also by the processing and use of zinc: the smelting of zinc sulfide ores, the manufacture and use of galvanized products in automobiles, plumbing, and so on, and as an additive in rubber tires and motor oil. Because of the importance of zinc in our technological society (1.6 million metric tons produced in the United States in 1968), it is going to be very difficult to halt the release of cadmium into the environment.

12–9 METALS—FRIEND OR FOE?

Metals are a basic requirement of the technological society to which we have become accustomed. However, we face some serious problems with metals. If the world continues to increase its use of metals at the present rate, we will soon run out of economically feasible supplies of several key metals. Even now, the supplies of some metals are probably not great enough to raise the entire world population to the American standard of living. Although new materials such as plastics have replaced metals in some applications, many metals have such unique properties that they cannot be replaced by any

---

[*]   Julian McCaull, "Building a Shorter Life," *Environment* **13** (7), 2–41 (1971).
[†]   Tucker, *The Toxic Metals.*
[‡]   McCaull, "Building a Shorter Life."

[§]   Ibid.
[||]   R. E. Carrol, "The Relationship of Cadmium in the Air to Cardiovascular Disease Death Rates," *Journal of the American Medical Society* **198**, 267–269 (1966).

materials presently developed. We owe it to future generations to conserve these vital resources by stopping wasteful uses of metals and by recycling the metals from discarded products. If we're not willing to do that, the least we can do is bury our trash in concentrated piles that are well marked, so that future generations can locate those very rich "ores" and process them.

Although many metals are vital nutrients, many are subtle poisons—toxic in such small amounts and often producing diseases so difficult to identify that the known poisonings may be only a portion of the total. To complicate matters, the metals often behave in strange ways in the environment—becoming concentrated in the food chain or transformed into highly toxic species, such as methyl mercury, after release. The enormous growth in the use of metals, as well as insecticides, plastics, and other products of our technological society, especially in the years since World War II, may have a great deal to do with the leveling off of the life expectancy curve of Fig. 1–3, despite continued advances in the prevention and cure of various diseases.

SUGGESTED READING

Harrison Brown, "Human Materials Production as a Process in the Biosphere," *Scientific American* **223** (3), 194–208 (Sept. 1970).

H. L. Cannon and H. C. Hopps, eds., *Environmental Geochemistry in Health and Disease* (Geological Society of America, Boulder, Colo., 1971).

J. J. Chisolm, Jr., "Lead Poisoning," *Scientific American* **224** (2), 15–23 (Feb. 1971).

E. S. Deevey, Jr., "Mineral Cycles," *Scientific American* **223** (3), 148–158 (Sept. 1970).

L. J. Goldwater, "Mercury in the Environment," *Scientific American* **224** (5), 15–21 (May 1971).

Hearings on "The Effects of Mercury and Other Toxic Metals on Man and the Environment," Subcommittee on Energy, National Resources, and the Environment of the Committee on Commerce, U.S. Senate, May–Aug. 1971 (U.S. Government Printing Office, Washington, D.C., 1971).

D. W. Jenkins, "The Toxic Metals in Your Future—and Your Past," *Smithsonian* **3** (1), 62–69 (1972).

B. H. Mahan, *University Chemistry*, 2nd ed. (Addison-Wesley, Reading, Mass., 1969).

D. H. Meadows, D. L. Meadows, J. Randers, and W. W. Behrens, III, *The Limits to Growth* (Universe Books, New York, 1972).

Linus Pauling, *General Chemistry*, 3rd ed. (W. H. Freeman, San Francisco, Calif., 1970).

J. J. Putman and K. W. Madden, "Quicksilver and Slow Death," *National Geographic* **142**, 507–527 (1972).

Anthony Tucker, *The Toxic Metals* (Ballantine Books, New York, 1972).

U.S. Bureau of Mines, Bulletin 650, *Mineral Facts and Problems* (U.S. Government Printing Office, Washington, D.C., 1970).

QUESTIONS AND PROBLEMS

1. Briefly describe or explain each of the following terms.

   a) ligand   b) ductile

   c) pig iron   d) thermal conductivity

   e) acid anhydride   f) bauxite

   g) amphoteric oxide   h) malleable

   i) magnetite   j) slag

2. a) Calculate the weight percent of Fe in Fe_2O_3.

   b) Assuming that the only reactions in a blast furnace are those of Eqs. (12–14) and (12–15), how many tons of coke (assumed to be pure carbon) are required to reduce one ton of Fe_2O_3?

3. a) What are the physical characteristics that distinguish metals from nonmetals?

   b) What is the major chemical characteristic that distinguishes most metals from nonmetals?

4. What kinds of electrical measurements would you make on a sample of a pure element to determine whether it was a metal or nonmetal, and how would you interpret them?

5. Chromic oxide, Cr_2O_3, is amphoteric. Write balanced chemical equations for the complete reaction of Cr_2O_3 with

   a) excess acid (H^+)

   b) excess hydroxide to form $Cr(OH)_4^-$

6. a) Write a balanced equation for the thermite reaction of chromic oxide, Cr_2O_3.

   b) What weight of Cr_2O_3 can be reduced to pure Cr by 1 kg of Al?

7. a) Why is it necessary to further process most pig iron from blast furnaces in a converter before the final product is made?

   b) What are the main functions of the processes that occur in the converter?

8. List the major uses of each of the following metals and describe the characteristics of the metal that make it most appropriate for the indicated uses.

   a) aluminum   b) iron

   c) copper   d) mercury

   e) gold   f) zinc

   g) titanium   h) cadmium

9. Explain, in terms of electronic structure, why metallic character increases down the columns of the periodic table and across the rows, from right to left.

10. Make a list of all of the environmental effects you can think of that are associated with the production, use, and disposal of the following metals.

    a) iron   b) zinc

11. As noted in Section 12–2, in the Christmas tree experiment, Cu goes into solution to replace Ag^+ ions. Since this reaction occurs spontaneously, it would be possible to obtain energy from the reaction by constructing a battery in which Ag is deposited on one electrode as Cu goes into solution from the other electrode (similar to the Cu–Zu battery discussed in Section 12–2). Sketch an arrangement for constructing such a battery, and indicate the positive terminal.

12. a) What was the significance of the discovery that inorganic forms of mercury can be converted into methyl mercury species by bacteria present in sediments?

    b) For what reasons is a given concentration of methyl mercury in water more hazardous for humans than the same concentration of inorganic mercury?

    c) What are the major effects of organic mercury compounds in the body?

*13. a) The solubility product of mercuric sulfide, HgS, is about 3×10^{-52} (moles/liter)^2. What is the solubility of Hg^{2+} (in moles/liter) in water containing S^{2-} ion at a concentration of $10^{-12}M$?

    b) Convert your answer in part (a) to μg Hg/l.

    c) Convert your answer in part (a) to ppm Hg (by weight).

    d) Suppose one lowered the pH of the solution in part (a). Would that increase or decrease the solubility of Hg^{2+}? Explain.

14. What are the lines of evidence suggesting that cadmium in our environment may have a sizable effect on the longevity of major segments of the U.S. population?

*15. It has been suggested that a major fraction of the Cd in urban atmospheres is released by the abrasion and heating of automobile tires, which contain about 50 ppm Cd. Assume that all the ~1.5 $\mu g/m^3$ Pb observed in city air results from the burning of gasoline containing about 2 g/gal Pb. Now estimate the amount of rubber worn from tires that have been driven 20,000 miles, and assume that all the Cd in the rubber is released to the atmosphere. Now compute the ratio of Cd to Pb emitted by one automobile and predict the atmospheric concentration of Cd from this source. The observed value in cities is typically about 0.01 to 0.02 $\mu g/m^3$. Do you account for this much from auto tires?

---

*   Indicates a difficult problem requiring greater mathematical skill than other problems.

13
THE CHEMISTRY OF CARBON COMPOUNDS

Plastic comprises a major portion of most present-day litter. (Photograph by Bruce Anderson.)

13–1 THE IMPORTANCE OF ORGANIC CHEMISTRY

In this chapter we take up organic chemistry, in essence the chemistry of compounds consisting mainly of carbon and hydrogen along with smaller amounts of elements such as oxygen, nitrogen, sulfur, phosphorus, and the halogens. Although carbon is but one out of the eighty-plus elements that occur naturally on earth, by far the majority of compounds present in our bodies and the biosphere have carbon as their most important element. Most of the consumer products of the chemical industry are organic compounds: synthetic fabrics, plastics, soaps, and detergents, packaging materials, and pharmaceuticals, to name but a few. Furthermore, some of our most serious environmental problems have arisen as the result of man's introduction of synthetic (man-made) organic compounds, such as pesticides, plastics, and polychlorinated biphenyls (PCB's) into the environment. Organisms are ill-equipped to decompose these "unnatural" materials, which thus remain in the ecological system for many years and often become concentrated in certain species with serious consequences.

In the early days of chemistry, the only organic chemicals known were those that occurred naturally in living systems. It was thought that organic chemicals were fundamentally different from inorganic chemicals in that the former could be formed only in processes involving the "vital force" present in living systems. In 1828, the German chemist Friedrich Wöhler successfully synthesized urea by heating an aqueous solution of ammonium cyanate, an inorganic chemical:

$$NH_4OCN \xrightarrow{heat} H_2N-\overset{\overset{\displaystyle O}{\|}}{C}-NH_2. \qquad (13-1)$$

ammonium cyanate   urea

The urea produced was identical to the well-known organic chemical urea, present in urine. This first synthesis of an organic chemical removed the "vital force" as a distinction between inorganic and organic compounds. Today the chemical industry synthesizes thousands of organic chemicals; some are the same as those made in living systems, but the majority are types unknown in nature.

13–2 SATURATED HYDROCARBONS

Carbon has four valence electrons and most of its compounds are predominantly covalent, i.e., based on bonds formed by the sharing of electrons. Carbon has a unique ability to form long chains of atoms bonded to each other, which accounts for the immense number and variety of compounds it forms in combination with such elements as H, O, N, S, P, and the halogens.

The hydrocarbons are compounds of just carbon and hydrogen. The term *saturated* means that the carbon atoms have their full capacity of attached hydrogen atoms; i.e., there are no double or triple bonds between carbon atoms. The simplest of the saturated hydrocarbons is methane, CH_4:

$$H\overset{\overset{\displaystyle H}{\times}}{\underset{\underset{\displaystyle H}{\times}}{\overset{\times}{C}}}H$$

Although the methane molecule appears flat in our diagram, we find that in three dimensions the hydrogen atoms get as far apart as possible by arranging themselves at the corners of a tetrahedron (a solid, each of whose four faces is a triangle); the carbon atom is in the center of the tetrahedron (see Fig. 13–1). Methane is a gas at room temperature and is the major component (about 82%) of the natural gas used in homes for cooking and heating.

We can make successively bigger hydrocarbon molecules by removing one hydrogen, attaching a carbon and filling in the rest of the hydrogens:

Ethane, C_2H_6:

$$H\overset{\overset{H}{\times}}{\underset{\underset{H}{\times}}{\overset{\times}{C}}}\overset{\overset{H}{\times}}{\underset{\underset{H}{\times}}{\overset{\times}{C}}}H \qquad H-\overset{\overset{\displaystyle H}{|}}{\underset{\underset{\displaystyle H}{|}}{C}}-\overset{\overset{\displaystyle H}{|}}{\underset{\underset{\displaystyle H}{|}}{C}}-H \qquad \text{or} \qquad H_3C-CH_3$$

Propane, C_3H_8:

$$H\overset{\overset{H}{\times}}{\underset{\underset{H}{\times}}{\overset{\times}{C}}}\overset{\overset{H}{\times}}{\underset{\underset{H}{\times}}{\overset{\times}{C}}}\overset{\overset{H}{\times}}{\underset{\underset{H}{\times}}{\overset{\times}{C}}}H \qquad H-\overset{\overset{\displaystyle H}{|}}{\underset{\underset{\displaystyle H}{|}}{C}}-\overset{\overset{\displaystyle H}{|}}{\underset{\underset{\displaystyle H}{|}}{C}}-\overset{\overset{\displaystyle H}{|}}{\underset{\underset{\displaystyle H}{|}}{C}}-H \qquad \text{or} \qquad H_3C-CH_2-CH_3$$

Although these are called "straight chain" hydrocarbons and appear as such above, we see in three dimensions that they are a series of tetrahedra connected at the corners (see Fig. 13–2). If we examine the carbon "skeleton" alone, it would appear to be staggered, or zigzag (see Fig. 13–3).

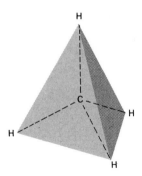

Fig. 13–1 Tetrahedral structure of methane, shown in two different ways. In the model shown at the right, the sizes of the balls are proportional to the sizes of the electron clouds around the hydrogen and carbon atoms.

After propane the next member of the series is butane, C_4H_{10}. Here we encounter a new phenomenon—we can make two different structures having the same formula, C_4H_{10}:

$$H_3C-CH_2-CH_2-CH_3 \qquad H_3C-\underset{|}{\overset{CH_3}{CH}}-CH_3$$

normal butane   isobutane

These are distinct compounds called *structural isomers*; i.e., they have the same formulas, but the atoms are arranged in different geometrical structures. We can imagine forming isobutane by pulling a hydrogen atom off the central carbon atom of propane and replacing it with a methyl group, i.e., a methane molecule with one hydrogen removed, as shown at the right.

propane   methyl group   isobutane

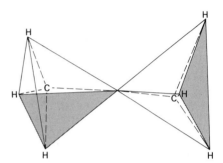

Fig. 13–2   Structures of ethane (above) and propane (below).

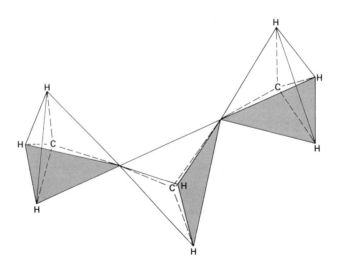

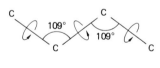

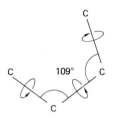

Fig. 13–3 Various shapes of "straight-chain" carbon skeletons can be made by rotating the single bonds at angles of 109°, as shown in the sketch. Models shown at the right show a 9-carbon molecule (nonane) stretched out as straight as possible (above) and twisted up (below).

Note that we cannot form a third isomer of butane by depicting the structure as

$$H_3C—CH_2—\underset{\underset{CH_3'}{|}}{CH_2}$$

since the four carbons are still in a "straight" chain; i.e., when viewed in three dimensions we see that this structure is the same as for n-butane (n = normal). Isobutane and n-butane have similar but slightly different properties: isobutane is more volatile, boiling at −12°C versus −0.5°C for n-butane.

The family of saturated hydrocarbons is called the alkanes. The name of each member of the series has the same ending, -ane, indicating that the structure is saturated with respect to hydrogen. The prefixes for the first four members, as used above, are meth-, eth-, prop-, and but-. For chains of five or more carbon atoms, Greek prefixes signifying the number of carbons are used: pent-, hex-, hept-, oct-, non-, dec-, etc. An alkyl group is an alkane molecule with one of the hydrogen atoms removed. Such a group has the same prefix as the corresponding alkane and the ending -yl, for example, methyl (CH_3—), ethyl (H_3C—CH_2—), and so on. We will sometimes indicate an alkyl group, in general, by the symbol R —.

For butane, only two structural isomers are possible, hence there is little difficulty in distinguishing between normal or straight-chain butane and the branched compound, isobutane (a carbon compound is termed "branched" if the carbon atoms do not follow each other in a chainlike fashion). However, as more carbon atoms are added, the variety of possible structures becomes so great that we must use a systematic procedure to describe them precisely. The International Union of Pure and Applied Chemistry (IUPAC) maintains guidelines for naming compounds. Many compounds that are used frequently also have common names, usually shorter but less precise than the IUPAC names. In order to acquaint you with both types of names, we will often give both names for various structures. For example, the two sets of names for the three possible pentane isomers are shown at the bottom of the page. (Note that

$$H_3C—CH_2—\underset{\underset{CH_3}{|}}{\overset{\overset{CH_3}{|}}{CH}}—CH_3$$

is the same as methyl-butane; it's just turned around, not another structural isomer.)

With the IUPAC system, a simple set of principles can be used to name a compound or to determine the structure of a compound from its name.

1. The longest chain of carbon atoms is considered the parent compound. For example, in isopentane the longest chain has four carbon atoms, hence the name butane for the parent compound.

2. The carbon atoms in this chain are then numbered to locate positions where a hydrogen has been replaced by an alkyl group or a functional group (e.g., a group containing atoms other than C and H). The numbering sequence should start at the end which will give the lowest numbers. For example, the following compound would be 2-methylpentane, not 4-methylpentane:

$$H_3C—CH_2—CH_2—\underset{\underset{CH_3}{|}}{CH}—CH_3$$

3. The complete name of the compound consists of: (a) the name of the parent compound, (b) the name of the attached alkyl or functional groups, and (c) the numbered positions of these groups.

$$H_3C—CH_2—CH_2—CH_2—CH_3 \qquad H_3C—\underset{\underset{CH_3}{|}}{\overset{\overset{CH_3}{|}}{CH}}—CH_2—CH_3 \qquad H_3C—\underset{\underset{CH_3}{|}}{\overset{\overset{CH_3}{|}}{C}}—CH_3$$

IUPAC:	pentane	methylbutane	dimethylpropane
Common:	n-pentane	isopentane	neopentane

The IUPAC names of the following complex hydrocarbons allow us to determine the structure of the compound easily:

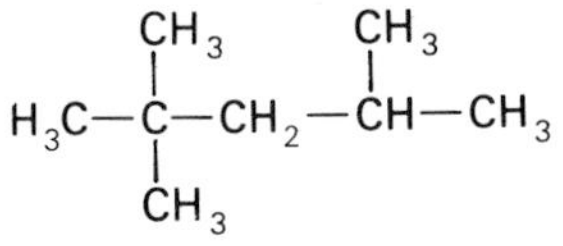

2,2,4-trimethylpentane
(isooctane)

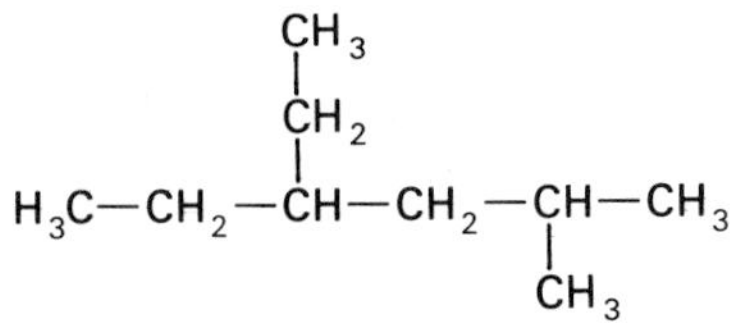

2-methyl-4-ethylhexane

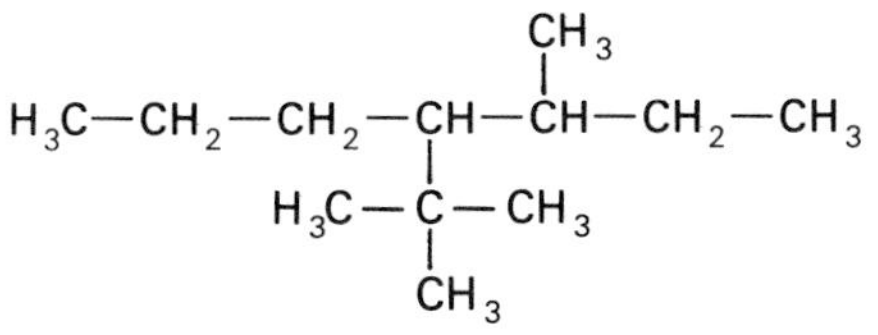

3-methyl-4-isobutylheptane

Halogen atoms, our first example of functional groups, can also substitute for hydrogen atoms (see Section 13–11) and likewise fall under the same naming rules:

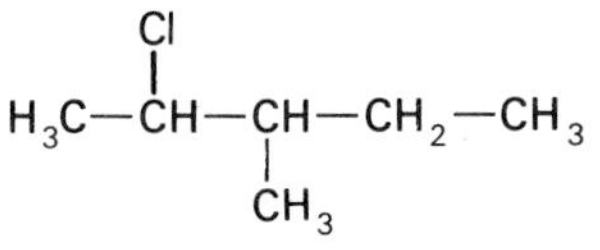

3-methyl-2-chloropentane

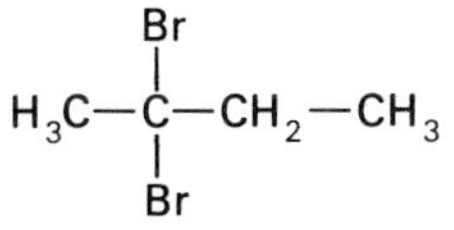

2,2-dibromobutane

The number of structural isomers increases sharply as we add more carbons (e.g., there are five hexanes, C_6H_{14}, and eight heptanes, C_7H_{16}). The general formula for the linear-chain saturated hydrocarbons is C_nH_{2n+2}, where n (n = number) can range from one (as in methane) to many hundreds or thousands. As the molecules get bigger, their compounds become more difficult to melt and evaporate; thus there is a general upward trend of melting and boiling points as the number of carbon atoms increases (see Table 13–1).

Alkanes can form rings as well as straight-chained structures; for example, the two compounds cyclopropane and cyclohexane are saturated hydrocarbons in ring form:

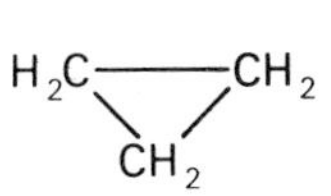

cyclopropane

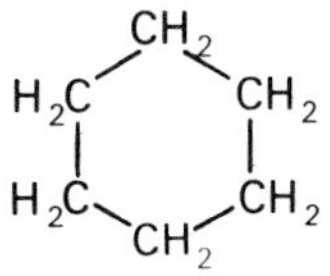

cyclohexane

The general formula for saturated hydrocarbons containing one such ring structure is C_nH_{2n}, two less than the corresponding straight-chain alkane, since one hydrogen must be removed from each end of the chain in order to form a ring.

The major use of hydrocarbons is in the form of natural gas, gasoline, lubricating oils, and other components of petroleum. Since hydrocarbon molecules are nonpolar, they are excellent solvents for most other organic compounds that

Table 13–1 Properties of Straight-Chain Alkanes

No. of carbons	Name	Number of isomers	Formula	Molecular weight	Melting point (°C)	Boiling point (°C)
1	Methane	1	CH_4	16.0	−182.5	−161.5
2	Ethane	1	C_2H_6	30.0	−183.0	−88.6
3	Propane	1	C_3H_8	44.1	−187.0	−42.1
4	Butane	2	C_4H_{10}	58.1	−138.0	−0.5
5	Pentane	3	C_5H_{12}	72.1	−130.0	36.1
6	Hexane	5	C_6H_{14}	86.2	−95.3	68.7
7	Heptane	9	C_7H_{16}	100.2	−90.6	98.4
8	Octane	18	C_8H_{18}	114.2	−56.8	125.7
9	Nonane	35	C_9H_{20}	128.3	−53.6	150.8
10	Decane	75	$C_{10}H_{22}$	142.3	−29.7	174.0
15	Pentadecane	4347	$C_{15}H_{32}$	212.4	10.0	270.5

are similarly non-ionized and nonpolar (in contrast with water as a solvent, Section 8–4). For example, hydrocarbons have been used as solvents for dry-cleaning materials that must not be washed in water. However, this is a dangerous procedure because of the high flammability and toxicity of the hydrocarbon solvents.

13–3 UNSATURATED HYDROCARBONS

Carbon atoms can be bonded to each other by double and triple bonds, as in ethene (common name, "ethylene") and in ethyne (commonly called "acetylene"):

Ethene (ethylene), C_2H_4:

Ethyne (acetylene), C_2H_2: H:C :: C:H   or   H—C≡C—H

Compounds with double bonds such as ethene are called *alkenes* and have the general formula, C_nH_{2n}, the same as cyclic saturated hydrocarbons. The systematic names of hydrocarbons with double bonds begin with the same prefixes as those listed for the alkanes in Table 13–1, but the suffix is -*ene* rather than -*ane*.

Hydrocarbons with triple bonds are called the *alkynes*, or the acetylene series. Their general formula is C_nH_{2n-2}. The alkynes are named using the prefixes in Table 13–1 and the ending -*yne*.

As discussed in Section 8–1, there is free rotation about single bonds. Figure 8–4 shows how the atoms attached to one carbon atom in a molecule such as ethane can rotate with respect to atoms or groups of atoms attached to the other carbon. There is free rotation about all the bonds in a long skeleton of carbon atoms held together by single C—C bonds. Thus because of the rotation and the 109° angle between single bonds of a carbon atom, a long chain of singly bonded carbon atoms can take many shapes, ranging from a zigzag about a straight line to a shape in which the chain is all "wrapped up" in itself (see Fig. 13–3). If very large groups of molecules are attached to nearby carbon atoms on a chain, they may prevent free rotation about the intervening single bonds, an effect called "steric hindrance."

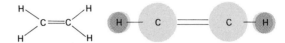

Fig. 13–4 The six atoms of ethene lie in a single plane, and there is no rotation about the double bond.

There is no rotation about double and triple bonds. All six atoms of ethene are in the same plane (see Fig. 13–4). Even if we replace a hydrogen atom from each end with carbon atoms, as in a segment of a long carbon skeleton, the six atoms remain in one plane and the double bond prevents rotation of the chain at that point. The four atoms of ethyne lie in a straight line. Since only one additional atom can be attached to each of the carbon atoms involved in the triple bond, the lack of rotation about the bond has little effect on the shape of a long skeleton of carbon atoms which includes a —C≡C— unit.

Unsaturated hydrocarbons can have structural isomers that differ in the location of their multiple bonds, as well as in the location of their attached alkyl groups. For example, in the four-carbon compounds with one double bond, there are two unbranched isomers and one branched*:

1-butene   2-butene   isobutene

All these hydrocarbons burn readily and many are used for fuels. Ethyne (acetylene) is used with oxygen gas in torches to obtain the very high temperatures needed to melt steel. Ethyne was also used for many years in the lamps on miners' helmets, since it can be generated right in the lamp by the reaction of solid calcium carbide and water:

$$CaC_2(s) + 2H_2O(l) \longrightarrow Ca^{2+}(aq) + 2OH^-(aq)$$
$$+ H—C≡C—H(g). \qquad (13–2)$$

All the volatile, short-chain hydrocarbons can be used as fuels in automobile engines or home heating units, but natural

---

*   Note that 2-butene has *cis* and *trans* isomers. See Sect. 8–1.

gas (mainly methane with smaller amounts of higher alkanes) is the best. It burns with a relatively cool flame, not the extremely hot flame of ethyne, and as a result produces only a small quantity of the various nitrogen oxides (more nitrogen oxides are produced in combustion processes at high temperatures). As we shall see in Chapter 14, nitrogen oxides are undesirable air pollutants and major components of photochemical smog.

The alkenes and alkynes are "unsaturated" in that they don't have the maximum number of hydrogen atoms possible, like the six hydrogens in ethane. There are series of unsaturated hydrocarbons similar to that of the saturated compounds of Table 13–1. Unsaturated hydrocarbons react much more readily with other compounds than do the saturated hydrocarbons. There are no excess electrons in a saturated hydrocarbon and the C—C and C—H bonds are quite stable. In order for a saturated compound to react with another species, a bond must be broken in the original molecule to open up a point of attachment for additional atoms. For example, if we add some reddish-brown liquid bromine to hexane, no reaction occurs unless the mixture is exposed to a high temperature or to ultraviolet light from a lamp or direct sunlight. These treatments cause a breaking of bonds and the substitution of one or more bromine atoms for hydrogen atoms in the hexane molecule:

$$H_3C-CH_2-CH_2-CH_2-CH_2-CH_3(l) + Br_2(l) \xrightarrow{UV}$$

normal hexane   bromine

$$H_3C-CH_2-CH_2-CH_2-CH_2-\underset{\underset{Br}{|}}{\overset{\overset{H}{|}}{C}}H(l) + HBr(g). \quad (13-3)$$

l-bromohexane   hydrogen bromide

This is known as a substitution reaction. In this reaction, the UV light apparently breaks Br_2 molecules into Br atoms which then attack the hexane molecules, removing hydrogen atoms to form HBr. The hexane radical then reacts with another Br_2 molecule to form the bromohexane compound.

Bromine reacts much more readily with alkenes than with alkanes because extra electrons are available at the double bond. If one mixes Br_2 liquid with hexene, for example, the reddish-brown color quickly disappears as the bromine reacts

with the unsaturated compound:

$$H_3C-CH_2-CH_2-CH_2-CH=CH_2(l) + Br_2(l) \longrightarrow$$

l-hexene   bromine

$$H_3C-CH_2-CH_2-CH_2-\underset{\underset{H}{|}}{\overset{\overset{Br}{|}}{C}}-\underset{\underset{H}{|}}{\overset{\overset{Br}{|}}{C}}-H(l). \quad (13-4)$$

1,2-dibromohexane

This is known as an addition reaction. We can readily distinguish between saturated and unsaturated hydrocarbon compounds by adding some bromine and noting whether or not the reddish-brown color quickly disappears (since the dibromoalkanes are colorless). The other halogens undergo similar addition reactions with unsaturated hydrocarbon chains. The amount of iodine, I_2, that will react with oils or fats is used as a standard measure of the degree of unsaturation of the hydrocarbon chains in its fatty acids (see Section 13–8).

Unsaturated hydrocarbons can be made saturated by hydrogenation, i.e., reaction with hydrogen gas under high pressure in the presence of catalysts:

$$H_2C=CH_2(g) + H_2(g) \xrightarrow[\text{high pressure}]{\text{catalysts}} H_3C-CH_3(g). \quad (13-5)$$

13–4 AROMATIC HYDROCARBONS

One of the most important structural units in organic chemistry is the benzene ring. Benzene itself has the formula C_6H_6 and the six carbon atoms are arranged in a ring with a hydrogen atom attached to each carbon. If one were trying to predict the bond structure of benzene on the basis of the octet rule the most logical choice would be the structure:

However, this structure does not agree with experimental observations. In most compounds, carbon atoms joined by single bonds are about 1.54 Å apart, whereas those of a double bond are about 1.34 Å apart. Thus if the structure shown above were correct, we would expect alternate long and short bonds around the ring. Instead, it is found that all the carbon-carbon bonds in the ring are of the same length, 1.39 Å, intermediate between single- and double-bond lengths. Likewise, if we made dichlorobenzene compounds in which two chlorine atoms were attached to adjacent carbon atoms, we would expect to have two distinct structural isomers, depending on whether the carbon atoms were joined by a single or double bond:

and

In fact, only one structural isomer is found (although other isomers can be made by moving a Cl atom to nonadjacent carbon atoms).

These and many other observations on benzene structures indicate that all the bonds between the carbon atoms of the ring are the same; i.e., instead of alternating single and double bonds, all of them might be described as being "1½ bonds." Another way of looking at it is to consider the true structure to be the "hybrid" of the structures

$\longleftrightarrow$

The benzene structure is thus another example of *resonance*, which we encountered in Section 8–1 in describing the structures of SO_2, SO_3, NO_3^-, and so on. In order to emphasize the fact that all the bonds are equivalent, we usually depict

the benzene structure as

which stands for the hybrid of the two structures shown above. Often the hydrogen atoms are left off, but it is understood that there is one hydrogen attached to each carbon atom unless otherwise indicated.

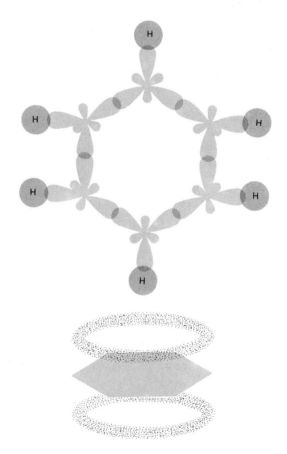

Fig. 13–5 The benzene ring structure.

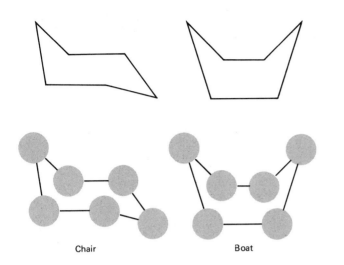

Chair   Boat

(Note that hydrogen atoms or other groups can be attached only at the indicated sites.) Structures in which the aromatic rings share one or more common edges are called "condensed-ring structures." An important source of these compounds is coal, from which they are obtained by distillation. Many of the compounds that have several condensed aromatic rings, especially those of the benzopyrene family having five rings, are carcinogenic (see Section 14–4):

pyrene   3,4-benzopyrene

If we consider each resonance form of the benzene ring to be made up of three, planar, ethene-like units, it is not surprising that all the atoms of benzene lie in a plane, with all the bond angles being 120° (see Fig. 13–5). By contrast all the bond angles in cyclohexane have approximately the tetrahedral value of 109°. The cyclohexane ring is thus "puckered" (nonplanar), existing in either the "chair" form or the "boat" form (see Fig. 13–6).

The benzene ring is a very stable structure that occurs in a wide variety of compounds called *aromatic compounds*. Originally the term referred to the fact that most compounds containing the structure had distinct odors, but now the term refers specifically to compounds containing the benzene ring structure. In some aromatic compounds, the benzene rings are "fused" together, as in naphthalene and anthracene:

The structure at the right is called "3,4" benzopyrene to indicate that the benzene ring is attached at positions 3 and 4 of the basic pyrene structure shown at the left.

Straight and branched-chain hydrocarbons may also be attached to aromatic rings, as in toluene (methyl benzene):

methyl benzene
(toluene)

All positions on the benzene ring are the same until one of the hydrogen atoms has been replaced. If another element

naphthalene   anthracene

or group is added it must be numbered as shown with the xylenes:

1,2-dimethyl benzene (ortho-xylene)   1,3-dimethyl benzene (meta-xylene)   1,4-dimethyl benzene (para-xylene)

Note that *ortho*, *meta*, and *para* refer to the three possible types of isomers of dimethyl benzene, as well as to the relationship between the sites on the benzene ring.

Many of the compounds based on the benzene ring are not naturally found in living substances. Benzene is poisonous if taken internally, as are many of its derivatives. Benzene settles in the liver, where it is stored since it does not easily take part in most biochemical reactions and cannot be easily removed from the body.

13-5 ALCOHOLS

Up to this point we have focused most of our attention on compounds consisting only of carbon and hydrogen. The alcohols are the first of a series of compounds that contain oxygen, in this case, the hydroxyl group —OH. The replacement of a hydrogen atom on a hydrocarbon molecule by oxygen, sulfur, nitrate, or other atoms or groups of atoms produces a more reactive center on the molecule. The replacement atom or group of atoms is called a *functional group* because of its effect on the chemical reactivity of the molecule. Double and triple carbon-carbon bonds are also sometimes called functional groups, as they are more reactive than C—H and C—C single bonds.

The general form of an alcohol is ROH, where R— is an alkyl group (see Section 13-2). Simple examples of alcohols are methyl and ethyl alcohol:

methyl alcohol (methanol)   ethyl alcohol (ethanol)

The systematic (IUPAC) names of the alcohols end in -*ol*, as with the two possible isomers of propanol:

1-propanol (*n*-propyl alcohol)   2-propanol (isopropyl alcohol)

Methyl alcohol can be made by the destructive distillation of hardwood, hence its industrial name, "wood alcohol." If taken internally in large amounts, methanol can cause blindness and death. Isopropyl alcohol is used mainly as "rubbing alcohol." Ethyl alcohol ("grain" alcohol) is produced by the fermentation of sugar, a reaction catalyzed by zymase, an enzyme mixture present in yeast (Chapter 15).

$$C_6H_{12}O_6(s) \xrightarrow{\text{zymase}} 2C_2H_5OH(aq) + 2CO_2(g). \qquad (13-6)$$

glucose   ethyl alcohol

Ethyl alcohol is the active ingredient in alcoholic beverages. These beverages can be made by fermentation of many fruits and grains, e.g., whiskey from wheat, rye, or barley, rum from molasses, wine from grapes, vodka from potatoes, hard cider from apples. In some cases, especially the fruits, the sugars occur naturally in the starting material, whereas in cereal grains and potatoes, starch must be converted to sugar before fermentation (see Chapter 15). Note that reaction (13-6) produces CO_2 gas. Normally the gas is allowed to escape, but for certain wines, the fermentation is done in bottles with a cork firmly wired on to trap the CO_2, producing a naturally carbonated wine called "champagne" or "sparkling wine."

The alcohol content of a beverage is indicated by the percent alcohol or the "proof," which is twice the percentage. Beers range from about 3 to 9% alcohol, wines from 13 to 19%, and whiskys, gin, and vodka from about 35 to 45% (70 to 90 proof). Natural fermentation stops when the alcohol content reaches about 15%, so in order to make stronger beverages, the natural products are distilled; i.e., they are heated to the boiling point and the vapor is condensed, as in Fig. 13-7. Since ethyl alcohol is more volatile than water, the vapors are much richer in alcohol than the boiling mixture is.

The alcohol content can be increased to a maximum of 95% by distillation.

Because of the high taxes on drinking-quality alcohol, nontaxed ethyl alcohol for uses other than drinking is "denatured." A toxic substance such as methyl alcohol is added to the ethyl alcohol to make it unfit for drinking. During the 1920s and early 1930s, in the Prohibition Era when alcohol consumption was unlawful, many rumors about the conversion of wood alcohol or denatured ethyl alcohol to "drinking" alcohol were in circulation. One persistent rumor had it that one could cut off both ends of a loaf of rye bread and let the alcohol trickle through the loaf. The alcohol draining through was supposed to be converted to ethyl alcohol or freed from its impurities. One wonders how many people were killed or blinded by such patent nonsense!

These are the type of illegal stills currently being uncovered in North Carolina. (Courtesy of the U.S. Department of the Treasury, Bureau of Alcohol, Tobacco, and Firearms.)

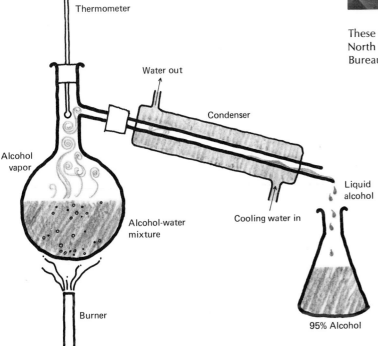

Fig. 13-7 Separation of an alcohol-water mixture by distillation.

Some alcohols such as ethylene glycol and glycerol have more than one hydroxyl group:

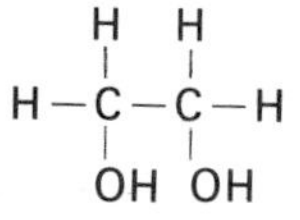

$$\begin{array}{cc} \text{H} \;\; \text{H} & \text{H} \;\; \text{H} \;\; \text{H} \\ \text{H} - \text{C} - \text{C} - \text{H} & \text{H} - \text{C} - \text{C} - \text{C} - \text{H} \\ \text{OH} \;\; \text{OH} & \text{OH} \; \text{OH} \; \text{OH} \end{array}$$

ethylene glycol   glycerol
(1,2-dihydroxyethane)   (1,2,3-trihydroxypropane)

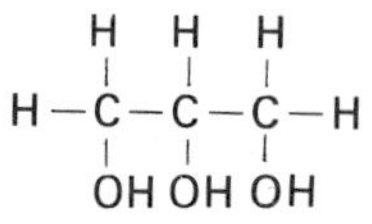

Ethylene glycol is used in automobile cooling systems as an antifreeze. A mixture of 60% ethylene glycol and water will not freeze at temperatures above $-40°C$.

Alcohols form weak hydrogen bonds, since they have a hydroxyl group containing a separation of charge. Because of their polar nature, alcohols are moderately good solvents for polar substances, although much less so than water. The alcohols are soluble in water to a very high degree. Ethyl alcohol, for example, is completely *miscible* with water, meaning that one can make solutions with any ratio of alcohol to water.

13–6 ETHERS

Ethers are compounds that contain two hydrocarbon groups connected by an oxygen atom. Ethers have the general formula ROR′, where R and R′ may be the same or different. A few common ethers are listed below.

Dimethyl ether
(methoxymethane) C_2H_6O: $H_3C - O - CH_3$

Methyl ethyl ether
(methoxyethane) C_3H_8O: $H_3C - O - \overset{\displaystyle H}{\underset{\displaystyle H}{C}} - CH_3$

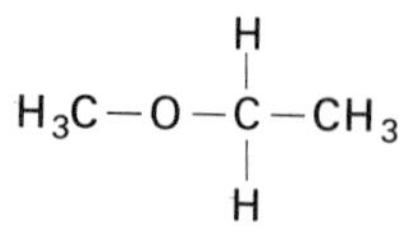

Diethyl ether
(ethoxyethane) $C_4H_{10}O$: $H_3C - \overset{\displaystyle H}{\underset{\displaystyle H}{C}} - O - \overset{\displaystyle H}{\underset{\displaystyle H}{C}} - CH_3$

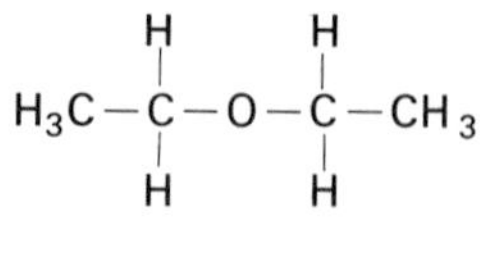

The ethers are structural isomers of the alcohols; for example, both ethyl alcohol and dimethyl ether have the empirical formula C_2H_6O. Their properties, however, are quite different because of the different placement of the atoms. The alcohols are more reactive because the oxygen is in a more exposed position at the end of the molecule than in the ethers. The ethers do not undergo hydrogen bonding and, therefore, are generally not very soluble in water and other polar solvents. The boiling points of the ethers are lower than alcohols of the same empirical formula. For example, of the two compounds having the formula C_2H_6O, ethyl alcohol is a liquid at room temperature and dimethyl ether is a gas.

Ethers can be produced by the dehydration (removal of water) of alcohols by hot concentrated sulfuric acid. The most common ether, diethyl ether, is produced from ethanol by the following reaction:

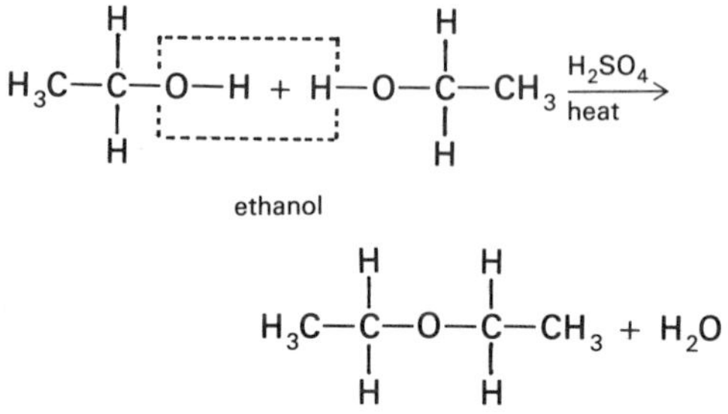

$$H_3C - \overset{\displaystyle H}{\underset{\displaystyle H}{C}} - O - H + H - O - \overset{\displaystyle H}{\underset{\displaystyle H}{C}} - CH_3 \xrightarrow[\text{heat}]{H_2SO_4}$$

ethanol

$$H_3C - \overset{\displaystyle H}{\underset{\displaystyle H}{C}} - O - \overset{\displaystyle H}{\underset{\displaystyle H}{C}} - CH_3 + H_2O \qquad (13-7)$$

diethyl ether

Diethyl ether is used extensively as an anesthetic during operations, but it is highly flammable and causes nausea in most patients, so that other types of anesthetic are being increasingly used. Diethyl ether is also a good solvent for fats, waxes, and many other water-insoluble organic substances, but its flammability and volatility make it hazardous to use in the laboratory.

13–7 ALDEHYDES AND KETONES

Aldehydes and ketones have the general structures

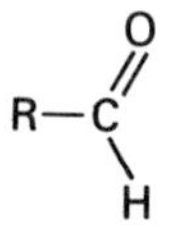

$$\begin{array}{cc} \overset{\displaystyle O}{R - C} & \overset{\displaystyle O}{R - C} \\ \quad\;\; H & \quad\;\; R' \end{array}$$

aldehyde 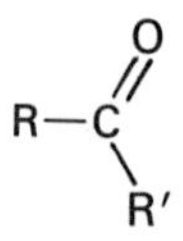   ketone

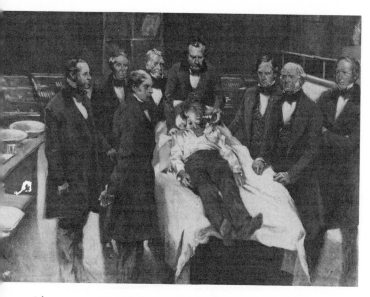

Ether was used in this first public demonstration of surgical anesthesia, which took place at Massachusetts General Hospital, Boston, in 1846. (Courtesy of National Library of Medicine, Bethesda, Maryland.)

where R and R′ are any alkyl radicals. The systematic names of the aldehydes end in -*al* and the ketones in -*one*. Some simple examples of compounds of these groups are given below:

H—$\overset{\overset{\text{O}}{\|}}{\text{C}}$—H   formaldehyde (methanal)

CH_3—$\overset{\overset{\text{O}}{\|}}{\text{C}}$—$CH_3$ acetone (propanone)

CH_3—$\overset{\overset{\text{O}}{\|}}{\text{C}}$—H   ethanal

CH_3—CH_2—$\overset{\overset{\text{O}}{\|}}{\text{C}}$—$CH_3$ butanone

(benzene ring)—$\overset{\overset{\text{O}}{\|}}{\text{C}}$—H   benzaldehyde

(benzene ring)—$\overset{\overset{\text{O}}{\|}}{\text{C}}$—(benzene ring)   diphenyl ketone (benzophenone)

Both aldehydes and ketones can be produced by the controlled, partial oxidation of alcohols. If the hydroxyl group is on the end carbon atom, an aldehyde is formed. If the

hydroxyl is not on an end carbon atom a ketone is formed:

$$H_3C—\overset{\overset{\text{H}}{|}}{\underset{\underset{\text{H}}{|}}{\text{C}}}—OH + [O] \longrightarrow H_3C—\overset{\overset{\text{O}}{\|}}{\text{C}}—H + H_2O, \qquad (13–8)$$

$$H_3C—\overset{\overset{\text{OH}}{|}}{\underset{\underset{\text{H}}{|}}{\text{C}}}—CH_3 + [O] \longrightarrow H_3C—\overset{\overset{\text{O}}{\|}}{\text{C}}—CH_3 + H_2O. \qquad (13–9)$$

The oxygen shown as $[O]$ can be supplied by various oxidizing agents, e.g., oxygen gas with copper as a catalyst or the dichromate ion $Cr_2O_7^{2-}$. If the oxidation is allowed to proceed too far, the aldehydes and ketones are converted to organic acids (Section 13–8) or destroyed.

Aldehydes and ketones are used heavily in the chemical industry as intermediates in the preparation of more complex organic molecules. They are also used as solvents for waxes, greases, and other solid organic materials. The most widely used aldehyde is formaldehyde, which is used in plastics (Section 13–13) and preservatives. A 37% solution of formaldehyde in water, called "formalin," is used to preserve biological samples.

The most common laboratory ketone, acetone, is used as a solvent for many organic materials, e.g., as fingernail-polish remover. Since acetone is quite volatile, it is used a lot around laboratories to clean and dry glassware. After flasks are washed with water, remaining water droplets can be rinsed out with acetone, which then evaporates very quickly. Although not as hazardous as ether, acetone is flammable and can cause serious laboratory fires if not used with care.

Molecules of sugar contain aldehyde or ketone functional groups. This fact is the basis of a standard test for sugar in urine, used in the diagnosis of diabetes. Aldehydes treated with Fehling's or Benedict's solution, basic solutions of copper sulfate, react to form a red precipitate of cuprous oxide:

$$R—\overset{\overset{\text{O}}{\diagup\!\!\|}}{\text{C}}—H \text{(aq)} + 2\,Cu^{2+}\text{(aq)} \xrightarrow{\text{NaOH}}$$

aldehyde   blue solution   (13–10)

$$R—\overset{\overset{\text{O}}{\diagup\!\!\|}}{\text{C}}—O^- \ Na^+ \text{(aq)} + Cu_2O\text{(s)}$$

salt of organic acid   red precipitate

on: copper is reduced from the
lehyde is oxidized to the anion

important *carboxylic acid* func-

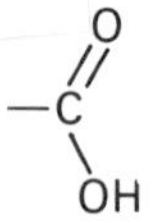

Examples of some simple organic acids are

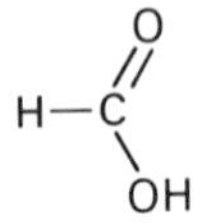

formic acid
(methanoic acid)

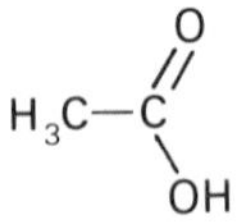

acetic acid
(ethanoic acid)

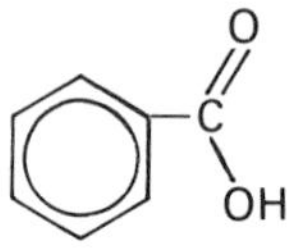

benzoic acid

Most organic acids are very weakly dissociated in water solution, e.g., acetic acid:

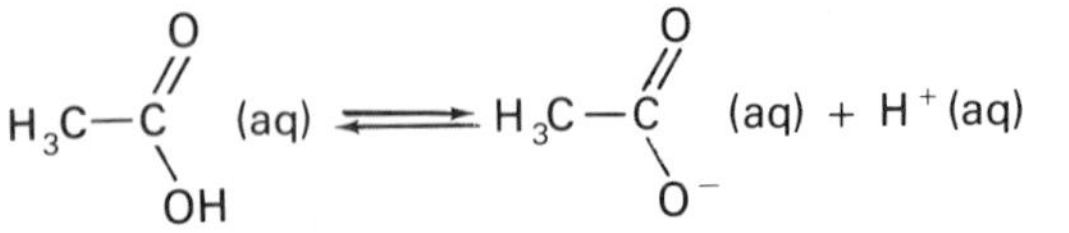

$$H_3C-C \overset{O}{\underset{OH}{\big\|}} \text{ (aq)} \rightleftharpoons H_3C-C \overset{O}{\underset{O^-}{\big\|}} \text{ (aq)} + H^+ \text{(aq)} \quad (13-11)$$

acetic acid   acetate ion

This is the same acid we used as an example of a weak acid in Section 8–4, but there it was represented by the empirical formula $HC_2H_3O_2$ instead of the structural formula given above.

Organic acids can be made by strong oxidation of alcohols or aldehydes. In Section 13–5 we noted that fermentation of apple cider converts sugars to alcohol. Low-temperature oxidation of the ethyl alcohol then yields acetic acid:

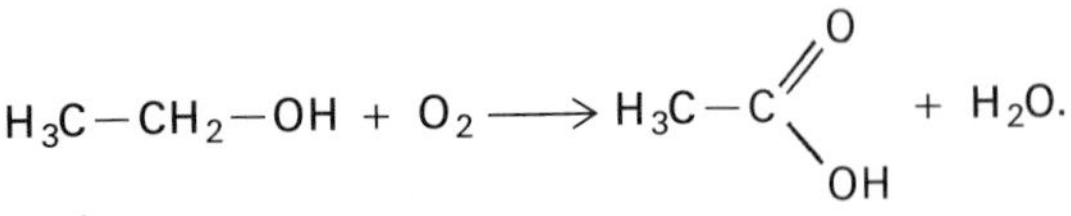

$$H_3C-CH_2-OH + O_2 \longrightarrow H_3C-C \overset{O}{\underset{OH}{\big\diagup}} + H_2O. \quad (13-12)$$

ethyl alcohol   acetic acid

Aside from some impurities that may give it flavor, vinegar is simply a dilute aqueous solution of acetic acid, and its characteristic odor is that of acetic acid. You may have

noticed that vinegar bottles often carry the label "cider vinegar," as the vinegar was made from cider by the chemical steps discussed above and in Section 13–5:

apple cider $\xrightarrow{\text{fermentation}}$ "hard" cider $\xrightarrow{\text{oxidation}}$ vinegar.
(sugar)   (alcohol)   (acetic acid)

The carboxylic acids of low molecular weight, which are soluble in water and dissociate weakly, have a sour taste (e.g., vinegar). The acids with one to three carbon atoms have sharp, acrid odors. Those with four to ten carbons have extremely unpleasant odors. Butyric acid, for example, is formed in rancid butter and is responsible for its characteristic odor. The *capr-* root of the six-, eight-, and ten-carbon acids (see Table 13–2) is derived from the Latin word for "goat," doubtless relating to their odor! Acids with twelve or more carbons are not very volatile, so have little odor.

Some of the more prominent organic acids with a single carboxylic acid group are listed in Table 13–2. Most such acids are derived from animal fats and oils from plants, and so these acids are referred to as *fatty acids*. Note that virtually all the important fatty acids have an even number of carbon atoms, a result of the fact that the biological synthesis involves the joining of two-carbon units. Most of the fatty acids have straight-chain alkyl groups. Many of the carbon chains are completely saturated with hydrogen atoms, but some contain one or more double bonds. At the same molecular weight, fats containing unsaturated alkyl groups have lower melting points than those that are saturated; therefore animal fats, which are mostly saturated, are solids, whereas most vegetable fats are liquids. In medical circles, a theory now in vogue has it that saturated fats in the diet contribute to cholesterol deposition in the blood vessels (hardening of the arteries) and in the gall bladder (as gall stones). For this reason, diets containing more "polyunsaturated" fats (such as safflower oil) and less animal fat are often suggested. Unsaturated fats can be "hydrogenated" to saturate the double bonds, converting vegetable oils to solids such as oleomargarine and vegetable shortening.

Some organic acids have two or more carboxylic acid groups, e.g., oxalic acid, which occurs in rhubarb leaves,

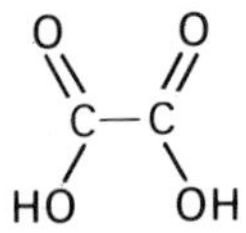

oxalic acid

Table 13–2   Important Monocarboxylic Acids (Fatty Acids)

| Name | | | | |
Common	IUPAC	Formula	No. of carbons	Major occurrence
Saturated				
Formic	Methanoic	HCOOH	1	Ants, bees
Acetic	Ethanoic	CH_3COOH	2	Vinegar
Butyric	Butanoic	C_3H_7COOH	4	Butter
Caproic	Hexanoic	$C_5H_{11}COOH$	6	Butter
Caprylic	Octanoic	$C_7H_{15}COOH$	8	Coconut oil
Capric	Decanoic	$C_9H_{19}COOH$	10	Coconut oil
Lauric	Dodecanoic	$C_{11}H_{23}COOH$	12	Palm kernel oil
Myristic	Tetradecanoic	$C_{13}H_{27}COOH$	14	⎫ Many animal fats
Palmitic	Hexadecanoic	$C_{15}H_{31}COOH$	16	⎬ and vegetable
Stearic	Octadecanoic	$C_{17}H_{35}COOH$	18	⎭ oils
Unsaturated				
One double bond:				
Palmitoleic	*cis*-9-hexadecanoic	$C_{15}H_{29}COOH$	16	Cod-liver oil
Oleic	*cis*-9-octadecanoic	$C_{17}H_{33}COOH$	18	Corn oil
Two double bonds:				
Linoleic		$C_{17}H_{31}COOH$	18	Linseed oil
Three double bonds:				
Linolenic		$C_{17}H_{29}COOH$	18	Linseed oil

and citric acid,

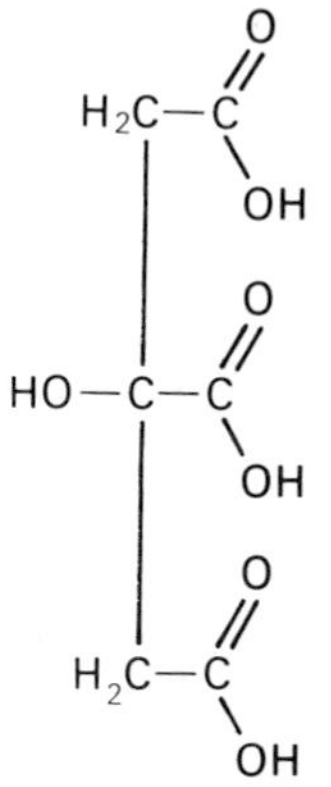

citric acid

Along with its sodium or potassium salts, citric acid occurs in citrus fruits (lemons, limes, oranges), giving them their sour taste. This important acid is also added to artificial citrus fruit drinks.

The reaction of organic acids with alcohols forms *esters*, which have the general structure:

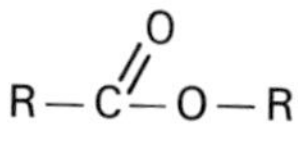

ester

When a mixture of the appropriate acid and alcohol are heated, a water molecule "splits out" and the remainder of the acid and alcohol molecules join together, thus forming the ester. The reaction between acetic acid and ethyl alcohol is typical:

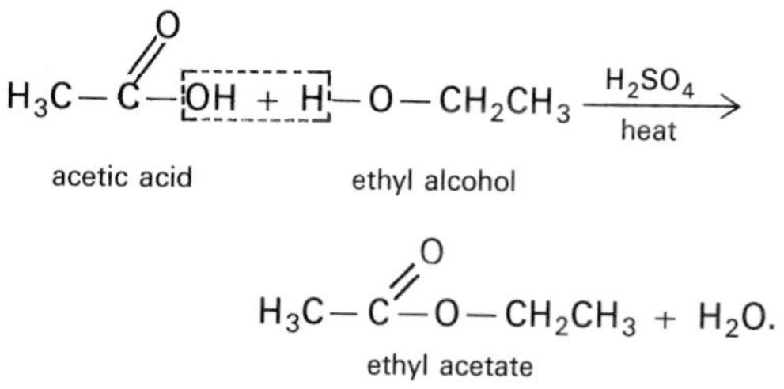

acetic acid   ethyl alcohol

$$H_3C-C\overset{O}{\diagup}O-CH_2CH_3 + H_2O. \qquad (13\text{–}14)$$

ethyl acetate

The sulfuric acid is used as a catalyst in many organic reactions.

The names of the esters are taken from those of their constituent alcohol and acid roots, with the suffix -ate, somewhat like the names of salts of organic acids, e.g.:

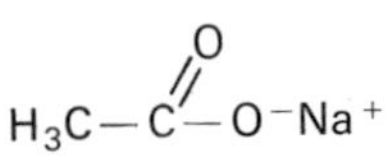

$$H_3C-\overset{\displaystyle O}{\overset{\|}{C}}-O^-Na^+ \qquad H_3C-\overset{\displaystyle O}{\overset{\|}{C}}-O-CH_2CH_3$$

Salt: sodium acetate 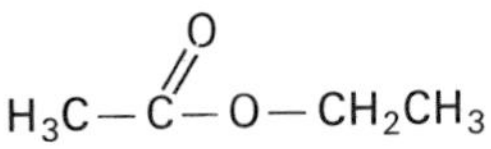   Ester: ethyl acetate

Some other simple esters are

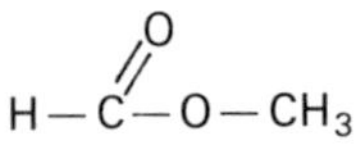

$$H-\overset{\displaystyle O}{\overset{\|}{C}}-O-CH_3$$

methyl formate

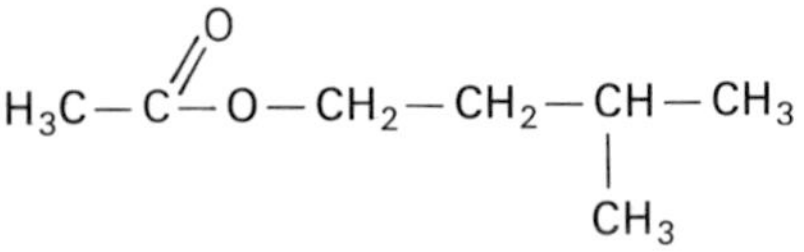

$$H_3C-\overset{\displaystyle O}{\overset{\|}{C}}-O-CH_2-CH_2-CH-CH_3$$
$$CH_3$$

isoamyl acetate

$$H_3C-CH_2-CH_2-\overset{\displaystyle O}{\overset{\|}{C}}-O-CH_2-CH_3$$

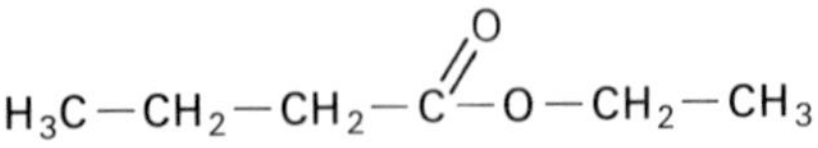

ethyl butyrate

Many esters have very sweet, pleasant odors and are responsible for the aromas of fruits and flowers. Some esters are used commercially as solvents, and a wide variety of them are used for artificial flavoring and perfumes. Isoamyl acetate is used as a banana flavoring, ethyl butyrate for pineapple, ethyl formate for rum, and amyl butyrate for apricot.

Two aromatic esters derived from salicylic acid have widespread household use: acetyl salicylic acid, the active ingredient in aspirin, and methyl salicylate, "oil of wintergreen":

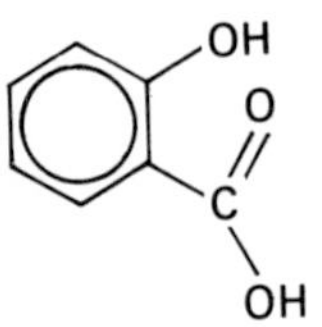

salicylic acid 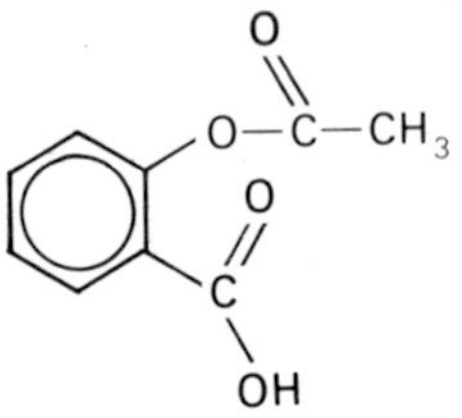   acetyl salicylic acid 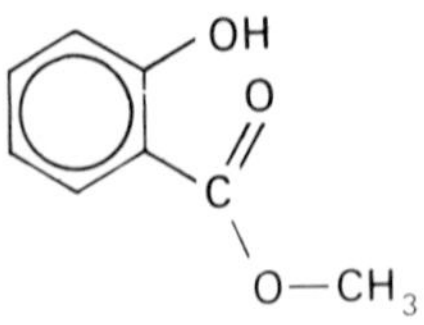   methyl salicylate

Aspirin is by far the most widely used medicine—20 billion pills per year in the United States! It has three main functions: reduction of swelling and inflammation, reduction of fever, and a general relief of pain. Despite the widespread use of aspirin, until recently we had almost no idea of the mechanism by which it acts upon the body. There is now some evidence that, in part, aspirin is effective because it reduces the concentration of prostaglandins in the body.[*] The prostaglandins are a class of biochemicals that seem to have very important effects in the regulation of many body processes and, among other things, they are under study as possible birth-control agents (see Chapter 15). Although aspirin is generally considered safe, hundreds of children annually become seriously ill or even die from overdoses of aspirin which they take themselves, especially fruit-flavored aspirins designed for children. Some persons suffer severe allergic reactions to aspirin, and fragmentary evidence indicates that aspirin taken by pregnant women may cause birth defects.[†]

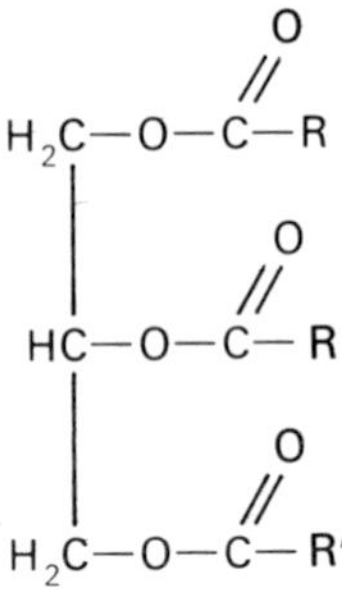

a fat   glyceryl tripalmitate   all palmitates

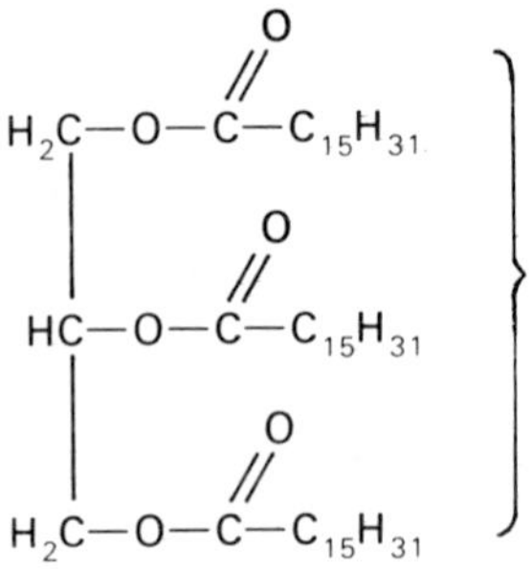

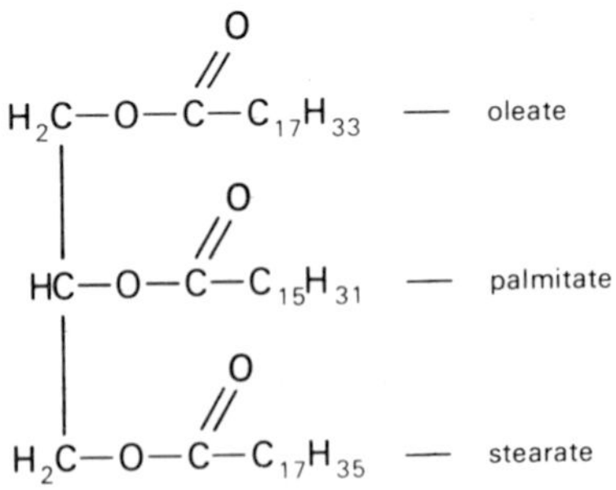

glyceryl oleopalmitosterate

[*]   A. L. Hammond, "Aspirin: New Perspective on Everyman's Medicine," Science 174, 48 (1971).

[†]   "Aspirin and Birth Defects: Fetal Cell Inhibition," Science News 100, 225 (1971).

Fats are a very important class of esters that are formed by the reaction between fatty acids and the trihydroxy alcohol, glycerol. The general formula for a fat and some examples are given in the preceding column.

In general, the three acid groups attached to the glycerol are different from each other, and they are either listed in Table 13–2 or are similar acid groups, with chain lengths ranging up to more than 20 carbon atoms. All fats and vegetable oils contain glycerol esters with a wide variety of attached acid groups.

Soap is made by the *saponification* of fats with sodium hydroxide. The sodium hydroxide splits the fat into glycerol and three molecules of the sodium salts of the attached fatty acids. For example,

$$
\begin{array}{c}
\text{H}_2\text{C}-\text{O}-\overset{\displaystyle \text{O}}{\overset{\|}{\text{C}}}-\text{C}_{17}\text{H}_{35} \\[2mm]
\text{HC}-\text{O}-\overset{\displaystyle \text{O}}{\overset{\|}{\text{C}}}-\text{C}_{17}\text{H}_{35} \quad + \quad 3\text{NaOH} \longrightarrow \\[2mm]
\text{H}_2\text{C}-\text{O}-\overset{\displaystyle \text{O}}{\overset{\|}{\text{C}}}-\text{C}_{17}\text{H}_{35}
\end{array}
$$

glyceryl tristearate

$$
\begin{array}{c}
\text{H}_2\text{C}-\text{OH} \\[2mm]
\text{HC}-\text{OH} \quad + \quad 3\text{C}_{17}\text{H}_{35}-\overset{\displaystyle \text{O}}{\overset{\|}{\text{C}}}-\text{O}^-\,\text{Na}^+ \qquad (13\text{–}15) \\[2mm]
\text{H}_2\text{C}-\text{OH}
\end{array}
$$

glycerol   sodium stearate
a soap

The cleansing action of soap molecules is dependent on the fact that the salt end of the molecule is ionic and the other end is a hydrocarbon. The hydrocarbon part dissolves readily at the surface of particles of grease and other organic materials that are not soluble in water. The carboxylate groups, $-\text{COO}^-$, remain outside the droplet, forming a layer of ionic material around the particle. These ionic layers, which are attractive to water molecules (*hydrophillic*), keep the organic materials suspended in the wash water and prevent

the organic globules from coalescing to form a heavy mass of water-repellant (*hydrophobic*) material that would settle on the clothes or dishes (see Fig. 13–8). In this way, soap lifts particles of organic material from clothes, dishes, or skin, suspending them in water and allowing them to be rinsed away. Since soap molecules are a natural type of material, they are easily decomposed in the environment. The major dis-

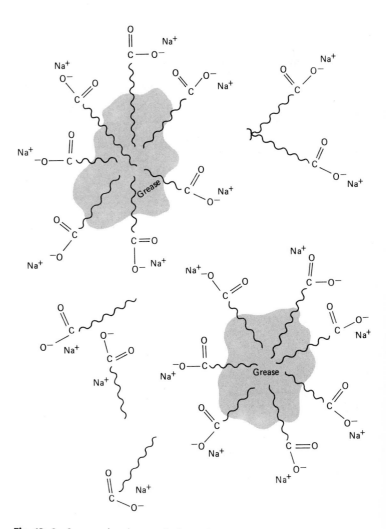

Fig. 13–8 Soap molecules attach themselves to grease droplets and keep the droplets suspended in water.

Soap is formed in this neutralizer tank when a stream of fatty acids is mixed with a stream of alkali. (Courtesy of Proctor and Gamble Company.)

Here we have written the nitric acid formula in a rather unusual way (normally it is "HNO_3") to emphasize the structure of the nitrate group. Nitroglycerin is a powerful explosive, very sensitive to vibration or shock as described in Section 10–2. In 1867, Alfred Nobel successfully stabilized nitroglycerin by absorbing it in siliceous earth to form dynamite, which can be transported without extreme danger. Nitroglycerin in the body causes dilation of the blood vessels and a decrease in blood pressure. People who work with nitroglycerin may have severe headaches because they absorb the compound through the skin. Pills containing small quantities of "nitro" are often given to people with heart conditions such as angina pectoris.

13–9 ORGANIC SULFUR COMPOUNDS

Sulfur is in the same group in the periodic table as oxygen, and can substitute for it in almost any position that oxygen normally takes. Compounds in which this occurs are called *thio-* compounds. For example,

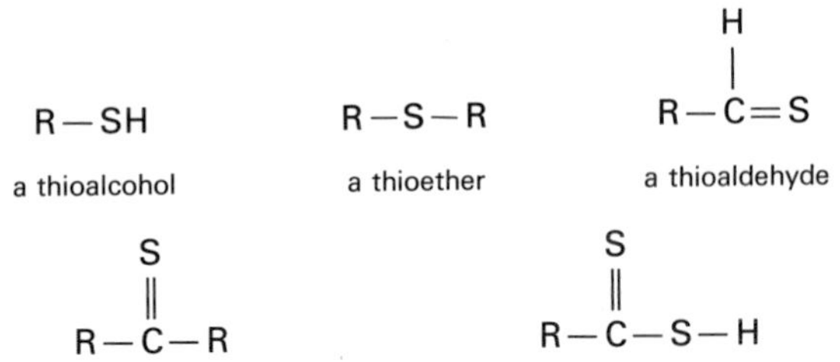

$$R-SH \qquad R-S-R \qquad R-\overset{\overset{\displaystyle H}{|}}{C}=S$$

a thioalcohol   a thioether   a thioaldehyde

$$R-\overset{\overset{\displaystyle S}{\|}}{C}-R \qquad\qquad R-\overset{\overset{\displaystyle S}{\|}}{C}-S-H$$

a thioketone   a dithioacid

advantage of soap is that although the sodium and potassium salts of the fatty acids are quite water soluble, the hard-water cations, principally Ca^{2+}, Mg^{2+} and Fe^{2+}, form insoluble salts such as Ca(stearate)$_2$ with the soap molecules. These insoluble precipitates form a scum on the water and leave deposits on clothes, skin, and wash basins (the "bathtub ring"). Synthetic detergents were developed to avoid this problem but they have caused environmental problems of their own.

Esters may also be made from inorganic acids and alcohols. An example is the reaction of nitric acid with glycerol to form the nitric ester glyceryl trinitrate, more commonly known as "nitroglycerin":

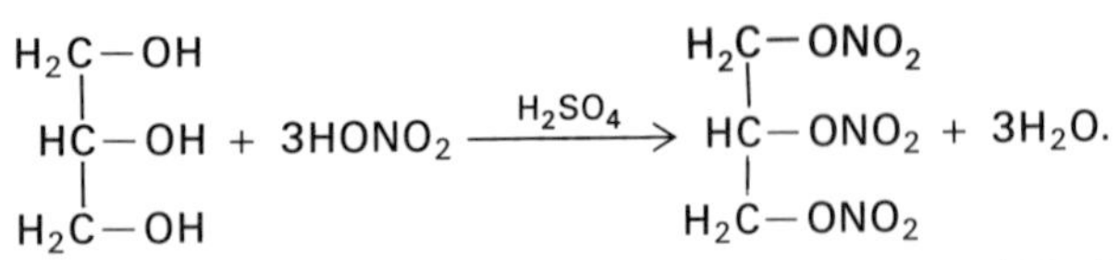

The thioalcohols, which are called *mercaptans*, have very powerful odors. The defensive agent of the skunk contains butyl mercaptan. Ethyl mercaptan is added in very low concentrations to natural gas, so that leaks can be detected by odor. A sensitive observer can detect it at concentrations as low as 0.0001 ppm in the air.

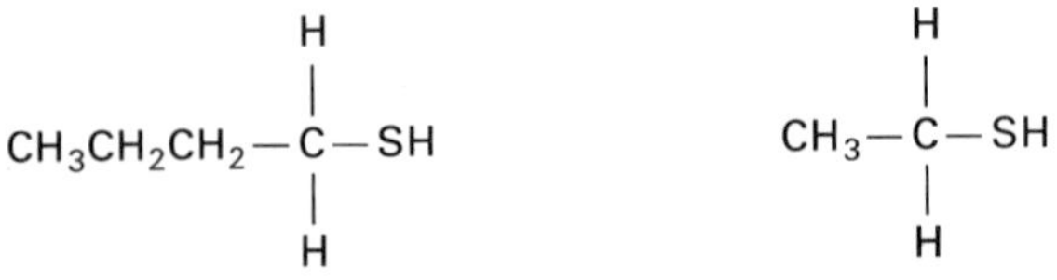

$$\begin{array}{c} H_2C-OH \\ | \\ HC-OH \\ | \\ H_2C-OH \end{array} + 3HONO_2 \xrightarrow{H_2SO_4} \begin{array}{c} H_2C-ONO_2 \\ | \\ HC-ONO_2 \\ | \\ H_2C-ONO_2 \end{array} + 3H_2O.$$

glycerol   nitric acid   nitroglycerin   (13–16)

$$CH_3CH_2CH_2-\overset{\overset{\displaystyle H}{|}}{\underset{\underset{\displaystyle H}{|}}{C}}-SH \qquad\qquad CH_3-\overset{\overset{\displaystyle H}{|}}{\underset{\underset{\displaystyle H}{|}}{C}}-SH$$

butyl mercaptan   ethyl mercaptan

Strong oxidation of a thio alcohol results in the formation of sulfonic acid:

$$5C_2H_5SH + 6MnO_4^- + 18H^+ \longrightarrow 5C_2H_5-\overset{\displaystyle O}{\underset{\displaystyle O}{S}}-OH$$

ethyl mercaptan

ethyl sulfonic acid

$$+ 6Mn^{2+} + 9H_2O. \qquad (13-17)$$

Sulfonic acids are related to sulfuric acid in structure, but they are not so strongly acidic. They are important in the production of synthetic detergents and ion-exchange resins used in water softeners and in the chemical laboratory to "de-ionize" water. An ion-exchange resin used as a water softener consists of a large, open, hydrocarbon network with many attached sulfonic acid groups. Before use, sodium hydroxide is passed through the ion-exchanger to neutralize the acid groups, i.e., replace the acid hydrogens with Na^+ ions. When hard water trickles through the ion-exchange resin, the weakly bound Na^+ ions are replaced on the sulfonate sites by hard-water ions such as Ca^{2+} and Mg^{2+}, which are more tightly bound to those sites (see Fig. 13–9). Water flowing out of the ion-exchanger contains Na^+ ions instead of hard-water ions, and so it can be used with soap for washing purposes.

Because of the problems of using soap in hard water, soap has gradually been replaced by synthetic detergents since World War II, until today soap accounts for only about 15% of the market (mainly as hand and bath soap bars). Until 1965, many detergents were of the alkyl benzene sulfonate (ABS) type, which has the general structure

alkyl benzene sulfonate, ABS

where R and R' indicate large alkyl groups of variable chain length. The ABS is made largely from reactions of petroleum products with sulfuric acid.

Although synthetic detergents are better than soap when we are washing with hard water, they have caused a number

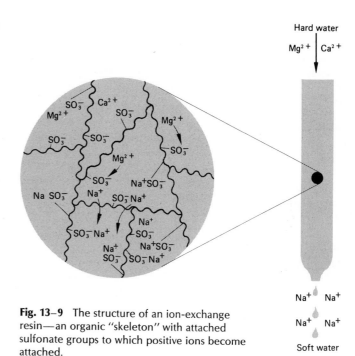

Fig. 13–9 The structure of an ion-exchange resin—an organic "skeleton" with attached sulfonate groups to which positive ions become attached.

of problems. The detergents do have less tendency to form precipitates with hard water ions than soap; however, in order to completely avoid precipitation, most household detergents contain sodium polyphosphates that form stable, soluble complexes with the hard-water ions. As noted in Section 11–6, the phosphates have added to our problems of eutrophication and to the decline of many bodies of water such as Lake Erie.

The ABS detergent, in particular, caused problems by not being *biodegradable*. We will encounter this problem frequently in our discussions of man-made organic chemicals. Over the billions of years of life on Earth, the ecological system has evolved in such a way that there are natural processes for breaking down and re-using all the naturally occurring materials: mountains are eroded away, rocks are broken up by rushing water, parts of them are dissolved, parts are converted to soil with nutrients needed by plants. After a plant or animal dies, the organisms and chemicals of the environment break down its structure and return its elements to the ecological cycle.

Think for a moment what would happen if a single, important destructive process stopped, e.g., if the limbs falling from trees no longer were rotted away by natural processes. Soon the ground in a forest would be so heavily littered with fallen wood that all small vegetation would be smothered out. Furthermore, the nutrients in the wood would not be returned to the soil and the trees would eventually starve.

Most man-made objects and materials are destroyed by natural processes. Even durable objects made of metal (except those of very inert metals such as gold or platinum) eventually rust and corrode. However, we are becoming increasingly aware of problems caused by several classes of organic chemicals for which there are only very slow, if any, natural breakdown processes in the environment. As we shall see below, many troublesome pesticides are of this sort. The structure of their molecules is so different from those of naturally occurring molecules that organisms cannot "digest" or break them down. Instead of decomposing, these unnatural substances may just wander around in the environment until they find a suitable place to stay—for pesticides, this place is in the fats of animals. Man-made chemicals may thus continue to accumulate at certain points in the environment until they reach levels significantly high enough to cause harm.

The ABS molecules are an example of such a nonbiodegradable substance. The salts of fatty acids in soap are so similar to the breakdown products of fatty acids that organisms can easily digest them and convert them into other biochemicals.

The structures of the ABS molecules are sufficiently different from those of fatty acids that organisms have great difficulty in breaking them down. In particular, the organisms that break down soap molecules readily attack straight-chain alkyl groups, but are unable to handle the branched chain of ABS (i.e., the two alkyl groups bonded to the carbon attached to the benzene ring). As a result, ABS detergents were not broken down by the usual processes in sewage-treatment plants or by organisms in streams after release of the sewage effluent. As the use of ABS detergents climbed, the problem became increasingly obvious to everyone: instead of rotting away, the ABS kept acting as a detergent in the streams. Creeks, rivers and lakes became choked with masses of foam and bubbles produced by the persistent detergent present throughout the nation's water system.

Because of public and governmental persuasion, the U.S. detergent industry in 1965 completed the switchover

Nonbiodegradable detergents caused this pollution in an Alabama stream in 1970. Scenes like this are becoming rarer, due to the increasing use of biodegradable detergents. (U.S. Dept. of Agriculture–Soil Conservation Service photo by Bill McCormick.)

from ABS to a different type of detergent, linear alkylbenzene sulfonate, LAS. The LAS is much more biodegradable than ABS, and so the problem of foaming on streams has now largely disappeared. The structure of LAS is of the form

linear alkyl benzene sulfonate, LAS

Although not so easily broken down in the environment as fatty acid salts, the absence of branching in the chain attached to the benzene ring makes LAS much more biodegradable than ABS, thus eliminating the problems of suds on rivers.

13–10 ORGANIC NITROGEN COMPOUNDS

Nitrogen can be found in many different forms in organic molecules. One of the simplest is the ammonium salt of an acid, for example, ammonium butyrate formed by the reaction of ammonia with butyric acid:

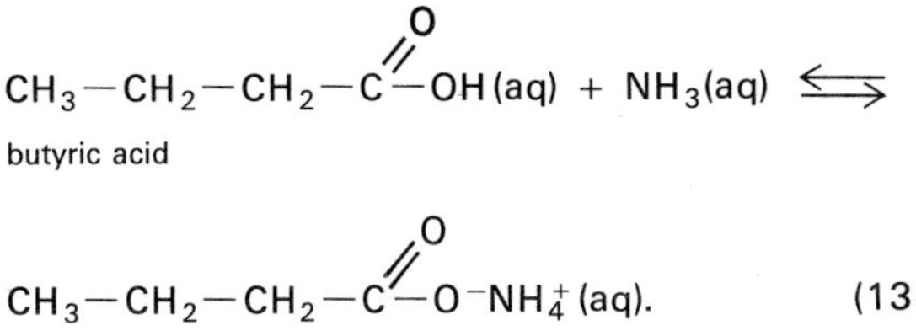

$$CH_3-CH_2-CH_2-\overset{\overset{\displaystyle O}{\|}}{C}-OH\,(aq) + NH_3(aq) \rightleftharpoons$$

butyric acid

$$CH_3-CH_2-CH_2-\overset{\overset{\displaystyle O}{\|}}{C}-O^-NH_4^+\,(aq). \qquad (13\text{–}18)$$

ammonium butyrate

Another class of compounds is the *amines*, which can be broken down into three types:

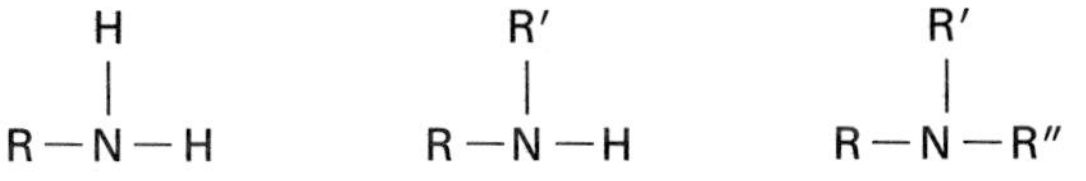

primary amine   secondary amine   tertiary amine

where R, R' and R'' stand for similar or different alkyl or aryl groups (the latter containing benzene rings). Some examples of typical amines are

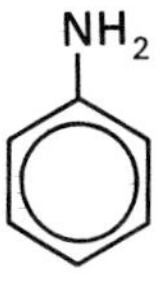

aniline

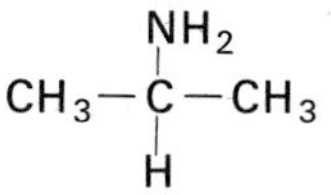

2-amino propane
(isopropyl amine)

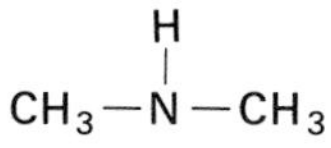

dimethyl amine

Other compounds containing amine groups are the amino acids, which are carboxylic acids in which there is an amino group ($-NH_2$) on the carbon atom next to the carboxylic group. The general formula is

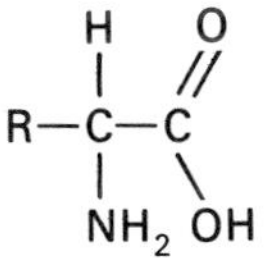

Examples of amino acids are glycine and tyrosine:

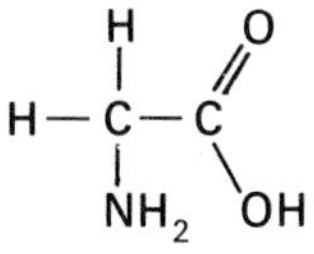

glycine 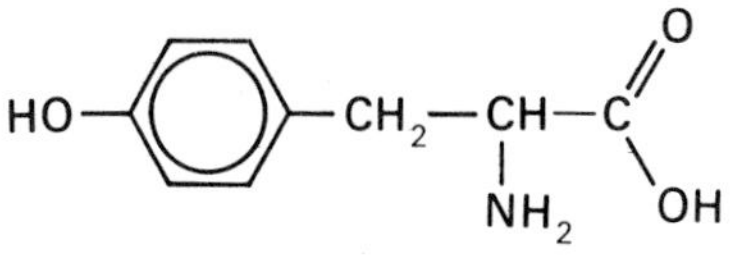   tyrosine

Amino acids are very important biochemicals, especially as proteins are made up of long chains of amino acids formed into coils (see Chapter 15).

Nitrogen atoms may also be a part of the ring structure in the aromatic and nonaromatic hydrocarbons called *heterocyclic* compounds, as the rings contain different kinds of atoms. Some examples of these compounds are given below:

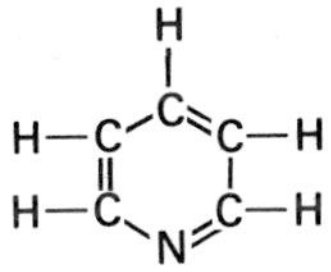

pyridine 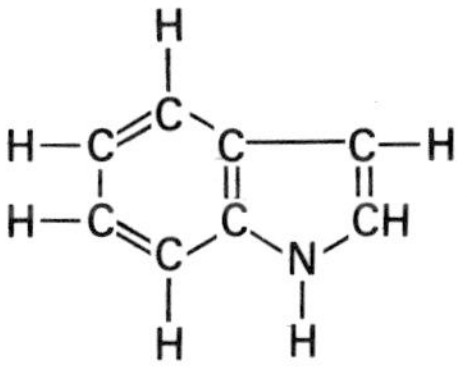   indole 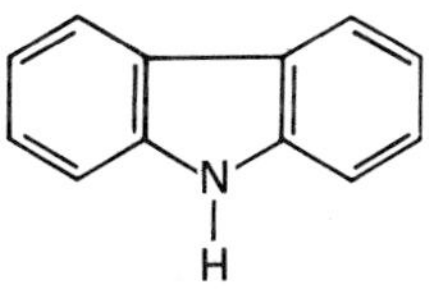   dibenzopyrrole

Of these compounds, pyridine, which has a strong, unpleasant odor, is used extensively in the chemical industry as a solvent and starting material for the synthesis of more complex compounds. Indole is present in raw sewage, causing its characteristic bad odor.

The basic indole and pyridine structures occur in many *alkaloids*, a class of nitrogen-containing compounds that occur in many plants. Some well-known alkaloids are the nicotine in tobacco, caffeine in coffee, morphine and heroin obtained from opium, and LSD (see Section 15–14).

Nitro compounds, such as trinitrotoluene (TNT), contain the $-NO_2$ group. This primary explosive is produced by the nitration of toluene with very strong nitric and sulfuric acids in the three separate steps outlined in the following diagram. This explosive is stable and safe to use, but it requires a strong detonator explosive to set it off. It is usually melted and cast into artillery shells or bombs.

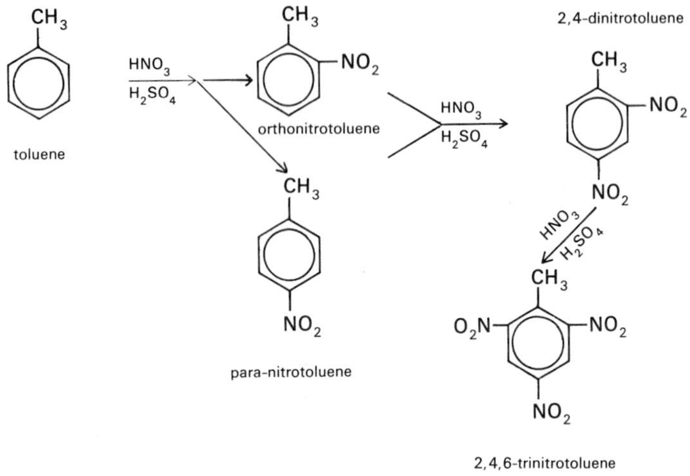

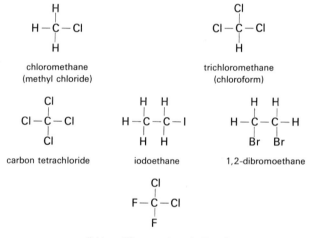

2,4-dinitrotoluene

orthonitrotoluene

para-nitrotoluene

2,4,6-trinitrotoluene

(13–19)

Another property of nitro compounds is that, upon reduction, they form diazo compounds which are important in some dyes:

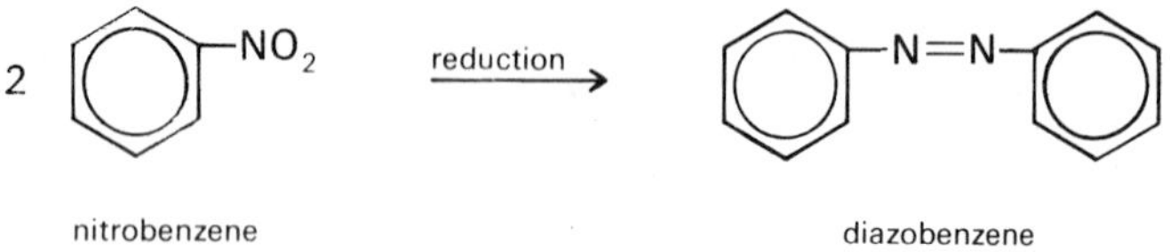

nitrobenzene   **diazobenzene**

(13–20)

Groups such as the diazo group ($-N=N-$) give molecules strong color characteristics. Examples of some dyes that contain nitrogen atoms are given below:

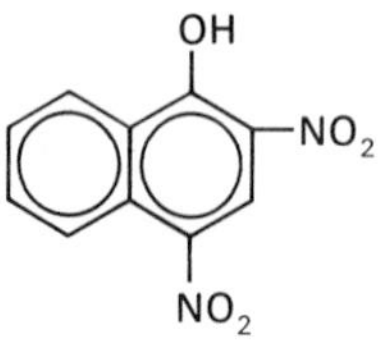

martinus yellow

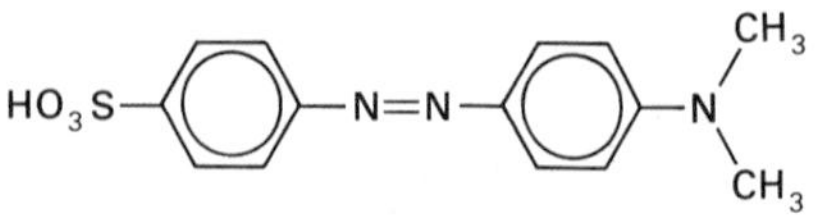

methyl orange

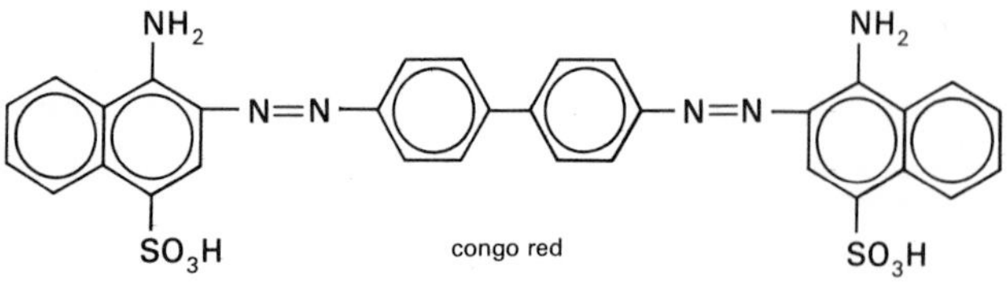

congo red

These dyes and indicators attach themselves to the cloth with the sulfonic acid or hydroxyl groups. The term *chromophore* has been given to the unsaturated groups responsible for the color. Some groups of chromophores are

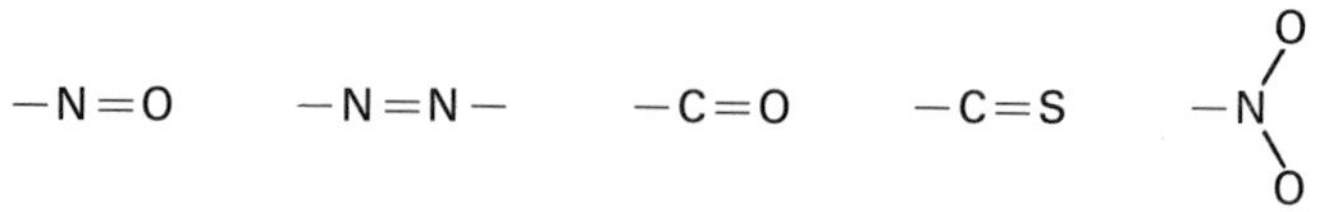

$$-N=O \qquad -N=N- \qquad -C=O \qquad -C=S \qquad -N\genfrac{}{}{0pt}{}{O}{O}$$

13–11 HALOGENATED HYDROCARBONS

Atoms of any of the halogens, Group VIIA in the periodic table, can replace one or more hydrogen atoms in organic compounds. These compounds, especially those containing chlorine and fluorine, are important in plastics, pesticides, anesthetics, and non-stick coatings (Teflon). Several examples of simple halogenated hydrocarbon compounds are shown below:

chloromethane (methyl chloride)

trichloromethane (chloroform)

carbon tetrachloride

iodoethane

1,2-dibromoethane

dichloro difluoromethane (a Freon)

Chloroform is used as a solvent for organic compounds, as an anesthetic sedative, and as a preservative medium for biological samples. Carbon tetrachloride is used extensively as a nonflammable dry-cleaning solvent and a starting material for the synthesis of other halogenated products. It should be used with care, as it is biologically hazardous: the body is unable to destroy or rid itself of carbon tetrachloride, which is thus stored in the liver or fatty tissues. Until a few years ago, carbon tetrachloride was used extensively in fire extinguishers. Today it has largely been replaced by CO_2 for that purpose, as CCl_4 reacts in flames to produce some phosgene, a deadly poison gas. The toxic effect of phosgene, $COCl_2$, results from its breakdown in the lungs to form hydrochloric acid, HCl.

Chlorinated hydrocarbon solvents released from dry-cleaning plants add to the problems of photochemical smog, because they react with components of the smog, such as nitric acid, to form chloropicrin:

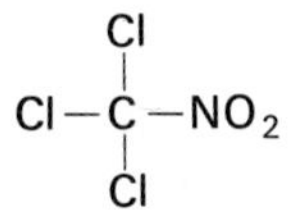

$$Cl-\underset{\underset{Cl}{|}}{\overset{\overset{Cl}{|}}{C}}-NO_2$$

chloropicrin

Chloropicrin is a lung irritant and a *lachrimator*, a compound that reacts with proteins to release acids that irritate the mucous membranes and make the eyes tear.

Chloropicrin and several other compounds, such as chloroacetophenone and bromoacetone, are components in the tear gas (actually aerosols of fine droplets or solids) used to quell riots:

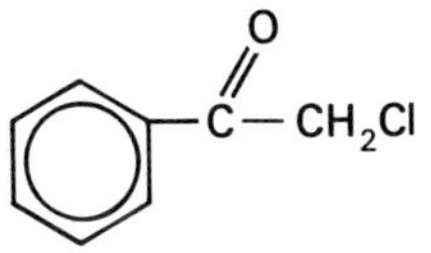

chloroacetophenone

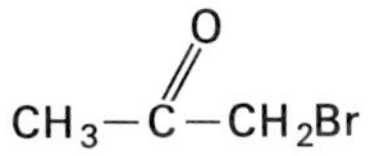

$$CH_3-\overset{\overset{O}{\|}}{C}-CH_2Br$$

bromoacetone

13–12 PESTICIDES AND POLYCHLORINATED BIPHENYLS

The debate on the benefits versus the hazards of pesticide use, probably more than any other single issue, has brought public attention to bear upon the world's environmental crisis. The beginning of widespread concern about man's degradation of the planet was initiated by the publication of Rachel Carson's *Silent Spring* in 1962, a book that exposed the many unfortunate side effects of the extensive use of pesticides, especially DDT and similar synthetic organic compounds.

Although one may think of pesticides only as the insecticide DDT, the term "pesticide" also includes herbicides (weed killers), fungicides, fumigants, rodenticides (rat poison), and so on. Before World War II, large amounts of pesticides containing inorganic elements such as copper, lead, arsenic, and mercury were mainly used. But during and after the war, chlorinated hydrocarbons such as DDT became

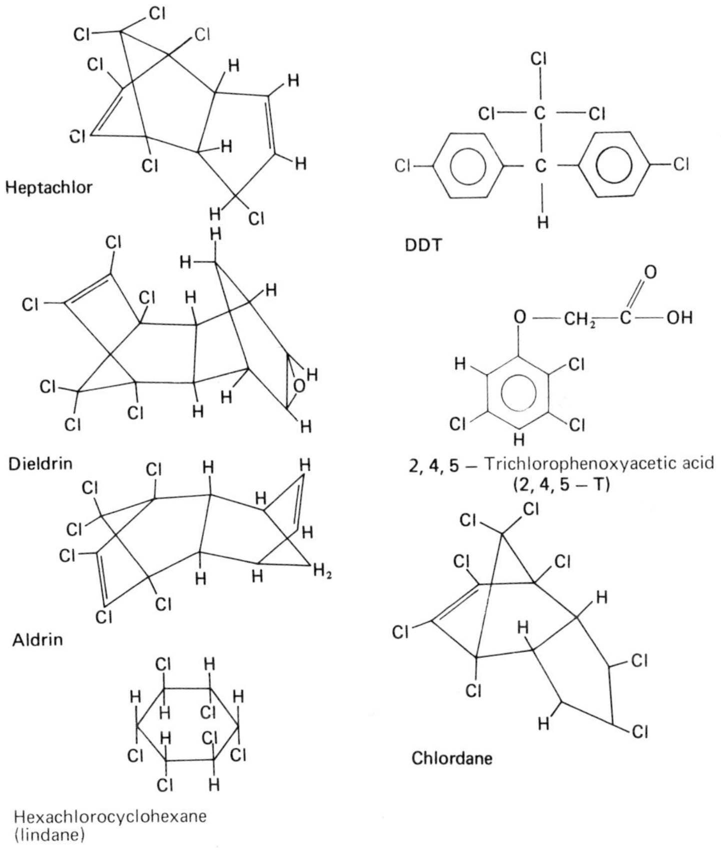

Fig. 13–10 Structure of several of the chlorinated hydrocarbon pesticides.

by far the most widely used pesticides, and it is these types that have caused serious environmental problems. Despite the problems they create, these modern pesticides have had some enormously beneficial effects: the control, by DDT and similar compounds, of insects bearing diseases such as malaria, and the increase of world food production by the use of weed killers and insecticides against pests that destroy crops.

Why do the chlorinated hydrocarbons cause problems in the environment? The most important reason is that these compounds, some of whose structures are shown in Fig. 13–10, are synthetic organic chemicals whose molecules are

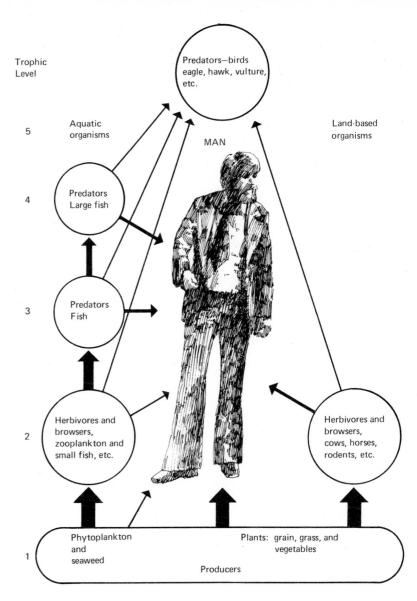

Trophic
Level

Aquatic
organisms

Land-based
organisms

5

MAN

Predators—birds
eagle, hawk, vulture,
etc.

4

Predators
Large fish

3

Predators
Fish

Fig. 13–11 A food web.

2

Herbivores and
browsers,
zooplankton and
small fish, etc.

Herbivores and
browsers,
cows, horses,
rodents, etc.

1

Phytoplankton
and
seaweed

Plants: grain, grass, and
vegetables

Producers

unlike those of any naturally occurring substances. In this way, the chlorinated hydrocarbons are similar to the non-biodegradable synthetic detergents discussed in Section 13–9. Few, if any, organisms of the biosphere are capable of breaking the pesticides down into simpler compounds* Furthermore, since the chlorinated hydrocarbons are non-

ionic and not very polar, they have extremely low solubility in water, e.g., about 1 ppb (part per billion) for DDT in water. When an animal ingests plants or animals containing such molecules, the water solutions of the body are unable to flush the pesticides out of the body or to degrade them; instead, major portions of the pesticides are stored up in the fat of the animal, in which the molecules are soluble. An animal or organism that feeds by passing large quantities of water and plant life through its body can therefore store up much larger concentrations of pesticides and residues in its body tissues than were present in the water or plants.

---

* In the environment, DDT molecules are modified slightly to form compounds called DDE and DDD which are termed "pesticide residues." These residues have properties very similar to those of DDT.

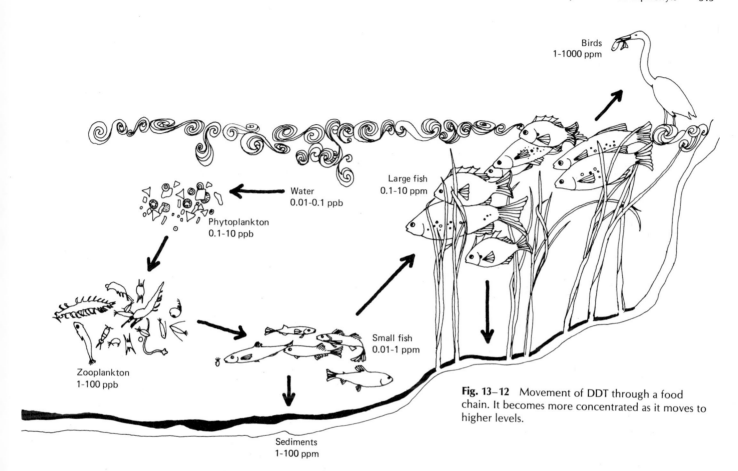

Birds
1-1000 ppm

Water
0.01-0.1 ppb

Phytoplankton
0.1-10 ppb

Large fish
0.1-10 ppm

Zooplankton
1-100 ppb

Small fish
0.01-1 ppm

Sediments
1-100 ppm

Fig. 13–12 Movement of DDT through a food chain. It becomes more concentrated as it moves to higher levels.

In any ecosystem such as a lake, there are generally a number of food "chains" that make up a food "web" (see Fig. 13–11). At the bottom of each food chain are the *primary producers*, which convert sunlight, water, and CO_2 into high-energy compounds needed by animal life, the *consumers* at higher levels of the food chain.* The various higher levels of the food chain are divided into *primary consumers*, which feed directly on the producers, *secondary consumers*, which feed on the primary consumers, *tertiary consumers*, and so on, up to a total of about five levels. As noted in Section 11–4, the transfer of energy and nutrients from one level of a food chain to the next higher one has an efficiency of only about 10%. Thus to provide sufficient food for the higher levels, the populations of the lower levels must have much greater total mass than levels above them. Animals of one level consume the much greater mass of animals or plants of the next lower level, in the process storing up the pesticides contained in the mass of the lower level. In this way, pesticides and their residues become increasingly concentrated at the higher levels of a food chain.

One of the food chains in Lake Michigan is an example of the concentration of DDT: Lake Michigan bottom muds are found to have a DDT concentration of 0.014 ppm, amphipods 0.41, fish 3 to 6, and herring gulls 99 ppm!* (See Fig. 13–12). Perhaps because man's diet is more diversified than that of

---

*   W. D. Russell Hunter, *Aquatic Productivity* (Macmillan, New York, 1971).

*   H. Egan, R. Goulding, J. Roburn, and J. O'G. Tatton, "Organic-Chlorine Pesticide Residues in Human Fat and Human Milk," *British Medical Journal* **2**, 66–69 (1965).

herring gulls and includes large portions of food from simpler chains (direct consumption of vegetables or of domestic animals fed on vegetation), the pesticide residues are not as concentrated in his body as they are in some other animals. Average concentrations of pesticide residues in body fat in 1965 varied from a high of about 26 ppm for people in India down to about 3 ppm in Alaska, with the United States population having about 12 ppm.*

What kinds of damages are caused by chlorinated hydrocarbons? Sometimes the pesticides are applied in such excessive amounts that they poison animals other than the intended species, causing massive kills of fish, birds, cats, etc. *Silent Spring* is filled with many examples of such overkills, resulting from airplane spraying or massive tree-spraying operations. Even when proper amounts are used, pesticides become more concentrated in species at the top of long food chains. The most susceptible species appear to be the predatory birds, such as the herring gulls mentioned above. Over the past several years, great declines in the populations of peregrines (duck hawks), bald eagles, ospreys, and pelicans have been observed.†

The major reason for the population decline that threatens the survival of many species is that the DDT in the birds' bodies upsets their calcium metabolism, so that their calcium-containing eggshells are abnormally thin. The thinner eggshells often break before the young are hatched. In addition, high concentrations of certain chlorinated hydrocarbons often cause a several-day delay of the hatching of the eggs.‡ Since the lifetime scales and migratory patterns of birds have evolved over thousands of years to provide the optimum survival chance for the offspring, a few-day delay decreases the survival rate. Conditions are no longer optimum for the young; e.g., the food supply may not be maximum when needed.

In comparison with the beneficial effects of pesticides, the loss of a few species of birds may seem a small price. However, many of the species removed may be the predators of undesirable animals or insects, whose populations may soar in the absence of their natural enemies.

The extinction of any animal species directly or indirectly

diminishes the quality of human life. But are there more direct threats to man caused by the chlorinated hydrocarbons present in his food and his body? The body fat of many people, including the average American, contains more DDT and residues than the 5 ppm limit generally used for edible fish! Even if the use of such pesticides is halted, their concentration in humans will continue to increase for several years, since it takes a long time for the pesticides to work their way up the food chains.§

Are the predatory birds just the early warning for a fate that man will later suffer? We are not likely to become extinct, but we do not know what kinds of subtle effects may result from long-term storage of chlorinated hydrocarbons in the body. Some test animals injected with DDT have shown an abnormally high incidence of tumors of the lungs, liver, and lymphoid organs. Human cancer patients have substantially higher DDT concentrations in their fat than people who do not have cancer. ‖ There is also evidence, again from animal studies, that DDT may cause female infertility. # This is not surprising, in view of the eggshell problem in birds: the disturbance of calcium deposition in birds is caused by a change in the concentration of the hormone *estrogen*, which plays a central role in the female reproductive cycle of animals and humans (Chapter 15).

Not only do the chlorinated hydrocarbons cause serious environmental problems, but they also lose their effectiveness as insecticides. The target insect species is never completely eliminated, even with massive spraying programs. There are always a few members of the insect population that have unusually strong resistance to the insecticide. Those that survive tend to produce offspring that, like themselves, have high resistance. After a few generations of survival against spraying, the entire population of the species will have evolved with much greater resistance to the insecticides—a case of "survival of the fittest." To control this hardier breed, we must greatly increase the dosage applied or switch to new types of insecticides, which may be effective for a few years until the insects develop resistance to it.

What is the answer to this serious environmental problem? To control diseases and feed the world population, man must

---

* A. S. Boughey, *Man and the Environment* (Macmillan, New York, 1971), Ch. 12.

† D. B. Peakall, "Pesticides and the Reproduction of Birds," *Scientific American* **222**, (4), 72–78 (April 1970).

‡ Ibid.

§ H. L. Harrison *et al.*, "Systems Studies of DDT Transport," *Science* **170**, 503–508 (1970).

‖ Peakall, "Pesticides and the Reproduction of Birds."

"Fertility: Effects of DDT," *Chemical and Engineering News*, April 14, 1971, p. 14.

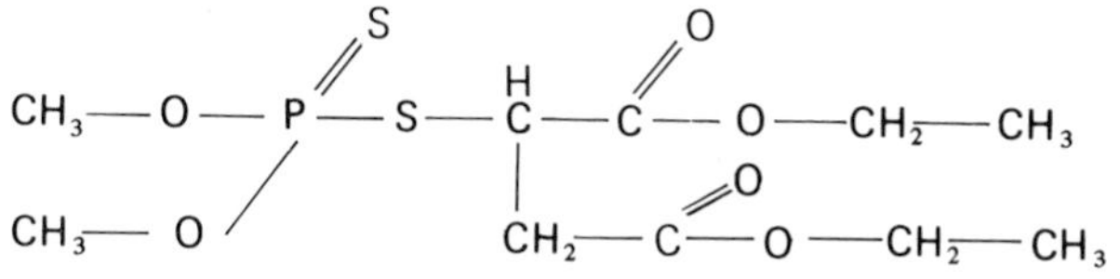

Fig. 13–13 Structures of malathion and parathion.

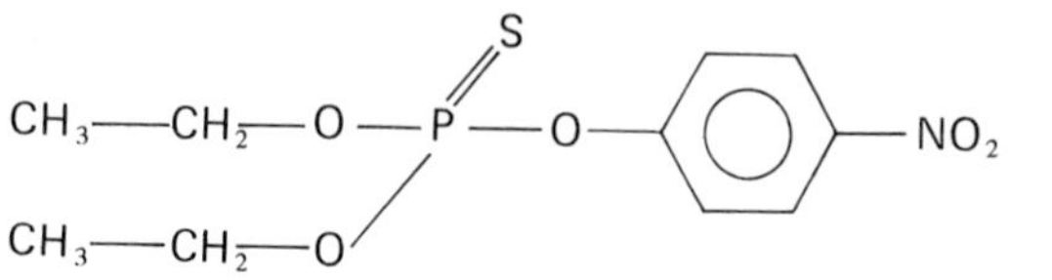

necessarily "unbalance the ecological system"—left to its own devices it would probably allow for more bugs and fewer people!* It seems clear that we must stop using DDT and similar compounds, but how can we control the insects? For one thing, we could use less persistent pesticides. There are several organic phosphorus insecticides, such as malathion and parathion, that break down in the environment a few days after application (see Fig. 13–13). However, there are a number of other imaginative possible solutions to the problem: control of entry of insects into the country, breeding food crops with increased resistance to insects, the introduction of predators of the target insect, the release of large numbers of sterilized male insects, and the use of insect sex attractants.† For each particular problem insect, any one or a combination of these strategies may prove to be effective, so we cannot say at this time which is the most likely to succeed.

Perhaps the most exciting new method of insect control is that of attractants. To attract each other for mating, insects of a given species emit compounds called *sex pheromones*, of very specific types. If we can identify and synthesize the pheromone for a particular species, we can place it in traps that kill the insect. These devices are used to eliminate gypsy moths. The moth was imported from France in 1868 by Professor Leopold Trouvelot of Medford, Massachusetts, for use in breeding experiments designed to produce improved species

of silkworms.‡ The experiments were a complete failure and, unfortunately, some of the moths escaped into the surroundings. Probably because of the absence of predators common to their natural habitat, the gypsy moth population gradually grew and, in recent years, exploded throughout the northeastern United States. The moths cause great damage to forests as they strip the leaves from trees. Massive spraying operations have been largely unsuccessful; because of the fear of damage from DDT, less persistent pesticides of the carbaryl type have been used. (See the following illustration.) This pesticide breaks down very quickly in the presence of water, so it must be applied at just the right point in the life of the moth to be effective. Variations in the time scale for individual moths allow many to escape the effects of the insecticide.

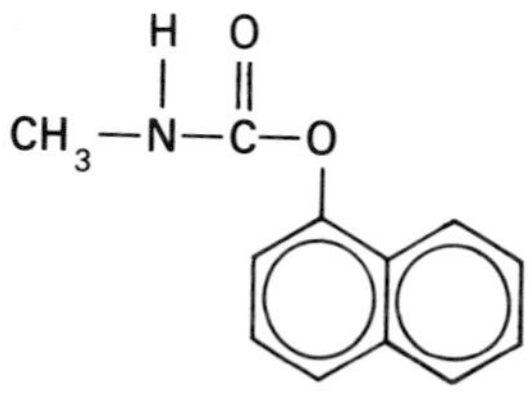

carbaryl

Two species of wasps that are predators of the gypsy moth have been introduced into the United States and have been somewhat effective in reducing the moth population, but ironically the population of the predators has been reduced by the spraying campaigns.§

The newest weapon in the fight against the gypsy moth is, as we have mentioned, its sex attractant (pheromone). Scientists at the U.S. Dept. of Agriculture in Beltsville, Maryland have been working on the problem for some time. They extracted the moth's sex attractant and identified it as *cis*-7,8-epoxy-2-methyloctadecane (see Fig. 13–14). ‖ At first the natural material was extracted and used as a sex attractant, but soon a cheaper method for artificial synthesis of the compound was developed. The synthetic material is being

---

* G. W. Irving, Jr., "Agricultural Pest Control and the Environment," *Science* **168**, 1419–1424 (1970).
† Ibid.

---

‡ A. D. Hinckley, "The Gypsy Moth," *Environment* **14** (2), 41–47 (1972).
§ Hinckley, "The Gypsy Moth."
‖ "Pheromones: Basis of Moth Control Methods," *Chemical and Engineering News*, Aug. 9, 1971, pp. 79–80.

A female gypsy moth lays her eggs on a tree twig on Cape Cod, Massachusetts (left.) (Courtesy of the U.S. Dept. of Agriculture.)

Hundreds of thousands of gypsy moth traps like this one are scattered throughout the eastern United States each year (right). The traps, which are baited with an artificial sex attractant, provide data needed to plan other ways of controlling the gypsy moth. (Courtesy of the U.S. Dept. of Agriculture.)

used in massive experiments to reduce the moth population. About 20 μg of the attractant are applied to the wick of a trap which contains a sticky substance that traps all moths attracted to it. In a second approach, the idea is to confuse the males by dropping papers impregnated with the attractant over wide areas. The male insects receive so many false signals that they cannot locate the females. The sex attractants are effective in areas of light infestation, but it is not yet known how useful they are in areas of high moth density, where there is an abundance of females available for mating.

Compounds that cause environmental problems similar to those caused by the chlorinated hydrocarbon insecticides

$$CH_3-(CH_2)_8-CH_2-CH-CH-CH_2-CH_2-CH_2-CH_2-CH \begin{array}{c} CH_3 \\ | \\ \\ | \\ CH_3 \end{array}$$

Fig. 13–14 Gypsy moth pheromone-cis-7, 8-epoxy-2-methyloctadecone.

are the *polychlorinated biphenyls*, PCB's. The basic structure of PCB's is the biphenyl molecule,

biphenyl   a PCB molecule

in which one or more chlorine atoms are substituted for the hydrogen atoms of the molecule. From the similarity of their structures to those of the pesticides in Fig. 13–10, it is not surprising that their behavior is similar. In fact, it has been hypothesized that some pesticide residues are converted into PCB's by sunlight.* The PCB compounds are liquids with low vapor pressures, high boiling points, low water solubility, stability against decomposition up to high temperatures (~800°C), good electrical insulating properties, and strong resistance to chemical attack. They are useful in a wide variety of applications: coolant-insulation fluids in electrical transformers, high-pressure hydraulic fluids, impregnation of cotton and asbestos insulation of cables, lubricants, additives to epoxy paints and coatings for wood, metal and concrete.† You have probably encountered PCB's in carbonless reproducing paper that is often used in business forms or charge slips that require multiple copies.

The same properties that make the PCB's useful are responsible for the environmental problems they create. Their low water solubility and high resistance to chemical breakdown cause PCB's to persist for long times in the environment and to be concentrated in the food chain, much like DDT and related pesticides. The PCB's in the environment mostly affect fish-eating birds.‡ Just like DDT, the PCB's cause thinning of eggshells and delayed hatching.

In their major applications, the PCB's are used in closed systems such as transformers in which there is no release of the material during normal operations. However, machinery using these "closed" systems occasionally spring leaks that

allow PCB's to be released to contaminate products being manufactured. In Japan in 1968, about 1000 people were poisoned by PCB's that got into rice oil used in cooking.§ The affected persons suffered skin discoloration and irritation, profuse sweating, weakness, and numbness, finally resulting in the death of five persons. These effects may have been caused by the PCB's themselves or by impurities, dioxins and benzofurans, contained in them. We've not had any similar ment mostly affect fish-eating birds.‡ Just like DDT, the PCB's but there have been several cases of contamination of fowl or eggs, by accidents that released PCB's into their feed. In separate instances, 50,000 turkeys in Minnesota, 88,000 chickens in North Carolina, and tens of thousands of eggs at various places have been seized and destroyed because of PCB contamination. ‖

Searches for adequate replacements for PCB's are being conducted, so far without much success. It would be a disaster today for the electricity distribution industry if a ban were placed on the use of PCB's in transformers. What is to become of the vast quantities of PCB's now stored in closed systems throughout the country? The major domestic supplier, Monsanto Corporation, has urged that PCB's from obsolete equipment be returned to them for incineration at 3000°C, a temperature high enough to decompose the PCB's.

13–13 POLYMERS AND PLASTICS

Polymers are extraordinarily long molecules, with molecular weights of tens of thousands or more, consisting of long chains of repeating or alternating chemical units called *monomers*. The many plastics that have become ubiquitous in modern society are examples of synthetic polymers. There are also many naturally occurring polymers, such as natural rubber, proteins, and cellulose.

Natural rubber is a polymer containing long chains of isoprene monomers:

isoprene   polyisoprene

(2-methyl-1,3-butadiene)

---

* T. H. Maugh, II, "DDT: An Unrecognized Source of Polychlorinated Biphenyls," *Science* **180**, 578–579 (1973).

† C. G. Gustafson, "PCB's—Prevalent and Persistent," *Environmental Science and Technology* **4**, 814–819 (1970).

‡ Ibid.

---

§ E. Carper, "Little-Understood PCB's Pose Hazard to Humans and Food," Washington *Post*, Dec. 27, 1971.

‖ "Controversy Continues over PCB's," *Chemical and Engineering News*, Dec. 13, 1971, pp. 32–33.

where *n* is a very large number. These long-chain molecules give rubber its properties of pliability, strength, and shape-retention: the molecules bend and stretch under stress, as in rubber tire. A segment of a molecule of natural rubber is shown in Fig. 13–15.

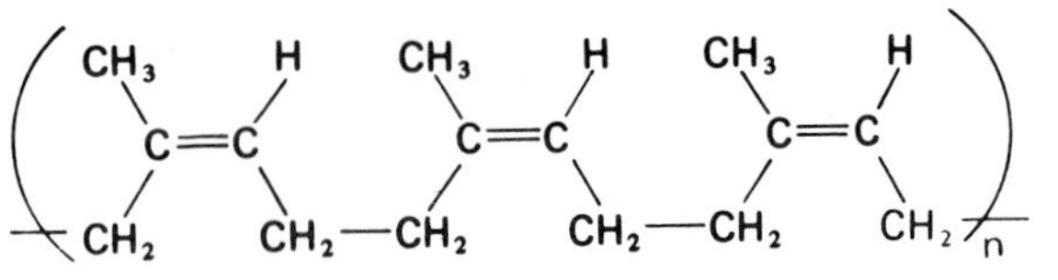

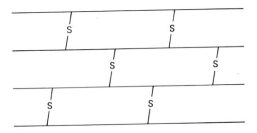

Fig. 13–15   Segment of a natural rubber molecule.

Pure natural rubber is not a very useful material, being hard and brittle at low temperature, and soft and tacky when warm. In 1839, Charles Goodyear discovered that he could develop cross links between polymer chains by adding sulfur, a process called "vulcanization," which gives the product much greater strength than the raw material (see Fig. 13–16).

Fig. 13–16   Cross-linked natural rubber.

Synthetic rubber products were developed between 1914 and the 1930s, but widespread use of them did not occur until World War II, when many countries were cut off from supplies of natural rubber in Southeast Asia.

Today many synthetic rubber products and other polymeric materials are produced on a massive scale. One of the most useful synthetic rubbers is butyl rubber, formed by the polymerization of isobutylene:

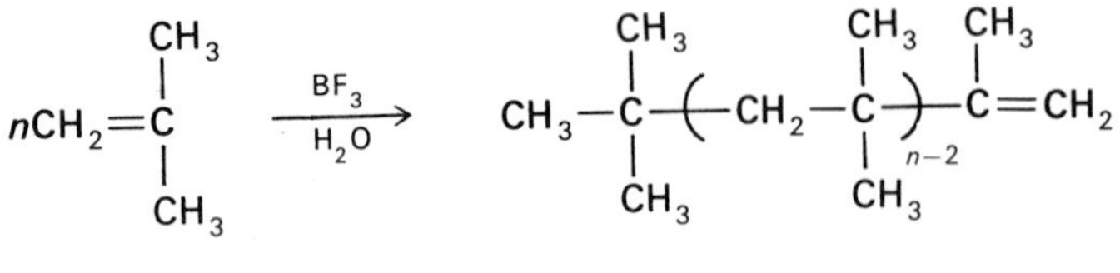

isobutylene   butyl rubber

(13–21)

A co-polymer between isobutylene and styrene forms Buna-S, the rubber used extensively in auto tires. In both structures, the *n* signifies a large number:

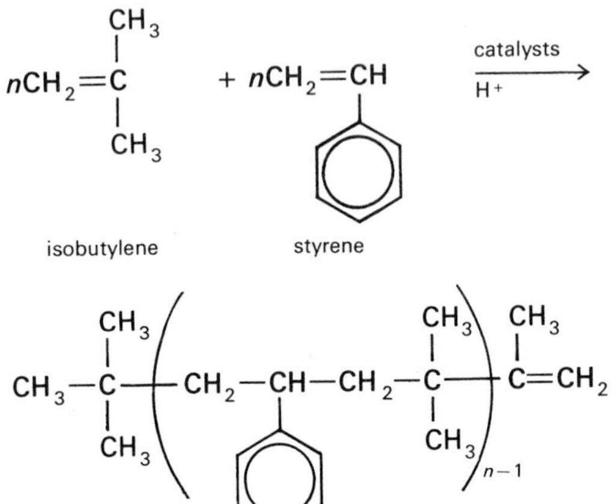

isobutylene   styrene

(13–22)

Buna-S

The importance of the shape and placement of the monomers is demonstrated with an isomer of natural rubber called gutta-percha, a hard, crystalline polymer. In this monomer, the methyl group and the hydrogen are on different sides of the double bond, whereas in the case of natural rubber (Fig. 13–15), they are on the same side:

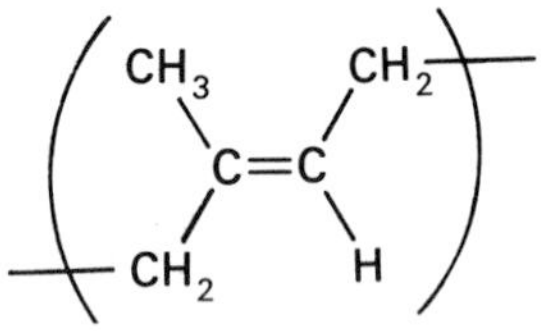

Gutta percha monomeric unit

Thus the natural rubber and gutta-percha molecules are *cis* and *trans* isomers respectively. (Recall from Section 8–1 that *cis* and *trans* are different isomers of the same compound.)

The development of synthetic products with the chemical regularity of natural rubber was made difficult because random mixtures of *cis* and *trans* isomers were obtained, yielding products with poor properties. In 1953, chemists Karl Ziegler of Germany and Giulio Natta of Italy found catalysts for the polymerization of rubber that yielded all *cis* or *trans* species. With these catalysts, it is now possible to make synthetic polyisoprene that is virtually identical to

$$nNH_2(CH_2)_6N\!\!\overset{H}{\underset{}{|}}\!\!\text{-}H + nHO\text{-}\underset{\|}{\underset{O}{C}}(CH_2)_4\underset{\|}{\underset{O}{C}}OH \longrightarrow \left(NH(CH_2)_6\underset{\|}{\underset{O}{N}}C(CH_2)_4\underset{\|}{\underset{O}{C}}\right)_n \qquad (13\text{–}23)$$

1,6-diaminohexane   1,6-hexanedioic acid   Nylon-66

natural rubber. Today about equal amounts of synthetic and natural rubber are in use. There are many types of synthetic rubber available with various properties for specific purposes, e.g., resistance to attack by oxidizing agents in photochemical smog (see Section 14–5). The starting materials for making synthetic rubber include large amounts of compounds produced by the catalytic breakdown of large molecules from petroleum (see Section 14–1).

One of the first synthetic polymers to be widely used was nylon, a product developed under the direction of Dr. Wallace H. Carothers of the DuPont Company in 1935.* The most common of the several types of nylons is Nylon-66, a copolymer of 1,6-diaminohexane and 1,6-hexanedioic acid, which are joined together by the splitting out of water molecules, as shown above, Eq. (13–23).

Polymers of this type are called *polyamides* because of the $-$NH$-$ units that occur regularly in the chain. In that respect nylon is similar to the natural protein polymers, including silk fibers, which are also polyamides. Most of the uses of nylon, as in hosiery and other fabrics, depend on its ability to be drawn into thin, strong strands that can be woven. Nylon was the first of a series of synthetic fibers used in clothing. More recent additions to this line of materials include Orlon (a polyacrylonitrile) and Dacron (a polyester). Modern synthetic fabrics or coatings on natural fibers have been designed to have various beneficial properties such as strength, stain resistance, water repellence, resistance to moths or fungi, and "drip-dry" and "permanent press" characteristics.

Americans live in an increasingly "plastic" world. One can scarcely buy a small item in a drugstore or hardware store that it is not encased in an indestructible plastic bubble attached to a large, colorful card. The use of plastics has grown enormously since World War II, as they have replaced many other materials and have made possible the manufacture of entirely new kinds of products.

One of the most widely used plastics is polyethylene,

which is used mainly for plastic containers. It is made from ethylene in the presence of a catalyst:

$$nCH_2\!\!=\!\!CH_2 \xrightarrow[H^+]{\text{catalyst}} \left(CH_2\text{–}CH_2\right)_n . \qquad (13\text{–}24)$$

ethylene   polyethylene

Polyvinyl chloride (PVC) is used extensively in such household products as Saranwrap, trash and sandwich bags, and films. PVC is chemically inert and inexpensive to produce. It is made from vinyl chloride monomers in the presence of catalysts:

$$nCH_2\!\!=\!\!\underset{|}{\underset{Cl}{CH}} \xrightarrow{\text{catalyst}} \left(CH_2\text{–}\underset{|}{\underset{Cl}{CH}}\right)_n . \qquad (13\text{–}25)$$

vinyl chloride   PVC

These are but a few examples of the many synthetic plastic compounds in wide use today. Other important ones are polystyrene, polymethyl methacrylate (called "Plexiglass" or "Lucite") and polyurethane (see Fig. 13–17).

One of the most chemically inert polymers is Teflon. This polymer has excellent heat qualities and is slippery to the touch, making it ideal in bearings on contact areas requiring a solid lubricant. Teflon is also used in nonstick frying pans and in chemical containers where its unusual properties are needed. The polymer is prepared from tetrafluoroethylene as shown below:

$$nCF_2\!\!=\!\!CF_2 \xrightarrow{\text{catalyst}} \left(CF_2\text{–}CF_2\right)_n . \qquad (13\text{–}26)$$

Teflon

Synthetic polymers have beneficial applications in modern society. They have replaced some of our other valuable natural resources in their previous uses. However, plastics and other polymers are not without their own set of problems for the environment. The polymer molecules themselves are for the most part chemically inert. However, most plastics contain additives that improve their properties. A major additive in plastics such as PVC are the "plasticizers"— usually phthalate compounds that lubricate the bending and

---

* Rayon, which predates nylon, is not truly a synthetic fiber. It is produced by chemical modification of the cellulose that occurs in natural materials such as wood (see Section 15–4).

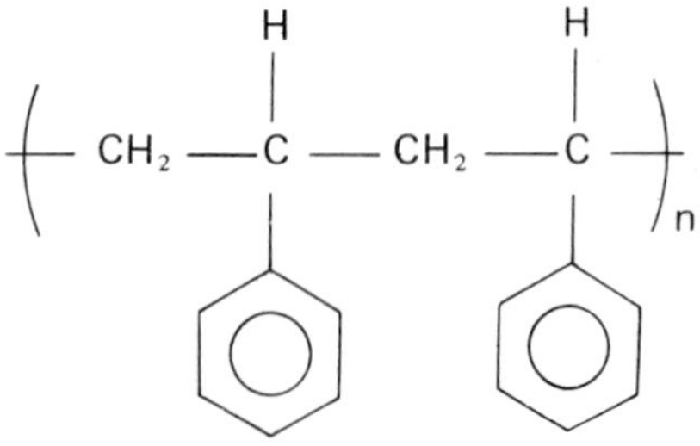

Polystyrene

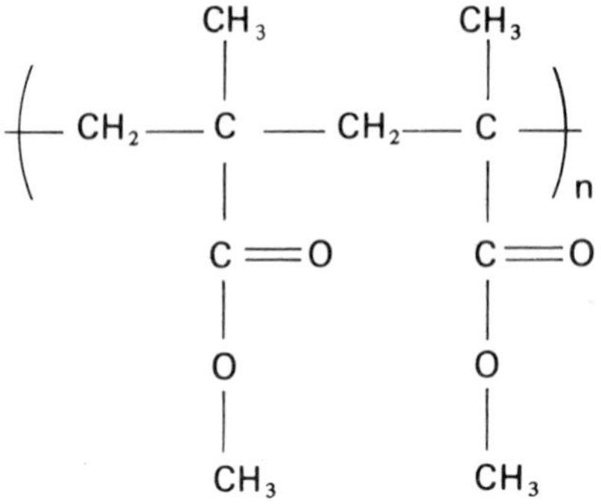

Polymethyl methacrylate

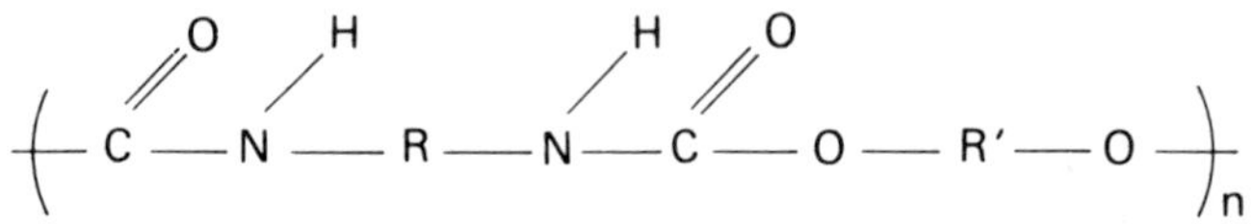

A polyurethane

Fig. 13–17 Structures of some important polymers—polystyrene, polymethyl methacrylate, polyurethane.

stretching of the material (see Fig. 13–18). Unfortunately, the phthalates are not chemically bonded to the polymer molecules, so they are sometimes released when the plastic is heated or brought in contact with fluids that dissolve the plasticizers. The characteristic smell of a new automobile is probably largely due to the evaporation of phthalates from plastics in the car, especially when it sits closed up in bright sunlight.* These compounds also cause much of the film that builds up on the inside of the car windows.

Because of the presumed chemical inertness of plastics, they have had extensive medical applications as storage bags for blood and tubing on devices such as kidney machines. It is becoming increasingly apparent that body fluids dissolve small amounts of phthalates from the plastics. Patients who have received transfusions of blood stored in plastic bags or who have been treated with kidney machines employing plastic tubing show observable concentrations of phthalates in their bodies.† The effects of the phthalates on humans are not fully known, although some affected patients have complained of nausea and abdominal pains, which subsided when plastics were replaced with other tubing. It is speculated, but

not proven, that the plasticizers may contribute to the fatal shock-lung syndrome that occasionally affects patients following massive blood transfusions.

The disposal of plastics causes problems in the environment. These synthetic organic compounds are similar to the chlorinated pesticides and PCB's in that they are nonbiodegradable. Plastic trash that litters the surroundings is virtually indestructable—ocean voyagers report seeing plastic objects floating throughout the seas far from their original sources. When plastics containing chlorine, such as PVC, are burned in municipal incinerators, the chlorine produces

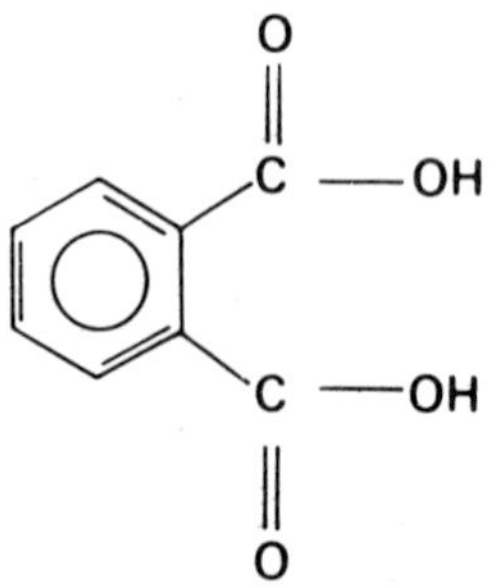

<hr/>

* K. P. Shea, "The New-Car Smell," *Environment* **13** (8), 2–9 (1971).
† Victor Cohn, "Plastics Residues Found in Bloodstreams," Washington *Post*, Jan. 18, 1972.

Ortho-phthalic acid

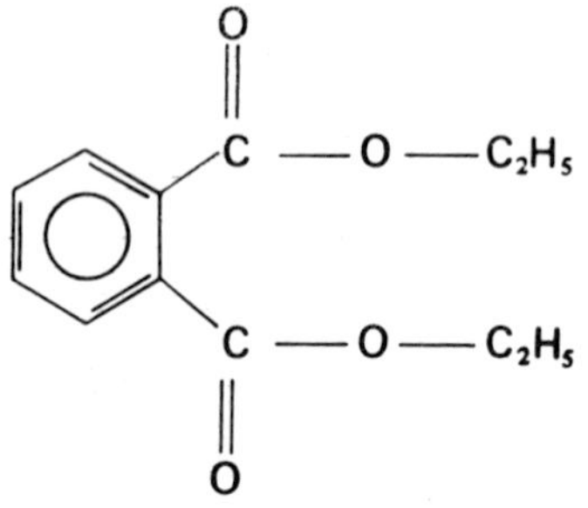

The ethyl ester of o-phthalic acid

Fig. 13–18 Phthalates.

strong hydrochloric acid, HCl, that is corrosive to the incinerator itself and an undesirable emission that is released from the stack unless the incinerator has good emission-control equipment.*

The plastics of our modern society have many beneficial applications, but the indestructible plastics themselves and their residues and additives are becoming spread throughout our environment and in our bodies. What will be the long-term effects of this burden of modern technology upon nature? We cannot say now, but the problem must be further investigated.

SUGGESTED READING

American Chemical Society, *Cleaning Our Environment—The Chemical Basis for Action* (ACS, Washington, D.C., 1969).

Rachael Carson, *Silent Spring* (Houghton Mifflin, Boston, Mass., 1962).

H. D. Embree and H. J. DeBey, *Introduction to the Chemistry of Life* (Addison-Wesley, Reading, Mass., 1968).

C. G. Gustafson, "PCB's—Prevalent and Persistent," *Environmental Science and Technology* **4**, 814–819 (1970).

C. B. Huffaker, "Biological Control and a Remodeled Pest Control Technology," *Technology Review* **73** (8), 30–37 (1971).

G. W. Irving, Jr., "Agricultural Pest Control and the Environment," *Science* **168**, 1419–1424 (1970).

J. McCaull, "Know Your Enemy," *Environment* **13** (5), 30–39 (1971).

D. B. Peakall, "Pesticides and the Reproduction of Birds," *Scientific American* **222** (4), 72–78 (April 1970).

J. I. Routh, D. P, Eyman, and D. J. Burton, *Essentials of General, Organic and Biochemistry* (W. B. Saunders, Philadelphia, 1969).

QUESTIONS AND PROBLEMS

1. Explain the following terms
   a) saturated and unsaturated hydrocarbons
   b) alkyl group
   c) aromatic compounds
   d) condensed ring structures
   e) formalin
   f) saponification of fats
   g) hydrophilic and hydrophobic
   h) a lachrymator
   i) *cis* and *trans* isomers

2. Draw structural formulas for the following molecules.
   a) both propyl alcohols   b) acetylene
   c) 2-methyl-pentane   d) 1,2-dichloro-pentane
   e) 3-ethyl hexanol   f) isooctane
   g) cyclopentane   h) benzene
   i) TNT   j) nitroglycerin
   k) benzoic acid   l) ethyl-*n*-propyl ether
   m) butanal   n) 2-pentanone
   o) sodium palmate   p) glyceryl trilaurate
   q) butyl sulfonic acid   r) 3-pentaneamine
   s) 2,2-dibromopropane

3. Draw the five structural formulas for hexane.

4. Write a balanced equation for the production of ethyl alcohol from sugar by fermentation.

5. Write equations for the oxidation of 1-butanol and 2-butanol to their respective aldehydes and ketones, naming the products.

6. Draw the structures and name the esters formed by the following reactants.
   a) propanoic acid and methyl alcohol
   b) acetic acid and isobutyl alcohol
   c) butyric acid and ethyl alcohol
   d) formic acid and isopropyl alcohol
   e) salicylic acid and *n*-propyl alcohol

7. What are some of the alternatives to DDT in the control of insects and pests?

8. What are some of the sources of the PCB's that we find in the environment?

9. Draw a portion (2 or 3 of the repeating units) of the polymer polypropylene.

*   C. E. Knapp, "Can Plastics Be Incinerated Safely?", *Environmental Science and Technology* **5**, 667–669 (1971).

14 PETROLEUM, THE AUTOMOBILE AND PHOTOCHEMICAL SMOG

When the Lucas Well came in in 1901, it opened up Texas as a major producing area for the oil industry. Modern equipment and conservation practices make oil gushers like this one a thing of the past. (From the Historical Photo Library, American Petroleum Institute, courtesy of Texas Mid Continental Oil and Gas Association.)

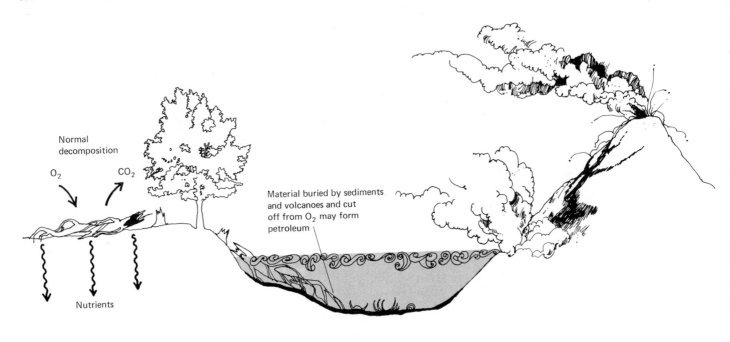

Normal decomposition

O_2 CO_2

Material buried by sediments and volcanoes and cut off from O_2 may form petroleum

Nutrients

14-1 PETROLEUM: AN IMPORTANT NATURAL RESOURCE

When plants and animals die, they decompose under the action of bacteria, fungi, and natural chemicals in the environment. Molecules of their cells are broken down into simpler structures and made available as nutrients for other living things in the biosphere. Carbon of the dead organic material is combined with oxygen and released to the atmosphere as carbon dioxide which may be used as a nutrient by photosynthetic plants (recall the carbon cycle discussed in Section 9-3). Over the past 600 million years, small fractions of the dead plant and animal material have been buried under rocks, soil or the sediments in water bodies. Sealed off from oxygen, the dead material was unable to undergo the usual decomposition (see Fig. 14-1). Trapped under overlying rocks and soil, the organic material was slowly transformed into fossil fuels, principally coal, natural gas, and petroleum.

The fossil fuels are an extremely important commodity in our technological society. Going back to Chapter 4, we can determine from Table 4-1 that about 95% of the energy used in the United States in 1968 was obtained from fossil fuels. Table 4-2 indicates that an estimated 75% of our energy needs will still be supplied by fossil fuels in the year 2000.

Table 14-1   Major Uses of Petroleum Products in the United States—1968

Use	Percent
Energy uses (fuels)	
Household and commercial	20.4
Industrial	9.8
Transportation	54.6
Electric power generation	3.8
Other	0.7
Non-energy uses	
Petrochemical feedstocks	5.2
Other	5.5

P. Meadows, ''Petroleum,'' in *Mineral Facts and Problems*, Bureau of Mines, U.S. Dept. of the Interior (U.S. Government Printing Office, Washington, D.C., 1970), pp. 147–182.

◄ **Fig. 14–1** Dead biological material exposed to air is eventually oxidized to CO_2, H_2O, and other compounds, but material shut off from oxygen by overlying rocks and soil may form "reduced" carbon compounds: the hydrocarbons of petroleum or the carbon of coal.

Although the ultimate reserves of coal are far greater than those of petroleum and natural gas, the latter fuels are more valuable and useful, as they can be recovered from the earth and transported more easily than the solid, bulky coal. Products derived from petroleum have high energy content per unit weight and perform well under the exacting conditions of high-performance engines, making them ideal fuels for the vehicles and airplanes of modern society. They supply about 95% of the energy used by U.S. transportation. Another major use of petroleum is in space heating of homes and other buildings. Oils are also used for lubrication, electric power generation, and asphalting of roofs and roads.

Table 14–1 shows that about 5% of the petroleum (called "petrochemical feedstocks") is used by the chemical industry in the synthesis of a wide variety of products: synthetic rubber, plastics, synthetic fibers, ammonia, fertilizers, etc. The use of petrochemicals has increased enormously since World War II, from 23 million barrels in 1947 to 253 million in 1968. Today petrochemicals account for about 85% of the organic chemicals produced by United States industry, and continued growth of the petrochemical industry is expected.[*] If we are to insure adequate future supplies of petroleum for these uses, we shall have to conserve petroleum more carefully.

With regard to petroleum, we may be living in a brief and unique period in the life of our planet. It has taken about 600 million years for the earth's reserves of petroleum to be formed. Negligible amounts of petroleum were consumed prior to this century. But since the invention of the automobile, the use of petroleum has skyrocketed: an estimated 227 billion barrels of petroleum had been consumed throughout the world up to the end of 1969, half of it during the century between 1857 and 1959 and the remaining half during the decade 1959–1969![†]

Estimates of the total world reserves of petroleum range between 500 and 2000 billion barrels.[‡] The annual world consumption is now about 15 billion barrels, but that figure will surely increase several-fold over the next three or four decades. One projection suggests that the world will have used 90% of the total petroleum reserves by about 2020 or 2030.[§] Perhaps by that time economical methods for recovery of petroleum from low-grade oil-shale deposits will have been developed, which would probably extend the supply for a few more decades. Even with oil-shale, unless enormous new petroleum deposits are discovered, we will have exhausted within one or two centuries all the petroleum formed near Earth's surface over 600 million years, depriving future generations of this valuable resource.

The United States consumes far more than its share of the world's petroleum: with about 6% of the world population, we use about one-third of the petroleum, nearly 6 billion barrels per year.[||] Fortunately, we have been well-endowed with petroleum deposits, mainly in Texas, Oklahoma, California, and off-shore areas of the Gulf Coast and Southern California. Although the United States is the world's largest oil-producing nation, we still have to import about 30% of our oil (mainly from Canada, Venezuela, and the West Indies).

During the winter of 1973–1974, the "energy crisis" discussed in Chapter 4 arrived in western Europe, Japan and, to a lesser extent, the United States. Severe petroleum shortages in Europe and Japan resulted from the partial embargo on oil from the Middle East in the wake of the Arab–Israeli War of Oct. 1973. Since those areas depend heavily on oil imports from the Middle East, the partial embargo had serious ramifications: bans on Sunday driving, gasoline rationing, electrical power shut-offs, and cut-backs to 3-day work weeks in industries in some countries. Although the United States depends much less on imports from the Middle East, there were also shortages here. Gasoline shortages appeared first during the summer of 1973 (*before* the Arab oil embargo) and, more severely, during early 1974. The supply of home heating oil was also quite uncertain. The reasons for the magnitude of the shortage in the United States are not clear. Since we imported relatively little oil, perhaps about 8% (reliable figures being almost impossible to obtain) from the

*   P. Meadows, "Petroleum."
†   M. K. Hubbert, "The Energy Resources of the Earth," *Scientific American* **224** (3), 61–70 (Mar. 1971).
‡   Ibid.

§   Ibid.
|| P. Meadows, "Petroleum."

Middle East, we would probably have had some shortages even without the embargo. Some critics have charged that the crisis was artificially created by industry collusion to force price increases.*

It has been suggested that apparent shortage of natural gas reserves is a result of the regulated price of gas.† Unlike other petroleum products, gas is considered a public utility and its price is controlled by government agencies. The gas industry indicates that the price is being held so low that they are not encouraged to seek and develop the use of additional gas deposits. If the price is deregulated, the cost of gas at the well head would probably triple. Since the cost of the raw gas represents only a portion of the expense of gas delivered to consumers, the price increase to the consumer would presumably not triple, but it would have to increase. It has been charged that gas suppliers have not been reporting the true amounts of their reserves, in order to bring about pressure for deregulation.‡

In part, the shortage of some fuels has stemmed from regulations designed to improve air quality. For example, limits on sulfur content of fuels that can be burned in many metropolitan areas has caused some consumers to shift from coal to gas or low-sulfur petroleum (see Section 10–8). It is frequently claimed that pollution-control devices on automobile engines have cut their efficiency. It's certainly true that post-1970 full-size automobiles exhibit poorer fuel economy than similar models of the mid-1960s. A few years ago, gasoline mileage of 17 and 20 miles per gallon was fairly common, but today values in the range of 10 to 15 mpg are more typical. Some measures taken to reduce auto emission do reduce efficiency (see Section 14–6); the EPA estimates that the average reduction of fuel economy is about 7% for 1973 automobiles. The real culprits, according to that agency, are the increased weight of automobiles and the accessories that demand extra power. Air conditioning, for example, cuts fuel economy by an average of 9% and up to 20% on hot days. Automatic transmissions decrease mileage by about 6%. A 500-lb increase in automobile weight causes a 14% decrease of economy.§

Regardless of the economic or technological reasons for the crisis of 1973–1974, it seems clear that the United States will need to obtain foreign oil in substantial amounts in the future. Domestic production peaked in 1970 and has declined since then. The discovery of a major new oil field in Alaska near Prudhoe Bay has received much attention. It is estimated that the field has reserves of about 10 to 20 billion barrels. Although that's only a 1.5 to 3 year supply for the United States, more reserves may be found as the field is developed. However, even when the Trans-Alaska Pipeline is in full operation in about 1979, it will supply only about 2 million of the 16 to 18 million barrels per day needed by the United States. Unless we take stringent measures to reduce our use of petroleum or develop other sources (e.g., oil shale), projections indicate that by 1985 we will have to import nearly half of our petroleum.‖

The world's largest known petroleum reserves are located in the Middle East (Saudia Arabia, Kuwait, Iran, and elsewhere) and in Northern Africa (Libya, Algeria, U.A.R.). Middle East oil fields supply most of Western Europe's and Japan's needs for oil. Over the next few years, the United States also will probably have to import substantial quantities of petroleum from that area unless we conserve oil more carefully or develop new sources.

Petroleum is not only a valuable natural resource, but its handling and use cause many of our most serious environmental problems, especially the oil slicks that frequently pollute water and the high concentrations of the urban air pollutants (CO, NO, NO_2, lead, and hydrocarbons) that result in large part from combustion of gasoline by motor vehicles, leading in extreme cases to the photochemical smog that often blankets Southern California cities.

14–2 PETROLEUM DEPOSITS AND CRUDE OIL

Petroleum is deposited in porous rock formations that are capped by strata of impervious rock. When an oil well is drilled through the gas-tight layers above the deposit, the natural gas usually present with the crude oil may have enough pressure to force the oil to the surface (a "gusher"), until the subterranean pressures decrease to the point that the oil must be pumped out (see Fig. 14–2).

---

*   J. L. Rowe, Jr., "FTC Blames Fuel Pinch on Oil Firms," *Washington Post*, May 12, 1973.

†   Thomas O'Toole, "The Energy Crisis," four-part series, *Washington Post*, Nov. 26–29, 1973.

‡   Morton Mintz, "Shortage of Natural Gas May Be Hoax, Hart Says," *Washington Post*, June 29, 1973.

§   "The Real Mileage Thief," *Science News* **102**, 418 (1972).

‖   Office of Emergency Preparedness, "The Potential for Energy Conservation" (U.S. Government Printing Office, Washington, D.C., 1972).

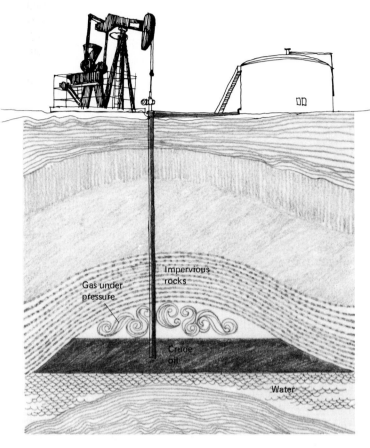

Fig. 14–2 A petroleum deposit tapped by an oil well.

Crude oil contains thousands of hydrocarbon compounds, most of them alkanes. The molecules range in size from methane, with a single carbon atom, to those having a hundred or more carbon atoms, such as those present in the asphalt and tar components of the oil. Properties and uses of the various fractions of the crude oil are summarized in Table 14–2.

Table 14–2 Hydrocarbon Fractions Obtained from Petroleum

Fraction	Number of carbons in each molecule	Boiling range °C	Uses
Gas	C_1–C_3	$-164°$ ⟶ $-20°$	Natural gas
Liquid petroleum gas	C_3–C_5	$-20°$ ⟶ $30°$	Bottled or LP gas
Petroleum ether (ligroin and naphtha)	C_5–C_7	$30°$ ⟶ $90°$	Solvent, dry-cleaning fluid
Gasoline (straight run)	C_6–C_{12}	$30°$ ⟶ $200°$	Internal combustion engine
Kerosene	C_{12}–C_{16}	$175°$ ⟶ $275°$	Jet fuel, camp stove and lamp fuel
Gas oil, fuel oil and diesel fuel	C_{15}–C_{18}	$250°$ ⟶ $400°$	Diesel engine fuel, furnace oil, cracking oils
Lubricating oils, greases, petroleum jelly	C_{16} and greater	$350°$ and up	Lubrication
Paraffin	C_{20} and greater	Melts $52°$–$57°$	Waxes, candles
Pitch and tar	—	Residue	Asphalt, residual fuels

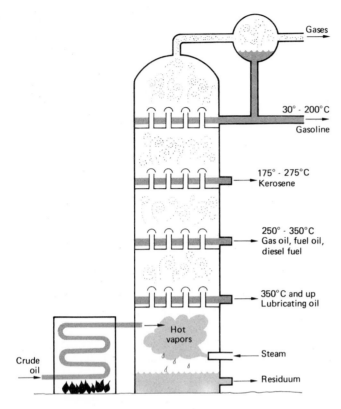

Fig. 14-3 Distillation column separates crude oil into fractions having different boiling points.

The natural gas that accompanies the crude oil is burned off unless there are facilities for handling and distributing it to consumers. Most of it produced in the United States is used, but a lot of it is still wasted in the big Middle East oil fields. Raw natural gas usually contains 50 to 99% methane, with small quantities of ethane, propane, and higher alkanes. It is often contaminated with N_2, CO_2 and H_2S (hydrogen sulfide) gases. Because of its bad odor and its corrosion of pipes, the H_2S is removed before the gas is used.

Natural gas is piped to homes, businesses, and industries, where it is used as a clean, cheap, convenient fuel for heating and cooking. The gas must be used with care, however, because of its hazards. Natural gas is not particularly toxic, but if it leaks into a room and expels the oxygen, a person can be asphixiated from lack of oxygen. The more serious danger of gas leaks is the threat that a spark will cause the

gas to explode and burn.* So that consumers can detect gas leaks, the companies add minute quantities of ethyl mercaptan (CH_3CH_2SH) to the gas. As noted in Section 13–9, ethyl mercaptan has an unpleasant skunklike smell that can be detected in small quantities.

The first step in the refining of crude oil is to separate it into components of various boiling points by distillation. The oil is heated in a furnace and then flashed (vaporized) into a fractionating column, as shown in Fig. 14–3. The more volatile, low-molecular-weight fractions move to the uppermost, coolest portions before condensing to form liquids. Any remaining gas is withdrawn at the top of the column. The heavier, less volatile fractions condense at lower, warmer levels of the column. The residuum left at the bottom of the column is quite nonvolatile and often a solid at room temperature. It is used in asphalt and tar or sold as cheap "residual" fuel (also called #6 oil or "Bunker-C") to large-volume users such as power plants or large central heating plants. Many undesirable impurities in the petroleum, such as sulfur and the metals vanadium and nickel, become concentrated in the residuum. Thus residual fuel is a "dirty" fuel and an important contributor to air pollution in cities in the northeastern United States, where it is used extensively (see Section 10–8).

Many products such as diesel fuel, kerosene, and home heating fuel can be used as they come from the distillation unit without much further processing, but a great deal must be done to the gasoline fraction. *Straight-run* gasoline from the distillation unit is of poor quality, with octane numbers in the 50 to 60 range, much below the 90 to 100 range required by modern engines. Other compounds must be added to the straight-run fraction to raise its quality. Furthermore, the demand for gasoline relative to the other petroleum components is much greater than the amount of straight-run fraction that occurs naturally. The market for gasoline accounts for about 40 to 45% of the total petroleum demand, but only about 25 to 35% of the crude falls into the gasoline boiling-point range. Thus portions of the other distillation products are converted into gasoline.

Processes used to convert portions of other fractions into gasoline are sketched in Fig. 14–4. These processes can

---

*   Methane is less dangerous than LP or bottled gas, since the former, with a molecular weight of 16, is less dense than air, whereas the latter is more dense and thus can become trapped more easily in a basement.

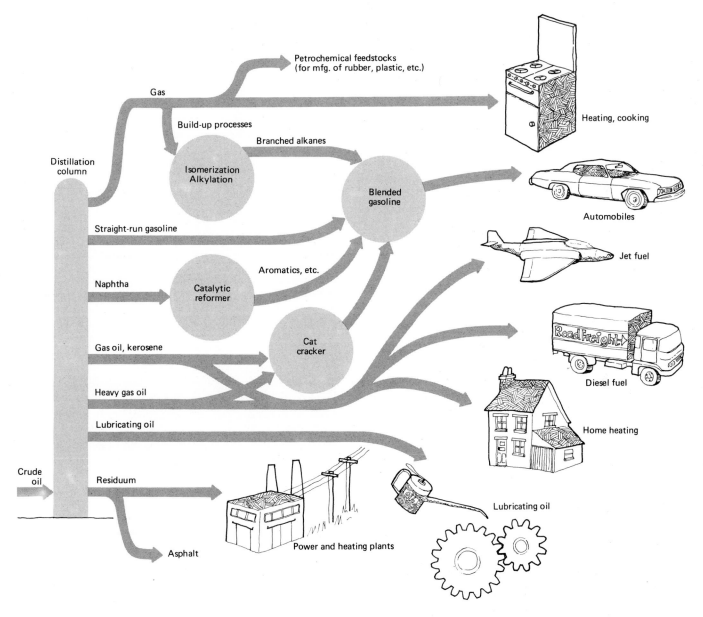

Fig. 14–4   Summary of the treatments and uses of various fractions from distillation of crude oil.

be divided into three classes: buildup, breakdown and change processes.* They are carried out at high temperature and pressure in the presence of catalysts such as platinum, aluminum chloride, sulfuric acid, and hydrofluoric acid. The nature of the catalyst is very important, as small modifications can greatly change the rates of the various reactions, thereby changing the yields of the products.

In the buildup processes, small molecules, mainly from the liquid petroleum gas fraction, are joined together to form the larger gasoline-type molecules (this process is called *polymerization*, which was discussed in Section 13–13), although here the molecules formed have only 6 to 12 carbon atoms. The more common buildup process today is *alkylation*, in which small alkene molecules are joined to isobutane, for example,

$$CH_2{=}CH{-}CH_3 + CH_3{-}\underset{\underset{CH_3}{|}}{CH}{-}CH_3 \longrightarrow CH_3{-}\underset{\underset{CH_3{-}\underset{\underset{CH_3}{|}}{C}{-}CH_3}{|}}{CH}{-}CH_3$$

propene	isobutane	2,2,3-trimethyl butane
	(methyl propane)	

$$(14{-}1)$$

The reaction illustrated is a particularly useful one, as the product is highly branched, rather than being an *n*-alkane. The branched compounds have higher octane numbers than normal compounds.

In the breakdown processes, the large molecules of the kerosene and distillate fuels are broken down into molecules of the gasoline size range. These processes are called *catalytic cracking*, and the units in which they are performed are termed "cat crackers."

Perhaps the most important processes for improving gasoline are those in which the structure of the molecules is rearranged. In *isomerization*, the purpose is to form new structural isomers of *n*-alkanes, converting them into highly branched structures. The more common process today is *catalytic reforming*, in which alkanes with six or more carbon atoms, e.g., the naphtha fraction, are converted into cyclo alkanes and, preferably, the aromatic compounds benzene, toluene, and xylene.

Products from these molecular modification processes are blended with straight-run gasoline to make the final

_____

* B. F. Greek, "Gasoline," *Chemical and Engineering News*, Nov. 9, 1970, pp. 52–60.

A catalytic reforming unit. Alkanes in the heater on the left go into four reactors (on the right) that contain platinum catalyst. The aromatic molecules produced there go to the tall distillation columns (center, rear) for separation into benzene, xylene, and so on. (Courtesy of the Shell Oil Co.)

product. The amounts of each that are added depend on many factors, including the characteristics of the engines in which the gasoline is to be used, the season of the year (which determines average temperature), and the altitude and climate of the area in which the gasoline is to be marketed. Other components are added in small quantities: tetraethyl lead, other compounds to prevent buildup of lead deposits, drying agents, and engine-cleaning compounds (detergents).

Fig. 14–5 Cycles of one cylinder of an internal combustion engine. Intake: (a) Piston moving down draws air/fuel mixture into cylinder through open intake valve. (b) Compression: With valves closed, rising piston compresses mixture. (c) Power: Spark ignites mixture which burns, releasing heat, causing expansion that forces piston down. Force is transferred to rotation of crankshaft. (d) Exhaust: Rising piston forces burned gases out through open exhaust valve. [From Joseph Priest, *Problems of Our Physical Environment* (Addison–Wesley, Reading, Mass., 1973) p. 232.]

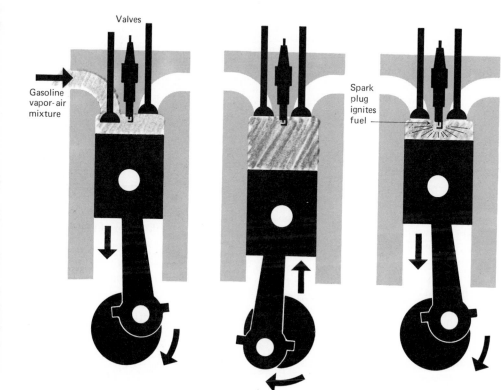

Intake stroke. A vapor-air mixture is pulled in from the carburetor through the open intake valve.

Compression stroke. The vapor-air mixture is compressed by the upward moving piston. Both valves are closed.

Power stroke. At the top of the compression stroke, the sparkplug ignites the vapor-air mixture, forcing the piston down to produce linear kinetic energy.

Exhaust stroke. At the bottom of the power stroke, the exhaust valve opens and the spent gas is forced out through the exhaust system.

14–3 THE INTERNAL COMBUSTION ENGINE

The internal combustion engine (ICE) converts the chemical potential energy of gasoline into heat and releases more moles of CO_2 and H_2O gases than the number of moles of O_2 consumed. Expansion of the hot gases in a cylinder yields mechanical energy in the form of a rotating crankshaft. The drive shaft transfers this energy to the wheels.

Gasoline is pumped from the tank by the fuel pump and forced into the carburetor, where it is mixed with a stream of air. The mixture of air and gasoline vapor is pulled into a cylinder, and where a piston compresses it to a much smaller volume (the ratio of initial to final volume being the "compression ratio"). The compressed gases are ignited by a spark from the sparkplug. The rapid combustion of the gasoline releases heat and gases. Expansion of the gases in the cylinder pushes the piston down and causes the crankshaft to rotate (see Fig. 14–5). Contrary to popular misconception, in a properly operating engine, the vapor does not explode, but burns smoothly as it releases the heat. Diesel engines operate in a similar way, but they do not use sparkplugs as the compression of the vapor heats the fuel enough to cause ignition.

The volatility of the gasoline must be controlled carefully. Oil companies change the composition of the gasoline throughout the year and from one area to another to adjust it to climatic conditions. In winter, a higher proportion of light, volatile components is needed, so the engine will fire when started cold; a small quantity of butane or other short-chain compounds may be added to handle the problem. The volatile components must be reduced in summer as they may boil in the fuel lines, reducing the flow of liquid gasoline, causing the car to stall because of a "vapor lock."

Gasoline quality is generally denoted by its *octane number*. The octane number (ON) of a gasoline is determined empirically by comparing the performance of a standard engine running on the gasoline with that of the same engine operating on a mixture of two pure hydrocarbons, *n*-heptane and isooctane.* *N*-heptane is a poor gasoline, defined to have an ON of zero. Years ago isooctane had higher quality than most gasolines and was defined to have an ON of 100. The ON of a gasoline under study is determined by the percentage of isooctane in the mixture that produces the same performance as the gasoline being tested. For example, if a gasoline performed the same as a 9%-heptane–91%-isooctane mixture, it would be assigned an ON of 91. The octane number of some representative hydrocarbons are given in Table 14–3.

It is not clear which chemical and physical properties determine the ON of the various compounds that make up gasoline, but it is probably related in part to the volatility of the compound. Branched hydrocarbons (which are more volatile than the unbranched kind) have higher ON's than normal hydrocarbons with the same number of carbons; for example, the ON of *n*-octane is −19, whereas that of isooctane is 100.† Some of the highest known ON's are those of

A technician pours gasoline into a special engine that determines the gasoline's antiknock qualities. (Courtesy of Exxon Company, U.S.A.)

aromatic compounds such as benzene (106), toluene (120), and xylene (107 to 118 depending on the relative placements of the two methyl groups):

toluene   xylene

An engine operating on gasoline of too low an ON "knocks" under a heavy load. The cause of knocking isn't certain, but it may be caused by a secondary wave of combustion. Instead of combustion preceding outward from the sparkplug in a well-controlled way, a second wave of combustion may be initiated somewhere else in the cylinder.‡ In any case, knocking is undesirable, causing the engine to lose power.

The ON of gasoline required by an engine for antiknock operation increases with the compression ratio of the engine.

---

*   As noted in Section 13–2, isooctane is the common name for the branched eight-carbon compound whose structure is

$$CH_3-\underset{\underset{CH_3}{|}}{\overset{\overset{CH_3}{|}}{C}}-CH_2-\underset{\overset{CH_3}{|}}{CH}-CH_3$$

† Don't be disturbed by ON's less than zero or greater than 100. The lower ones simply indicate a lower level of performance than that obtained from *n*-heptane. The higher ones indicate performance superior to that of isooctane. The values outside the range 0 to 100 are determined by adding tetraethyl lead to the compound being tested or to the isooctane.

‡   V. R. Gutman, Letter to the Editor, *Chemical and Engineering News*, Aug. 17, 1970, pp. 6–7.

Table 14–3 Octane Numbers of Some Representative Hydrocarbons

Compound	Octane Number
n-Octane	−19
n-Heptane	0 (by definition)
n-Hexane	25
2-Octene	56
2,4-dimethyl hexane	65
Propane	97
2,2,4-trimethyl pentane (isooctane)	100 (by definition)
Benzene	106
Xylenes	107–117
Toluene	119

From Greek, "Gasoline."

Over the years, as the quality (ON) of gasoline has improved, it has been possible to raise the compression ratio of the engines so that they produce greater power for their weight. At a compression ratio of 8/1 (typical of engines that use "regular" gasoline), the gasoline must have an ON of about 91. Higher performance engines with ratios of 10/1 or more need gasoline with an ON of 98 or better ("premium" grade).

A major objective of the catalytic processes discussed in the preceding section is the production of high-octane products, especially highly branched alkanes and aromatics from reforming processes. However, the cheapest way in which to raise the ON number is by adding tetraethyl lead (TEL) or tetramethyl lead (TML) to the gasoline.

$$CH_3-CH_2-\underset{\underset{CH_3}{\overset{|}{\underset{|}{CH_2}}}}{\overset{\overset{CH_3}{\overset{|}{CH_2}}}{Pb}}-CH_2-CH_3 \qquad CH_3-\underset{\underset{CH_3}{|}}{\overset{\overset{CH_3}{|}}{Pb}}-CH_3$$

tetraethyl lead, TEL   tetramethyl lead, TML

The action of the lead compounds is not understood, but it is found that the addition of three grams of lead per gallon raises the ON about seven units. Lead from combustion of the alkyl lead compounds also acts as a lubricant to the valves, preventing the miniature spot welds that can form between the valve and its seat.* However, as discussed below and in

_____

*   Greek, "Gasoline."

Section 12–8, the lead emitted in the exhaust is a serious air pollutant and has harmful effects on the car itself. Lead deposits can foul the sparkplugs, and the so-called "lead-acids" attack metals in the exhaust pipe and muffler, causing them to corrode. These acids are actually HCl and HBr formed by the combustion of ethylene dichloride and ethylene dibromide, which are added along with the lead alkyl compound in order to remove lead from the engine by forming volatile lead halides such as $PbCl_2$, $PbBr_2$, and $PbBrCl$.

14–4 AIR POLLUTANTS FROM INTERNAL COMBUSTION ENGINES

Carbon Monoxide

Vehicles powered by ICE's are a major source of several air pollutants (see Table 10–3), of which one is carbon monoxide, CO. When hydrocarbons are burned in an abundant supply of oxygen, the carbon forms carbon dioxide, CO_2, and the hydrogen forms water vapor, e.g.:

$$2C_8H_{18}(l) + 25O_2(g) \longrightarrow 16CO_2(g) + 18H_2O(g). \quad (14–2)$$

These are "clean," natural emissions, except insofar as CO_2 contributes to the "greenhouse effect" discussed in Chapter 9. But if the supply of O_2 gas is inadequate, some of the carbon atoms are oxidized only to CO instead of CO_2. The amount of air drawn into an automobile cylinder per amount of fuel, i.e., the air/fuel (A/F) ratio, is determined by the adjustment of the carburetor. The A/F ratio must be at least 14.5 in order to provide enough O_2 for complete combustion under ideal conditions. Most automobiles operate with A/F between 11.5 and 14.5; that is, the mixture is usually too "rich" in gasoline relative to air. The richer mixture generally promotes better engine performance, but at the cost of incomplete combustion. The exhaust gas ranges from about 0.5% CO for A/F = 14.5 to nearly 8% CO when A/F = 11 (see Fig. 14–6).

Carbon monoxide is a poisonous gas that reduces the oxygen-carrying ability of the blood. In an oversimplified way, we can consider the blood a conveyor belt, as depicted in Fig. 14–7. Oxygen in the lungs becomes attached to the hemoglobin (Hb) of the red cells of the blood. Hemoglobin consists of four units of heme, a large organic molecule centered about an iron atom, connected to the globin, which consists largely of proteins (see Section 15–6). The hemoglobin carrys oxygen throughout the body via the blood-

Pasadena Freeway at rush hour. (Photograph by Los Angeles County Air Pollution Control, courtesy of Environmental Protection Agency.)

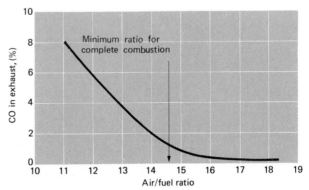

Fig. 14-6 Percentage of CO in exhaust for various air/fuel ratios. [From U.S. Dept. of Health, Education, and Welfare, National Air Pollution Control Administration, *Air Quality Criteria for Carbon Monoxide*, AP-62 (U.S. Government Printing Office, Washington, D.C. 1970).]

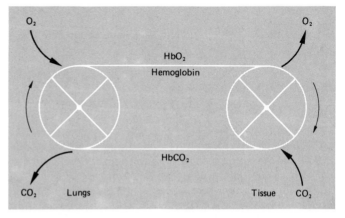

Fig. 14-7 Conveyor belt analogy shows the role of hemoglobin in the blood, transferring gases between lungs and tissue.

stream. When the blood reaches tissue that needs oxygen, the O_2, which is rather weakly bound to the Hb, is detached. Waste products from the tissue, especially CO_2 (which can also become attached to Hb), are carried by the blood back to the lungs where they are exhaled from the body.

If we breathe air containing carbon monoxide, the CO molecules become attached to the Hb with about 200 times the strength of the attachment of O_2.* Although the air may contain far less than 1% CO (compared with the 21% O_2 of the atmosphere), a person's blood may have several percent of the Hb in the HbCO form. Furthermore, unlike O_2, which readily "gets off" the Hb, the CO is firmly attached, requiring several hours for full removal. Thus the CO ties up some of the Hb in the HbCO form, reducing the oxygen-carrying capacity of the blood.

If a person is subjected to a very high concentration of CO (several thousand ppm), such as by running an engine in a closed garage or by driving a car with a faulty exhaust system, death occurs rapidly. (When Hb becomes about 60% HbCO, the brain can no longer get the O_2 that it needs). The effects of breathing air containing certain concentrations of CO for various times are summarized in Fig. 14–8. The Hb of a person exposed to 100 ppm of CO for four hours would be converted to about 10% HbCO form, reducing the O_2 supply to the tissue to about 90% of its normal value. The person would probably have a headache and somewhat reduced mental acuity because of oxygen deprivation of the brain. A person's ability to do arithmetic is reduced even at a 2% HbCO concentration. The symptoms become more serious for a person who is smoking tobacco or marijuana.† As the concentration of CO increases in the air breathed by a person, the HbCO concentration increases until death occurs at about 60% HbCO.

The yearly average atmospheric CO concentrations in Chicago, Los Angeles, and Washington, D.C., are about 14, 10.6 and 4.4 ppm, respectively,‡ which would seem to present no significant problem. However, yearly averages over large

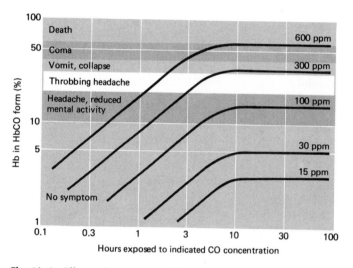

Fig. 14–8   Effects of exposure to CO at several concentrations for various lengths of time. (From W. Agnew, *Proc. Roy. Soc.* **A307**, 153, 1968; courtesy of General Motors Corporation.)

regions are deceptive. Over short intervals near busy highway intersections or in tunnels, CO concentrations are much higher than average. For example, the concentration of CO on FDR Drive in New York City, a partially enclosed highway but not a tunnel, reaches 85 ppm during peak traffic periods.§

What are the effects of chronic exposure to sublethal doses of CO? It is not known if there is any long-term effect on healthy people. Over a period of hours after a person is removed from a high CO environment, the CO gradually becomes detached from the Hb and no long-term damage is apparent.‖ However, there may be a more subtle effect on commuters, policemen, toll-booth attendants, etc., who are

---

*   National Academy of Sciences—National Academy of Engineering, "Effects of Chronic Exposure of Low Levels of Carbon Monoxide on Human Health, Behavior and Performance," 1969.

†   Cigarettes, like gasoline, are not burned in a sufficient stream of oxygen to convert them completely to harmless CO_2 and H_2O vapor. Enough CO is produced that an average pack-a-day smoker maintains about 5% HbCO in his blood. (The moral for students: Don't smoke during exams!)

‡   NAPCA, *Air Quality Criteria for Carbon Monoxide.*

---

§   P. C. Wolf, "Carbon Monoxide," *Environmental Science and Technology* 5, 212–214 (1971).
‖   National Academy of Sciences, "Effects of Carbon Monoxide."

exposed to high CO levels for long periods. The level of HbCO in their blood, especially if they are smokers, may rise high enough to reduce their alertness, and it may increase their fatigue and mental irritation enough that their accident rates are noticeably increased. It is known that automobile accident rates are nearly twice as great for smokers as for nonsmokers, but no definite link to carbon monoxide has been established.*

Increased CO levels probably do present hazards to people who already have serious respiratory or circulatory conditions. Survival rates of heart-attack victims brought to Los Angeles hospitals during periods of high CO concentration are significantly less than during periods of low CO concentration.† A general correlation between mortality and CO concentrations in the Los Angeles area has also been observed.‡

Hydrocarbons

Another major auto emission is hydrocarbons. These pollutants are either the gasoline molecules themselves, which may, for example, be evaporated from the fuel tank, or the partially reacted products of the combustion of hydrocarbons. For example, a molecule of octane, C_8H_{18}, may be only partially oxidized to CO_2 and H_2O, leaving an unburned hydrocarbon fragment such as methane, ethene (C_2H_4), or ethane. In a typical older automobile (pre-1965) with no pollution control devices, about 9% of the hydrocarbons emitted are evaporated from the fuel tank, 9% are emitted from the carburetor, 20% from the crankcase "breather," and the remaining 62% from the exhaust pipe. This typical auto would emit about 570 grams of hydrocarbons per day. The reason for hydrocarbons in the exhaust is the same as for CO: there is insufficient O_2 in the cylinders to cause complete combustion. The amount in the exhaust decreases with increasing air/fuel ratio.

A wide variety of hydrocarbons result from incomplete combustion of the gasoline. The highest-yield product is methane, but smaller amounts of hundreds of compounds containing greater numbers of carbon atoms are found. Most are alkanes, but alkenes (also called olefins), cycloalkanes, aromatic compounds, and even condensed-ring compounds are found.

Two major atmospheric problems are caused by hydrocarbons. First, reaction of hydrocarbons (especially the olefins) with nitrogen oxides in the presence of sunlight produces harmful substances present in smog (discussed in more detail below). Second, some emitted hydrocarbons are carcinogenic to animals and probably to humans. The most important carcinogens appear to be the complex aromatic compounds, especially those similar to benzopyrene which have condensed ring structures. There is little of these compounds in crude oil and straight-run gasolines, but some are formed during the catalytic processes used to produce the other components of blended gasoline. Condensed-ring compounds are also formed in the engine during combustion. Gasolines containing large amounts of aromatic compounds as replacements for alkyl lead additives produce greater amounts of these carcinogenic substances. Blended gasolines produce about ten times as much carcinogenic material in an engine than does pure isooctane.§ When a high-octane gasoline made of 50% xylene and 50% benzene is burned, the carcinogen production is three times greater than from ordinary gasoline.

Automobiles are not the only man-made sources of organic carcinogens. Any time an organic compound is burned in a limited supply of oxygen, some organic carcinogens are formed. That's probably one of the main reasons for the fact that cigarette smoking greatly increases one's chances of contracting lung cancer. When fat from a steak drips onto glowing charcoal and burns, organic carcinogens are released and some deposit on the steak.

The atmospheric concentration of benzopyrene in most American cities ranges between 0.001 and 0.01 $\mu g/m^3$.‖ Although the concentration is 10,000 times smaller than those of low-molecular-weight hydrocarbons (see Table 14–4), Benzopyrene is thought to be a serious life-shortening con-

*   Andrew Wilson, "Smoker Motorist Is Accident Prone," Washington *Post*, Jan. 31, 1973.
†   National Academy of Sciences, "Effects of Carbon Monoxide."
‡   A. C. Hexter and J. R. Goldsmith, "Carbon Monoxide: Association of Community Air Pollution with Mortality," *Science* 172, 265–267 (1971).

§   D. Hoffman and E. L. Wynder, "Respiratory Carcinogens: Their Nature and Precursors," in *International Symposium on Identification and Measurement of Environmental Pollutants*, B. Westley, ed. (National Research Council of Canada, Ottawa, 1972), pp. 9–16.
‖   U.S. Dept. of Health, Education, and Welfare, National Air Pollution Control Administration, *Air Quality Data from the National Air Sampling Networks and Contributing State and Local Networks*, 1968 ed. (U.S. Government Printing Office, Washington, D.C., 1968).

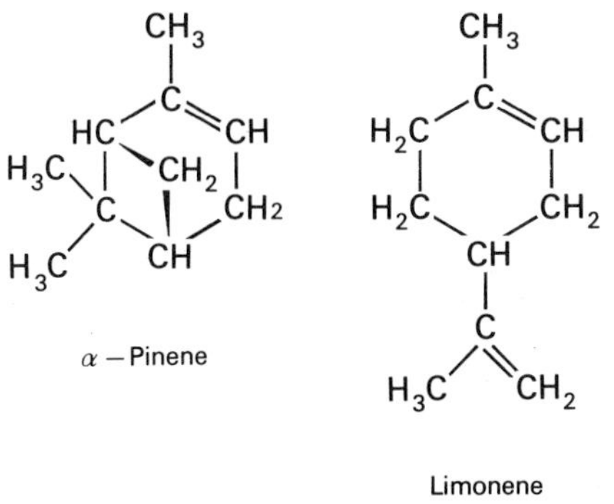

Fig. 14–9   Structure of two terpenes.

Table 14–4   Concentrations of Several Hydrocarbons in the Los Angeles Atmosphere

Compound	Formula	Concentration $\mu g/m^3$	Compound	Formula	Concentration $\mu g/m^3$
Methane	CH_4	1600	2-Butene	C_4H_8	5
Ethane	C_2H_6	} 120	n-Pentane	C_5H_{12}	63
Ethene	C_2H_4		Nonane	C_9H_{20}	} 28
Ethyne	C_2H_2	} 110	Decane	$C_{10}H_{22}$	
Propane	C_3H_8		Toluene	C_7H_8	115
Propylene	C_3H_6	18	Xylene	C_8H_{10}	73
n-Butane	C_4H_{10}	110			

Computed from data of A. P. Altshuller, W. A. Lonneman, F. D. Sutterfield, and S. L. Kipczynski, "Hydrocarbon Composition of the Atmosphere of the Los Angeles Basin—1967," *Environmental Science and Technology* **5**, 1009–1016 (1971).

taminant. The benzopyrene breathed by a person every day in an average-size American city is equivalent to that of smoking one-third of a pack of cigarettes, but it could easily be as high as a pack per day in heavily polluted cities such as Los Angeles or New York.*

Concentrations of some hydrocarbons in the atmosphere of the Los Angeles basin are listed in Table 14–4. Note that there is more methane than all other hydrocarbons combined. Even in Los Angeles, not all hydrocarbons result from gasoline combustion. There are many other man-made sources: escape of volatile hydrocarbons from oil and gas wells and refineries, evaporation of solvents from industrial and dry cleaning plants, and the combustion of oil in furnaces and power plants. Furthermore, there are many natural hydrocarbon sources. On a global basis, man's activities account for only about 15% of the atmospheric hydrocarbons.† Deposits of decaying matter, as in swamps or marshes, liberate great quantities of methane, which is sometimes called "swamp gas." Some forms of vegetation such as pine forests emit *terpenes*, compounds of the sort contained in turpentine (used as a paint thinner). It is terpene emission that causes the "smokiness" of the Great Smoky Mountains. Alpha-pinene is the principal terpene in turpentine and limonene is the main terpene in lemons, oranges, and certain oils derived from plants and fruits (see Fig. 14–9).

As in the case of sulfur gases (see Chapter 10), the natural sources of hydrocarbons are sufficiently spread out and far enough away from population centers that they cause little threat to life. Figure 14–10 shows the sources of many atmospheric hydrocarbons.

---

* E. Edelson and F. Warshofsky, *Poisons in the Air* (Pocket Books, New York, 1966).

† American Chemical Society, *Cleaning Our Environment—The Chemical Basis for Action* (ACS, Washington, D.C., 1969).

Fig. 14–10 Hydrocarbons in the atmosphere originate from various natural and man-made sources.

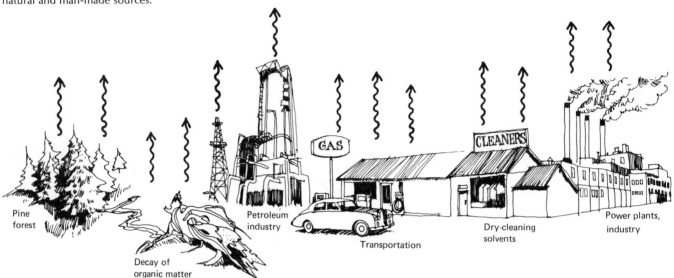

Pine forest

Decay of organic matter

Petroleum industry

GAS

Transportation

CLEANERS

Dry-cleaning solvents

Power plants, industry

Nitrogen Oxides

Perhaps the most troublesome air pollutants from internal combustion engines (ICE's), and the most difficult to control, are the nitrogen oxides produced by the reaction between N_2 and O_2 of the air drawn into the cylinders. In Section 11–2 we noted that, at normal atmospheric temperature, the reaction

$$N_2(g) + O_2(g) \longrightarrow 2NO(g) \qquad (14–3)$$

proceeds to an insignificant extent. However, since the reaction is endothermic (requires heat), it proceeds much farther at the high temperatures attained in the cylinder (about 2800°C), and so the exhaust gases may contain several thousand ppm of NO. If the carburetor is adjusted to a "leaner" mixture (higher air/fuel ratio) to obtain more complete oxidation of hydrocarbons, the mixture burns at a higher temperature and produces even more NO (see Fig. 14–11). Thus any measures that we might take to decrease CO and hydrocarbon emissions may increase the NO. Diesel engines produce smaller amounts of NO than ICE's, but greater amounts of CO and hydrocarbons.

Nitrogen oxides play an important role in the formation of photochemical smog. Motor vehicles are the major source of nitrogen oxides in many areas, but *any* high-temperature

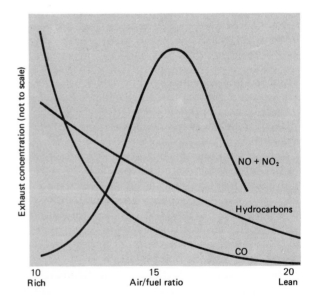

Fig. 14–11 Relative concentration of NO in exhaust as a function of air/fuel ratio.

process in the presence of air produces NO. In Table 10–3, we saw that on a nationwide basis, only about half of the man-made nitrogen oxides originate from transportation. The remaining half comes from stationary sources such as incinerators, power plants, and home heating units. Even if NO from automobiles is completely eliminated, many areas will still have high NO concentrations.

Lead

Alkyl lead compounds TEL or TML are generally added to gasoline to raise its octane number. Although we may think of lead as a minor additive, on the average about 2 g of lead is added to each gallon. If one drives an automobile about 10,000 mi/yr on leaded gasoline, the gasoline used would contain about one kilogram of lead. In a year's time, enough alkyl lead is burned by 25 cars to make a 2″ × 4″ × 8″ lead brick! The average American annually uses 2 lb of lead as gasoline additives.

About half of the lead is immediately emitted in the exhaust as very small particles. Most of the remainder is deposited in the exhaust system and muffler, from which it may later flake off as much larger particles. The former constitutes the major source of atmospheric lead in most urban areas which typically have average atmospheric lead concentrations of one to two $\mu g/m^3$. In areas of high traffic density, the atmospheric lead concentration is even greater. For example, the concentration in downtown San Diego is frequently greater than 7 $\mu g/m^3$.*

At concentrations of 1 to 2 $\mu g/m^3$, the lead constitutes about 1 to 2% of the atmospheric particulate material in a city. This lead, attached to particles small enough to be deposited in the lungs, is a major contributor to the body burden of lead in urban residents, accounting for about 25 to 35% of the daily buildup. As discussed in Section 12–8, lead is a highly toxic element. The average American urban resident carries enough lead in his body to be near the threshold of symptoms of lead poisoning, so any unusual exposure (such as living next to a busy intersection) could cause lead poisoning.

Fine particles of lead settle to the ground very slowly, but they can be brought down by rain or snow, or they may stick to surfaces with which they come in contact (especially leaves of vegetation). Much of the atmospheric lead depo-

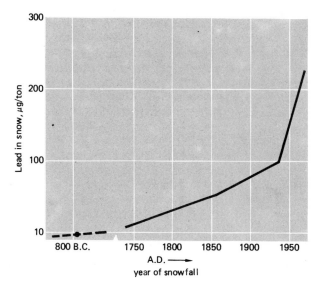

Fig. 14–12 Concentration of lead in layers of snow deposited in Greenland at various times. (After Patterson and Salvia, "Lead in the Modern Environment.")

sited in these ways and the larger particles that flake from the exhaust systems of automobiles is deposited with the dust of city streets. In part, the lead poisoning of inner-city children results from the lead in the dust of the streets. It is becoming increasingly evident that we must remove lead from both interior paints *and* gasoline.

The lead from automobiles and other activities of man is not simply a localized pollutant. Professor Clair Patterson of Caltech and his coworkers have demonstrated that man has polluted the atmosphere of the entire Northern Hemisphere with lead. They went to regions near the North and South Poles to collect snow from various depths below the present surface that was deposited as long ago as 800 B.C.† Analysis of the snow from various ages showed that near the North Pole, the concentration of lead in the snow has increased more than 40-fold since 800 B.C. (see Fig. 14–12). There is a definite upswing of the curve after the middle 1920s, when leaded gasolines were introduced. The lead concentration in South Pole snow was nearly constant in layers deposited

---

*   T. J. Chow and J. L. Earl, "Lead Aerosols in the Atmosphere: Increasing Concentrations," *Science* **169**, 557–580 (1970).

†   C. C. Patterson and J. D. Salvia, "Lead in the Modern Environment, How Much is Natural?," *Scientist and Citizen* **10**, 66–79 (1968).

over a wide span of years; the maximum value was about equal to the 800 B.C. value of the North Pole region. These observations are explained by the fact that most of man's lead-producing activities, especially the use of large amounts of leaded gasolines, occur in the Northern Hemisphere. The amount of lead mined and introduced into the environment by man each year is more than 100 times the amount that is naturally leached from rocks and soil each year and added to the world's oceans.

14–5 PHOTOCHEMICAL SMOG: THE "LOS ANGELES SYNDROME"

Automobile emissions are the major pollutants responsible for the *photochemical smog* that first appeared in Los Angeles and now occurs with increasing frequency in other cities. The severe air pollution that appeared in Los Angeles during and following World War II was rather a surprise, since the area had little of the industrial and heating activities (e.g.,

Denver is another city subject to the Los Angeles syndrome. Note how the smog is confined to altitudes lower than the mountains. (Charles E. Grover photo courtesy of National Air Pollution Control Administration.)

Fig. 14–13   The Los Angeles Basin.

soft-coal burning) typical of older polluted cities.* Now that we know more about photochemical smog, we can see that Los Angeles has an abundance of the components necessary for smog. First, the ring of mountains around Los Angeles forms a cup-shaped basin (with sides several thousand feet high) whose only open side is on the Pacific Ocean. Because of the terrain and the location of the city at the interface of land and sea, Los Angeles has temperature inversions about 320 days per year. Air pollutants are confined to altitudes below the height of the surrounding mountains, so they can only slowly diffuse out of the Los Angeles basin horizontally via passes through the mountains (see Fig. 14–13).

Enter the next component—the automobile. Los Angeles is a new, rapidly growing city, highly suburbanized and designed for the automobile, with a massive freeway system

and little public transportation. The high density of traffic produces vast quantities of CO, hydrocarbons, and nitrogen oxides, all of which are trapped in the city air below the inversion.

The third component of smog formation is sunlight, which Los Angeles has nearly every day. Figure 14–14 shows the cycle for formation of smog. This cycle has the following features.†

1. During the morning rush hour, the concentration of NO from automobiles builds up to high levels.

2. Some of the NO is oxidized to NO_2. (A low percentage of the NO is converted to NO_2 by reaction with oxygen in automobiles or other combustion sources before the hot gases leave the sources.) In Section 11–2 we noted that the reaction

*   A. J. Haagen-Smit, "Air Conservation," *Science* **128**, 869–878 (1958).

†   S. J. Williamson, *Fundamentals of Air Pollution* (Addison-Wesley, Reading, Mass., 1973), Chapter 10.

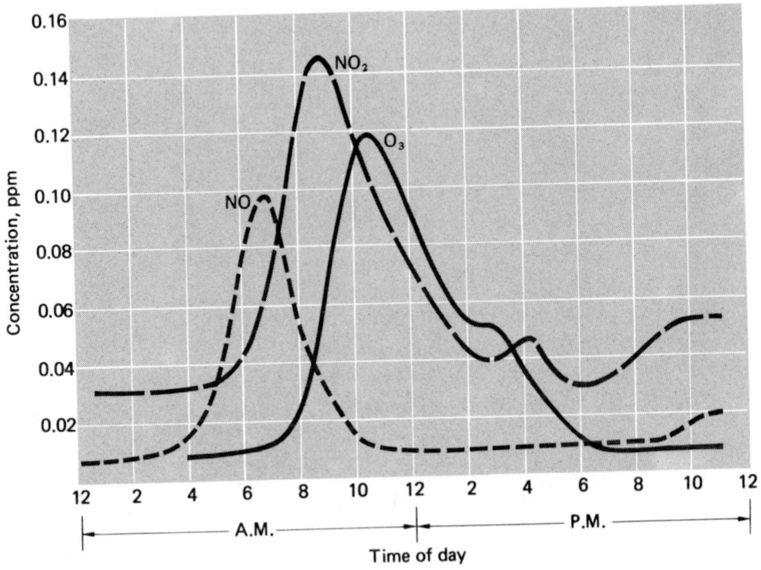

Fig. 14–14 Typical concentrations of NO, NO_2, and O_3 throughout the day in Los Angeles. [From U.S. Dept. of Health, Education, and Welfare, National Air Pollution Control Administration, *Air Quality Criteria for Photochemical Oxidants,* AP-63 (U.S. Government Printing Office, Washington, D.C., 1970).]

of O_2 with NO to form NO_2 is one step in the production of nitric acid, so we might suppose that the same reaction would occur in the atmosphere. However, the NO, even in a polluted atmosphere, is at such a low concentration that the reaction is too slow to explain the observed rate of NO_2 formation.

3. NO_2 absorbs UV light from the sun and dissociates to form nitric oxide and oxygen atoms:

$$NO_2 + UV\ photon \longrightarrow NO + O. \tag{14–4}$$

4. Free oxygen atoms are quite reactive and some react quickly with O_2 molecules to form ozone:

$$O + O_2 \longrightarrow O_3. \tag{14–5}$$

5. Ozone reacts rapidly with NO to form more NO_2:

$$O_3 + NO \longrightarrow NO_2. \tag{14–6}$$

If reactions (14–4) through (14–6) were the only ones occurring in the polluted atmosphere, no smog would be formed. Note that the set of reactions is circular: the NO_2 formed in (14–6) would simply react according to (14–4) and then go

through the cycle again. All three species, NO, NO_2, and O_3, would build up during the day as pollution sources continued to add more NO to the atmosphere, but to produce smog, other components are needed.

6. One of those components is carbon monoxide. Probably the major reaction that oxidizes NO to NO_2 is a three-step sequence or *chain reaction* involving CO.* First, we have the reaction of CO with a hydroxyl radical, $\cdot OH$

$$CO + \cdot OH \longrightarrow CO_2 + H \cdot \tag{14–7}$$

Radicals are unstable, highly reactive species having unpaired electrons. The hydrogen atom reacts quickly with oxygen in the presence of a third molecule M (generally N_2):

$$H \cdot + O_2 + M \longrightarrow HO_2 \cdot + M \tag{14–8}$$

Then the hydroperoxyl radical, $HO_2 \cdot$, reacts with NO:

$$HO_2 \cdot + NO \longrightarrow NO_2 + \cdot OH \tag{14–9}$$

---

* Hiram Levy II, "Normal Atmosphere: Large Radical and Formaldehyde Concentrations Predicted," *Science* **173**, 141–143 (1971).

Note that reaction (14–9) regenerates an ·OH radical that can be used in (14–7) to start the cycle over again. The net reaction of the chain is just the sum of the three reactions:

$$CO + O_2 + NO \longrightarrow CO_2 + NO_2. \qquad (14-10)$$

Thus CO may play an important role in the formation of photochemical smog, as well as being itself an undesirable pollutant. The $HO_2·$ produced by the reactions with CO may have other important reactions with pollutants, such as the oxidation of SO_2 to SO_3 (see Section 10–8.)

7. Another essential sequence of smog-forming reactions involves the hydrocarbons, especially the highly reactive olefins. They react with oxygen atoms from reaction (14–4) to form radicals:

$$O· + R \longrightarrow R'O· + \text{possibly other products such as } R''.$$

hydrocarbon   radical
$$(14-11)$$

In this equation, R' and R" represent different alkyl groups. As an example, ethene can react with oxygen atoms:

$$(14-12)$$

8. Once hydrocarbon radicals are formed, many complex reactions can occur. The radicals react with O_2, O_3, NO_2, or additional O atoms to form various types of organic compounds, such as aldehydes and ketones, and various peroxyacyl nitrates (PAN's), of which a prominent member is peroxyacetyl nitrate:

peroxyacetyl nitrate

9. The brown color of smog is produced by the NO_2 gas, and the hazy appearance of smog by particles in the atmosphere. Many of the pollutant gases eventually end up on particles, particularly the high-molecular-weight hydrocarbons and the nitrogen species, which are finally deposited as nitrates.

A highly oversimplified diagram of the major smog-forming atmospheric reactions is shown in Fig. 14–15.

The primary automobile exhaust components are not responsible for much of the damage caused by photochemical smog. The major culprits are the secondary products, i.e., ozone, PAN, aldehydes, ketones, etc., formed by reactions of the primary pollutants in the presence of sunlight.

Fig. 14–15 Simplified scheme of some important reactions in formation of photochemical smog. Hc is an abbreviation for "hydrocarbons."

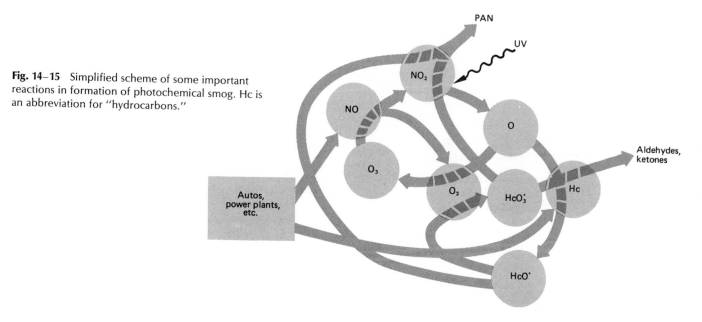

(a) & (b) Two New York City scenes recorded one day apart. The heavy smog is a result of heavy smoke concentrated by a stagnant air condition. (Photo courtesy of *New York Daily News.*)

(c) A visual illustration of clear conditions and an intense smog formation in Los Angeles. (United Press International Photograph.)

Photochemical smog is different from the London or typical eastern United States air pollution. The latter is created mainly by SO_2 and particulate matter and is a "reducing" form of pollution, particularly since SO_2 can be oxidized to SO_3. On the other hand, the major components of photochemical smog are oxidizing agents, e.g., ozone, NO_2, and PAN. Most of the damage of photochemical smog is a result of the oxidation of substances brought about by these secondary pollutants.

The intensity of photochemical smog is measured by the atmospheric concentration of photochemical oxidants (mainly ozone) capable of oxidizing I^- in a potassium iodide solution to I_2. An area is said to have smog on any day that the oxidant concentration is greater than 0.15 ppm for an hour or longer.* Cities of Southern California have the most frequent smog. Between 1964 and 1967, 41% of the days in Pasadena, California were smoggy; Los Angeles had 30%, San Diego, 5.6%, and Washington, D.C., had only about 1%. Although smog is most serious in Southern California, no area can safely ignore the problem. With increasing traffic density in urban areas, smog frequency will increase in all cities unless pollution controls are used on the offending engines.

Photochemical oxidants attack many materials. Ozone attacks rubber products, breaking bonds that link together the long rubber molecules (polymers) that give rubber its strength and elasticity (see Section 13–13). Ozone degradation is much faster when the rubber is under stress. One can make a simple "smog detector" by bending a thick strip of common rubber into a hairpin shape, tying the ends together, and exposing it to the atmosphere until it cracks and eventually breaks in two.† On a very smoggy day, cracks start to appear within a few minutes. Ozone takes a heavy toll in

---

* NAPCA, *Air Quality Criteria for Photochemical Oxidants.*

† Haagen-Smit, "Air Conservation."

damage to rubber products, especially automobile tires, wiper blades, and so on.

High smog levels irritate the eyes, nose, and throat, apparently because of the PAN and similar compounds. Asthma, bronchitis, and other respiratory ailments are aggravated by smog, although there is little apparent long-term damage or increased mortality in regions of excessive smog. Athletes exhibit poorer performance on days of high smog. Vegetation is strongly affected by smog. Leaves or needles of citrus and pine trees, flowers and vegetables are damaged by the oxidants. There is a dramatic difference between the damaged trees on the Los Angeles side of the mountains and the almost unaffected ones on the other side.

14–6 CONTROL OF EMISSIONS FROM INTERNAL COMBUSTION ENGINES

Automobile emissions must be significantly reduced or we will have to find other sources of transportation energy. The federal government and the state of California have established emission standards that will become increasingly stringent for the next few years, as shown in Table 14–5. An option was written into the Clean Air Act of 1970 (which specified the federal standards) for the Director of the EPA to delay implementation of the 1975 and 1976 standards if it appears that auto manufacturers, despite strenuous efforts, are unable to meet the standards by those model years. Debate between the EPA and United States auto makers over implementation of the standards has been quite vigorous and, as it turns out, an embarrassment for American technology when it is compared with Japanese and German technology. The United States auto manufacturers convinced EPA that they would not be able to meet the 1975 standards with the use of catalytic converters (described below), the only approach they seem to have pursued strongly. However, two Japanese autos (using Wankel and "statified-charge" engines) and two German models with diesel engines demonstrated that they could meet the standards.

In April 1973, William Ruckelshaus, Director of EPA, announced a relaxation of the 1975 standards because of the difficulties of the United States auto companies.* According to his ruling, 1975 autos sold throughout most of the United States will have to meet only half of the planned reduction of

Table 14–5   Automobile Emission Standards

	Hydro-carbons, g/mi	Carbon monoxide, g/mi	Nitrogen oxides, g/mi
Average uncontrolled vehicle	11.0	80.0	4.0
Federal standards*			
1972–1974	3.4	39.0	3.0
1975	0.41	3.4	3.0
1976	0.41	3.4	0.4
California standards			
1974	1.5	23.0	1.3

* Standards specified in the Clean Air Act of 1970. However, the EPA was given the option of delaying implementation of 1975 and 1976 standards by one year. The 1975 standards have been partially delayed (see text).

CO and hydrocarbon emissions. Cars sold in California in 1975 will have to meet two-thirds of the originally required reduction. The complete reductions will go into effect in 1976. There is a possibility that the 1976 standards for nitrogen oxide emissions will be relaxed somewhat. Data obtained since 1970 suggest that the original requirements may have been more stringent than necessary, as they were based on a method of measuring atmospheric nitrogen oxide concentrations so that they appeared higher than the true values.†

American manufacturers have considered two basic methods for control of CO and hydrocarbons: (1) the manifold reactor and (2) the catalytic converter. The former, attached to the exhaust manifold, is a large chamber in which the hot exhaust gases are mixed with more air to oxidize most of the CO and hydrocarbons. It is a sort of "after burner," but unlike those used on jet engines, it does not increase the power of the engine.

The second approach is the use of a catalytic converter, a chamber in the exhaust system containing a catalyst that speeds the oxidation of CO and hydrocarbons to CO_2 and water vapor. The catalyst, which might consist of a noble metal such as platinum or palladium, would be supported on a porous ceramic material through which the exhaust would pass.

The basic ideas for the oxidation of CO and hydrocarbons are fairly well developed, so the building of practical

---

* G. C. Wilson, "Auto Makers Win Delay on Clean Air," Washington *Post*, April 12, 1973.

† EPA Moves on Curbs for Plant, Auto Emissions," *Chemical and Engineering News*, June 11, 1973, p. 2.

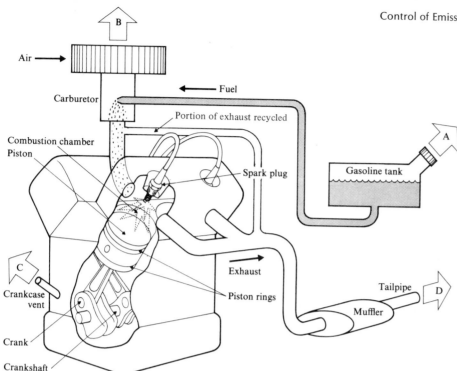

Fig. 14–16 Recirculation of a portion of exhaust gases back into the engine to dilute the air/fuel mixture, reduce combustion temperature, and decrease NO emission. [From Samuel J. Williamson, *Fundamentals of Air Pollution* (Reading, Mass.: Addison-Wesley, 1973), p. 314.]

systems is largely an engineering task. This is not the case for NO control. A good bit of scientific investigation will have to be done before serious engineering work can even begin. The NO emissions can be reduced by recirculating about 10 to 20% of the exhaust gases back through the cylinders (see Fig. 14–16). This dilutes the air/fuel mixture with unreactive gas, lowering the combustion temperature and, thus, the amount of NO formed.* However, recirculation is not a satisfactory solution, as the fraction recirculated must be controlled carefully to avoid engine stalls and generally poor performance.

A better solution would perhaps be the development of materials that catalyze the reduction of NO to N_2. Many catalysts are under study, such as Cu-Ni, Pd, Pt, and compounds of rare earths with Pb, Mn, and O. If satisfactory catalysts for both reduction of NO and oxidation of CO and hydrocarbons are developed, they would probably be placed

in series in the exhaust system (see Fig. 14–17). The reduction catalyst would be placed first so that CO could be used as the reducing agent:

$$2CO(g) + 2NO(g) \xrightarrow{\text{catalyst}} 2CO_2(g) + N_2(g) \quad (14\text{–}13)$$

The actual reactions that occur in catalytic converters are probably quite complex. In many cases, nitrogen appears to be reduced all the way to ammonia:†

$$CO + H_2O \longrightarrow CO_2 + H_2, \quad (14\text{–}14)$$

$$5H_2 + 2NO \longrightarrow 2NH_3 + 2H_2O, \quad (14\text{–}15)$$

followed by the decomposition of ammonia:

$$2NH_3 \longrightarrow N_2 + 3H_2. \quad (14\text{–}16)$$

The nature of the catalyst and the temperature of the gases must be controlled carefully to insure complete decomposi-

*   M. Malin and C. Lewicke, "Pollution Free Power for the Automobile," *Environmental Science and Technology* **6**, 512–517 (1972).

†   R. L. Klimisch and K. C. Taylor, "Ammonia Intermediacy as a Basis for Catalyst Selection for Nitric Oxide Reduction," *Environmental Science and Technology* **7**, 127–131 (1973).

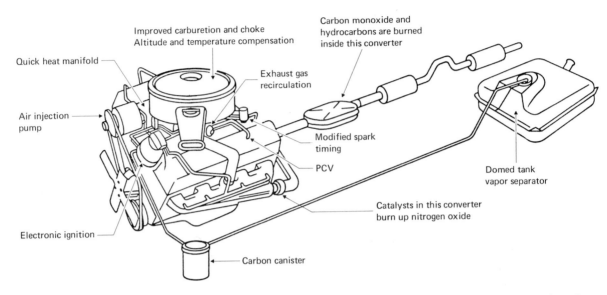

Fig. 14–17 Engine equipped with two catalytic converters to remove first NO, and then CO and hydrocarbons. [From Joseph Priest, *Problems of Our Physical Environment* (Addison-Wesley, Reading, Mass, 1973), p. 241.]

tion of the ammonia. Ammonia allowed to pass on to the oxidizing chamber would be oxidized back to NO.

Although it may be possible to clean up the emissions from conventional ICE's with catalytic converters, there are many uncertainties. Converters may work when first installed, but what about a dirty catalyst after 20,000 miles of use? The Environmental Protection Agency insists that the control devices must achieve maintenance-free operation for 50,000 miles—a difficult challenge! Most controls will require the engines to be kept well adjusted, something that most car owners are lax about. Even with pre-1973 automobiles, proper tuning and adjustment of all engines could reduce CO and hydrocarbon emissions by about 30%.* Furthermore, emission-control adjustments and gadgets generally reduce the engine's performance or decrease fuel economy, so owners may purposely remove the devices or change adjustments to improve performance, while increasing emissions. Thus any control strategy will depend upon frequent compulsory inspection and adjustment of the engines and severe fines for removal of control devices.

Lead will probably be removed from gasolines of the future because of the toxicity of the lead emissions themselves and because lead deposits ruin most emission control catalysts. EPA will require most gasoline stations to make available at least one grade of lead-free gasoline by July 1974.† In January 1975, according to proposed regulations, gasolines produced by refineries would be limited to an average of 2.0 g/gal lead, and in January 1978 the limit would drop to 1.25 g. Today the removal of lead is not an unmixed blessing, as it is usually replaced by aromatic compounds that boost the octane but increase the emission of organic carcinogens and the olefins that participate in smog-forming reactions. As a short-range measure, the auto industry is today producing nearly all engines with low enough compression ratios that they can use "regular" gasoline with octane numbers in the 91 to 93 range, thus reducing the need for lead or aromatics. When control devices for hydrocarbons are in use, lead compounds could be safely replaced with aromatics, as the organic carcinogens would mostly be oxidized to CO_2 and water vapor.

---

*   Revel Shinnar, "System Approach for Reducing Car Pollution," *Science* **175**, 1357–1360 (1972).

†   "EPA Lead-Free Gas Rules Refuel Controversy," *Chemical and Engineering News*, Jan. 8, 1973, p. 10.

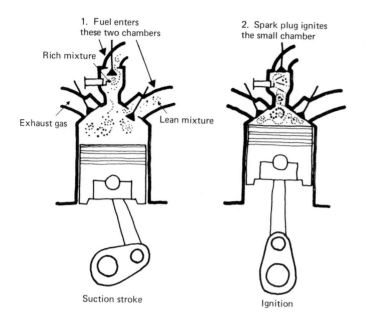

1. Fuel enters these two chambers

Rich mixture

Exhaust gas

Lean mixture

Suction stroke

2. Spark plug ignites the small chamber

Ignition

3. Fire spreads to rest of chamber, pushing piston down

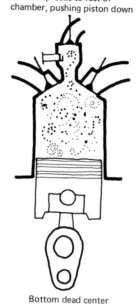

Bottom dead center

Fig. 14–18 Honda's stratified charge permits a leaner air-gasoline mixture by the use of two combustion chambers. The small chamber is ignited first, and the fire spreads from there to fuel inside the cylinder. This arrangement also ensures that most impurities are burned up inside the engine.

The Japanese manufacturers of the Honda and Mazda have come up with solutions to the emission problem that appear to be more satisfactory than the catalytic converters. The Honda employs a piston engine that operates with a "stratified charge" in a mode that the manufacturer calls "compound vortex controlled combustion" (CVCC). As shown in Fig. 14–18, the spark plug is contained in a small chamber above the main cylinder chamber. Two different air/fuel mixtures (the "charges") are supplied to the two chambers by separate carburetors. The major charge, fed to the main chamber, is a very lean mixture with air/fuel mixtures toward the right-hand side of the graph in Fig. 14–11.* It's almost impossible to ignite such a lean mixture with a spark plug—that's the reason for the small chamber. The charge supplied to it is rich enough to be ignited by the spark and the flame from that chamber ignites the main charge. Already the Honda CVCC engine has met the original 1975 standards for CO and hydrocarbons and it will apparently be able to meet the 1976 nitrogen oxide standards with little difficulty.

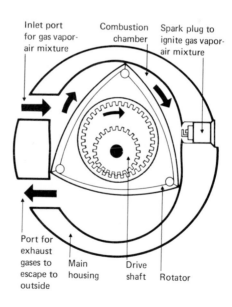

Inlet port for gas vapor-air mixture

Combustion chamber

Spark plug to ignite gas vapor-air mixture

Port for exhaust gases to escape to outside

Main housing

Drive shaft

Rotator

Fig. 14–19 The Wankel engine. [From Joseph Priest, *Problems of Our Physical Environment* (Addison-Wesley, Reading, Mass., 1973), p. 248.]

The Mazda automobiles can meet the 1975 standards with a very unusual type of ICE, the Wankel engine (see Fig. 14–19). The three-chamber rotary "piston" Wankel engine has fewer moving parts, runs more smoothly, and yields more power per unit volume than conventional ICE's.* The Wankel-equipped Mazdas have been quite popular in the United States since their introduction, although they are less efficient (19% versus 25%) than conventional ICE's in converting the chemical energy of gasoline into mechanical energy.† Also, an uncontrolled Wankel emits considerably greater amounts of CO and hydrocarbons than conventional ICE's. The Mazda was able to meet the 1975 standards for these pollutants by attaching a thermal reactor to the engine. At the higher exhaust temperatures of the Wankel engine, the CO and hydrocarbons are readily oxidized by additional air without the need for a catalyst. In late 1974, General Motors will market Chevrolet Vegas powered by similar engines. Despite their popularity, the Wankel engines are probably not so good a solution to the emissions problem as the stratified-charge approach, especially in view of gasoline shortages, because of the lower efficiency of the Wankel.

14–7 ALTERNATIVES TO THE CONVENTIONAL INTERNAL COMBUSTION ENGINE

Before considering other types of engines, let us consider the automobile in relation to other transportation modes used in metropolitan areas, where pollution problems are the most serious. The automobile is an inefficient and expensive method for moving large numbers of people. Most of the energy is used to move two tons of steel and very little to move the person himself. In Fig. 14–20, note that an automobile with one passenger in urban traffic has a "net propulsion efficiency" (NPE), the number of passenger miles per gallon of fuel consumed, of about 7. Thus its efficiency is about one-sixth that of an urban bus and one-seventeenth that of a two-level commuter train! (Perhaps surprisingly, the most efficient "people movers" are low-powered motorcycles and heavily loaded VW microbuses.) Fuel consumption within urban uses could be greatly reduced by use of buses and

Bicyclers commuting to work in Cambridge, Massachusetts. (Courtesy of The Cambridge Chamber of Commerce.)

trains. Automobiles are much more expensive to operate in a city than most people imagine. Considering the cost of gasoline, oil, tires, maintenance, licenses, insurance and parking, which the motorist pays, and the costs of the highways and the land they occupy, the estimated cost of urban driving is about one dollar per mile, much greater than the highest mass transportation fares.‡ The growing use of trucks for hauling freight is also inefficient. The efficiency of 40-ton trucks is about 50 ton-miles of cargo per gallon, whereas

*   D. E. Cole, "The Wankel Engine," *Scientific American* **227** (2), 14–23 (Aug. 1972).
†   C. M. Summers, "The Conversion of Energy," *Scientific American* **224** (3), 148–160 (March 1971).

‡   R. H. Gilluly, "Nitrogen Oxides, Autos, and Power Plants," *Science News* **101**, 252–253 (1972).

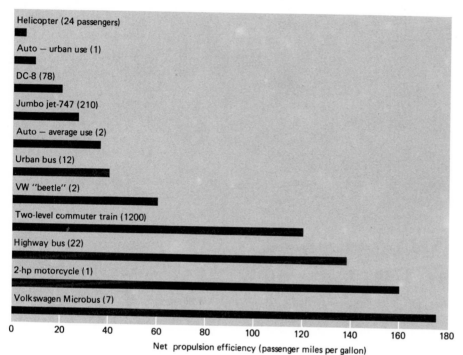

Fig. 14–20 Net propulsion efficiency for several modes of travel. The figures for number of passengers which are cited for the several transport modes are not the total capacity of the unit—but the average load on which the calculations are based. [From R. A. Rice, "Systems Energy and Future Transportation," *Technology Review* **74** (3), 31–37 (1972). Reprinted by permission.]

that of large freight trains and inland barge tows is about 200 or more.

Despite future improvements of public transportations, we will probably have heavy automobile and truck traffic in cities for many years. Are there mobile energy sources that would emit fewer or more easily controlled pollutants than the conventional ICE at a competitive cost? The answer is a guarded "yes". There are some interesting possibilities, but one must remember that a large fraction of the U.S. economy is "wedded" to the gasoline-burning ICE. A shift to a radically different mobile energy source would cost enormous sums for new machine tools, fuel processing and distribution systems, retraining of mechanics, etc., resulting in sizable em-

ployment dislocations. Strenuous attempts to fix up the ICE to meet future standards will doubtless be made before alternatives are considered. Possible competitors of the conventional ICE fueled by gasoline are the following.

ICE fueled by natural or LP gas. With minor additions (mainly a different fuel tank), ICE's can be converted to use natural or LP (liquefied petroleum) gas. They run well and emit much smaller amounts of pollutants. Some large fleets of industrial and government cars, trucks, and buses have been switched to gas, but the shortage of gas and the distributional problems will probably prohibit widespread use by one- or two-car owners.

Modern rapid transit station in Oakland, California. (Courtesy of Bay Area Rapid Transit.)

Gas turbines.   The gas turbine can use a variety of fuels. The incoming air/fuel mixture is compressed and warmed by passing near the hot outgoing exhaust gases before it is ignited and allowed to expand in the turbine that turns the drive shaft. Prototypes meet the 1975 CO and hydrocarbon standards. They emit less NO than present-day ICE's, but not as little as the 1976 standard; however, developers feel they can meet the NO standard by that time. Gas turbines are simpler than ICE's—they have fewer moving parts, require no cooling system, and run more smoothly. However, they respond more slowly to increased fuel supply than ICE's do, thus exhibiting a "hesitation" problem. Despite this difficulty, they may be a strong competitor to the ICE if modifying the latter to meet the 1975–1976 emission standards proves unreasonably expensive.

Steam engines.   In order to burn properly in the cylinders of a conventional ICE, both when the engine is cold and after it warms up, gasoline must be within narrow ranges of volatility, heat release, etc. The simpler operations of gas turbines and steam engines make their fuel requirements much less demanding. Steam engines can use any of a variety of fuels to heat the water or other working fluid to produce a high-

pressure vapor capable of expanding in a piston or turbine. Emissions from steam engines are much less than those from ICE's: a 220-hp steam turbine built by Lear Motors Corporation and installed in a bus in 1972 met the 1975 and 1976 standards.*

Modern steam engines are quite different from those used in Stanley Steamers early in this century. One doesn't have to build a fire under them to warm them up nor does one have to stop every few miles to "take on" water. The working fluid is in a closed system and seldom has to be recharged. Future improvements may make the steam engine very competitive with the ICE in a few years.

Battery-powered electric cars.   Electric cars powered by batteries do not look very promising today. The batteries are heavy and expensive and commonly have ranges of only 100 miles or less between recharges. They are completely clean in operation but, of course, the power plant that generates the electricity to charge the batteries may be a major pollution source. Research is under way to develop batteries that can store more energy per unit weight, but electric cars probably

---

*   Malin and Lewicke, "Pollution Free Power for the Automobile."

won't become practical soon, except in specialized applications.

Fuel cells. In the long range, fuel cells are probably the most desirable mobile energy sources. Instead of converting chemical energy to heat and the heat to mechanical energy, as engines do, the fuel cell converts chemical energy directly to electricity that is used in electric motors to power the vehicle. As discussed in Section 4–5, the conversion of heat to mechanical energy is inherently inefficient because of fundamental limitations dependent on the operating temperature of the engine. There are no similar limitations on the conversion of chemical to electrical to mechanical energy, although in practical systems the efficiency would be less than 100%. A fuel cell is similar to a battery: compounds of high potential chemical energy react to produce more stable, lower-energy chemicals, the energy difference being given off as electrical current.

In Section 8–4 we discussed the fuel cells used in space capsules. In that case the fuels are H_2 and O_2 gases, which are combined to give water, yielding electricity in the process. Hydrogen-oxygen fuel cells for automobiles have been built, but they contain expensive catalysts and the production, storage, and distribution of the hydrogen would be a problem. It could be made from natural or petroleum gas or by electrolysis of water, but fuel cells that could use common fuels such as gas or petroleum products directly would be preferable. Prototypes of such fuel cells exist, but they are not economically practical today.* If inexpensive catalysts can be found, fuel cells may become very practical, clean energy sources. Since they operate at low temperatures they produce no nitrogen oxides. Their only products are harmless gases such as CO_2 and H_2O vapor. Someday we may also have fuel cells in our homes to supply electricity by the reaction of natural gas or other petroleum products with air, thus removing the inefficiency of electrical supplies from steam-electric plants, which dump half to two-thirds of their heat energy into rivers as thermal pollution.

14–8 OIL SPILLS

Another serious environmental problem associated with petroleum is oil spillage. Some of the more spectacular oil

---

*   Terri Aaronson, "The Black Box," *Environment* **13** (10), 10–18 (1971).

The 1969 oil leak in the Santa Barbara Channel. In this photo, the oil has spread to an area 1500 feet in diameter. Later it spread to, and polluted, nearby shores. (Courtesy of the Environmental Protection Agency.)

spills were the shipwreck of the oil tanker *Torrey Canyon* off the southwestern coast of England in 1967, the leak of an offshore oil well in the Santa Barbara Channel in 1969, and the fire and gusher from an offshore platform in the Gulf of Mexico off Louisiana in 1970.†

Large spills make headlines, but they represent only a small fraction of the petroleum released to the seas. Estimates of the annual worldwide spillage range from one to ten million metric tons (118,000 tons were carried by *Torrey Canyon*).‡ Most of the oil is spilled in the minor tanker col-

---

†   W. R. Young, "Possible Solutions to Oil Spillage, A Growing Problem," *Smithsonian* **1** (8), 18–27 (1970).

‡   Max Blumer, "Submarine Seeps: Are They a Major Source of Open Ocean Oil Pollution," *Science* **176**, 1257–1258 (1972).

lisions that occur almost daily, losses in transfer of oil between tankers and port facilities, and deliberate flushing of the tanks at sea. After delivery of oil, tankers often have no cargo for the return trip. The tankers would ride high in the water and be rather unstable, so they frequently pump water into the tanks to act as ballast. Just before re-entering port, the water is flushed into the sea, along with any oil in the tanks. Although this procedure is generally illegal many ships continue to do it. Unfortunately, most oil spillage occurs in coastal areas, which have the highest productivity of aquatic life in the oceans.

Most oil components are less dense than water and insoluble in it. Thus the oil speads out in a thin "slick" on top of the water. Some of the lighter petroleum fractions evaporate quickly. If high winds agitate the surface, emulsions of oil in water are formed (like vinegar-and-oil salad dressing). Aside from the toxicity of some oil components, an oil slick may have more subtle effects: oil prevents exchange of oxygen and other gases between water and the air, and the slick may severely reduce sunlight penetration of the water.

In a large near-shore spill many marine organisms are killed by the toxic components of the oil. Oil on the feathers of birds makes them unable to fly and swim properly, so that they die. Beaches are coated with smelly, tarlike oil residues. These are the obvious short-range effects. What are the long-term effects? There are conflicting reports. A few months after some of the major spills, marine biologists have examined the flora and fauna of the affected areas and reported that "things are getting back to normal." These optimistic reports may be misleading. Few areas have been well enough studied *before* the incident that the investigators know what the *normal* population of the various species should be. Also, after organisms in the area have been killed, organisms from areas not affected may move in, suggesting to the investigators that the area has suffered no ill effects.

One of the most carefully studied incidents was a minor spill of 600 tons of fuel oil in Buzzards Bay, Massachusetts, in September 1969. Scientists at nearby Woods Hole Oceanographic Institution made careful studies of the effects of the spill over a two-year period, with some rather distressing findings.*

---

* M. Blumer and J. Sass, "Oil Pollution: Persistence and Degradation of Spilled Fuel Oil," *Science* **176**, 1120–1122 (1972); see also Lawrence Meyer, "Oil Spills Heavily Damage Sea Life," *Washington Post*, July 24, 1970.

Thousands of water birds are killed each year by oil spills in Narragansett Bay, Rhode Island. (Courtesy of the Federal Water Pollution Control Administration, U.S. Dept. of the Interior.)

1. Oil concentrations in the bottom sediments decreased noticeably over the two years, but were still well above normal.

2. By eight months after the spill, pollution had spread to 5000 acres of bottom waters.

3. Fish, crabs, and shellfish were killed in the area, and in the most heavily polluted areas of Wild Harbor River, almost no animals survived.

4. By nine months after the spill, many areas had not been repopulated, but by two years many familiar species had begun to reappear.

5. Although some species cleanse themselves of oil when moved to clean waters, shellfish do not. They may retain so much oil as to be unfit to eat because of their oily taste.

6. Some of the oil components are carcinogenic and, after incorporation into the bodies of aquatic species, may become concentrated in species of higher trophic levels (as some pesticides do; see Section 13–12).

Thus even a minor spill can have serious effects on aquatic life over a sizable area for many months or several years.

What are the long-term global effects of oil pollution of the oceans? In many remote areas of the ocean, marine biologists frequently find globs of tarlike material (containing toxic hydrocarbons) in their nets when trying to collect small aquatic animals. The long-term effects of these ubiquitous globs of oil and the petroleum residues deposited in bottom sediments and in the bodies of organisms are not known.

What can be done to alleviate the damage caused by oil spills? The most important thing is to prevent their occurrence. More stringent regulations must be devised for the drilling and operation of offshore oil wells and the transportation of petroleum.

Some of the most active exploration and development of new petroleum deposits are taking place in offshore areas. The development of the Alaskan oil field presents special hazards because of the fragile environment created by permafrost on the land, and ice in the Arctic seas.

Once an oil spill has occurred, the most important steps are early detection of the spill, shutting off the source of the spill if possible, and confining the oil to a small area. Every minute of delay in confining the oil may increase the cleanup cost by hundreds of thousands of dollars. Confinement measures now in use are not adequate. Various types of physical barriers ("booms") or chemicals can confine slicks on calm waters, but most accidents occur on rough seas, in which these techniques are generally ineffective.

Cleanup measures are almost as ineffective. After the more volatile hydrocarbons have evaporated, it's almost impossible to burn the oil because of the cooling effect of the underlying water. In the *Torrey Canyon* incident, large amounts of detergent were thrown into the slick to disperse the oil. Detergents do disperse the oil, making it appear that the mess has been cleaned up, but by keeping the oil droplets suspended in the water for a long time, the detergents may do more long-range harm than good. Also, the detergents themselves are toxic to many microorganisms. In some cases dense particles such as chalk dust have been dropped on the slicks. The particles absorb oil and sink to the bottom. The crudest, but in some cases most effective treatment has been to spread straw on the slick. The straw absorbs the oil and is then gathered up and burned or buried. This produces some air pollution, but at least the oil is removed from the water.

One imaginative approach is the culturing of species of microorganisms that are capable of biodegrading the petroleum.* The microorganisms, along with supplies of essential nutrients, can be dropped into the slick and allowed to consume the oil. This promising approach should be studied carefully and developed for widespread use if successful.

We must begin to take more effective measures to prevent and clean up damage from oil spills, as the world use and transport of petroleum is growing and, with it, the threat of more extensive oil pollution in the seas. In its time, *Torrey Canyon* (with a capacity of 118,000 metric tons) was a large tanker, but now tankers twice that size are not uncommon and tankers of 500,000 or 1,000,000 tons may be in use in the future.†

SUGGESTED READING

L. C. Bliss, "Why We Must Plan Now to Protect the Arctic," *Science and Public Affairs* **26** (8), 34–38 (1970).

D. H. M. Bowen, "Los Angeles: The Uphill Fight against Photochemical Smog," *Environmental Science and Technology* **5**, 394 (1971).

D. H. M. Bowen, "The Drive to Control Auto Emissions," *Environmental Science and Technology* **5**, 492–495 (1971).

P. P. Craig and E. Berlin, "The Air of Poverty," *Environment* **13** (5), 2–9 (1971).

B. F. Greek, "Gasoline," *Chemical and Engineering News*, Nov. 9, 1970, pp. 52–60.

---

* "Microorganisms Consume Oil in Test Spills," *Chemical and Engineering News*, Sept. 7, 1970, pp. 48–49.
† Julian McCaull, "The Black Tide," *Environment* **11** (9), 2–16 (1969).

A. J. Haagen-Smit, "Air Conservation," *Science* **128**, 869–878 (1958).

J. B. Heywood, "How Clean a Car," *Technology Review* **73** (8), 21–29 (1971).

M. H. Hyman, "Timetable for Lead," *Environment* **13** (5), 14–23 (1971).

D. B. Luten, "The Economic Geography of Energy," *Scientific American* **224** (3), 164–175 (March 1971).

M. Malin and C. Lewicke, "Pollution-Free Power for the Automobile," *Environmental Science and Technology* **6**, 512–517 (1972).

J. McCaull, "The Black Tide," *Environment* **11** (9), 2–16 (1969).

C. M. Summers, "The Conversion of Energy," *Scientific American* **224** (3), 148–160 (March 1971).

U.S. Dept. of Commerce, "Automotive Fuels and Air Pollution," (U.S. Government Printing Office, Washington, D.C., (1971).

U.S. Dept. of Health, Education, and Welfare, National Air Pollution Control Administration, *Air Quality Criteria for Carbon Monoxide*, AP–62, March 1970; *Air Quality Criteria for Hydrocarbons*, AP–64, March 1970; *Air Quality Criteria for Photochemical Oxidants*, AP–63, March 1970 (U.S. Government Printing Office, Washington, D.C., 1970).

QUESTIONS AND PROBLEMS

1. Briefly define or explain the following terms:

    a) straight-run gasoline   b) "cat cracker"

    c) carburetor   d) compression ratio

    e) octane number   f) air/fuel ratio

    g) photochemical oxidant   h) olefin

    i) TEL   j) net propulsion efficiency

    k) catalytic converter   l) diesel engine

2. What are the major constituents of

    a) bottled or LP gas

    b) natural gas

    c) kerosene

3. a) Why is it necessary to add volatile components to gasoline in winter?

    b) What problems do these components often cause on a hot spring day?

4. What is the percentage of lead by weight in tetraethyl lead?

5. a) Write a balanced equation for the combustion of heptane, C_7H_{16}.

    b) How many moles of O_2 are required for complete oxidation of each mole of heptane?

    c) What is the weight ratio of O_2 to heptane for the oxidation?

    d) The air/fuel ratio is computed in terms of weight. Compute the air/fuel ratio needed for combustion of heptane in an engine. (*Hint*: Air is about 21% O_2 by volume and has an average molecular weight of 29.)

6. Suppose your carburetor is adjusted properly at sea level, but you are planning a vacation to the mountains that will involve driving mostly at altitudes of several thousand feet. In order to maintain proper engine performance with the same gasoline, should you adjust the carburetor to a richer or leaner mixture? Explain.

7. Suppose a sample of gasoline gives the same performance in a test engine as a mixture of *n*-heptane and iso-octane in which the heptane content is 12%.

    a) What is the octane rating of the gasoline?

    b) About how much Pb would one have to add per gallon to produce "regular" gasoline?

8. a) Why is a fuel cell–electric motor car theoretically more efficient than an ICE?

    b) To conserve energy resources, why would we be better off using natural gas fuel cells to produce electricity right in the home than obtaining it from steam-electric plants?

    c) If the ICE is only 25% efficient, what happens to its "thermal pollution"; that is, where does the wasted heat go?

9. Which of the following measures decrease emissions of hydrocarbons and CO from internal combustion engines?

    a) richer air/fuel mixture

b) recirculation of some exhaust gases back into the cylinders

c) thermal reactor or "after burner"

d) catalytic oxidizing converter

e) catalytic reducing converter (explain your answer to this one).

10. Why does Los Angeles have so much more smog than most other cities?

11. a) What is the drawback of replacing lead additives with aromatic hydrocarbons in gasolines, for today's automobiles?

b) Why will it be safe to do so in post-1974 cars?

12. Let's see if we can make a reasonable estimate of the atmospheric lead concentration caused by a city's automobiles. Consider the city of 10 million people described in Problem 15 of Chapter 10. Let's assume the conditions given in part (a) of that problem. Furthermore, assume that there are about 3 million automobiles, each of which goes about 10 miles per day, using one-half gallon of gasoline containing 2 g Pb/gal.

a) If all the Pb in the gasoline used is emitted to the atmosphere and remains suspended there, what is the atmospheric lead concentration (in $\mu g/m^3$) at the end of a 24-hr period of stagnant air?

b) Now allow a 5 km/hr wind to blow across the city as in part (b) of Problem 10–15. What is the equilibrium concentration of Pb in the atmosphere?

c) Compare your answer in part (b) above with the typical atmospheric Pb concentrations found in U.S. cities. Is it reasonable to assume that most atmospheric lead originates from automobiles?

d) Your calculations throughout this problem were based on certain assumptions. Which of these assumptions are most subject to error, thus affecting the conclusion you reached in part (a)?

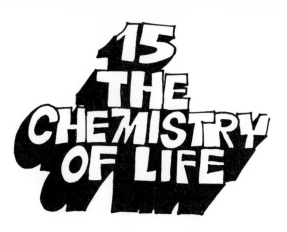

15 THE CHEMISTRY OF LIFE

Francis Crick (r) and James D. Watson (l) with model of DNA. (Courtesy of James D. Watson.)

15–1 INTRODUCTION

The chemistry of life, or "biochemistry" as chemists call it, is an area in which the classical fields of chemistry and biology meet. It can equally well be called "molecular biology." Biochemistry is the study of the structures and reactions of the thousands of compounds involved in life processes. Biochemistry is the most complex area of chemistry. You may have thought that the processing of iron ores described in Chapter 12 and the refining of petroleum discussed in Chapter 14 were complicated processes, but not even the most complex chemical factory devised by man approaches the chemical complexity of one of the cells of your body. Living things represent the most efficient, sophisticated, compact chemical "factories" ever known. How, for example, does the body of a teenager take in its daily quota of hamburgers, french fries, and malted milks, break down these foods into simpler chemical compounds, distribute these compounds throughout the body where they are rearranged and joined together to form new structures needed for maintenance and growth of the body, and then collect the debris from the body for elimination? How do cells of the body know when to divide and multiply into new cells having the same characteristics as the original cell? When the body is afflicted by disease or a wound, how does the body protect itself and repair the damage? These processes involve thousands of different chemical compounds.

When we compare the nervous system to man-made electronic computers, the efficiency and complexity of the biological systems become even more impressive. Despite great advances in computer technology, the greatest computer ever built is almost insignificant by comparison to a human brain weighing little more than a kilogram. A computer can perform mathematical operations millions of times faster than a person, but think of some of the things the nervous system can do: it can cause your arm to reach out smoothly and touch an object, no small feat if you've ever seen a mechanical robot move! The brain can translate signals from the retina of the eyes into three-dimensional color images. It can translate a series of frequencies detected by the ear into thoughts, whereas a computer can only perform operations for which it has been programmed by a person. When it comes to storage capacity, the brain really wins out. The largest computers have storage capacities of about one million "words" (e.g., numbers or instructions), but some experts believe that the brain stores up all the signals it receives.

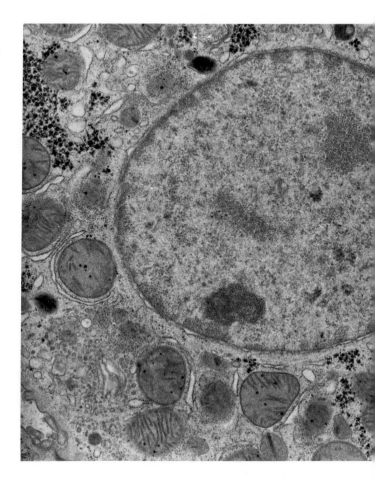

The chemical processes of our bodies involve enormously complex sequences of reactions. Not surprisingly, the details of these processes are far from completely understood. Nevertheless, great progress has been made in our understanding of biochemical structures and reactions, so that we now have a rough outline of the processes that occur in the body. The first half of this century might well be termed the "Golden Age of Physics," because so many breakthroughs in understanding the structure of molecules, atoms, and nuclei were made. By the same virtue, we may now be in the midst of a golden age of biochemistry. The next few years of research

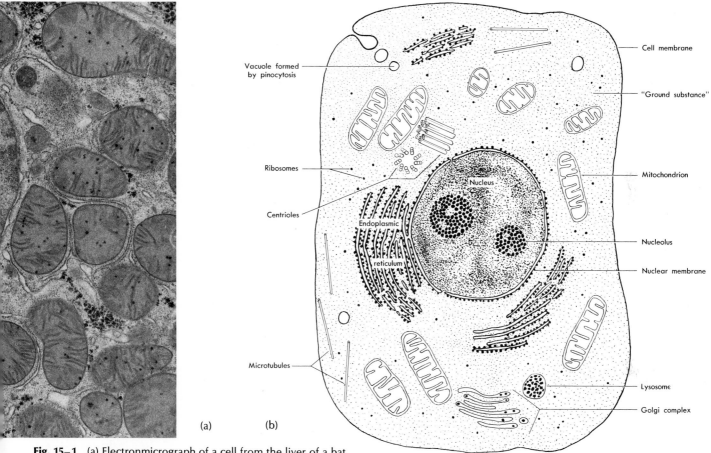

Fig. 15–1 (a) Electronmicrograph of a cell from the liver of a bat. Magnification about 22,000. (Courtesy of Keith Porter.) (b) Major features of a typical animal or human cell.

may bring such increased understanding of chemical processes in cells that we will be able to devise cures for diseases such as cancer, sickle-cell anemia, and multiple sclerosis.

15–2 CELLS

All living matter is made up largely of small units called *cells*. Cells vary greatly in size, from diameters less than 1 μm (micrometer) in some bacteria up to the 75 mm diameter of an ostrich egg! In humans, cells typically are about 100 μm across. Figure 15–1 shows that cells are surrounded by a

membrane, and the matter inside (*protoplasm*) is generally made up of two parts: a dense *nucleus* near the center of the cell and the remaining material, which is called *cytoplasm*. Present in the cytoplasm are various subcellular units collectively referred to as *organelles*, including the *mitochondria*, *ribosomes*, *lysosomes*, and *endoplasmic reticulum*.

Most activities of the cell are thought to be controlled by the nucleus. The nucleus contains little blobs of material called *chromatin*, whose name derives from its ability to take up the color of dyes added to the cell. When a cell is getting ready to divide, the chromatin comes together to form rod-

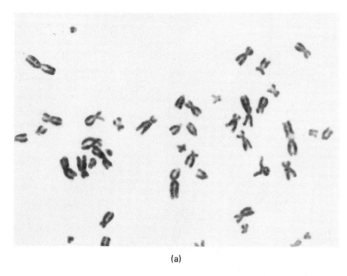

(a)

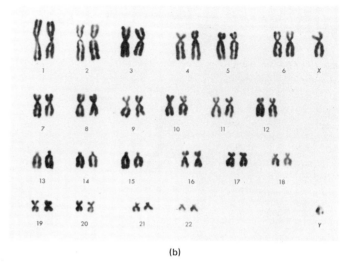

(b)

Fig. 15–2   (a) The chromosomes of a human cell as they are arranged shortly before dividing. [Courtesy of Drs. T. T. Puck and J. H. Ijis.] (b) Homologous pairs of chromosomes (photo prepared by cutting a photo into pieces containing one chromosome each and then rearranging the pieces). [Courtesy of Dr. James L. German, III.]

Nerve cells from the spinal cord of an ox. (From John W. Kimball, *Biology*, Reading, Mass., Addison-Wesley, 1965, p. 344.)

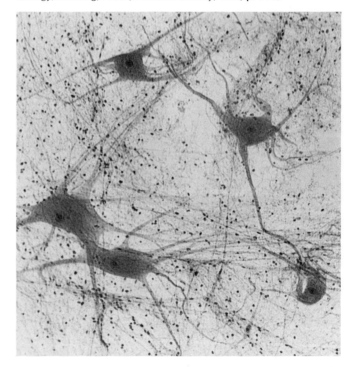

like structures called *chromosomes*. Normal human cells contain 46 chromosomes that line up as 23 pairs of chromosomes prior to division. Then each chromosome divides into two, one for each of the new cells (see Fig. 15–2). Each chromosome, which is about 10 μm long in humans, contains a large number of *genes*. The genes carry all the genetic information about the individual they inhabit—the "blueprint" that indicates how tall he should grow, what color his eyes are, whether his nose is shaped like his mother's or father's, etc. The genetic information is "encoded" in the structure of enormous molecules of *deoxyribose nucleic acid*, or DNA for short, that is present in the genes.

There are great differences in cells from one species to another and from one part of the body to another in a particular species. Cells of bacteria or yeast are completely self-sufficient and can exist as individual units if provided with nutrients. Most cells in our bodies have very specialized functions and properties and thus can exist only in a community with other types of cells. Some cells, such as the red blood cells of our bodies, have no nuclei and thus cannot divide. New cells have to be manufactured, as, for example, the red blood cells are in the marrow of our bones. Brain cells do not divide—a baby is born with all the brain cells he will ever have. Human sex cells, the sperm of the male or the egg of the female, have only 23 chromosomes, so that the fertilized egg receives half of its 46 from each parent. Unlike animal cells, some plant cells contain chlorophyll and other compounds that allow them to perform photosynthesis.

15–3 CLASSES OF BIOCHEMICAL COMPOUNDS

Most biochemical molecules are much larger than any we have previously considered except for the polymers discussed in Chapter 13. In fact, many biochemicals are naturally occurring polymers built up as long chains of simple molecular units. Many have molecular weights in the tens of thousands and values as high as one billion are not unknown. The essential element here, as in organic chemistry, is carbon, which has the unique property of forming long chains of like atoms bonded together to form the "backbone" of these large molecules.

Carbohydrates derive their name from the fact that some of their empirical formulas can be expressed as an equal number of carbon atoms and water molecules as, for example, in the case of the sugar glucose, $C_6H_{12}O_6$, which could be written as $C_6(H_2O)_6$. However, these compounds are not hydrates of carbon and they contain no water molecules as such. The basic units of carbohydrates are simple sugars containing three to six carbon atoms. Many sugars, including common table sugar (sucrose), consist of two six-carbon sugars linked together. The cellulose and starch that are important structural components of plant cells are large polymeric structures consisting of glucose sugar units. Glucose is important as a source of energy in animal bodies. It circulates in the blood and is stored up in long-chain molecules of glycogen.

Lipids are compounds that are soluble in nonpolar solvents such as ether and chloroform, but not in water. The most abundant lipids are the *triglyceride fats*, esters formed between a glycerol molecule and three molecules of long-chain fatty acids, as described in Section 13–8. Fats have very high energy content per unit weight. When our bodies take in more food energy than needed, the excess energy is mostly stored as fat. Other important lipids are cholesterol and some key hormones.

Proteins are polymers composed of amino acid units (Section 13–10). There is an infinite variety of different types of protein molecules that can be made up from different arrangements of the twenty-odd amino acids that occur in biological systems. Proteins are essential components of all biological materials. About 20% of the weight of the blood and muscles consists of proteins, and about 7% of the brain. Hair and silk are almost pure proteins. Many of the important classes of biological substances discussed below are simply special types of proteins.

Enzymes are proteins that are highly specific catalysts for biochemical reactions. There is a particular enzyme for nearly every type of reaction that occurs in the body. At the low temperature of the body, nearly all reactions between biochemical molecules would proceed very slowly unless catalyzed by enzymes. Thus, in general, only those reactions occur for which there is a corresponding enzyme. The control of the body's chemistry, therefore, is largely dependent on the amounts of the various enzymes present.

Vitamins have no particular type of chemical structure or a specific function as a group. They are simply substances required in very small amounts but which cannot be manufactured in the body. Plants are self-sufficient if supplied with water, sunlight, CO_2, and the essential chemical elements, because they synthesize all the chemicals for their growth. On the other hand, animals must eat plants (or other animals that have consumed plants) to obtain chemical energy and the basic building blocks (e.g., sugars, amino acids) needed for synthesis of more complex biochemicals in the body. As the various animals species evolved, they often found it inefficient to incorporate into their bodies mechanisms for the synthesis of chemicals needed in very small amounts that were readily available in their diets. Thus man has come to depend on his food for the vitamins he requires. If a person has a reasonably well-balanced diet, there is usually no difficulty in obtaining vitamins, but if his diet is limited in some way, he may suffer various vitamin-deficiency diseases such as beriberi, pellagra, or scurvy. Some vitamins are *coenzymes*, compounds that must be present for certain enzymes to function.

Hormones are chemicals secreted by the body's glands for the chemical control of various body functions or for the digestion of food. For example, the hormone *adrenalin*, which is released into the blood when a person is faced with a sudden fear or emergency, initiates many reactions throughout the body to give it the unusual amounts of energy needed to react to the emergency. Many sex functions are regulated by the release of hormones. Some hormones are proteins and others are steroids (a special class of lipids).

You may have noticed that we have not attempted to make any distinction between living and nonliving substances on a molecular scale. Common sense tells us that a brick is dead and a growing tree is alive, but can we make similar distinctions between molecules? In the early 1800s it was thought that only biological species, imbued with a "vital force," could synthesize organic compounds (see Section 13–1). That idea was discarded when Wöhler synthesized urea from inorganic compounds in 1824. Today, as far as can be determined, there is no fundamental difference between the molecules of living systems and those of nonliving materials. The molecules of biochemistry are generally very large and complex, but these enormous molecules appear to obey the same physical laws that apply to simpler molecules.

Even when we consider entities larger than molecules, it is difficult to make a clear definition that will separate the living from the nonliving. It is tempting to base such a definition on the ability of living species to *replicate* themselves, e.g., some cells divide into two new cells identical to the original. But how do we distinguish between inorganic salts, which always form crystals of a particular shape under a given set of conditions, and which are clearly nonliving, and *viruses*, which we might consider to be alive? Outside the cells of other living species, viruses appear to be quite "dead." Although they consist of nucleic acids surrounded by proteins, they are unable to reproduce themselves or perform other operations typical of live species. However, when they invade a cell, they use the material of the host cell to make new "copies" of themselves. Typically about half an hour after a virus enters a cell, the cell bursts open to release about two hundred new virus particles.

Many common diseases (e.g., measles, mumps, polio, influenza, and perhaps the common cold and cancer) are caused by viruses that multiply in the body until they produce conditions that are the symptoms of the disease. The presence of the invading viruses causes the body to form *antibodies*, biochemical species that kill the viruses and bring the disease to a conclusion. In many cases, the antibodies are retained permanently in the body, preventing a person from having the disease again (e.g., mumps, measles). In other cases, such as a cold, the antibodies give only temporary immunization. Vaccination for a disease causes one to have the disease in a very mild form, but sufficient to produce the antibodies that give one immunity from it.

15–4 CARBOHYDRATES

Carbohydrates are the simplest of the large, polymeric biochemical molecules. The building blocks, or *monomers*, of the carbohydrates are the simple sugars, also called *monosaccharides*. The monosaccharides have the general formula $C_x(H_2O)_x$, where x can have values from 3 to 7, but 5 and 6 are the most common. The main functional group of a sugar molecule is the carbonyl group (C=O), making the molecule an aldehyde or a ketone, depending on its location in the molecule. As the names of sugars have the suffix "ose," these two classes of sugars are called the *aldoses* and the *ketoses*. The remaining bonds of the carbon atoms are taken up by hydrogen atoms or OH groups. Examples of two simple three-carbon sugars ("trioses") are the following.*

D-glyceraldehyde
an aldose

dihydroxy-acetone
a ketose

The most common monosaccharides in nature are the pentoses and hexoses, like those shown in Fig. 15–3. Fructose, the sweetest monosaccharide, is found predominantly in fruits. Galactose appears in milk as a part of the double sugar lactose, whose name is derived from the Greek root for milk. Glucose is the most common monosaccharide, being present in the blood, muscles, and certain organs, as well as serving as the monomer for the polymers cellulose, starch, and glycogen. The pentose, D-ribose, is an important part of the structure of nucleic acids. Under most circumstances the pentoses and hexoses exist as ring structures formed by an oxygen "bridge," as shown in Fig. 15–4.

Monosaccharides exist in nature in small amounts. For example, human blood contains about 0.8 mg of glucose per ml. Cells obtain energy from glucose by oxidizing it to CO_2 and H_2O. When a person is too sick to take food orally, he is fed intravenously with a solution of glucose in water.

---

* The "D" of D-glyceraldehyde indicates that the OH of the carbon atom next to the carbonyl group (the middle carbon in this case) is on the right-hand side. The corresponding left-handed molecule is called the L form. The D and L structures of a particular compound are mirror images of each other. One form cannot be converted to the other by rotations about the single bonds. This is not obvious from the flat structures written on paper, but you can see this easily by constructing three-dimensional molecular models. The naturally occurring sugars on Earth are nearly all of the D form.

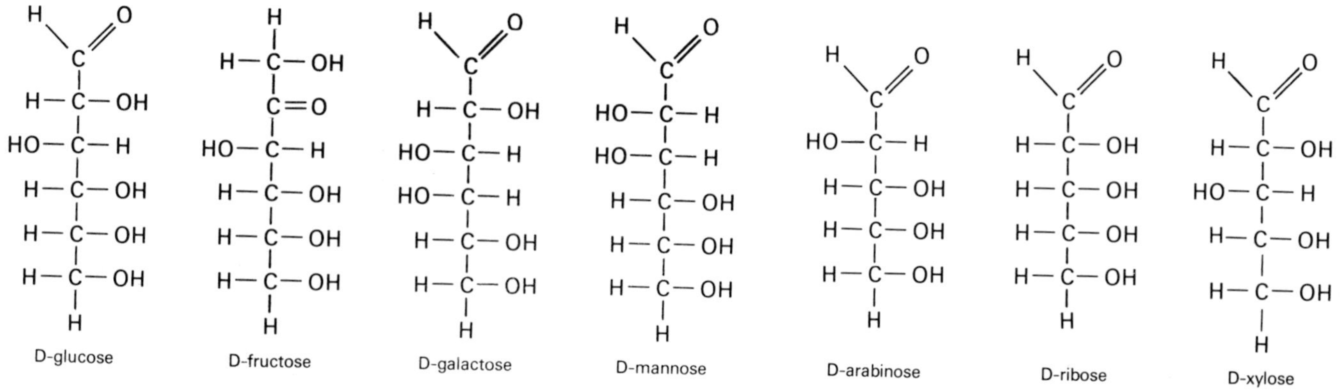

Fig. 15–3 Some representative pentoses and hexones.

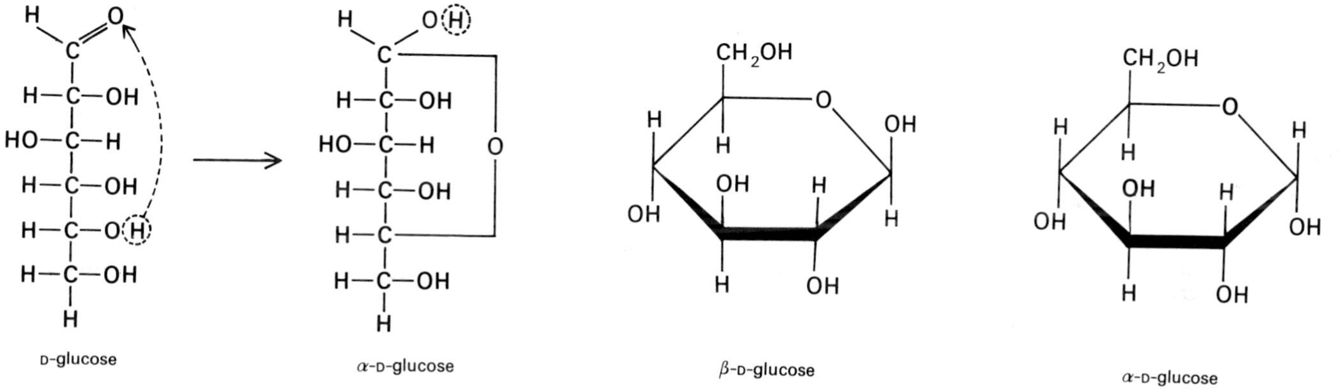

Fig. 15–4 An open chain sugar can form a ring structure via an oxygen bridge. The α and β indicate whether the OH group on the carbon next to oxygen in the ring is below or above the ring.

Sugars often occur as disaccharides, i.e., two monosaccharides linked together by an oxygen bridge. Figure 15–5 shows how a water molecule is "split out" to form the oxygen bridge when glucose and fructose combine to make common table sugar, sucrose. The structures of two other common disaccharides, maltose and lactose, are shown in Fig. 15–6. Maltose, an intermediate product in the breakdown of starch, is found in sprouting barley, from which the "malt" of malted milks is obtained. Lactose occurs in milk and is not particularly sweet.

The monosaccharides, particularly glucose, are stored in the bodies of plants and animals in the form of enormous polymers called *polysaccharides*, which may have molecular weights between 40,000 and five million amu. The principal polysaccharides are starch, cellulose, and glycogen, which are made up entirely of glucose units. Starch, which occurs in two forms, *amylose* and *amylopectin* (see Fig. 15–7), is a major constituent of most people's diets as it is abundant in cereal grains. It is a valuable nutrient because its glucose units contain much of the energy the plants receive from the sun and utilize by photosynthesis.

Glycogen, which has a structure similar to that of amylopectin, is the storage form of glucose in animals, mainly in muscles and the liver. Fats have higher energy content

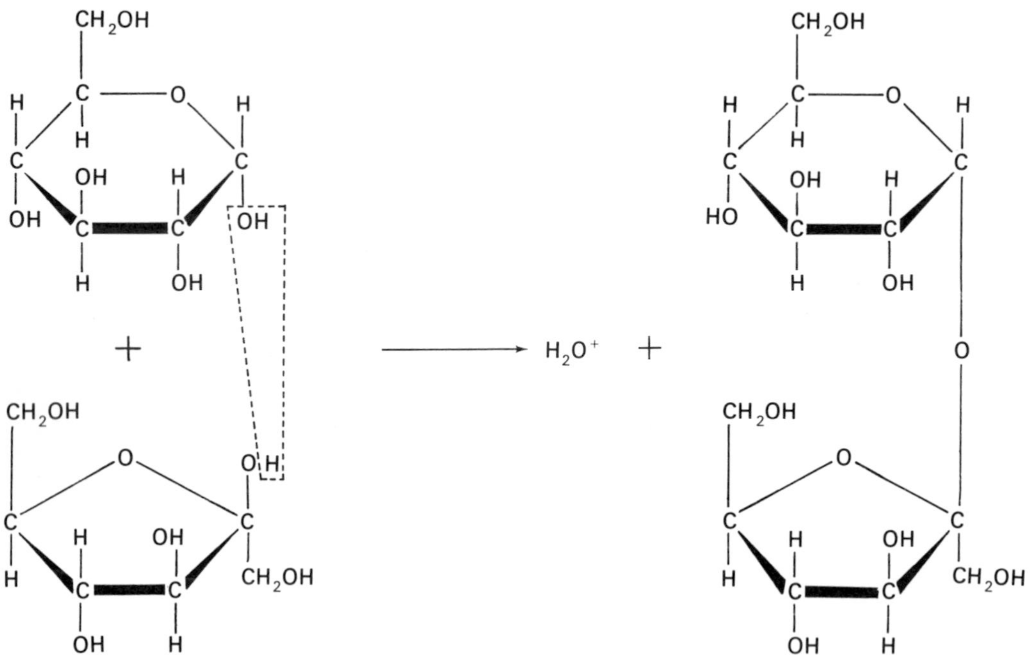

Fructose

Sucrose

Fig. 15–5 Sucrose is formed by the joining of glucose and fructose molecules, with the "splitting out" of water.

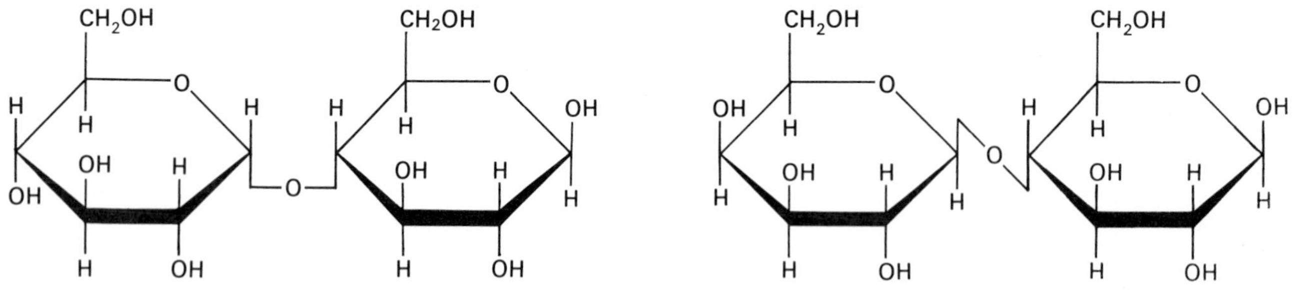

β-maltose

β-lactose

Fig. 15–6 Maltose and lactose.

per unit weight and are thus a more efficient form of energy storage. However, the glycogen can be broken down and converted to energy more quickly than fats. The two forms of energy storage are somewhat analogous to a checking account versus a savings account.

Cellulose forms the supporting structure (or "bones") of many plants. As shown in Fig. 15–8 the oxygen bridges between the glucose units are different from those of glycogen and starch. Enzymes of the digestive systems of most animals, including man, are unable to break the bonds of

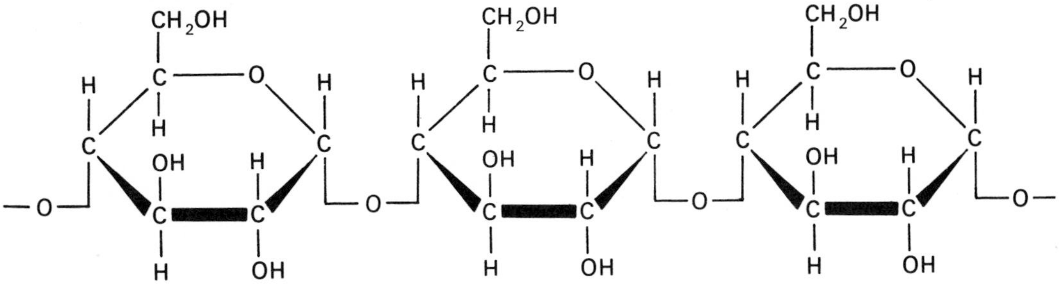

Amylose

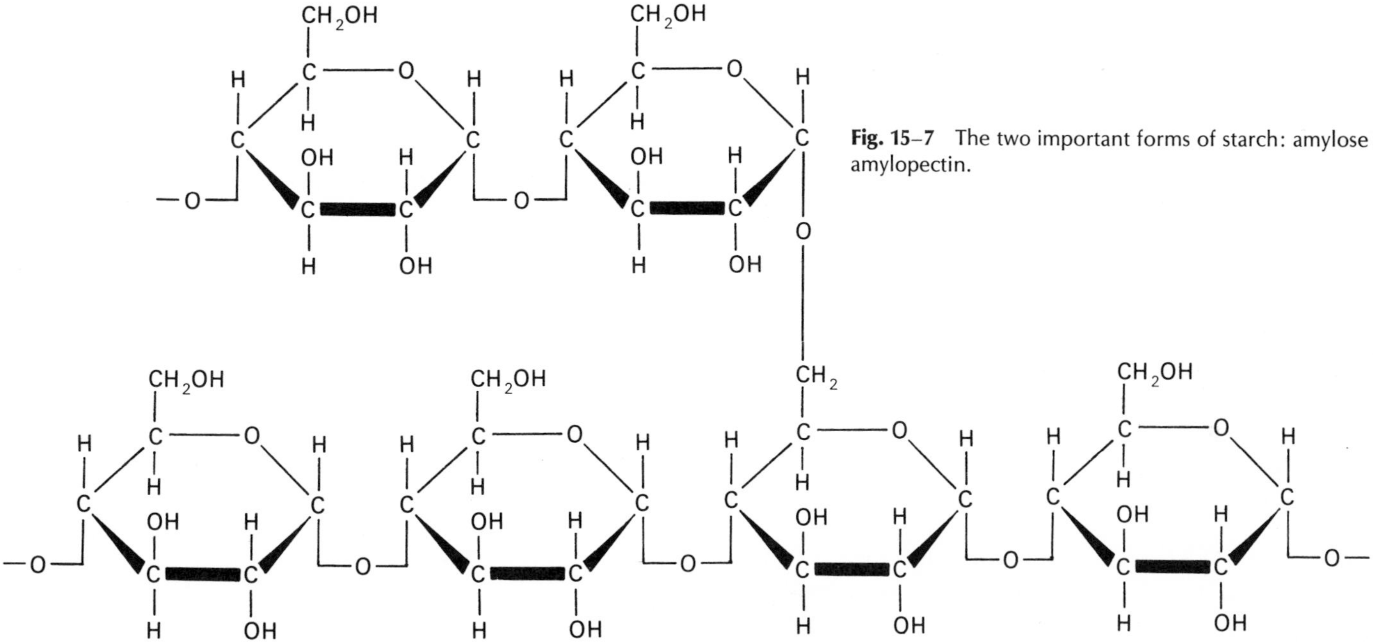

Fig. 15-7 The two important forms of starch: amylose and amylopectin.

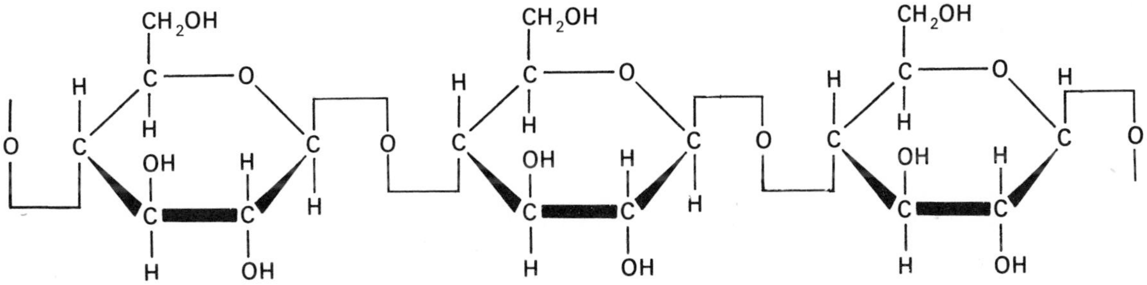

Amylopectin

Fig. 15-8 Cellulose.

Cellulose

cellulose; thus those species are unable to obtain direct nutritional value from cellulose. Ruminant animals (cows, sheep) and termites have microorganisms in their digestive tracts that are able to break down cellulose, allowing these species to utilize its glucose. The major sources of cellulose are wood, paper made from wood pulp, and the fibrous parts of plants. Foods such as celery, which consist mainly of cellulose, are considered "roughage" because of their limited food value for humans.

Carbohydrates (wheat) for human consumption in an early stage of processing. (Courtesy of the U.S. Food and Drug Administration.)

15–5 LIPIDS

The term *lipids* is often used interchangeably with "fats," but the term applies to all biological molecules that are soluble in nonpolar solvents such as ether or chloroform and insoluble in water. The lipids include waxes, steroids, terpenes, and phosphatides, among others.

As discussed in Section 13–8, animal fats and plant oils are esters, usually *triglycerides*, made by the reaction of three long-chain fatty acid molecules with a molecule of glycerol (see Fig. 15–9). The fatty acids prominent in the triglycerides of various plant and animal materials are listed in Table 13–2.

The major function of fats is the efficient storage of energy: the energy content of fats is about 9 cal/g versus 4 cal/g for carbohydrates and proteins. (Think how much heavier a person would be if he stored all excess energy as carbohydrates rather than fats!) Thin layers of fat just beneath the skin also insulate the body against extremes of temperature.

The transport of insoluble fats and oils in the aqueous fluids of the body is managed with the help of *phospholipids* (see Fig. 15–10). Since their molecules have both polar and nonpolar features, they keep globules of fatty acids suspended in water in the same way that detergents "float" particles of grease from dirty dishes (see Section 13–8). If the energy of the fatty acids is needed somewhere in the body, it is released when the fatty acids are oxidized to CO_2 and H_2O. Otherwise, the fatty acids and glycerol may be reassembled and stored as fatty tissue.

Another class of lipids is the *steroids*. The basic unit of the steroids is the fused ring structure shown in Fig. 15–11. Many biochemical molecules contain this basic structure. One is *cholesterol*, which for reasons that are not clear has a vital function in the nervous system. One-tenth of the solid

$$
\begin{array}{c}
\text{H} \\
| \\
\text{H}-\text{C}-\text{O}-\overset{\overset{\displaystyle \text{O}}{\|}}{\text{C}}-\text{C}_{17}\text{H}_{35} \\
| \\
\text{H}-\text{C}-\text{O}-\overset{\overset{\displaystyle \text{O}}{\|}}{\text{C}}-\text{C}_{17}\text{H}_{35} \\
| \\
\text{H}-\text{C}-\text{O}-\overset{\overset{\displaystyle \text{O}}{\|}}{\text{C}}-\text{C}_{17}\text{H}_{35} \\
| \\
\text{H}
\end{array}
$$

Tristearin
a fat

Fig. 15–9 A typical triglyceride (fat) molecule.

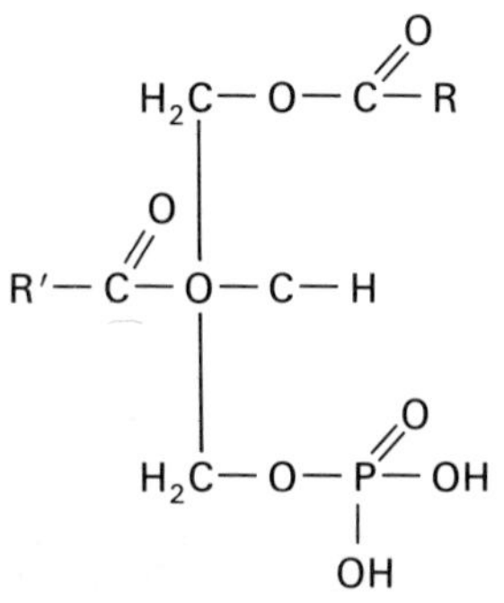

phosphatidic acid

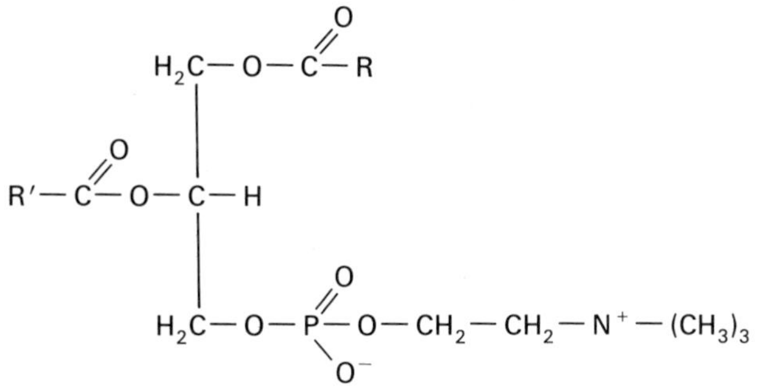

a lecithin

Fig. 15–10 Examples of phospholipids.

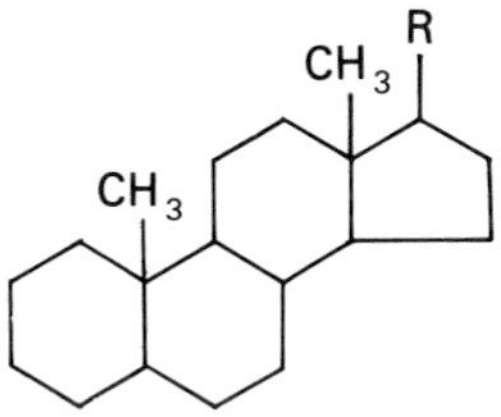

Steroid nucleus

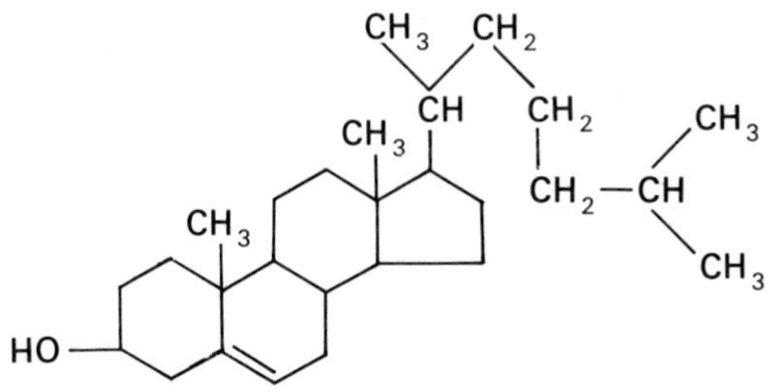

cholesterol

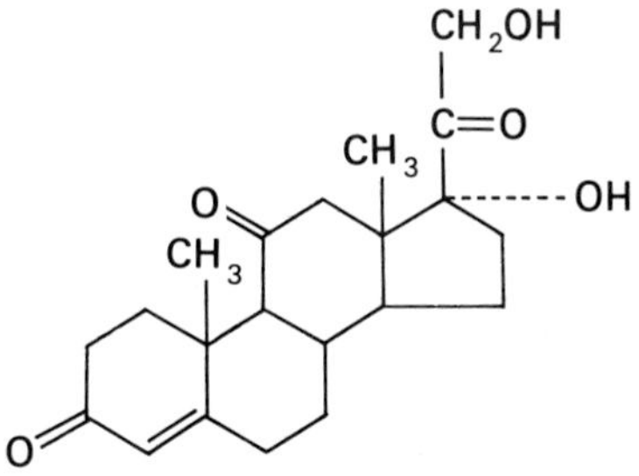

cortisone

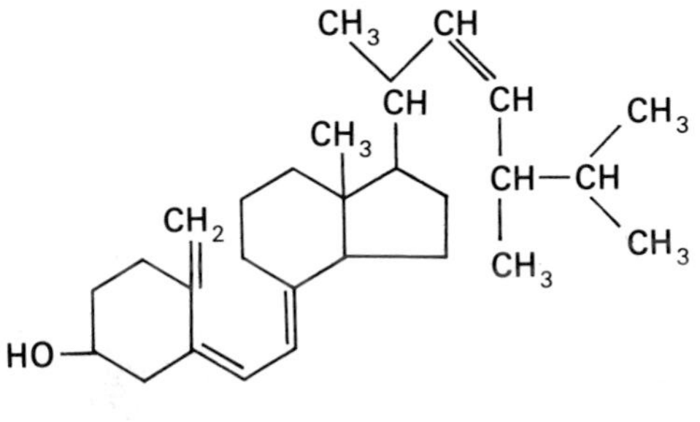

vitamin D$_2$

Fig. 15–11 Some important steroid compounds: the steroid "nucleus" along with cholesterol, cortisone, and one of several forms of Vitamin D.

β Carotene

Vitamin A

Fig. 15–12 Beta carotene and vitamin A, which is derived from β-carotene.

matter of the brain is cholesterol. It is also secreted in the bile from the liver. Excess cholesterol in the body sometimes forms stones in the gall bladder. Also, deposits of cholesterol in the arteries cause "hardening of the arteries." Since there is cholesterol in eggs and animal fat, some doctors suggest that one should limit one's intake of these sources of lipids. However, at present, opinion seems to be that the cholesterol synthesized in the body itself is the more critical factor.

Cortisone is a hormone used as a medicine for various conditions. It gives relief to many sufferers of arthritis, a disease in which the joints become swollen and painful to move. For a while it was considered a wonder drug, but it must be used with caution because of undesirable side effects such as increased susceptibility to tuberculosis and infections.

Many sex hormones contain the steroid nucleus (see Section 15–11 below). Vitamin D has a steriodlike structure, but with a break in one of the four rings. The terpenes discussed in Section 14–4 are examples of plant lipids. A closely related class of compounds is the *carotenes*, orange pigments present in some green plants and vegetables such as carrots.

Bacon fat is one form in which human beings consume lipids. ▶ (Courtesy of the U.S. Dept. of Agriculture.)

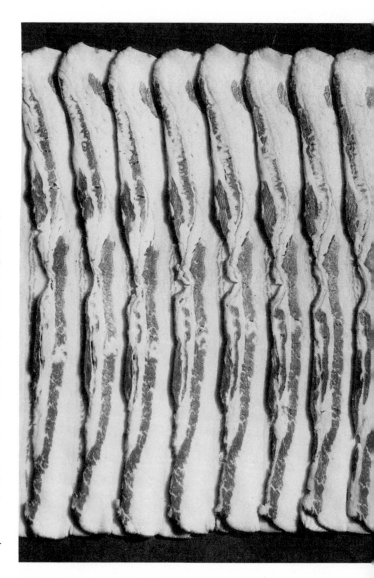

During the summer, the green chlorophyll in leaves covers up the color of the carotenes, but after a frost, the green color disappears, leaving the red and yellow colors of autumn leaves. When β-carotene is split apart and an OH group added to the end of its chain, vitamin A is formed (see Fig. 15–12).

15–6 PROTEINS

Proteins are probably the most important class of biochemicals. Certainly they are an essential component of our diets. We have seen that lipids and carbohydrates are good sources of energy, but if they are in short supply, the body can metabolize proteins in ways that release energy. However, that is not normally their major function. There are no other substances that can take the place of proteins.

Like cellulose, starch, and glycogen, the proteins are long-chain polymers. In the former, the monomers are all glucose molecules, but proteins are made up of about 20 different kinds of amino acid molecules. Thus, whereas all glycogen molecules are pretty much the same, there is an infinite variety of proteins, with widely varying properties that result from combining the amino acids in various ways. Glycogen's only major function is to store energy, but the wide range of proteins fill a broad range of functions. All enzymes and about half the hormones are special types of proteins.

Structures of some of the amino acids are shown in Fig. 15–13. Table 15–1 lists 20 amino acids commonly found in the body; note that ten of them are listed as "essential." All amino acids in the bodies of animals are derived directly or indirec-

America's favorite form of protein on the hoof. (Courtesy of the U.S. Dept. of Agriculture.)

tly from plants. The proteins consumed are broken down (metabolized) into amino acids and some types of amino acids can be synthesized from others. But in most animals there are some amino acids that cannot be synthesized in the body, although all twenty-odd of them are necessary. Human bodies are unable to synthesize the ten essential acids, so these must be present in the foods we eat.

Table 15–1 Amino Acids Commonly Found in Proteins of the Body

Essential		Nonessential	
Name	Abbreviation	Name	Abbreviation
Arginine	Arg	Alanine	Ala
Histidine	His	Asparagine	Asn
Isoleucine	Ileu	Aspartic acid	Asp
Leucine	Leu	Cysteine	Cys
Methionine	Met	Glutamic acid	Glu
Phenylalanine	Phe	Glutamine	Gln
Threonine	Thr	Glycine	Gly
Tryptophan	Try	Proline	Pro
Valine	Val	Serine	Ser
Lysine	Lys	Tyrosine	Tyr

Aliphatic Amino Acids

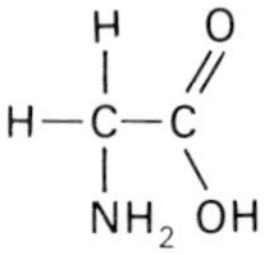

glycine

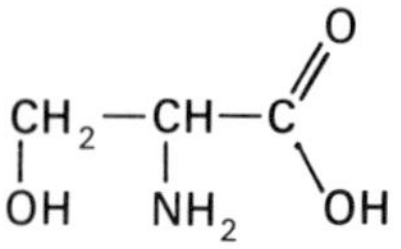

serine

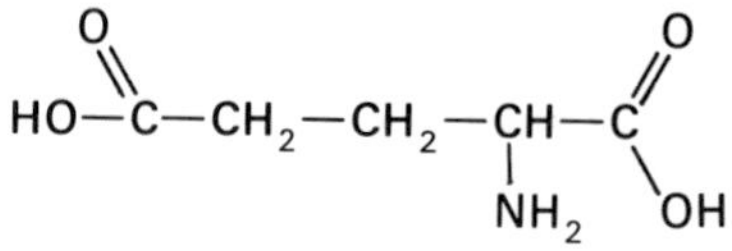

glutamic acid

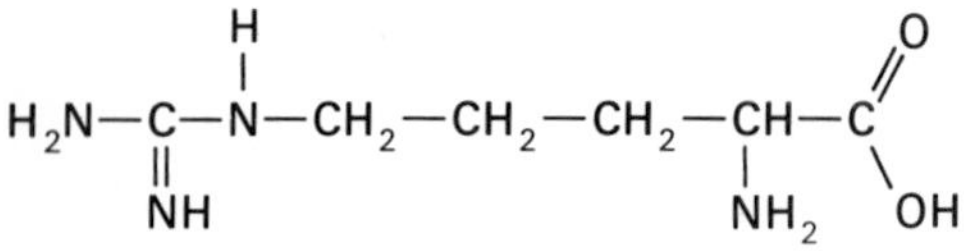

arginine

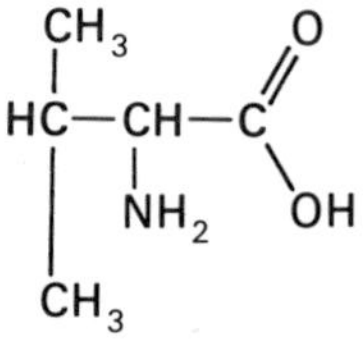

valine

Sulfur-containing Amino Acids

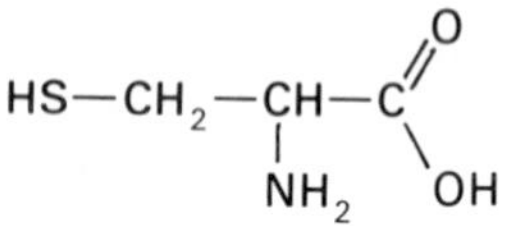

cysteine

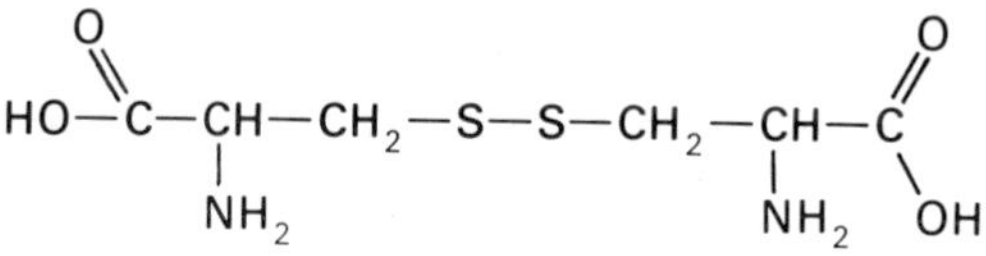

cystine

Aromatic Amino Acids

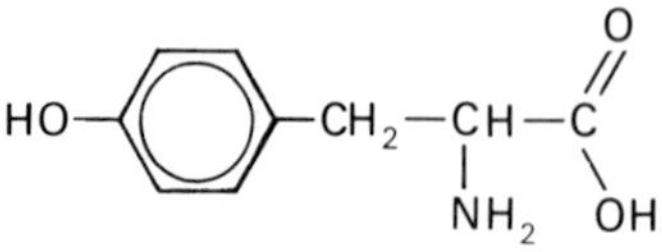

tyrosine

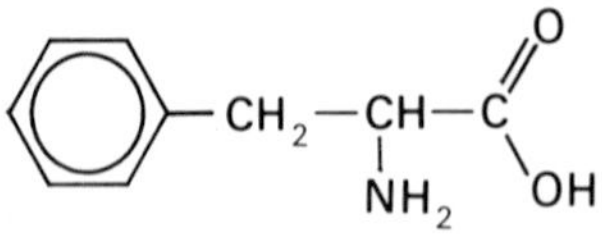

phenylalanine

Heterocyclic Amino Acids

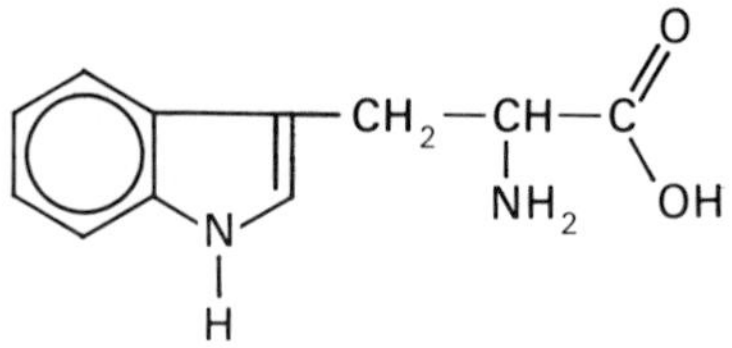

tryptophan

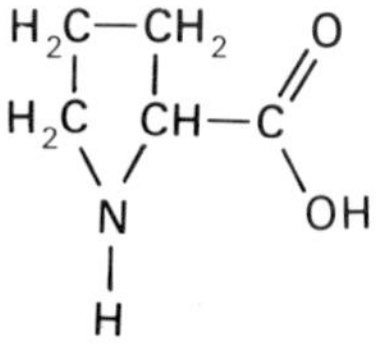

proline

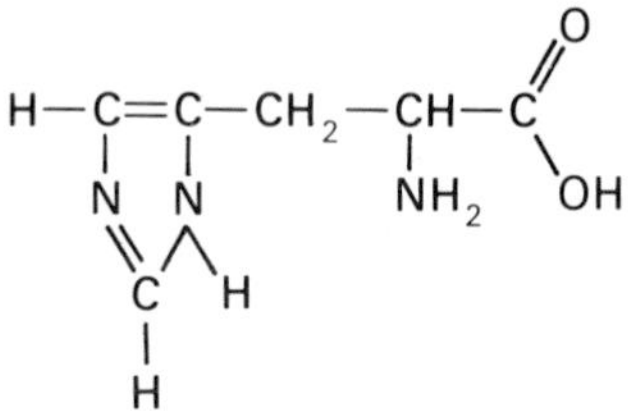

histidine

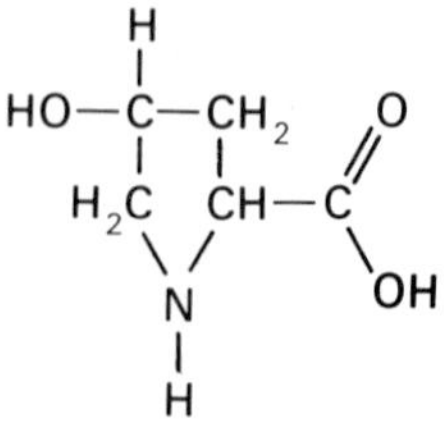

hydroxyproline

Fig. 15–13 Several representative amino acid structures.

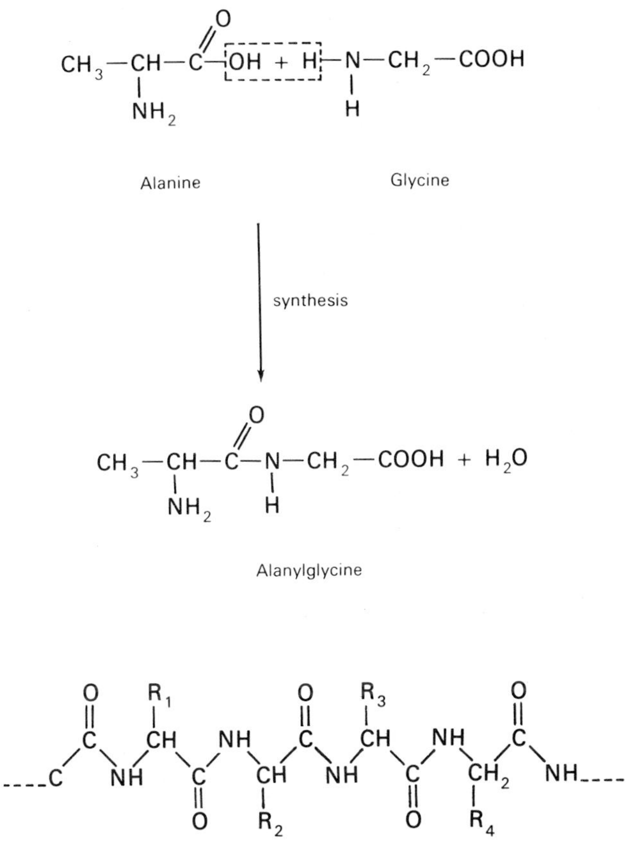

Alanine   Glycine

synthesis

Alanylglycine

A typical peptide

Fig. 15–14 Various amino acids are linked together by the splitting out of water to form peptides (proteins).

A person's diet must supply not only the total amount of protein needed, but also the required amounts of each of the essential amino acids in the protein. Ordinary corn, a major staple for many people around the world, is generally low in lysine and tryptophan, but efforts are being made to correct this deficiency by introducing high-lysine corn (see Section 11–4).

Proteins are built up when amino acids are joined together by the splitting out of water molecules as shown in Fig. 15–14. When several amino acids are joined together, the structure is referred to as a *polypeptide*. The bond formed

between the amino acid units, as in Fig. 15–14, is called the *peptide bond*. Proteins are simply very large polypeptides. Molecular weights of proteins cover a range from a few thousand amu up to about four million amu. The proteins of egg albumin, with molecular weights of about 35,000 amu, are a fairly common size. Proteins are made up almost entirely of only five elements, in the following proportions.

Element	% Weight in protein
Carbon	53
Hydrogen	7
Oxygen	23
Nitrogen	16
Sulfur	1

In *fibrous* proteins, the long chains of amino acids tangle around each other in a rather random way. These proteins are strong, tough, and flexible. The collagen of tendons and cartilage is a fibrous protein. So is the keratin of skin, nails, hooves, feathers, and hair.

The more common proteins belong to the class called *globular* proteins, in which the amino acids are linked together not only by the peptide bonds, but also by hydrogen bonds. In 1951, Linus Pauling made one of the most brilliant breakthroughs in modern science when he determined the structure of protein molecules.* Starting with the chain of amino acid units linked together by peptide bonds, as in Fig. 15–14, he sought to construct a three-dimensional model of proteins. In doing so, he assumed that the lengths of the various bonds and the angles between bonds would be the same as those of simple molecules containing the same elements. For example, a single bond between two carbon atoms has a length of about 1.54 Å in most compounds. When four atoms are attached to a carbon atom by single bonds, the angles between bonds are about 109°. Structures in which all bond angles and lengths have their usual values are much more stable than those in which the angles or lengths are forced to have unusual values, such as the 60° angle of the −C−C−C− bonds in cyclopropane. Pauling assumed that molecules as common as proteins would have structures that would accommodate all atoms comfortably and thus be of high stability.

---

*   L. Pauling, R. B. Corey, and R. Hayward, "The Structure of Protein Molecules," *Scientific American* **191** (1), 51–59 (July 1954).

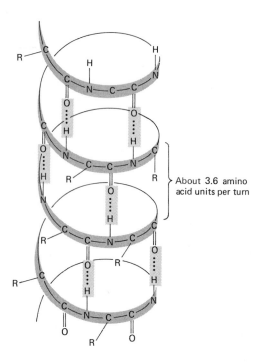

About 3.6 amino acid units per turn

Fig. 15–15 The α-helical structure of most proteins as discovered by Linus Pauling.

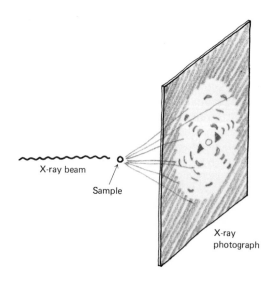

X-ray beam

Sample

X-ray photograph

Given the assumed lengths and angles of all the bonds, Pauling had to solve a three-dimensional "jigsaw puzzle." In solving it he also looked for any features that would give additional stability to the structure, which turned out to be hydrogen bonds (see Section 8–3). The protein structure deduced by Pauling is shown in Fig. 15–15. The atoms of the repeating sequence $-N-C-C-N-C-C-N-$, etc., along the backbone of the polypeptide chain form a helix. Not only does this structure have all the correct bond lengths and angles, but it has the added stability of hydrogen bonds between the $=O$ unit of one amino acid group and a hydrogen atom attached to a nitrogen atom of the amino acid one "turn" above or below on the helix. Also, the structure is quite general, since the $R-$ groups which are different for the various amino acids stick out of the skeleton and don't usually have much effect on the structure.

Despite the logic and beauty of the Pauling model, it would be of no value if it did not agree with the various experimental observations that have been made on proteins. Pauling used many such observations in constructing his model, and predictions based on it have been confirmed by many experiments performed since the model was proposed. For example, one can determine some features of the structure by allowing a beam of x rays to strike a sample of protein. The intensity of x rays scattered at various angles to the beam are recorded on a photographic film placed near the sample (see Fig. 15–16). The scattered x rays produce a pattern of intense spots on the film at points where there is constructive interference between the waves of x rays scattered from different points on the molecule (see Section 7–1). By a mathematical analysis of the pattern of dots one can determine certain structural features of the molecule.

In addition to the peptide bonds and hydrogen bonds, protein structures may also be held together by bonds or attractions between the "R" parts of the amino acids sticking out from the helical backbone. Several of the amino acids contain sulfur and, in some proteins, a disulfide link $(-S-S-)$ is formed between the sulfur atoms of two amino acid units, as in the structure of insulin shown in Fig. 15–18. The protein strands of hair contain a lot of the amino acid cysteine, which has an $-SH$ group at the end of the $R-$ group.

◀ **Fig. 15–16** Information about the structures of proteins and other biochemical molecules is obtained from the locations of spots on a photographic film when x rays are scattered by the sample.

1 2 3 4 5 6
Two strands of
protein connected by
disulfide bonds
S S S S S S
S S S S S S
Break
bonds
1 2 3 4 5 6

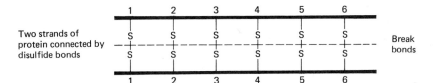

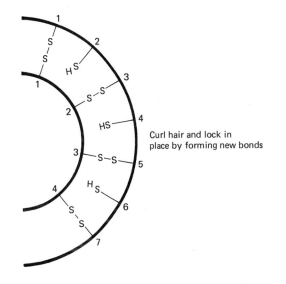

Fig. 15–17 In the "permanent" curling of hair, disulfide bonds between protein strands are broken, the curl forces one strand to slide relative to the other, and the curl is locked in place by the new disulfide bonds.

Curl hair and lock in
place by forming new bonds

Disulfide bonds between separate protein strands in the hair strengthen its structure. The first step of a "home permanent" involves application of compounds that break the disulfide links. After the protein strands have been displaced to give curl to the hair, another lotion is applied to reform the bonds and "lock in" the curl (see Fig. 15–17).

Although protein structures are enormously complex, the sequences of amino acids along the chains and the three-dimensional structures of several proteins have been determined by sophisticated analytical and physical measurements. One well established protein structure is that of the hormone *insulin*. We see in Fig. 15–18 that the insulin structure consists of two chains of 21 and 30 amino acids joined together by two disulfide links.

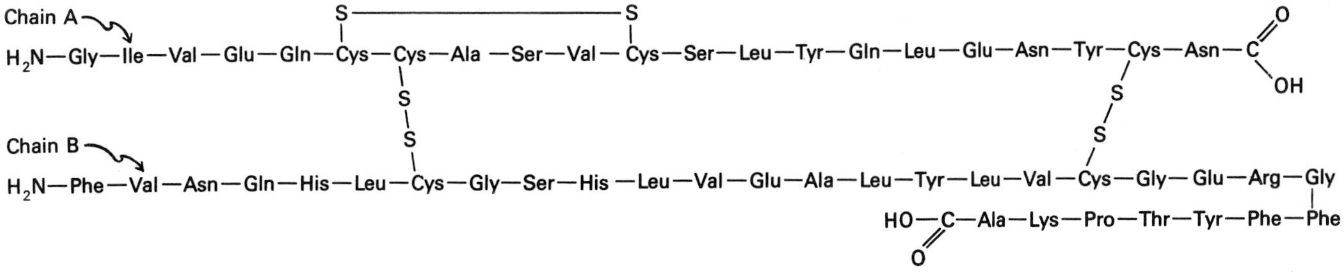

Chain A
H2N—Gly—Ile—Val—Glu—Gln—Cys—Cys—Ala—Ser—Val—Cys—Ser—Leu—Tyr—Gln—Leu—Glu—Asn—Tyr—Cys—Asn—C
O
OH
Chain B
H2N—Phe—Val—Asn—Gln—His—Leu—Cys—Gly—Ser—His—Leu—Val—Glu—Ala—Leu—Tyr—Leu—Val—Cys—Gly—Glu—Arg—Gly
HO—C—Ala—Lys—Pro—Thr—Tyr—Phe—Phe
O

Fig. 15–18 The protein insulin, a hormone.

Fig. 15–19   Heme, the iron-containing portion of hemoglobin.

Heme ~ Globin ~ Heme
Heme   Heme

Hemoglobin

Heme

Fig. 15–20   Model of myglobin, closely related to hemoglobin, showing the heme units "tucked into" the surrounding protein chains. (b) In this drawing of the myoglobin molecule, the dark portion is the heme group. [From Arthur L. Williams, Harland D. Embree, and Harold J. DeBey, *Introduction to Chemistry* (Addison-Wesley, Reading, Mass., 1968), p. 481.]

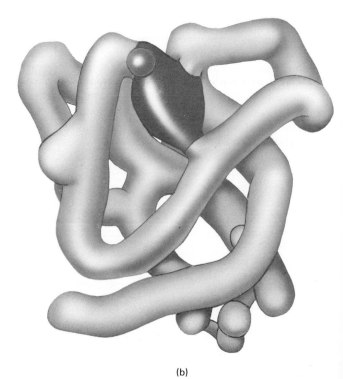

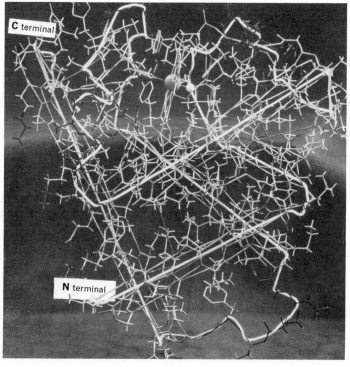

(a)

(b)

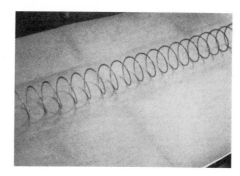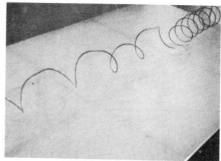

Fig. 15–21 The denaturation of proteins is similar to the stretching of the toy "slinky" beyond its elastic limit. [From Arthur L. Williams, Harland D. Embree, and Harold J. DeBey. *Introduction to Chemistry* (Addison-Wesley, Reading, Mass., 1968), p. 378.]

The helices of globular proteins generally don't continue indefinitely in a straight line, but bend at various points along the chain. Bends often occur when a proline or hydroxyproline occurs in the chain. In these amino acids the amine groups and the carbon to which it is attached are members of a five-membered ring. The ring structure disrupts the helix and causes a bend. In the hemoglobin molecule, for example, which has a molecular weight of 68,000, each of the four heme groups is tucked into one of four protein chains of the molecule (see Fig. 15–19). Each chain has about 150 amino acid units, with some helical parts in between the bends.* One of the protein chains, with its heme group, is similar to the models of myoglobin shown in Fig. 15–20.

The structure of proteins is altered by chemical reactions. Heat or treatment with mild chemicals causes *denaturation*, a term used to describe changes short of rupture of the peptide bond. The boiling of an egg causes the albumin protein to be denatured. The denaturation of a protein is similar to the overstretching of a "slinky" to the point where it cannot return to its original structure (see Fig. 15–21). In the case of the coiled protein, the hydrogen bonds that hold the helix together are stretched beyond the breaking point.

15–7 SICKLE-CELL ANEMIA: A MOLECULAR DISEASE

The properties and behavior of various proteins depend sensitively on the order of the amino acids in their chains. Nevertheless, one might suppose that an occasional mistake in the order of amino acids in molecules having hundreds of units wouldn't cause much difference in the protein. However, there are tragic examples of just such molecular mis-

takes. One of the best known involves the hemoglobin of red blood cells. As noted in Section 14–4, oxygen is attached to hemoglobin for transport throughout the body by the bloodstream. Persons suffering from the hereditary disease *sickle-cell anemia* have an error in their genetic code that causes a mistake in one of the 150 amino acids of their hemoglobin protein chains: a valine amino acid unit is inserted in place of glutamic acid (see structures in Fig. 15–13). After giving up its oxygen, this abnormal form of hemoglobin (designated HbS) is much less water-soluble than deoxygenated normal hemoglobin, HbA.† This change of solubility causes HbS to crystallize, distorting the wall of the red blood cells. The blood cells so affected transform from the normal disk shape to the sickle shape that gives the disease its name (see Fig. 15–22).

The sickled cells don't perform their proper oxygen-carrying role. Also, they cause "log jams" in the blood vessels. These effects cause the person great pain in the abdomen and the joints. The decreased oxygen-carrying capacity causes chronic anemia, weakness and drowsiness.‡ Most sufferers die at an early age.

Although scientifically the disease is well understood, there is a great deal of public confusion about sickle-cell anemia. Since it is a genetic disease, its incidence is determined by the genetic code passed on to the person by his parents. For each genetic trait, the child receives one gene from the father and one from the mother. If both genes are the same, the child will have the given trait. But what if the genes are different?

---

* M. F. Perutz, "The Hemoglobin Molecule," *Scientific American* **211** (5), 64–76 (Nov. 1964).

† H. J. DeBey, *Introduction to the Chemistry of Life: Biochemistry* (Addison-Wesley, Reading, Mass., 1968), Ch. 11.
‡ See the four-part series by Victor Cohn on the "Sickle-Cell Controversy," Washington *Post*, Nov. 1972, 12–15.

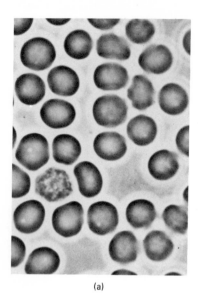

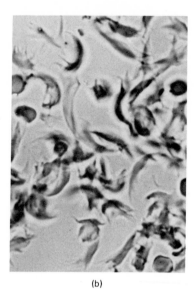

(a)   (b)

Fig. 15–22   (a) Normal red blood cells and (b) sickle cells as viewed with a microscope. (Photograph courtesy of Dr. A. C. Allison.)

	Father	Mother
Parents' genes	MM	MM
Vision	M	M
Children's genes	All MM	
Vision of children	All normal	

Father   Mother

MM   Mm

M   M

MM   Mm   MM   Mm

All normal

Fig. 15–23   Inheritance of genes for normal vision and extreme myopia from parents having various combinations of vision genes. M = normal vision, m = myopia.

Parents' genes	Mm	Mm
Vision	M	M
Children's genes	MM Mm mM mm	
Vision of children	Normal Myopic	
	75% 25%	

Family of four children

No. of normal children	No. of myopic children	Likelihood %
0	4	0.4
1	3	4.7
2	2	21.2
3	1	42.1
4	0	31.7

Parents' genes	Mm	mm
Vision	M	m
Children's genes	Mm Mm mm mm	
Vision of children	Normal Myopic	
	50% 50%	

Father   Mother

mm   mm

m   m

All mm

All myopic

According to classical genetics, there are two types of genes, dominant and recessive. When there is a conflict between genes the dominant one determines the trait. Consider the hereditary disease of extreme myopia (nearsightedness). It is a recessive trait, whereas normal vision is dominant. A child with one gene for normal vision and one for myopia would have normal vision. He would have myopia only if both genes were of the recessive myopic type.

The genes obtained by the child are determined by mathematical probability. There is an equal chance of getting either of the two genes from the father and either of the two from the mother. Figure 15–23 shows some typical applications of genetics to the prediction of normal vision and myopia in children. The dominant normal-vision genes are labeled M and the recessive myopia genes by m. When the parents have the combinations Mm-Mm and Mm-mm, some of the children are likely to have myopia and some normal vision. For Mm-Mm, 75% of the children are expected to have normal vision and 25% myopia. That doesn't mean that all such families with four children will include three children with normal vision and one with myopia. That's the most likely result, but the probability of obtaining exactly that result is only 42%. There is a small (0.4%) but finite probability that all of the children will have myopia.

Sickle-cell anemia, like myopia, is carried by a recessive gene. Although it occasionally occurs in white people, here in the Western Hemisphere sickle-cell anemia is primarily a disease of blacks. An estimated 8 to 12% of American Negroes carry the recessive sickle-cell trait. This causes little difficulty unless both parents carry the HbS gene. In that case the statistics are just like those worked out in Fig. 15–23, for an Mm-Mm couple. In a family of four there is a 42% chance that one child will have the HbS-HbS gene combination that will cause him to have sickle-cell anemia. There is a 68% probability that one or more of the four will have it and only a 32% chance that none will. About one out of every 400 black children born in this country has sickle-cell anemia.

There is a great deal of confusion between "sickle-cell anemia" and "sickle-cell trait" in the public mind. Carriers of the recessive HbS gene can be identified by a simple blood test. Mass blood-testing programs have been performed to identify carriers in order to counsel them about the chances of having children with the disease if they marry carriers. However, the information is sometimes not made clear and carriers may fear that they have the disease. Just as tragically, some institutions and companies discriminate against the carriers as if they had sickle-cell anemia: some insurance companies charge higher premiums and prospective employers occasionally feel they would be high-risk employees.* The fact is that most carriers lead completely normal lives. There is not a complete dominance of HbA genes over HbS genes, so carriers have some sickle-cells. But generally they would produce no observable effect on the person except in an extreme emergency situation where one needed every bit of oxygen-carrying capacity, e.g., in a sudden loss of pressure of a high-flying aircraft.

Since sickle-cell anemia is a very serious condition and its sufferers don't often have children, one might wonder why the trait has not died out. First, most children of sickle-cell carriers do not have the disease but many are carriers of the trait. Second, there is a correlation between the sickle-cell gene and resistance to malaria such that carriers have unusually high resistance to malaria. Thus, evolution has selectively enhanced the number of carriers in areas where malaria is a serious, prevalent disease. In certain areas of Africa today, an estimated 45% of the people are carriers.† Although our understanding of sickle-cell anemia is quite good, there is no known cure or prevention for it, except genetic counseling of potential parents. Recently it has been found that sufferers can obtain temporary relief during the sickle-cell crisis typical of the disease by injections of urea or certain other nitrogen compounds.

Sickle-cell anemia is not the only genetic disease that is prevalent primarily among members of a particular race or other population group. Another blood disease, thalassemia (or "Cooley's disease") mainly affects Greeks and Italians. Tay-Sachs disease, a degenerative disease of the nerves which dooms children to death within a few years, mostly affects Jews. Phenylketonuria, a metabolic disorder, has an unusually high incidence among the Irish. Genetic diseases are no respecters of social class. A famous example of a genetic disease is hemophilia, whose sufferers lack the clotting factor in the blood and may bleed to death from injuries that would be minor cuts for an ordinary person. Hemophilia was prevalent among members of the royal families of Europe who were descendents of Queen Victoria, who was a carrier of hemophilia. (Like color-blindness, hemophilia is a sex-linked genetic disease—although women can be carriers, only males commonly have the disease.

---

*   Victor Cohn, "Sickle-Cell Controversy."
†   H. J. DeBey, Introduction to the Chemistry of Life: Biochemistry (Addison-Wesley, Reading, Mass., 1968), Ch. 11.

Although there are no cures for the genetic diseases, a number of them, including Tay-Sachs disease and Down's syndrome (mongolism) can be detected in the fetus during early pregnancy by examining some of its cells in amniotic fluid withdrawn from the mother's womb.* These tests make it possible for couples who are carriers of such diseases to have normal children and avoid the risk of giving birth to an afflicted child, if they chose to abort the fetus in the case of a positive test. Methods for obtaining blood samples from the fetus will probably be developed soon, making possible the identification of many more diseases, including sickle-cell anemia.

15–8 ENZYMES

There are thousands of different chemical compounds in the body, each capable of reacting with many others. How does the body ensure that only the desirable reactions occur? The secret of control is largely found in the enzymes. Most reactions between ions in solution take place quite rapidly, but those involving non-ionic species, as most biochemicals are, generally occur much more slowly at the temperature of the body. Enzymes are catalysts that speed up the rates of biochemical reactions. Catalyzed reactions are so much faster than other reactions that generally only those reactions for which there is a catalyst occur to an appreciable extent. Thus, for nearly every type of reaction that occurs in the body, there is a corresponding catalyst.

Unlike metallic catalysts such as platinum and nickel, most enzymes are highly specific; that is, each one catalyzes only reactions of a certain type. Some break up starch molecules into polysaccharides, some only split disaccharides into monosaccharides. Some are so specific that they break up proteins only by attacking the link between two particular amino acid units. One highly specific enzyme is glutamate decarboxylase, which catalyzes the removal of one carboxylic acid group from glutamic acid:

The enzyme is so specific that it does not catalyze the same reaction of aspartic acid,

which differs from glutamic acid only by having one fewer CH_2 group in the chain.

Other enzymes are not so specific, and can be "fooled" into using molecules of slightly different structure than those for which they are designed. This "trick" is used in the control of certain bacterial diseases with sulfanilamide and similar "sulfa drugs." An enzyme in the bacteria catalyzes the synthesis of the vital compound folic acid, from starting materials that include para-aminobenzoic acid (PABA). As shown in Fig. 15–24, the sulfanilamide structure is very similar to that of PABA, and so the enzyme tries to use it in the synthesis of folic acid. However, sulfanilamide does not react to form folic acid and so the bacteria starve and die, or they are so weakened that the body's defenses can control them and bring about recovery from the disease. Sulfa drugs have been particularly effective in combating pneumonia.

Since our bodies also need folic acid, why don't sulfa drugs harm the body just as they do bacteria? The answer is that our bodies don't synthesize the folic acid—we must obtain it in our food; thus it is classified as a vitamin.†

How do enzymes catalyze reactions and what makes them so specific? The enzymes are proteins made up from one hundred to several thousand amino acid units. These protein molecules are constructed in particular shapes (held in place by strategically placed disulfide bonds and similar links) that are ideal for attachment to molecules involved

$$\underset{\text{glutamic acid}}{\text{HO}\diagup\overset{O}{\underset{\|}{C}}-CH_2-CH_2-\overset{H}{\underset{\underset{NH_2}{|}}{C}}-\overset{O}{\underset{\diagdown OH}{C}}} + H_2O \xrightarrow{\text{enzyme}} \text{HO}\diagup\overset{O}{\underset{\|}{C}}-CH_2-CH_2-\underset{\underset{NH_2}{|}}{CH_2} + H_2CO_3 \qquad (15–1)$$

*   "Genetic Manipulation—A Fresh Problem," *Technology Review* **75** (5), 60–61 (1973).

†   Some folic acid is synthesized by bacteria living in the intestines.

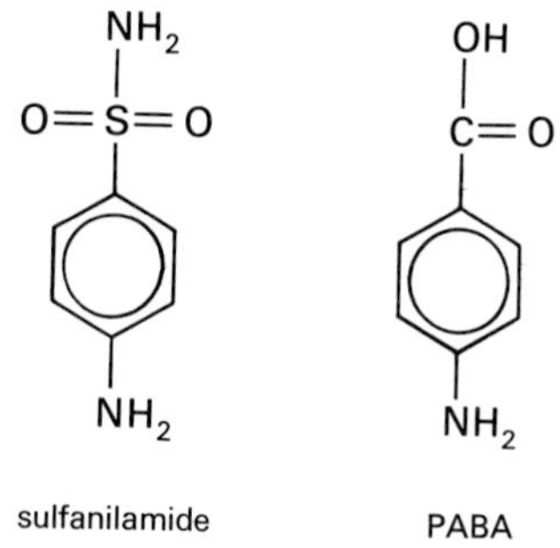

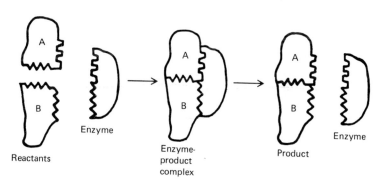

Fig. 15–24 Sulfamilamide (a Sulfa drug) has a structure so similar to that of part of the PABA molecule that enzymes attempt to use the former in synthesis of folic acid.

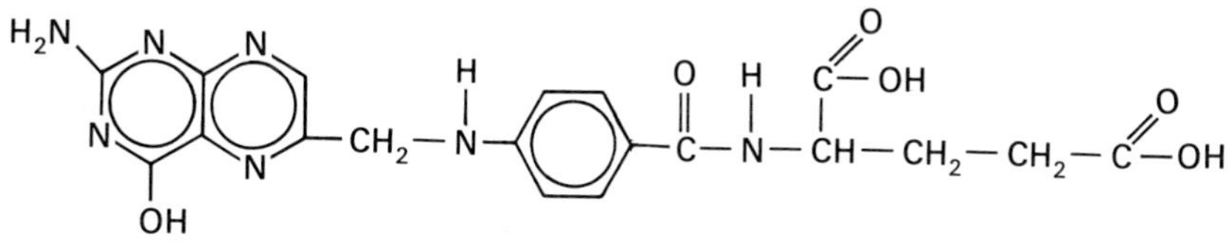

sulfanilamide   PABA

folic acid

Fig. 15–25 A schematic diagram of the way in which an enzyme catalyzes a reaction between molecules A and B. The enzyme holds the two molecules together in an optimum configuration for the reaction.

Reactants   Enzyme   Enzyme-product complex   Product   Enzyme

in the reaction to be catalyzed. Suppose an enzyme is to catalyze a reaction between species A and B to form AB molecules. As shown schematically in Fig. 15–25, the enzyme has particular slots designed to hold the A and B molecules close together in the right orientation for them to react with each other. When the enzyme encounters an A molecule, it may hold it in place until it encounters a B molecule. As soon as both are in place, they react quickly and leave the enzyme free to accept new A and B molecules.

Enzymes are more or less specific depending on how completely the A and B molecules fit into the slots. Apparently the enzyme that manufactures folic acid does not envelop PABA molecules completely enough to be able to distinguish between it and sulfanilamide. Despite their great complexity,

the three-dimensional structures of some enzymes have been determined.*

Like other catalysts, enzymes are not used up in the reactions. As soon as the products leave the enzyme molecule, it is available to work on additional reactants. Some enzyme molecules are able to induce reactions between several thousand pairs of reactant molecules per minute. Thus, although enzymes are extremely important regulators of chemical activity in the body, very small amounts are needed to accomplish the task. In a body weighing 70 kg, the total weight of each type of many important enzymes is

* D. C. Phillips, "The Three-Dimensional Structure of an Enzyme Molecule," *Scientific American* **215** (5), 78–90 (Nov. 1966).

probably less than one microgram! We noted in Chapter 12 that some of the more toxic elements are harmful in microgram amounts, suggesting that their mechanism of toxicity involves reactions with compounds that are present in the body in minute quantities. In many instances these highly toxic metals achieve their effects by interfering with the actions of enzymes, such as lead's interference in the manufacture of red blood cells (see Section 12–8).

The rates of enzyme-catalyzed reactions depend on a number of factors. Some can take place effectively only at pH levels that exist in certain regions of the body. The rates of nearly all reactions, including those catalyzed by enzymes, are sensitive to temperature. For most reactions, a temperature increase of about 10°C causes the rate to double. The rates of enzyme-catalyzed reactions, however, do not rise indefinitely with temperature. Since the enzymes are proteins, they can be denatured at about 45°C, thereby losing their effectiveness. We are able to preserve foods for long times in refrigerators and freezers because decomposition reactions become quite slow at low temperatures. Despite the marvelous properties of enzymes, they cannot make a reaction occur unless the reaction would occur spontaneously given sufficient time. That is, enzymes cannot supply the energy needed to drive reactions opposite to their natural direction.

Enzymes are named by adding the suffix-ase to the root of the name of the substrate, the substance upon which the enzyme acts, or the root of the word describing the reaction catalyzed. Enzymes that catalyze the breakup of proteins, carbohydrates and lipids are called proteases, carbohydrases, and lipases, respectively. The oxidoreductases catalyze oxidation and reduction reactions. Some important enzymes involved with metabolism of foods are known by non-systematic common names, e.g., pepsin, trypsin.

Many enzymes require the presence of a nonprotein compound in order to perform their catalytic activity. This other compound which "activates" the enzyme is called a coenzyme. Many of the vitamins are either themselves coenzymes or they react to form coenzymes. Many of the essential metals needed by the body in trace amounts, as discussed in Chapter 12, are part of coenzyme molecules.

15–9 NUCLEIC ACIDS: THE HEREDITARY MATERIAL OF CELLS

We have seen that proteins, specifically enzymes, control much of the chemistry that occurs in the body. But how are the proteins and enzymes themselves constructed? What controls the order in which different amino acids are added to the peptide chains?

As discussed in Section 15–2, the genes on the chromosomes of cell nuclei carry the genetic information that determines the characteristics of an individual. They exert this influence over the body by governing the construction of proteins within the cells. Molecules in the chromosomes contain "blueprints" or "patterns" that determine the make-up of the proteins that are manufactured. This information, the genetic code, is obtained from the chromosomes of the male sperm and female egg of the parents of the individual. Throughout the billions of cell divisions that occur during the lifetime of the offspring, the code is preserved by the chromosomes, which divide simultaneously with the cell.

Chromosomes are generally made up of proteins and deoxyribonucleic acid (DNA). The DNA serves as the template that carries the genetic code in its structure.

The quest to determine the structure of DNA was one of the most exciting developments of post-World War II science. DNA was not only the object of extraordinary scientific curiosity, but the knowledge of it is of great practical significance. Among other things, this knowledge will be helpful in the search for cures for cancer-type diseases. Cancer is a condition in which the mechanism regulating cell division becomes defective, causing certain cells to multiply in a wild, uncontrolled fashion. Knowledge of the structure and reactions of DNA in cells will surely be of immense help in the conquest of cancer and other diseases. This knowledge may also make it possible to correct genetic "molecular" diseases such as sickle-cell anemia.

By the early 1950s a number of facts about the chemical composition of DNA were known. It was known to contain three types of constituents: a five-carbon sugar called deoxyribose, a phosphate linkage, and nitrogen-containing ring compounds called "bases." There are four types of bases, two of them (adenine and guanine) containing the purine nucleus and the other two (cytosine and thymine) containing the pyrimidine nucleus (see Fig. 15–26). These three types of constituents can be linked together to form nucleotides, as shown in Fig. 15–27, which were thought to be subunits of DNA.

How could one use the scanty information about this enormous molecule, having a molecular weight in the millions, to determine its structure? Fortunately for future scientists and interested nonscientists, one of the discoverers of the structure, Professor James Watson (now at Harvard

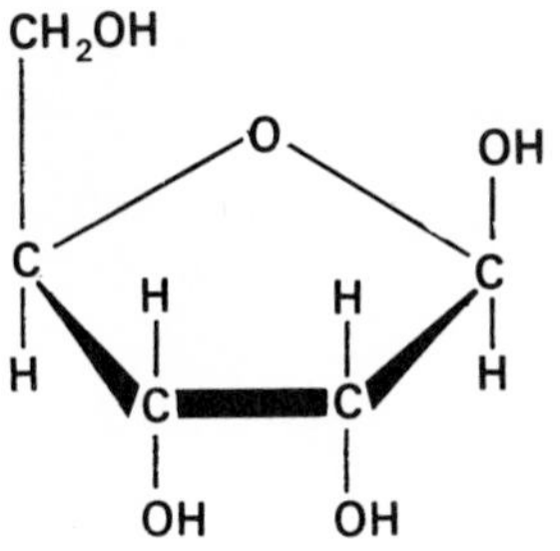

β-D-ribose

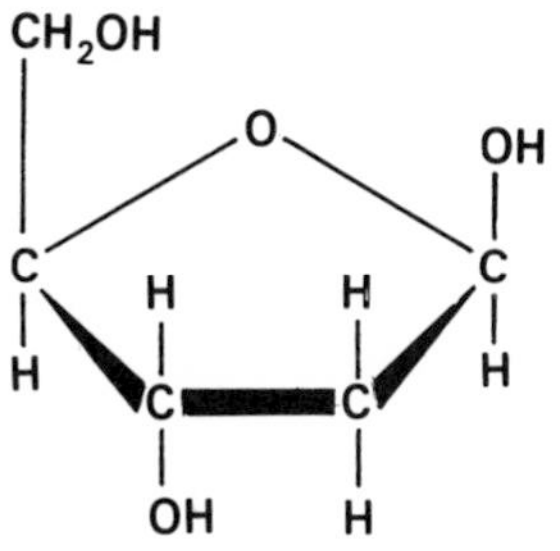

β-2-deoxy-D-ribose

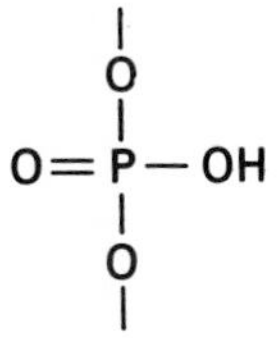

phosphate linkage

Fig. 15–26 The constituents of DNA: deoxyribose, two pairs of bases containing the pyrimidine and purine structures, and phosphate linkages.

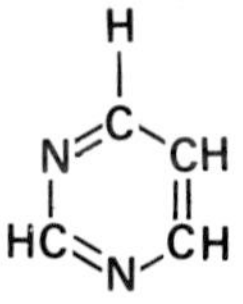

Pyrimidine 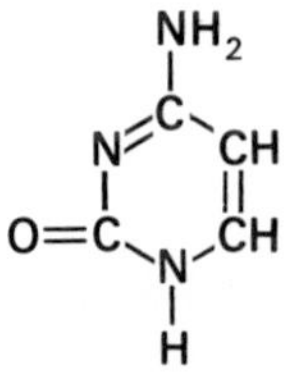   Cytosine 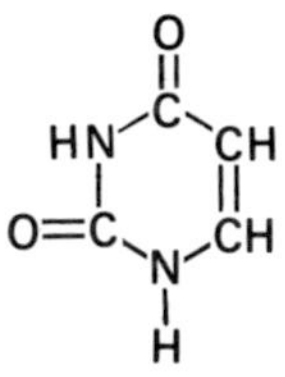   Uracil 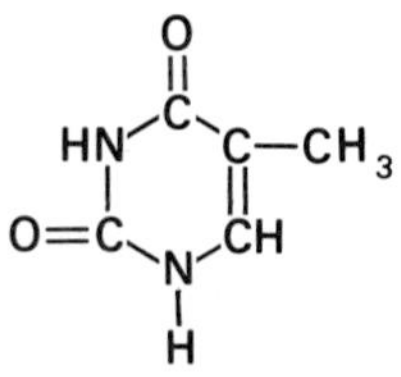   Thymine

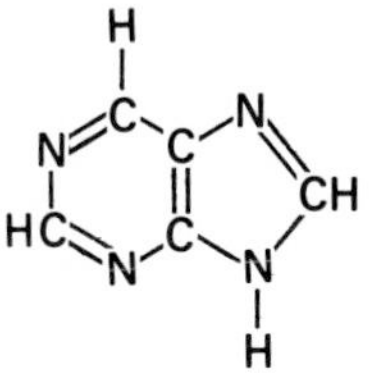

Purine 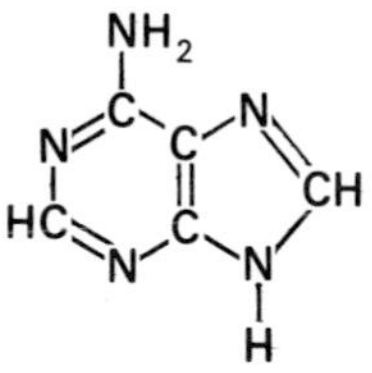   Adenine 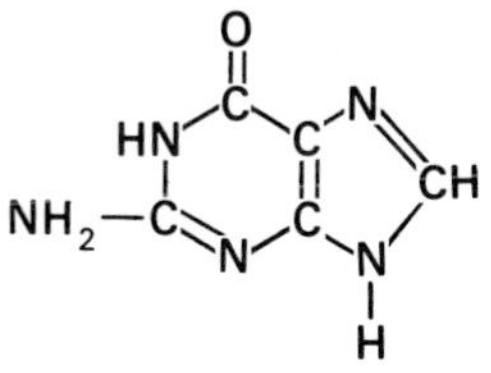   Guanine

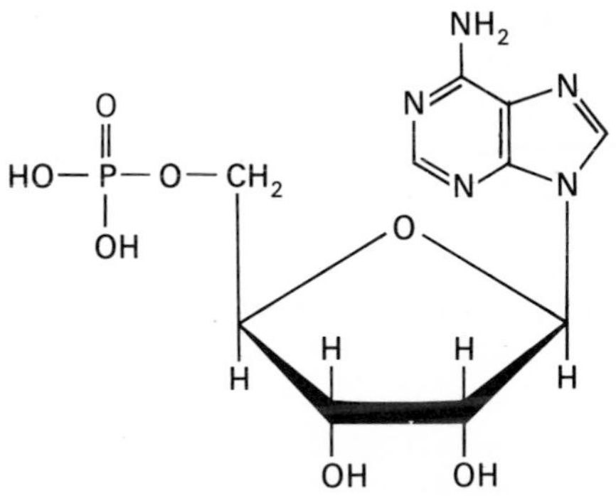

Fig. 15–27 Example of a nucleotide—adenosine monophosphate, AMP.

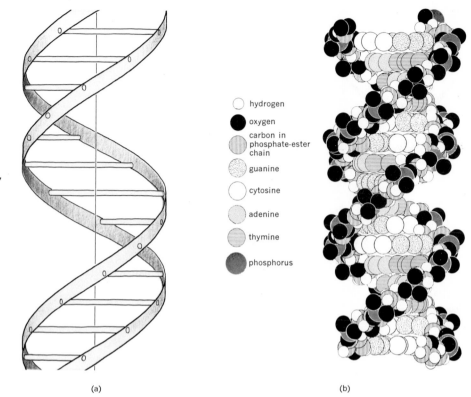

Fig. 15–28 The double helix "backbone" of DNA is connected by pairs of bases. [From Arthur L. Williams, Harland D. Embree, and Harold J. DeBey, *Introduction to Chemistry* (Addison-Wesley, Reading, Mass., 1968), p. 507.]

hydrogen

oxygen

carbon in phosphate-ester chain

guanine

cytosine

adenine

thymine

phosphorus

(a)   (b)

University and the Cold Spring Harbor laboratory on Long Island), has written a fascinating and candid account of the discovery.* Watson and Francis Crick, then at the Cavendish Laboratory in Cambridge, England, attacked the problem by the method previously used by Linus Pauling in his discovery of the structure of proteins.

Watson and Crick had available the chemical information noted above, as well as the following chemical clue concerning the nitrogen-containing bases: in DNA from various organisms, the proportions of the bases differ but always in such a way that the number of adenine molecules equals the number of thymines, and the number of guanine molecules is the same as the number of cytosine molecules.

Models of the skeletons of the pentose units and phosphate linkages were made of wire; the flat bases were originally cut from cardboard and, later, from sheet metal. It was a complex, three-dimensional puzzle to fit the nucleotides to-

gether into a stable structure. Watson and Crick had the good fortune to be in close touch with Maurice Wilkins and Rosalind Franklin at Kings College in London, who were taking x-ray scattering pictures of DNA at the same time that Watson and Crick were constructing their models.

In March 1953, Watson and Crick succeeded in building a model that explained all observations on DNA, including the latest x-ray data from Wilkins and Franklin.† The structure of this complex, all-important biochemical molecule had been solved! The molecular architecture of DNA is so beautiful and deceptively simple that, as in the case of many scientific discoveries, one wonders why it was so difficult to construct. But most scientific advances appear much easier in retrospect than they do to someone struggling to achieve them.

---

*   J. D. Watson, *The Double Helix* (Atheneum, New York, 1968).

†   In 1962, Watson, Crick, and Wilkins received the Nobel Prize for Physiology and Medicine for the discovery of the structure of DNA.

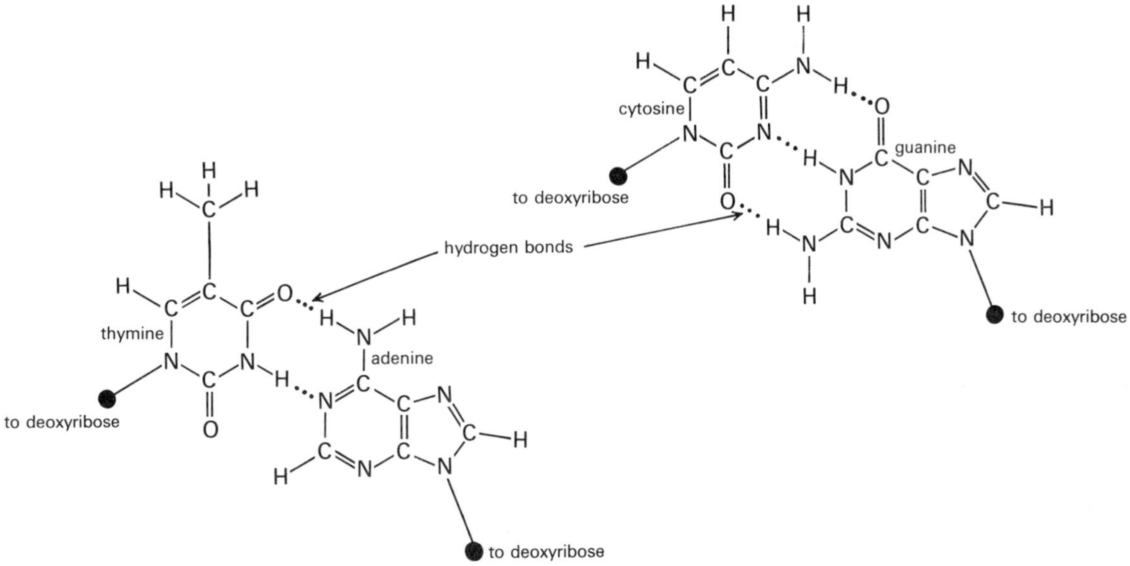

Fig. 15–29   The base pairs of DNA are held together by hydrogen bonds.

The structure of DNA is helical, like that of proteins, but in DNA two helices are wound together.* The helices consist of alternating five-carbon sugars (deoxyriboses) and phosphate linkages. Each sugar of one helix is linked to a sugar unit of the other helix by a pair of bases. The key to understanding DNA is the way in which the base pairs are fitted between the helices. Since the four bases have somewhat different structures, as shown in Fig. 15–26, how can pairs of the various bases be fitted smoothly between the helices without causing bulges in the double helix or gaps between the bases? The secret of success was Watson's discovery that if adenine (A) is always paired with thymine (T) and guanine (G) with cytosine (C) as shown in Fig. 15–29, the distance between the connection points to the sugars of the helices is the same for both pairs. Thus either base pair can be fitted smoothly between the sugars of the helices. Just as in the Pauling model of proteins, hydrogen bonds play an important role in holding together the DNA structure.

The A-T pair is joined by two hydrogen bonds and the G-C pair by three hydrogen bonds.

With knowledge of the structure of DNA, we can understand the way in which DNA functions. The genetic code is contained in the order of the bases along either of the strands of the double helix. For example, if we assign the numbers 0, 1, 2, 3 to the bases A, T, G, C respectively, we can consider a group of bases as representing a number in the base-4 system. The sequence TCAG would represent the base-4 number 1302 (see Fig. 15–30). This is similar to the representation of numbers in a computer, but most computers operate with the binary (base-2) number system, which has only two symbols, 0 and 1. Crick estimates that enough information can be encoded by the several centimeters of DNA in each human cell to fill 1000 large textbooks![†]

Since DNA is the "hereditary material" of cells, it must preserve the genetic code through the division of cells. Shortly before a cell starts to divide, the DNA double helix apparently starts unwinding, breaking apart at the relatively

---

* F. H. C. Crick, "The Structure of the Hereditary Material," *Scientific American* **191** (4), 54–61 (Oct. 1954).

† Ibid.

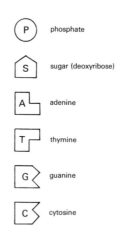

P   phosphate

S   sugar (deoxyribose)

A   adenine

T   thymine

G   guanine

C   cytosine

Fig. 15–30 The order of base pairs in DNA carries information analogous to numbers in the base-4 system.

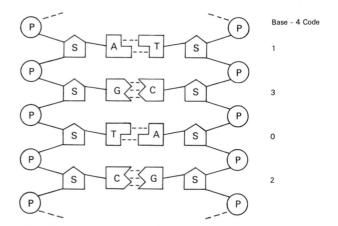

weak hydrogen bonds between the bases at the center of the molecule.* As soon as a section of each strand is unwound, nucleotides present in the cell material start to form new strands on the original sections. The new strands must be complementary to the original strands; e.g., at every point where there is an A base on the original strand, a T base must be built into the new strand in order for the new double helix to fit together properly (see Fig. 15–31). Thus each of the old strands serves as a template that forces the new strand to be just like the complementary old strand. Prior to cell division, each DNA molecule becomes two molecules, each of which contains one new strand and one old one.

The DNA molecule must have bends and loops to fit into the cell nucleus because the molecule is often centimeters in length, several hundred times as long as the cell. These long helices, having about ten million turns, must divide in periods of about twenty minutes. Thus the strands must unwind at the astounding rate of about 15,000 turns per minute! Furthermore, if the genetic code is to be faithfully preserved, these rapid divisions must be accomplished with an extremely small margin for error. It has been estimated that bacterial genes may be duplicated about one hundred million times before there is a fifty percent chance of error.†

How does the genetic code of DNA govern the manufacture of enzymes and other proteins? This is accomplished with the aid of ribonucleic acid (RNA), which translates the

* John Cairns, "The Bacterial Chromosome," *Scientific American* **214** (1), 36–44 (1966).

† P. C. Hanawalt and R. H. Haynes, "The Repair of DNA," *Scientific American* **216** (2), 36–43 (Feb. 1967).

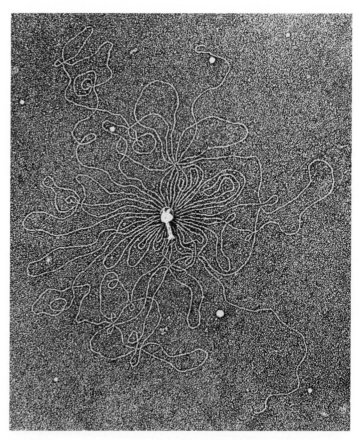

This virus, a T₂ *E. coli* bacteriophage, consists almost entirely of DNA. Magnification 97,000 X. (Courtesy of A. K. Kleinschmidt.)

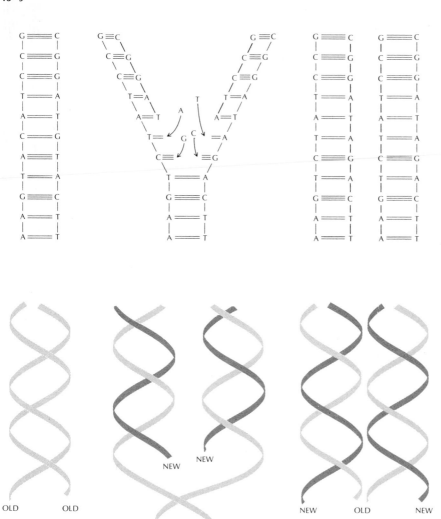

Fig. 15–31 The replication of a DNA molecule. The original strands unwind and new strands, with complementary bases, form on each old strand.

message of DNA into the structures of proteins. The structure of RNA is similar to that of a single strand of DNA except that the five-carbon sugar of RNA, ribose, has an — OH group where there is a hydrogen atom in the deoxyribose of DNA. Also RNA has the base *uracil* in place of the thymine of DNA.

There are several types of RNA that perform different functions. One is messenger RNA (m-RNA), which carries information from the DNA to the ribosomes, where proteins are synthesized. When the m-RNA is synthesized, a small section of DNA is unwound to serve as the template for the

Fig. 15–32 RNA being made, using the pattern of a partially unwound strand of DNA. The building of a protein by m-RNA and t-RNA. (a) The DNA chains unwind temporarily, exposing the sequence of nitrogenous bases. The separation allowing this unwinding occurs at the union of the paired bases. The messenger RNA is built on the template from the DNA subunits, located in the nucleoplasm of the cell. The sequence of bases on the completed messenger RNA will bear a definite relationship to the DNA sequence. This relationship is like that of an automobile fender to the mold in which it was shaped. At this stage it is theoretically possible to determine the DNA sequence by examining the RNA, just as the design of the fender tells a great deal about the shape of the mold. After the completion of this step, the role of the DNA in protein synthesis is essentially finished. (b) The completed messenger RNA molecule leaves the DNA template. The RNA passes out of the cell nucleus and moves to the ribosomes. These are located along the endoplasmic reticulum of the cell, where they have ready access to entering raw materials. The enlarged view here shows the substitution of the base uracil for thymine. (c) Meanwhile, transfer RNA molecules become bonded to free amino acids within the cell cytoplasm. These amino acids may have been derived from proteins which had previously been digested and absorbed. The type and sequence of the unpaired triplet of bases at the end of each transfer RNA molecule determines the kind of amino acid to which it attaches. (d) Molecules of transfer RNA, with their amino attached, enter the ribosome. The type and arrangement of the unpaired bases in the triplet now determine where each molecule of transfer RNA fits onto the messenger RNA. The attachment of transfer RNAs side by side on the messenger RNA automatically brings the amino acids which they carry into close contact. Here, the transfer enzyme can unite them to form protein chains. Thus the genetic code of the DNA molecule establishes the sequence of amino acids along the protein, which, in turn, determines the protein's characteristics. [From J. J. W. Baker and G. E. Allen, Matter, Energy, and Life (Reading, Mass.: Addison-Wesley, 1965).]

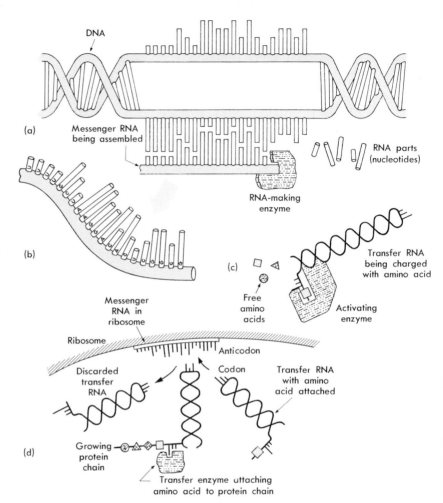

RNA (see Fig. 15–32). The bases attached to the RNA must be complementary to those of the corresponding DNA section. After the m-RNA is synthesized, it moves from the nucleus of the cell to a ribosome, where it serves as the pattern for the manufacture of protein.

Once the m-RNA is in place, *transfer-RNA* (t-RNA) brings amino acids to the ribosome (see Fig. 15–32). There is a particular type of t-RNA for each of the twenty-odd amino acids.* The amino acid is attached at one end of the t-RNA molecule, and at the other end a series of three bases form a *codon*, a code that indicates the type of amino acid that it carries. The codon of the t-RNA attaches to three com-

---

* For an excellent discussion of this topic, see R. W. Holley, "The Nucleotide Sequence of a Nucleic Acid," *Scientific American* **214** (2), 30–39 (Aug. 1966).

plementary bases (the *anticodon*) of the m-RNA in the ribosome. The amino acid is attached to the growing peptide chain. Then another molecule of t-RNA brings the next amino acid indicated by the next three-base anticodon of the m-RNA. In this way the sequence of bases on a strand of DNA determines the order of the three-unit anticodons on the messenger-RNA. These in turn determine the order in which transfer-RNA molecules bring up the appropriate amino-acid units for addition to the growing peptide chain.

15–10 HORMONES

Another class of compounds that control many of the body's functions is the hormones. They are manufactured in the endocrine glands and secreted into the bloodstream. Different kinds of hormones help digest food, control the body's growth rate and sexual development, regulate the female menstrual cycle, and marshall the body's defense mechanisms to react to emergencies. These are but a few of the important functions performed by hormones. Some hormones are proteins and others are lipids, in particular, steroids. Some of the hormones consist of a single compound whereas others include several closely related compounds. Although the structures of many hormones have been determined, little is known about the chemical reactions by which they cause their effects.

The hormone *insulin* is the protein whose structure was shown in Fig. 15–18. Secreted by the pancreas from the "islets of Langerhans," it helps control the concentration of the energy compound, glucose, in the bloodstream. The normal concentration of glucose is about 70 to 90 mg/100 ml blood, although it is higher shortly after meals. If the concentration of glucose becomes much too high or too low, conditions called *hyperglycemia* or *hypoglycemia*, respectively, are the result. A person suffers severe symptoms and may even die in extreme cases. These conditions can result from insufficient or excessive secretion of insulin.

In some persons, the pancreas produces excessive amounts of insulin in response to consumption of a carbohydrate-rich meal. This may drop the glucose level to the point where one suffers some symptoms of hypoglycemic shock such as weakness and trembling, but it is rarely fatal.

The more serious insulin problem is diabetes, caused by the failure of the pancreas to produce insulin. If untreated, diabetes is fatal. Fortunately, the disease can be treated by injection of insulin obtained from the pancreas of cows or hogs that are slaughtered. (It can't be taken by mouth because it would be broken down by digestive processes.) Even with

insulin, a diabetic has to watch his diet with some care. A well-operating pancreas adjusts its insulin output to the amounts and types of food taken in, but the insulin level of a diabetic is fixed by the dose of his injection. The danger for a diabetic is that the amount of insulin injected will be too great for the amount of carbohydrates ingested. This may cause the person's blood glucose level to drop so low that he suffers hypoglycemic shock. If he feels it coming on, he can usually ward it off by quickly eating some candy or other sugar-rich food. If he goes into a coma, it may require an injection of sugar to use up the extra insulin.

Another important and interesting hormone is *epinephrine*, more commonly called "adrenalin." Have you ever found that under great stress, fear, or excitement you were able to run faster or exert more strength than you are normally capable of? These effects are caused by epinephrine, which is released into the bloodstream by the adrenal glands (located above the kidneys) when a person is faced with some kind of emergency (see Fig. 15–33). The entire bloodstream normally contains only about five nanograms (5×10^{-9} g) of epinephrine, but during emergencies, the amount may increase a thousand-fold.

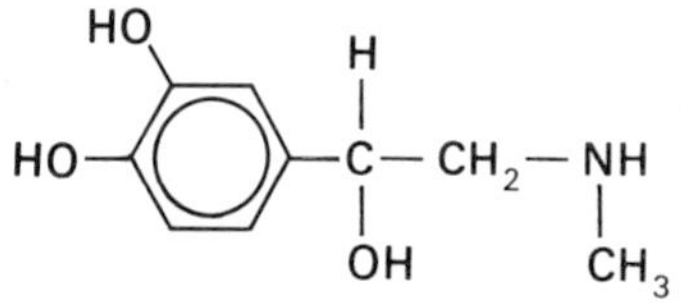

Fig. 15–33 Epinephrine (adrenalin).

Epinephrine is a much simpler compound than most biochemicals, but it performs many functions to mobilize the body's defense mechanisms. Among other things, the epinephrine causes the heart to beat faster in order to deliver glucose and oxygen to the muscles. It dilates blood vessels leading to the arms and legs to aid the supply of these essentials. In order to concentrate the flow of blood to those areas where it is most needed, the hormone constricts the vessels carrying blood to less essential areas such as the digestive tract. For this reason, one may suffer an "upset stomach" when the "adrenalin has been flowing."

Some of the most important hormones are those secreted by the *pituitary* gland, which is often considered the "master gland" of the body because of the many activities it directs. This gland weighs only about one-half gram in human beings, but it secretes at least 25 kinds of hormones, all of them proteins. It is located at the center of the skull at eyebrow level, where it is well protected by the skull.

One important hormone secreted by the pituitary gland is somatotrophin, the so-called "growth hormone." A defective pituitary that produces too little or too much of this hormone causes dwarfism or gigantism. Another is adrenocorticotrophic hormone (ACTH), a protein containing 23 amino acids. ACTH is released when the body is under stress and causes the adrenal glands to secrete adrenalin, as discussed above.

15-11 SEX HORMONES AND BIRTH CONTROL

Hormones from the pituitary gland are also involved in the control of the female menstrual cycle. This complex cycle is governed by five different kinds of hormones, three from the pituitary gland and two from the ovaries. Understanding this cycle will help us understand the effects of contraceptive pills, which have such enormous potential for preventing overpopulation.

1. At the start of the normal 28-day cycle, the pituitary begins to release *follicle-stimulating hormone* (FSH) into the blood stream (see Fig. 15–34). The FSH causes a follicle in one of the ovaries to start developing; the follicle contains an immature egg.

2. As it develops, the follicle begins to release *estrogen* hormones. The estrogens cause the inner lining of the uterus to expand and grow spongy with extra red blood cells, in preparation for a possible pregnancy. The estrogens also affect the production of the pituitary gland as follows.

They cause the pituitary to shut off the flow of FSH (to keep the follicle from growing too large).

They stimulate the production of two other hormones, luteinizing hormone (LH) and lactogenic hormone (LTH).

3. The LH causes ovulation (ejection of the egg from the follicle) on about the fourteenth day of the cycle, thus shutting down the production of the estrogens (see Fig. 15–35).

4. Since it is necessary for the uterus to continue expanding to prepare for a possible pregnancy, another hormone must take over control of that function from the estrogens. The production of this hormone is begun when the LTH stimulates the reorganization of the old follicle into a yellow body called the corpus luteum, which is located at the ruptured site on the ovary from which the egg was ejected.

5. The corpus luteum synthesizes the fifth hormone in the cycle, *progesterone*. Progesterone has several important functions.

It suppresses FSH production in the pituitary, thus inhibiting development of a new follicle.

It causes continued development of the uterine lining.

It causes slight enlargement of the breasts, in preparation for milk production.

It inhibits uterine contractions.

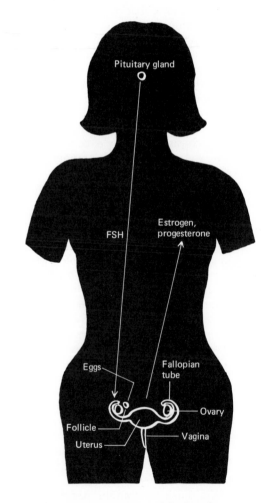

Fig. 15–34 The major components of the female reproductive system.

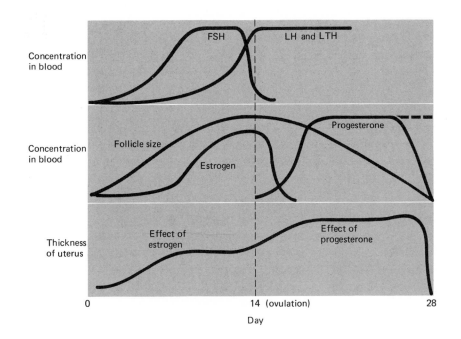

Fig. 15–35 Levels of hormones and size of the uterus during a typical menstrual cycle.

The egg ejected by the ovary is "caught" by one of the Fallopian tubes, which carries it to the uterus. If the egg is fertilized, the corpus luteum continues to produce progesterone, which is needed for the duration of the pregnancy.

If the egg is not fertilized, the corpus luteum degenerates and stops producing progesterone, causing the uterus to shrink. Starting on about the 28th day of the cycle, a small amount of blood flushes the unfertilized egg out of the uterus (the monthly "period") and the progesterone level drops to zero. The pituitary gland is then able to restart the cycle by producing FSH again.

Contraceptive methods depend on preventing the fertilization of the egg. At about the midpoint of the menstrual cycle, that is, about the fourteenth day, the stage is set for fertilization. If sexual intercourse causes one of the thousands of male sperm cells to reach the egg, there is a good chance that the egg will be fertilized and pregnancy will result.

Despite considerable research on fertility and birth control, there are many uncertainties about the timing of events at this stage. For example, for how long a time is the egg prepared for fertilization? A few hours or a couple of days? How long can the male sperm survive in the uterus after intercourse if it arrives before the egg does? To these uncertainties about the "average" time scale there is added the variation of the cycle in a particular woman. Although the female menstrual cycle is about 28 days, it may be considerably greater or less than that for a particular female. Furthermore, especially for girls not too far beyond puberty, the length of the menstrual cycle may vary greatly from one cycle to another. All these uncertainties of timing militate against the attempts by many couples to prevent pregnancy by the "rhythm method," that is, by avoiding intercourse for a few days at the midpoint of the menstrual cycle.

Birth-control pills contain compounds that are similar to the estrogens and progesterone. The hormones themselves are not very effective when taken by mouth, but slightly different structures can be used which mimic the effects of the actual hormones. Several types of compounds are used in preparations made by the various pharmaceutical companies. A common artificial estrogen is *mestranol* whose structure is similar to that of *estradiol*, the most active of the estrogens. Compounds such as *norethindrone* are used to duplicate the effects of progesterone (see Fig. 15–36). From the structures of these hormone substitutes, it is not obvious why they react so similarly to the corresponding

mestranol   estradiol   testosterone   norethindrone   progesterone

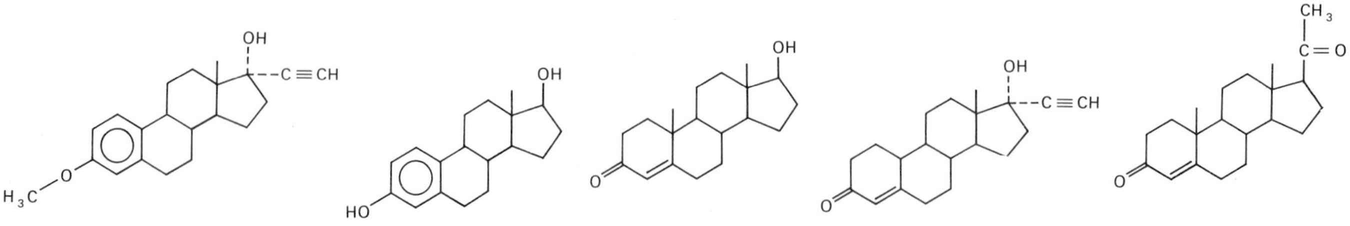

Fig. 15-36 Female hormones estradiol and progesterone along with typical substitutes mestranol and northindrone. Also, the major male hormone testosterone, which is quite similar in structure to estradiol.

natural compounds. Despite its great structural similarity to estradiol, the hormone *testosterone* shown in Fig. 15-36 is the major male hormone, being secreted by the testes and stimulating the development of many male characteristics.

The idea behind the use of birth-control pills is to keep the levels of artificial estrogen and/or progesterone high enough at most times to suppress the release of FSH by the pituitary gland. This prevents follicle development and ovulation. Thus the use of birth-control pills mimics the conditions during pregnancy, during which high levels of progesterone prevent ovulation. However, the pills are taken only for 21 days out of the 28-day cycle, in order to allow menses to occur as usual.

Two different types of pills are used. In the "combined" type, the pills taken for the 21-day period contain both estrogen- and progesterone-like compounds, whereas in the "sequential" type, pills taken the first 15 days contain only the estrogen substitute and the remaining six contain only the progesterone substitute. One might suppose that the sequential pills would be superior, as they more closely duplicate the natural hormone sequence. However, no clear-cut difference in the effectiveness or freedom from side-effects of the two types has been generally observed, although for a particular individual, one may prove more satisfactory than the other.

Birth-control pills are highly effective in preventing pregnancies if taken as directed. However, like any artificial hormonal substance, they can have undesirable side effects. Since the artificial hormones mimic the conditions of preg-

nancy, it is not surprising that they often cause weight gains (from increased water retention) and uncomfortable swelling of the mammary glands. One of the more serious side effects is an increase of the clotting factor, fibrinogen, in blood, increasing the danger of fatal internal blood clots. However, the small chance for serious illness or death from use of the pills must be balanced against the significant risks associated with pregnancy. Djerassi has estimated that among one million users (aged 20-34 years) of birth-control pills, during one year about thirteen would die from causes related to the pills and only one from complications of the expected 5000 pregnancies among the group.* Among a similar group using the "rhythm" method, about 240,000 would become pregnant and, of these, about 55 would die. Decisions regarding the use of birth-control pills must be made individually in consideration of the risks and the possible discomforts associated with use of the pills.

In spite of their effectiveness, present-day birth-control pills are not totally satisfactory as a solution to the population problem of underdeveloped nations. They are relatively expensive and the user must take them on a carefully prescribed schedule, a difficult task in areas where educational levels are low.

For widespread use it would be desirable to have a chemical contraceptive that could be administered much less frequently (perhaps by implanting a slowly dissolving

---

*   Carl Djerassi, "Prognosis for the Development of New Chemical Birth-Control Agents," Science **166**, 468-473 (1969).

capsule beneath the skin) or an "abortifact" that would have to be taken only after intercourse, the so-called "morning-after pill." Although a great deal of research on chemical contraceptives is under way, it may be many years before an ideal contraceptive is developed and tested.

One class of compounds under study is the *prosta-glandins*, hormone-like compounds that occur in various body tissues (see Fig. 15–37).* Preliminary studies have shown the prostaglandins to be effective in inducing abortion within the first few weeks of pregnancy, suggesting that they may be effective morning-after pills. Although they cause nausea and diarrhea if taken orally, they may be effective in the form of vaginal suppositories. Other workers are attempting to develop chemical contraceptives involving the use of proteinlike pituitary hormones of the LH and FSH types.†

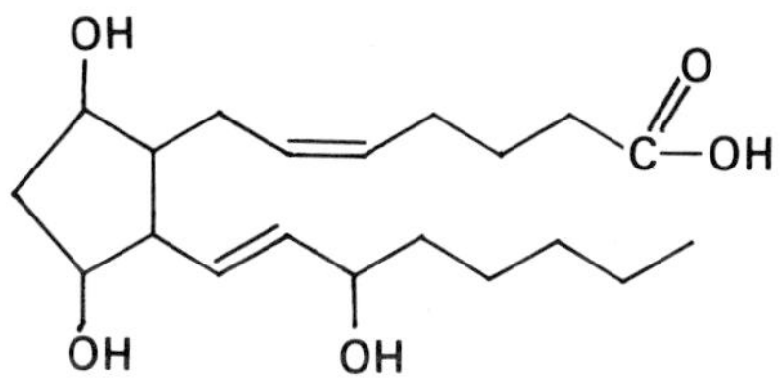

Fig. 15–37 A typical prostaglandin.

Although its molecular structure is quite different from that of estradiol, the artificially synthesized compound *diethylstilbestrol* (DES) exhibits physiological activity some-what like that of the estrogens (see Fig. 15–38). During the late 1940s and early 1950s, it was administered to many pregnant women who had difficulty carrying fetuses to term. The DES was supposed to prevent miscarriages, but its effectiveness was never clearly demonstrated. Recently it has been found that the daughters of those women have an unusually high incidence of vaginal cancer.‡ Some doctors, particularly at college health centers, have been prescribing DES as a "morning-after" contraceptive despite its question-

able effectiveness and the apparent risk of cancer in daughters of the users.§

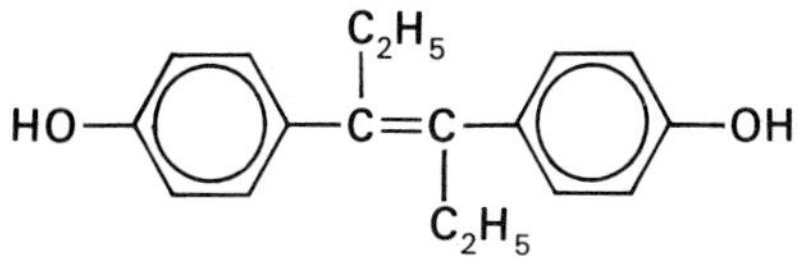

Fig. 15–38 Diethylstilbesterol, DES.

Since 1954, large quantities of DES have been used as an additive to cattle and sheep feed as it considerably increases their rates of growth. Because of its carcinogenic properties, the U.S. Food and Drug Administration banned the use of DES in feed after January 1, 1973. Cattle feeders were still permitted to implant DES in steers 120 or more days prior to marketing. However, traces of DES were still detected in livers of the implanted animals. Thus in April 1973, all use of DES in animals was halted. ‖

15–12 VITAMINS

Years ago, only two vitamins, A and B, were known. Vitamin A is soluble in nonpolar solvents such as lipids, and vitamin B is soluble in water. Since that time, it has been found that vitamin B is actually a set of compounds, subsequently designated by number, such as B_1, B_2, B_{12}, and so on. In addition, several other vitamins have been discovered, such as C (water-soluble) and D, E, and K (all lipid-soluble). Now it is common to designate vitamins by their chemical names (ascorbic acid, thiamine, riboflavin, etc.) rather than by letter, but both designations are used in this discussion.

People whose diets are deficient may suffer diseases resulting from the lack of vitamins, conditions referred to as *avitaminoses*. Today in the United States, most people are well enough nourished to avoid these problems. The more prevalent condition is probably *hypervitaminosis*, too much of some vitamin. The combination of vitamin supplements in many commercial food products (e.g., milk, bread) and excessive "pushing" of vitamin pills by overzealous mothers

*   B. J. Culliton, "Something for Everyone", *Science News* **98**, 306–307 (1970).
†   J. Arehart-Treichel, "Birth Control in the Brave New World," *Science News* **103**, 93–95 (1973).
‡   Morton Mintz, "Drug Linked to Cancer in Daughters," *Washington Post*, Oct. 27, 1971.

§   Morton Mintz, "'Morning After' Pills," *Washington post*, Dec. 12, 1972.
‖   Morton Mintz, "U.S. Ends Use of DES as Livestock Stimulant," *Washington Post*, April 26, 1973.

causes some children to consume excessive amounts of lipid-soluble vitamins, particularly vitamin D.

Vitamin A

As shown in Fig. 15–12, vitamin A is closely related to β-carotene. Two vitamin A molecules can be obtained by splitting the β-carotene molecule in the middle and adding the appropriate —OH and —H groups at the "split end." The body can obtain vitamin A directly from foods or convert β-carotene to the vitamin. Both vitamin A and β-carotene are present in many yellow and orange vegetables, which derive their colors from these compounds. They are also present in many green plants, where the characteristic color is masked by the green chlorophyll.

Vitamin A is particularly important for good vision, especially acute night vision. The rods of the retina of the eye, which are vital for seeing in dim light, contain a compound, retinene, quite similar to vitamin A and derived from it. Although we suffer little deficiency of vitamin A in the United States, there are areas in which it is a problem. An estimated one million persons in India suffer blindness because of vitamin A deficiency, a particularly tragic situation since the blindness could be prevented by a vitamin supplement costing only a few cents per year per person.

The B Vitamins

The B vitamins are water soluble and, as shown in Fig. 15–39, most contain at least one nitrogen atom in a ring structure. Most of the B vitamins act as coenzymes in the body.

Deficiencies of vitamin B_1, or *thiamin*, lead to a condition called "beriberi," a weakness of muscles and a disorder of the nervous system. It occurs primarily among those whose diet consists mainly of polished rice. The outer coating of rice kernels contains thiamin, but it also contains oils that can become rancid during storage. To avoid the spoilage, the coating is frequently removed. People subsisting on this polished rice suffer beriberi unless they have other thiamin sources. The same is true of the white flour that is made from wheat after removal of the darker outer layer of the kernels. For this reason, thiamin is added to most commercial white breads in this country.

Although it doesn't have a number, *nicotinamide* (commonly called "niacin") is one of the B vitamins. It is a coenzyme important in catalyzing reactions in which food is oxidized and the energy content stored in the form of ATP

(see the following section). Niacin deficiency causes a disease called "pellagra." In the past, this disease affected many poor people of the southern United States whose diets consisted mainly of corn products and other foods low in niacin.

Although the B vitamins cannot be synthesized by the human body, many are produced by bacteria in human intestines. Generally, enough of those B vitamins are absorbed through the intestinal walls to satisfy our needs, even if our diets are deficient in them. However, the intestinal bacteria need the amino acid tryptophan as a starting material for synthesis of niacin. Unfortunately, corn, which is deficient in niacin, is also deficient in tryptophan (see Section 11–4), and so the intestinal bacteria are unable to make up the niacin deficiency on a diet consisting mainly of corn.

Vitamin B_{12}, or *cobalamin*, contains a metal atom, cobalt, located at the center of a large molecule similar to that of heme. Cobalamin plays an important role in the formation of hemoglobin. People deficient in vitamin B_{12} suffer a condition known as "pernicious anemia." Anemia is a condition in which a person's hemoglobin level in the blood is abnormally low. The common form of anemia can be corrected with iron supplements or foods such as liver that are high in iron. Pernicious anemia is not so easily controlled and the disease used to be fatal. It does not result from the lack of vitamin B_{12} in food, but from the body's inability to absorb it from food. Injections of cobalamin, however, take care of the problem.

The folic acid discussed above in connection with sulfa drugs (see Section 15–8 and Fig. 15–24) is also classified as a B vitamin. One practical note regarding the nutritional aspects of the B vitamins should be made. Recall that they are among the water-soluble vitamins. This means that if foods containing B vitamins are cooked in water, a considerable fraction of them may dissolve in the water and be lost. From the standpoint of vitamins, it is thus preferable to avoid cooking food in water.

Vitamin C (Ascorbic Acid)

Vitamin C, or *ascorbic acid*, is also water-soluble. It has a structure much like that of a sugar (see Fig. 15–40). Deficiencies of this vitamin cause scurvy, whose symptoms include weight loss, weakness, frequently bleeding gums and skin, and loosening of teeth. Ascorbic acid is present in many fresh vegetable and fruits, especially citrus fruits. In the days of sailing merchant ships, sailors who were cut off from supplies

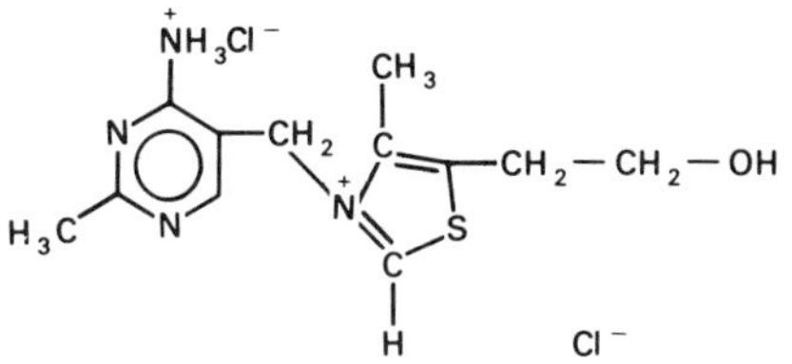

Thiamine (Vitamin B_1)

thiamine chloride

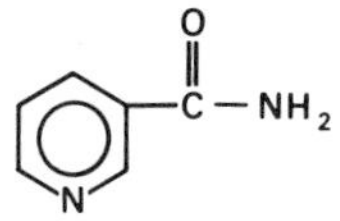

Nicotinamide (niacin)

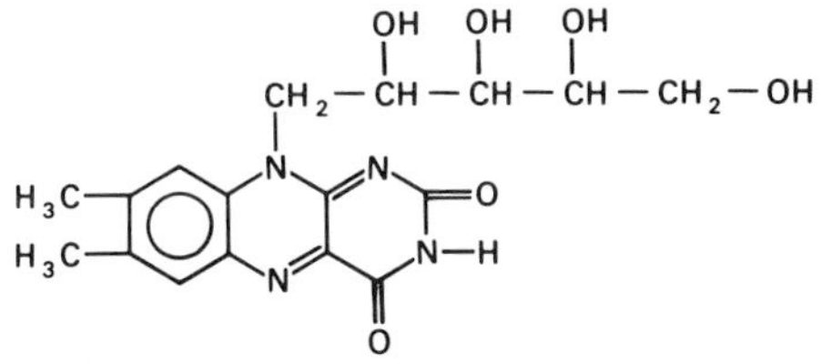

Riboflavin (Vitamin B_2)

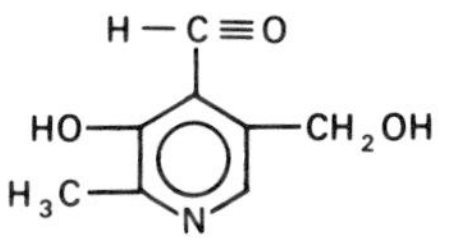

Pyridoxal (Vitamin B_6)

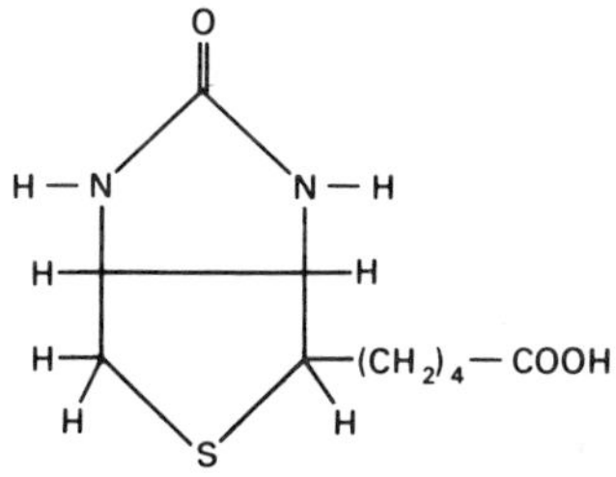

Biotin

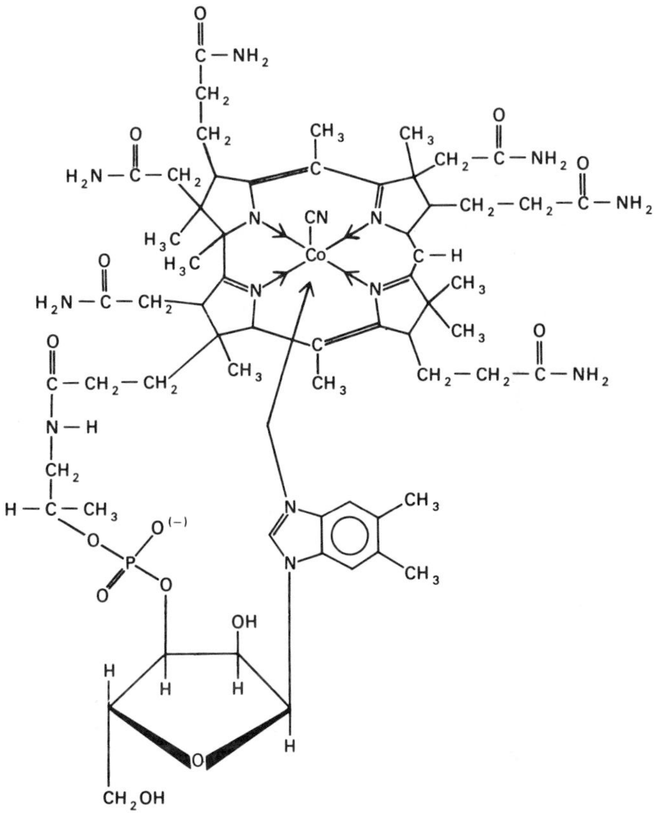

Cobalamin
(Vitamin B_{12})

Fig. 15–39 Several B Vitamins: B_1, B_2, niacin, B_6, biotin, and B_{12}.

of fresh fruits and vegetables were frequent sufferers. Then British sailors discovered that the disease could be prevented by consuming limes and lemons during long voyages. They thus acquired the nickname "limeys," which is still applied to the British today.

According to the National Academy of Sciences, a person requires about 60 mg of ascorbic acid per day, far more than the amounts of most other vitamins. For example, the daily human requirement of thiamin is about 1 mg, riboflavin (B_2)

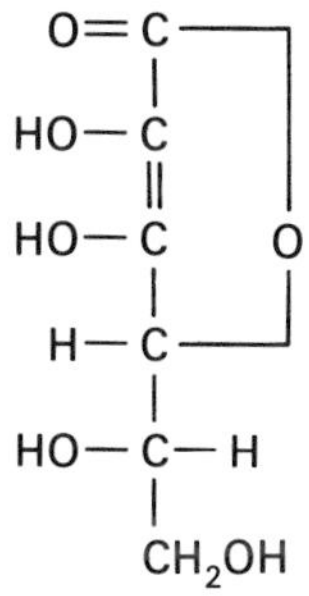

Fig. 15–40 Ascorbic acid, vitamin C.

Valencia oranges ripening in Florida. (Courtesy of the U.S. Dept. of Agriculture.)

recommended by the NAS is much too low.* On the basis of a wide variety of experiments with people and animals, he suggests that the true value is in the neighborhood of 1 g/day or more. Furthermore, he feels that daily doses up to 5 g/day are beneficial in warding off colds or ameliorating their effects, as well as promoting generally good physical and mental health. His suggestions have been hotly disputed by many other scientists and medical researchers. Support of Pauling's point of view has come from Man-Li Yew of the University of Texas, who studied effects of vitamin C on growth, wound healing, and resistance to surgical stress in guinea pigs.† Yew found that the guinea pigs needed 5 mg per 100 g of body weight per day to grow well, heal quickly, and recover from stress. If children have similar requirements, they need about 1.5 g per day.

Other recent evidence suggests that vitamin C helps remove cholesterol from the body, thus helping to prevent hardening of the arteries and other diseases of the circulatory system.‡ Regardless of their value, excesses of vitamin C are probably not harmful, as vitamin C is water-soluble and the excess is flushed out of the body.

Vitamin D

As can be seen from its structure in Fig. 15–11, vitamin D is closely related to the other steroid compounds. Actually there are several compounds, differing in the nature of the R— group, which act as vitamin D. One of them is formed when photons of sunlight strike molecules of 7-dehydro-cholesterol, a compound that occurs just beneath a person's skin. Vitamin D plays an important role in the deposition of calcium and phosphate minerals in growing bones. Deficiencies of vitamin D in children, whose bones are in a formative state, lead to rickets—a softness and possible deformation of bone structure. In order to prevent rickets, foods such as cereals, milk, and bread are often irradiated with ultraviolet light, which converts the steroid ergosterol into a type of vitamin D. The oil from fish livers is also high in vitamin D. Children of previous generations frequently were given daily doses of cod-liver oil to make sure they got enough vitamin D, especially during winter in high latitudes when sunlight was at a minimum.

1.5 mg, niacin 10 to 16 mg, and cobalamin, only about 1 μg. Vitamin C is very fragile, being easily destroyed by heat or by oxidation when exposed to the air. Thus fruits or vegetables containing vitamin C should be eaten raw, if possible.

A great deal of controversy surrounds vitamin C today. Professor Linus Pauling feels that the 60-mg daily amount

*   Linus Pauling, *Vitamin C and the Common Cold* (W. H. Freeman, San Francisco, 1970).
†   "A Plus for Pauling and Vitamin C," *Science News* **103**, 290 (1973).
‡   "Vitamin C and Heart Attacks," *Science News* **103**, 106 (1973).

Vitamin D is harmful in excessive amounts. Too much of it causes calcium and phosphate minerals to be deposited too rapidly, especially in the joints, causing stiffness and pain. With all the irradiated food and vitamin pills available, children of the United States probably suffer more from excess vitamin D than from deficiencies.

Vitamin D probably played a major role in the development of skin colors of the races.* Persons living in equatorial areas such as Africa were better protected against excessive vitamin D production by the abundant sunlight if they had dark skin. Because of their enhanced chances of survival to parenthood, dark-skinned people had more children than light-skinned people. Thus races originating in that area acquired dark skins, but as people moved to higher latitudes those with lighter skins had a better chance of survival because they could make maximum use of the limited sunlight available.

Vitamin E

Vitamin E has long been controversial (see Fig. 15–41). Many claims have been made for its enhancement of sexual potency and virility. It is essential for reproduction in rats, but it has not been clearly demonstrated to be essential in humans. Recently it has been used an an underarm deodorant. Unlike the use of many other ingredients in deodorants, which are often ineffective, this use of vitamin E has some basis in fact, since it is known to act as an antioxidant of lipids. Thus it may prevent oxidation of lipids in perspiration that form compounds with unpleasant odors.

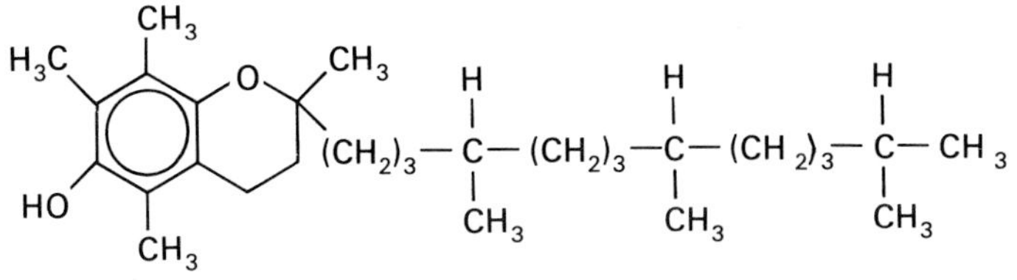

α-tocopherol

Fig. 15–41 Vitamin E.

_____

*   A. S. Boughey, *Man and the Environment* (Macmillan, New York, 1971), Ch. 3.

15–13 METABOLISM

Metabolism includes the processes by which large molecules in food are broken down into simpler units that are absorbed into the body and transported to appropriate cells. There they are reassembled into new large molecules or further broken down into CO_2 and H_2O in order to release energy needed by the body. Finally, the waste products and non-digestible portions of the food (e.g., cellulose) are eliminated from the body.

The very large molecules of carbohydrates, proteins, and lipids must first be broken down into smaller molecules. In each case, this bond breaking involves *hydrolysis*, i.e., the breaking of a bond with the addition of a hydrogen and an OH group from water to the new ends of ruptured bonds as shown in Fig. 15–42. Hydrolysis is catalyzed by appropriate *hydrolyzing enzymes* added to the food in the mouth, stomach, or small intestine. Food is not usable by the body until it has been absorbed through the walls of the small intestine into the bloodstream or lymph system. For the most part, only monomeric units resulting from hydrolysis of the polymers can penetrate the intestinal wall.

Digestion starts in the mouth, where the food is mixed with *saliva* by the salivary glands (see Fig. 15–43). Saliva contains an enzyme, *salivary amylase*, that catalyzes the hydrolysis of starch into smaller polysaccharides called *dextrins*. Food remains in the stomach for several hours. The stomach secretes gastric juice, which contains both hydrochloric acid and another hydrolyzing enzyme, *pepsin*. The high acidity of the stomach (pH 1 to 2.5) shuts off the hydrolysis of starch, but the acid and pepsin catalyze the hydrolysis of proteins.

Although we tend to think of the stomach as the main site of digestion, most of the "action" occurs in the small intestine. After the partially digested food enters the small intestine, it is acted upon by several hydrolyzing enzymes secreted by the pancreas or the intestine. After hydrolysis of the large biochemical molecules, the resulting fatty acids, sugars, and amino acids are absorbed through the intestinal wall.

Under the influence of insulin, the liver helps regulate the level of glucose in the blood by acting as a glucose storage bank. When the level is high, the liver withdraws glucose, forming it into glycogen, the storage form of glucose in animals. When the glucose level drops, the liver breaks up glycogen and returns glucose to the blood.

The blood carries glucose to cells throughout the body where it may also be stored as glycogen until energy is

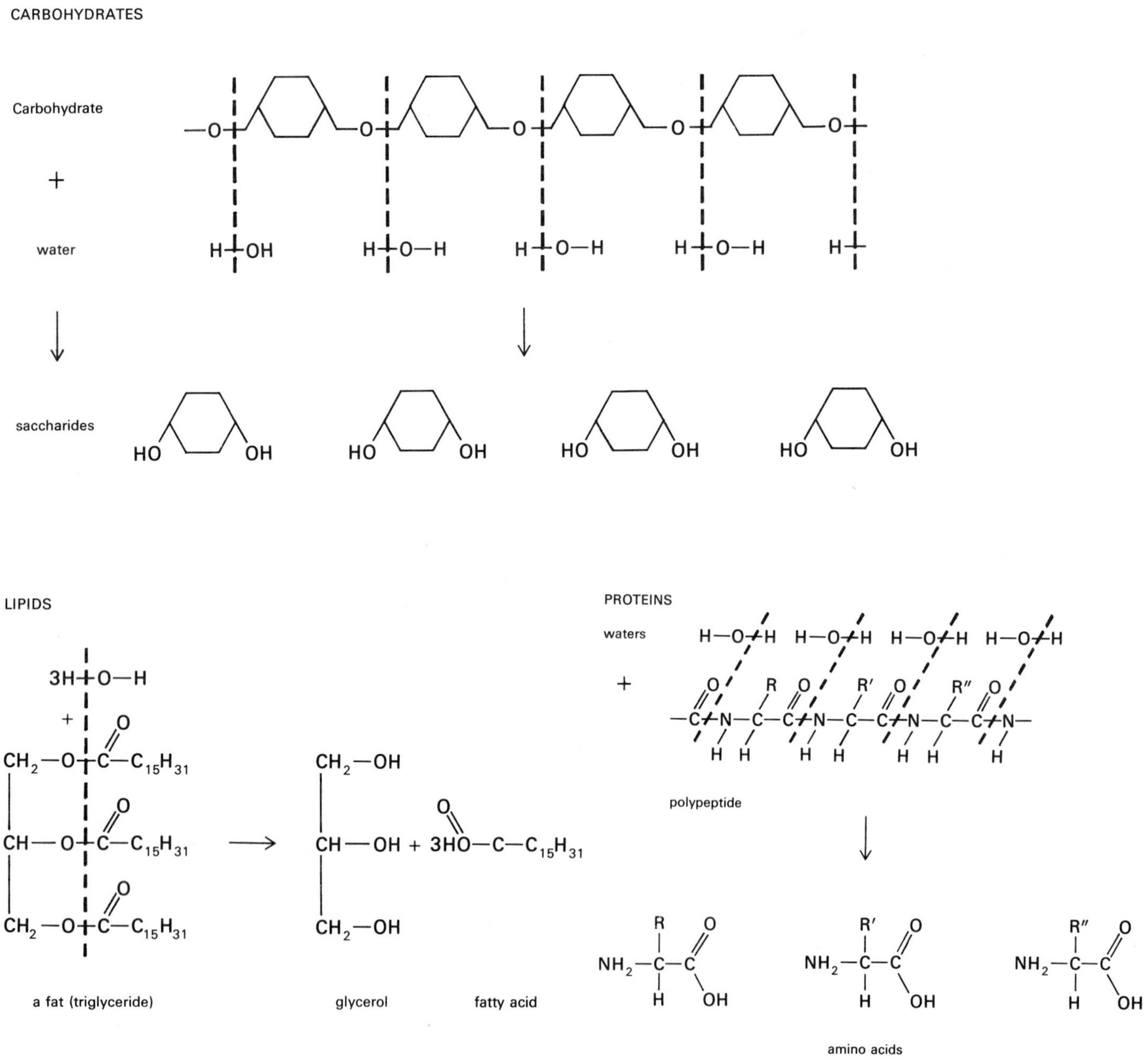

CARBOHYDRATES

Carbohydrate

+

water

↓   ↓

saccharides

LIPIDS

$3H{+}O{-}H$

+

$CH_2{-}O{+}C{-}C_{15}H_{31}$
     $\parallel$
     O

$CH{-}O{+}C{-}C_{15}H_{31}$
     $\parallel$
     O

$CH_2{-}O{+}C{-}C_{15}H_{31}$
     $\parallel$
     O

⟶

$CH_2{-}OH$

$CH{-}OH + 3HO{-}C{-}C_{15}H_{31}$
              $\parallel$
              O

$CH_2{-}OH$

a fat (triglyceride)   glycerol   fatty acid

PROTEINS

waters $H{-}O{/}H$ $H{-}O{/}H$ $H{-}O{/}H$ $H{-}O{/}H$

+

polypeptide

↓

amino acids

Fig. 15–42 Hydrolysis of carbohydrates, lipids, and proteins, is the first step of digestion.

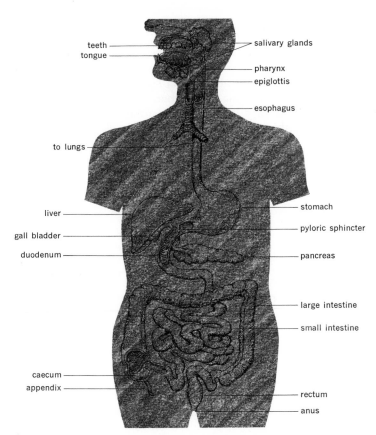

Fig. 15–43 Major features of the digestive tract. [From John W. Kimball, *Biology*, 2nd ed. (Addison-Wesley, Reading, Mass., 1969), p. 182.]

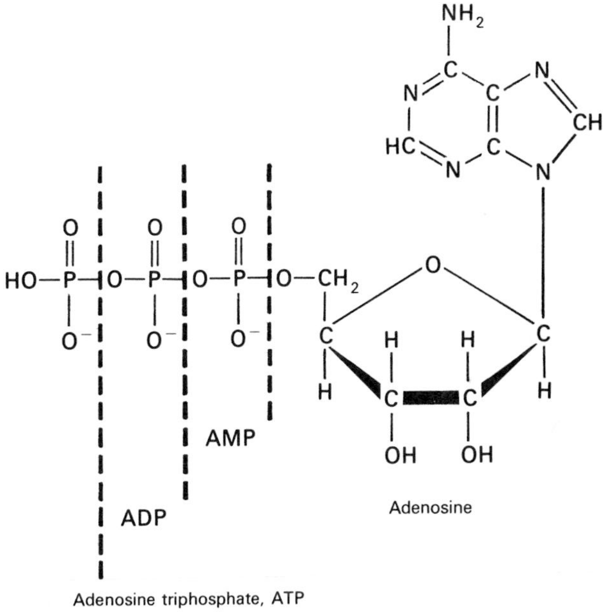

Fig. 15–44 Structures of ATP, ADP, and AMP.

$$ATP \xrightarrow{\text{hydrolysis}} ADP + H_3PO_4$$

$$ADP \xrightarrow{\text{hydrolysis}} AMP + H_3PO_4$$

$$AMP \xrightarrow{\text{hydrolysis}} \text{Adenosine} + H_3PO_4$$

needed. The energy of glucose is released by combining it with oxygen to form carbon dioxide and water:

$$C_6H_{12}O_6 + 6O_2(g) \longrightarrow 6CO_2(g) + 6H_2O + 4 \text{ kcal/g.} \quad (15–2)$$

We could carry out this reaction in the laboratory by burning the glucose in oxygen gas to release the energy. Although this is the net reaction by which glucose releases energy in the body, the body could not stand to have the glucose actually "burned" in its cells. Instead, reaction (15–2) is carried out in the mitochondria of the cells via a complicated series of reactions, each of which releases a small amount of the energy stored in glucose, rather than in a burst of fire.

Most of the energy released in each of these steps is used to convert *adenosine diphosphate* (ADP) into *adenosine tri-*

phosphate (ATP) (see Fig. 15–44). The energy of one molecule of glucose is sufficient to convert 38 molecules of ADP into ATP:

$$C_6H_{12}O_6 + 6O_2 + 38 \text{ ADP} + 38H_2PO_4^- \longrightarrow$$
$$6CO_2 + 6H_2O + 38 \text{ ATP}^-. \quad (15–3)$$

The smaller units of energy stored in the form of the phosphate bond of the third phosphate of ATP are of convenient size to be used for energy needed in other biochemical reactions.

The ATP released by the mitochondria is used as the energy source for biochemical reactions in other parts of the cell. For example, the conversion of ATP back to ADP supplies the energy needed for muscle contractions. The end products of carbohydrate metabolism are carbon dioxide and water. The carbon dioxide is carried by the blood to the lungs, from which it is exhaled.

Before lipids can be hydrolyzed, the large globs of insoluble fatty materials must be broken down into smaller globules that provide greater surface area for reactions to occur. This is done by *bile salts* produced by the liver and stored in the gall bladder. They act like detergent molecules, surrounding the lipid droplets with hydrophilic groups that keep the globules in suspension. Because of this important function of bile salts, a person whose gall bladder functions poorly or has been removed is likely to suffer considerable digestive discomfort after eating a meal rich in fats or oils.

The lipid fragments, i.e., glycerol, fatty acids, cholesterol, etc., pass through the intestinal wall into the bloodstream or the lymph system. Lymph is a clear fluid somewhat like blood, but without red blood cells; it circulates throughout the body through the lymphatic system, which is similar to the blood circulatory system. The lymphatic system connects to the bloodstream near the base of the neck, allowing additional lipids to flow into the blood.

Fats not immediately needed for energy are stored in fatty deposits throughout the body. The body can store energy in the form of glycogen only in limited amounts. When the storage capacity for glycogen is used up, additional carbohydrates are converted to fats for more efficient energy storage (9 kcal per gram versus 4 for carbohydrates). The body cannot convert fats back into glycogen or glucose, but when the stored energy is needed, the fats are oxidized to form carbon dioxide and water. Just as in the case of glucose oxidation discussed above, the body "dismantles" the fatty acid molecules piece by piece. The energy released in the process is used to manufacture ATP, converting the stored energy into the body's "currency form" of energy. The fatty acid molecules are subjected to repeated cycles of chemical reactions, each of which removes two carbon atoms from the chain. The oxidation of palmitic acid, which contains 16 carbon atoms, causes formation of 130 ATP molecules. The products of fatty acid oxidation are again carbon dioxide and water.

The amino acids resulting from protein hydrolysis undergo several types of metabolic reaction. Many amino acids are simply transported to the cells, where they are used as building blocks for the synthesis of enzymes and other proteins needed by the body. This is particularly true of the essential amino acids, which cannot be synthesized from other compounds in the human body.

Amino acids in excess of those needed for protein synthesis are metabolized by complicated reaction sequences similar to those of the carbohydrates and lipids. These processes release ammonia, which must be eliminated from the body as it is toxic to humans at levels above 5 mg/100 ml in the blood. Ammonia and carbon dioxide are combined in the liver to form urea, which is a much less toxic form of nitrogen than ammonia, occurring at levels between 20 and 40 mg/100 ml in the blood of normal persons. The net reaction for urea synthesis is

$$2NH_3 + CO_2 \longrightarrow H_2N-\overset{\displaystyle O}{\overset{\displaystyle \|}{C}}-NH_2 + H_2O. \qquad (15-4)$$

The urea is transported by the bloodstream to the kidneys, from which it is excreted in urine. An average adult excretes about 30 g of urea daily.

15-14 MEDICINES AND DRUGS

From a scientific point of view, it is frustrating to discuss medicines and drugs, because one rarely knows the chemical mechanisms by which the compound produces the observed effects. For example, the most widely used medicine on the market is aspirin. It is well established that aspirin generally reduces fever, swelling, and pain, but we have only the foggiest notions of how it accomplishes these effects. Recent evidence suggests that aspirin prevents synthesis of prostaglandins by body tissue.* In some cases, we do know the mechanisms by which medicines work (e.g., the sulfa drugs, Section 15-8), but those are the exceptions.

The chemical and pharmaceutical industries have produced some truly miraculous medicines for the cure of diseases that were formerly often fatal. Prominent among these are the sulfa drugs and the *antibiotics*, of which penicillin is the most widely used. Structures of several antibiotics are shown in Fig. 15-45. The antibiotics are produced by various molds and yeasts, although synthetic methods of producing some of them, penicillin particularly, have now been worked out. Hundreds of antibiotics have been made, but most are as damaging to the body as to the bacteria they are supposed to fight, so they cannot be used to treat infection. Even the antibiotics that are used can cause severe side effects. Many

---

* A. L. Hammond, "Aspirin: New Perspective on Everyman's Medicine," *Science* **174**, 48 (1971).

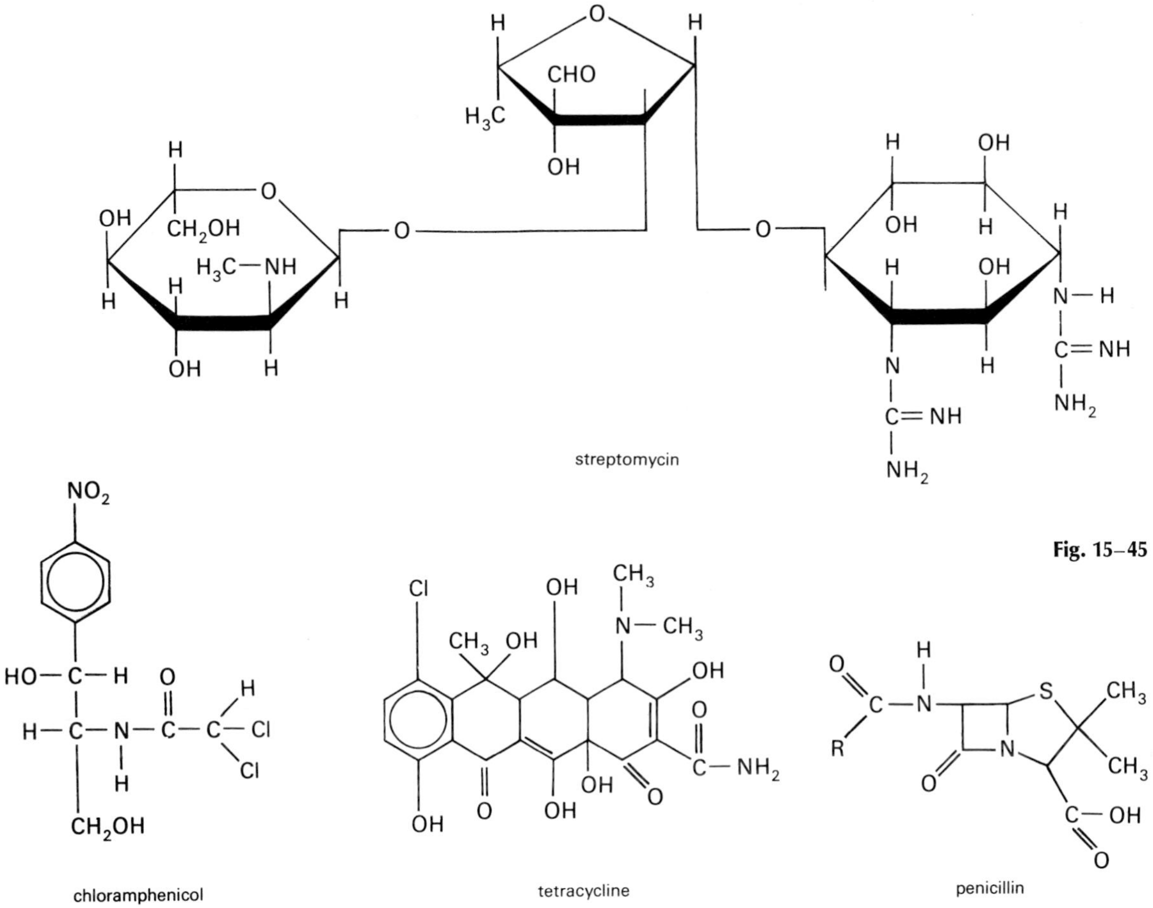

streptomycin

chloramphenicol

tetracycline

penicillin

Fig. 15–45 Several antibiotics.

people, for example, have developed a sensitivity to penicillin. In extreme cases, penicillin causes such a violent reaction that it is fatal.

All medicines and drugs cause some kinds of undesirable side effects in some persons. Therefore, the prescription of a medicine is usually a matter of weighing the benefits of the medicine against the danger or certainty of unwanted reactions. A particular problem of medicines and drugs is the frequently observed *synergistic effect* of two or more substances taken simultaneously; that is, the combined effect of the substances is often greater than the sum of the effects when the substances are taken separately. The adverse reaction of alcohol and barbiturates taken simultaneously is well known and often fatal.

There is no clear-cut distinction between most medical drugs and "street drugs." Many street drugs were originally developed for medical use and are still used for those purposes, but an estimated 50% of the amphetamines and barbiturates manufactured for medical use end up in illicit markets.* Other drugs, such as heroin, were developed for possible medical use but rejected because of the very effects that make them attractive to drug sellers (i.e., their strongly addictive properties).

The various classes of drugs exhibit a variety of chemical structures. About the only thing that they have in common is that nearly all of them except marijuana contain nitrogen, frequently as part of a ring structure. The reason for the prominence of nitrogen is not clear. Perhaps some drugs cause their effects by substituting for amino acids (which also contain nitrogen) in chemical reactions of the nervous system. Also, many drugs are *alkaloids* or chemical derivatives

---

* See the four-part series on drugs that appeared in *Chemical and Engineering News* from October 26 to November 16, 1970.

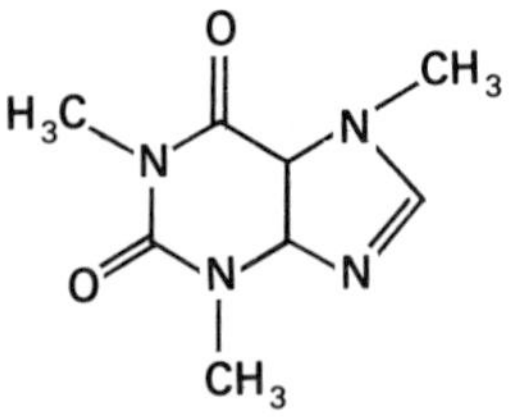

Fig. 15–46 Caffeine and nicotine.

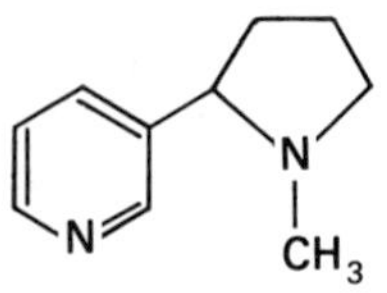

of alkaloids. Alkaloids are nitrogen-containing compounds found mostly in plants. The reason for their presence there is not clear, as they are not essential for growth of the plant. Perhaps they protect the plant by causing discomfort or death of animals that eat the plant. Common examples of alkaloids are the caffeine in coffee and the nicotine in tobacco (see (Fig. 15–46).

Some of the mildest drugs are the nonprescription sleeping pills and tranquilizers, most of which contain a combination of scopolamine and methapyrilene hydrochloride (see Fig. 15–47). Methapyrilene hydrochloride is quite similar in structure to the antihistamines, which are used to treat allergies. Despite the similarity of the scopolamine structure to that of the narcotic cocaine, scopolamine's effects are generally much milder than those of cocaine. There is considerable uncertainty as to the effectiveness of the sleeping aids and tranquilizers of these types and some concern about their adverse effects, especially those of scopolamine. Some users have suffered temporary mental conditions similar to those of schizophrenia.* The fact that a medicine can be obtained without a prescription does not mean that it is completely harmless.

Structures of stronger tranquilizers, obtainable only by prescription, are shown in Fig. 15–48. Meprobamate is used

_____

*   Suzanne Dean, "Study Links Sleep Pills, Delusions," Washington *Post*, Aug. 21, 1972.

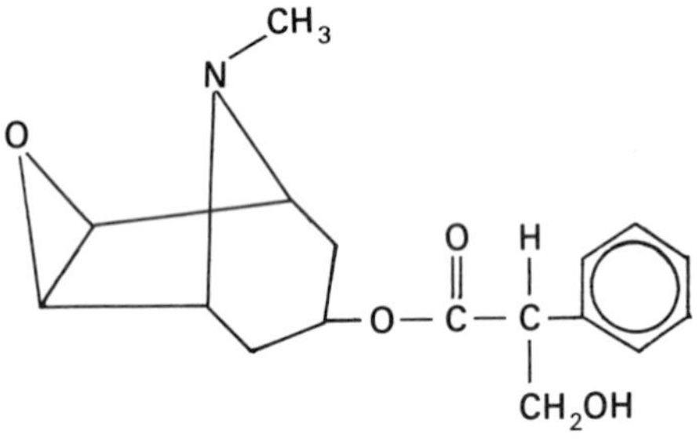

scopolamine

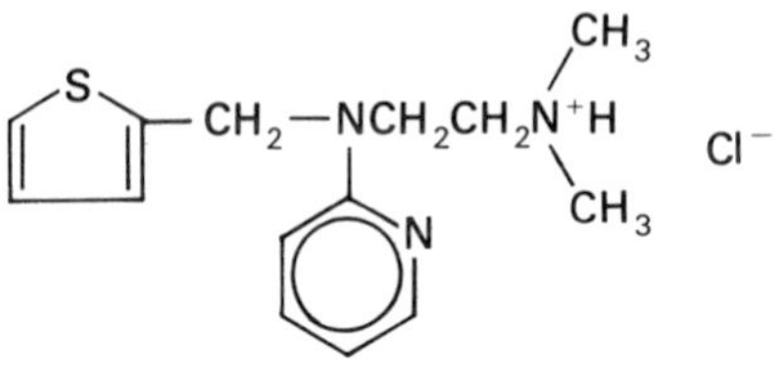

methapyrilene hydrochloride

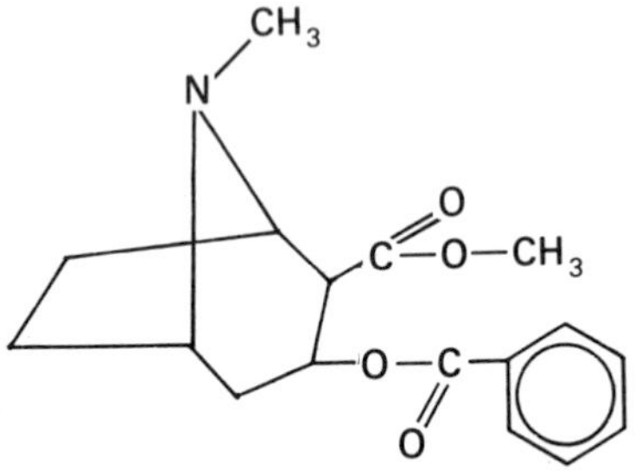

cocaine

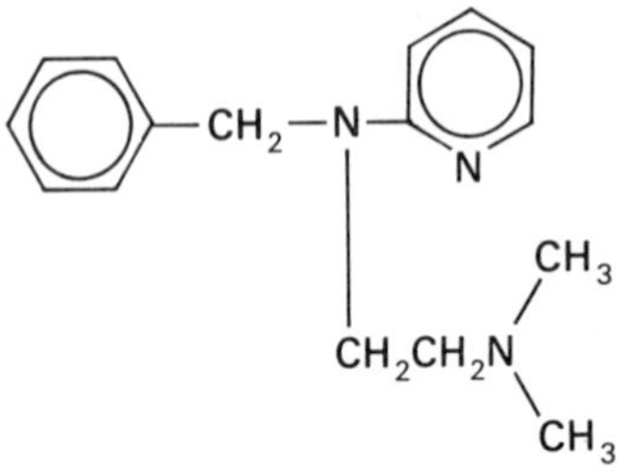

pyribenzamine

Fig. 15–47 Scopolamine and methapyrilene hydrochloride are contained in many over-the-counter sleeping aids and tranquilizers. Also shown is the structure of cocaine, a narcotic, and pyribenzamine, an antihistamine.

in tranquilizers such as Miltowns for persons who are over-anxious but not mentally ill. Reserpine and chlorpromazine are quite strong tranquilizers used to calm severely ill mental patients who are on the verge of becoming violent. Chlorpromazine was originally synthesized in an attempt to produce an improved antihistamine (note similarity to pyribenzamine in Fig. 15–47), but was soon found to be a strong tranquilizer. Reserpine is an alkaloid present in an extract from the roots of the plant *rauwolfia serpentina*. Here we first encounter a "mind-altering" drug containing the indole structure (see Section 13–10). This structure occurs in many hallucinogenic compounds, including LSD. The indole structure is also contained in the amino acid tryptophan (see Fig. 15–13). Indole and closely related compounds are produced by the breakdown of proteins in the large intestine; they cause the characteristic odor of feces.

The tranquilizers we have discussed don't appear to be used heavily in illicit drug traffic. (However, they may be abused, even under prescription.) Some of the most widely abused drugs, both legally and illicitly, are the synthetic chemicals *amphetamines* and *barbiturates*, the so-called "uppers" and "downers" (see Fig. 15–49). The barbiturates, variations on the structure of barbituric acid (which is not pharmacologically active), are useful as sedatives in doses of 10 to 20 mg, but repeated consumption of increasing amounts can lead to addiction. Habitual users can take up to 5000 mg (fifty sleeping pills) per day! *

Amphetamines are "psychic energizers" or "pep pills." They are often prescribed to give depressed persons more energy and a feeling of well-being. In view of their similarity to the structure of the hormone epinephrine (adrenalin, see Section 15–10), it is not surprising that the amphetamines

_____

* Four-part series on drugs in *Chemical and Engineering News*, Oct. 26 to Nov. 16, 1970.

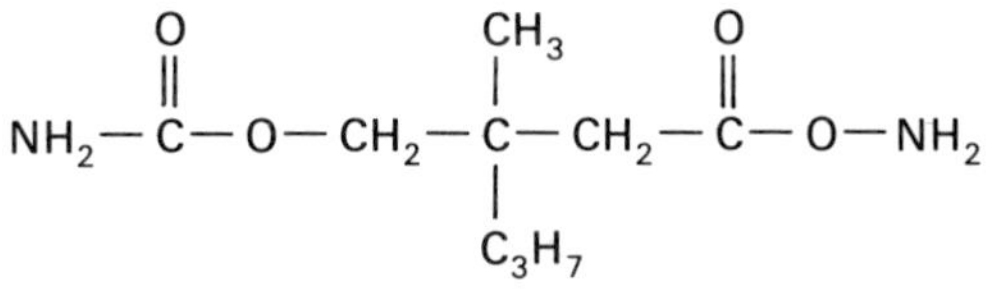

meprobamate (Equanil, Miltown)

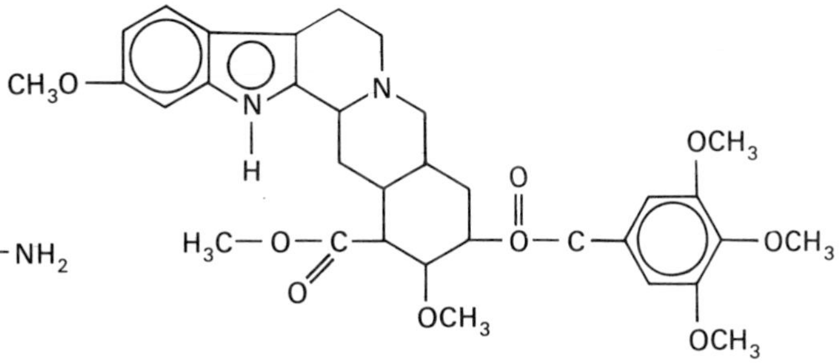

reserpine (Serpasil)

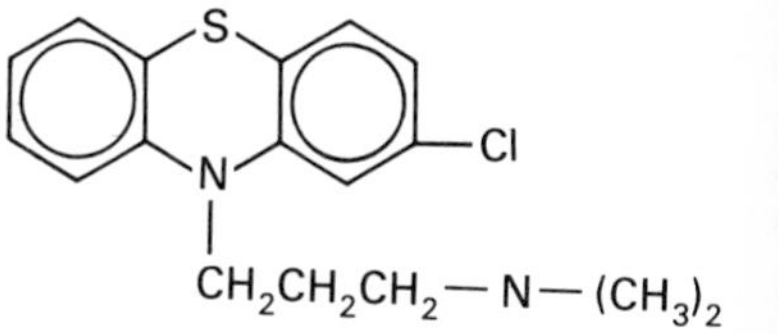

chlorpromazine (Thorazine)

Fig. 15–48 Structures of some of the strong tranquilizers. The darkened portion of the reserpine molecule is the indole structure present in many drugs.

Fig. 15–49 Barbituric acid and several barbiturates.

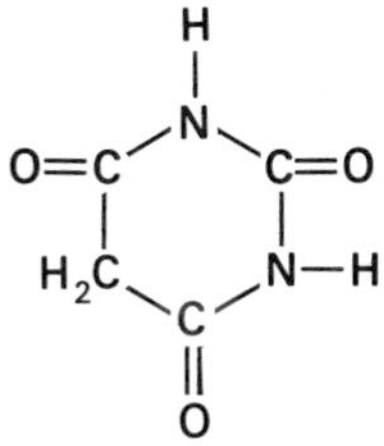

barbituric acid

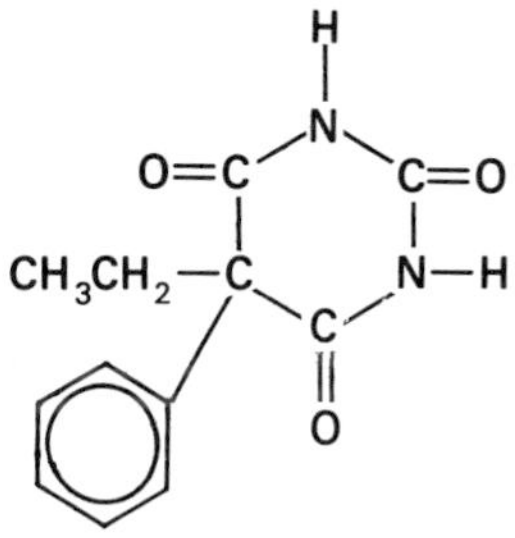

phenobarbital

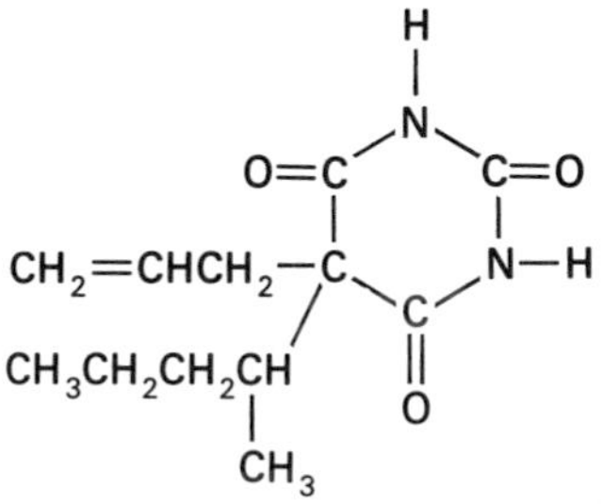

secobarbital

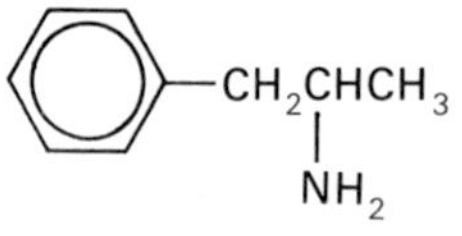

amphetamine
(Benzedrine)

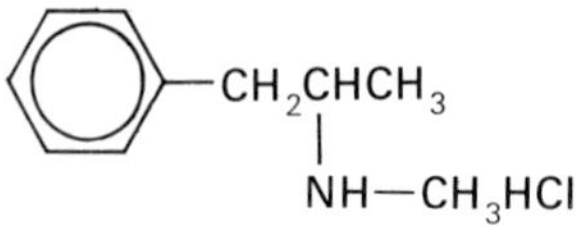

dextro-desoxyephedrine
(Dexedrine)

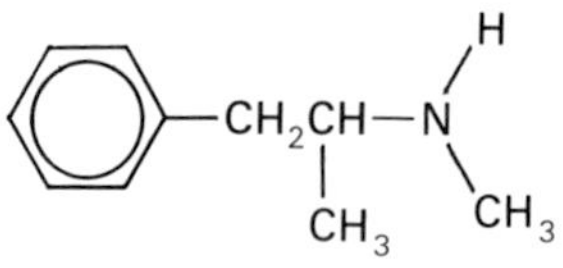

methamphetamine
("speed")

Fig. 15–50 Several amphetamines.

have these effects (see Fig. 15–50). Just as a person often loses his appetite when the "adrenalin is flowing," the amphetamines cause a loss of appetite. They are often prescribed in weight-loss programs, although their effectiveness is questionable and often overshadowed by their negative aspects. Amphetamines are widely abused. People for whom they are prescribed often take increasing amounts of them for the "lift" they give. College students or long-distance drivers often use "Bennies" (Benzedrine) or "Dexies" (Dexedrine) to stay awake for long periods, sometimes for several days. Some people get so hooked on pills that they completely control their mental states with uppers in the morning and downers at night.

Many of the uppers and downers sold illicitly are diverted from those produced by pharmaceutical companies for medical use. It is estimated that more than half of the eight billion amphetamine pills manufactured each year end up on the street.* In addition, large quantities of methamphetamine ("crystal," "speed," or "splash") are made privately by a rather simple two-step reaction. This is the main material used by "speed shooters."

Barbiturates and amphetamines are very dangerous drugs when abused. Heavy users of barbiturates can easily, in their soporific state, make a mistake and take too many sleeping pills, especially if also using alcohol. They go into a coma, and many die if not attended. The long periods of a high state of tension produced by large doses of amphetamines are very hard on the body. You've doubtless had a feeling of great fatigue, perhaps even shakiness, after a prolonged crisis during which your adrenalin level has been high. Imagine how exhausted the body is when the effects of a large dose of speed begin to wear off. During the high, the user may suffer hallucinations, but the most dangerous period is pro-

bably the "crash" experienced when the drug wears off. The motto of the drug culture, "Speed kills," is certainly accurate! In view of the widespread abuse of amphetamines and the ineffectiveness of some of them for weight-control programs, the U.S. Food and Drug Administration recalled certain types of injectable amphetamines and combination pills in 1973.† Also, the Bureau of Narcotics and Dangerous Drugs set production quotas that would limit 1973 production to 60% or less of that in 1972.‡

The latest "medicine" to hit the illicit drug market in quantity is methaqualone, known by names such as "Quaaludes" or "Sopors" (see Fig. 15–51). Originally billed as a nonbarbiturate, nonaddictive sleeping pill in 1965 when it was introduced, methaqualone has not turned out to be quite so innocuous.§ Large quantities of methaqualone pills are in circulation in the illicit market, and hospitals are beginning to see many cases of overdose. It is physiologically addictive and causes severe problems when suddenly withdrawn. From animal studies it appears likely that methaqualone can cause birth defects if taken by a woman during pregnancy. At present, methaqualone is not under control

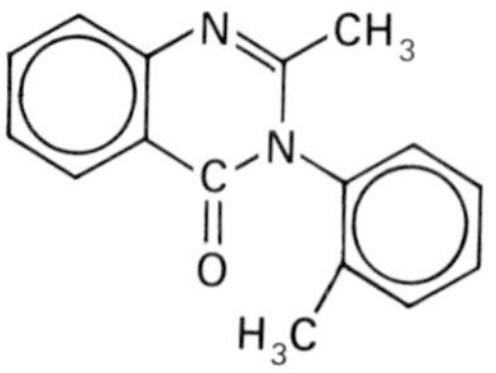

Fig. 15–51 Methaqualone.

---

* Ibid.

† "U.S. to Recall Most Types of Amphetamine Diet Drugs," Washington *Post*, April 3, 1973.
‡ W. L. Claiborne, "U.S. to Cut Production of 'Speed'," Washington *Post*, April 4, 1973.
§ Daniel Zwerdling, "Methaqualone: The 'Safe' Drug that Isn't Very," Washington *Post*, Nov. 12, 1972.

of the Bureau of Narcotics and Dangerous Drugs, but it probably will be controlled soon.

The most widely used hallucinogenic drug ("mind expander") is marijuana, obtained from hemp plants. The chemistry of marijuana is very complex, but it appears that its most active component is *tetrahydrocannabinol*, or THC (see Fig. 15–52). The amounts of THC and related cannabinol compounds vary greatly among various strains of the hemp plant and the locales where it is cultivated. The strongest effects are produced by marijuana from India, whereas that obtained from the plants that grow wild in the midwestern United States is quite weak.* The use of marijuana induces a mild state of euphoria and causes a distortion of a person's perception of time, so that things appear to happen more slowly than they actually do; e.g., an object dropped may appear to float down to the floor.

Fig. 15–52 Tetrahydrocannabinol (THC), the active ingredient of marijuana.

In using many drugs, the user has to keep increasing the dose to achieve the same "high." Marijuana is not like that. In carefully controlled tests in a neutral surrounding, first-time users felt much less intense subjective effects of a given dose than experienced users.† The effects of marijuana on behavior are somewhat similar to those of alcohol. The performance of first-time users in intellectual and psychomotor tests is impaired, but the performance of experienced users is much less impaired, and that of some users is even improved with marijuana.‡ In one recent study, the reaction times of persons under marijuana influence in simulated driving conditions were slowed down somewhat, as they are by consumption of alcohol.§

* K. E. Schultes, "Hallucinogens of Plant Origin," *Science* **163**, 245–254 (1969).
† A. T. Well, N. E. Zinberg, and J. M. Nelson, "Clinical and Psychological Effects of Marihuana in Man," *Science* **162**, 1234–1242 (1968).
‡ Ibid.
§ O. J. Rafaelsen et al., "Cannabis and Alcohol: Effects on Simulated Car Driving," *Science* **179**, 920–923 (1973).

Marijuana plant. (Photograph by permission of Harold S. Sweet.)

As yet there is no clear evidence of the long-term effects of marijuana on health. Marijuana can be habit-forming, but continued use does not appear to cause physiological dependence that would cause great physical stress upon withdrawal. There is no evidence that marijuana causes physical effects that lead to use of hard narcotics; however, the social surroundings in which it is consumed may encourage one to try other drugs.

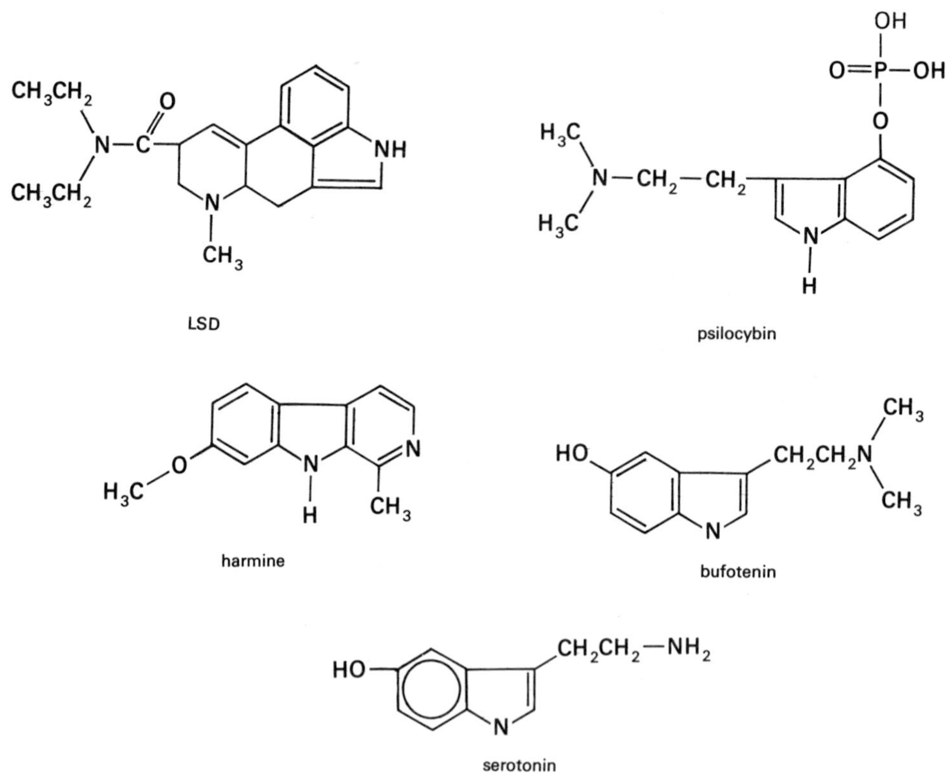

Fig. 15–53 The alkaloids LSD, psilocybin, harmine, bufotenin. Serotonin occurs naturally in the brain.

D-lysergic acid diethylamide (LSD) is an extremely powerful hallucinogen. Doses of only 20 to 30 μg are sufficient to cause psychological responses in some persons, and users seeking a "trip" generally take only 200 to 250 μg.* Such a dose causes spectacular hallucinations for 12 hours or more. LSD is produced by simple chemical treatment of lysergic acid, an alkaloid that occurs naturally in the ergot fungus of wheat or rye. As shown in Fig. 15–53, the LSD structure contains the indole ring structure, as do several other prominent alkaloid hallucinogens, such as *psilocybin* (present in hallucinogenic mushrooms), *harmine*, and *bufotenin*. The latter two occur in a number of Central and South American plants.

The details of the way in which hallucinogens affect the brain are not fully understood. However, it is thought that they interfere with the action of *serotonin*, which occurs naturally in the brain and plays an important role in the transmission of signals between the brain's nerve cells. Note the structural similarity between the hallucinogens and serotonin (which contains the indole structure) in Fig. 15–53.

LSD is a very dangerous hallucinogen. Under its influence, many users have done much harm to themselves, often ending in permanent disability or death. Some have suffered terrifying "bad trips" and ended up in mental institutions. Many have had recurring trips long after the LSD was taken. The long-term physiological effects of LSD are uncertain. The observation of "broken" chromosomes in the cells of LSD users has aroused concern about possible genetic effects of the hallucinogen or damage to fetuses. Observation of birth defects and deformities in offspring of laboratory animals and humans who have consumed LSD suggest increased rates of birth defects and spontaneous abortions, but the data aren't

---

* Four-part series on drugs in *Chemical and Engineering News*, Oct. 26 to Nov. 16, 1970.

conclusive.* Recent evidence suggests that LSD reduces the body's immunity to diseases. Since the LSD structure is similar to that of the amino acid tryptophan, it is suggested that LSD substitutes for tryptophan when the body is attempting to manufacture antibodies (which are protein structures). This would disrupt the growing peptide chain, probably preventing construction of the antibody.† Although questions remain about some of these possible harmful effects of LSD, its dangers are well enough established that one would be foolish to take it for thrills.

Some of the most severe drug problems in the United States are caused by the *narcotic analgesics* derived from the opium poppy. Opium contains about twenty types of alkaloids including *codeine* (used in cough syrups) and *morphine*. Morphine has been used for years as a powerful analgesic (pain killer). It is the "hypo" so often given to hospital patients. Despite its beneficial properties in relieving severe pain, morphine is addictive. Patients recovering from illnesses that require extended use of morphine often find it difficult to function without the drug.

Slight chemical alterations of biochemically active molecules usually modify their effects. In the hope of producing a drug with the pain-killing action of morphine, but without its addictive characteristic, chemists made various derivatives of morphine by changing some of the side chains of the basic structure. One of the products obtained was heroin, whose structure is quite similar to that of morphine (see Fig. 15–54). Heroin does have the pain-killing power of morphine, but it induces a state of great euphoria and is much more addictive than morphine.

In recent years, the synthetic product *methadone* has been used to relieve addicts of the need for heroin. Methadone satisfies much of the desire for heroin and prevents withdrawal symptoms, which are quite severe when a person's body has become physiologically dependent on frequent doses of heroin. However, it is not a "cure" for heroin addiction, as one must keep taking methadone indefinitely to stay off heroin. Methadone itself is abused and many deaths from overdoses have occurred. However, a person on methadone is able to function fairly normally and hold a job, whereas many of those on heroin are unable to. Also, the methadone is distributed at little or no cost, so that one is not forced into a life of crime to support the habit.

---

* T. H. Maugh II, "LSD and the Drug Culture: New Evidence of Hazard," *Science* **179**, 1221–1222 (1973).
† Ibid.

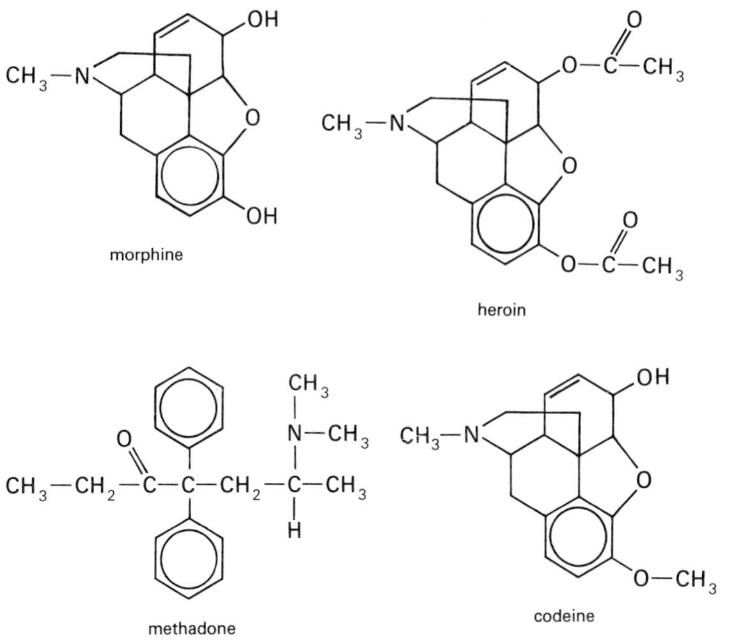

Fig. 15–54 Heroin, morphine, codeine, and methadone.

SUGGESTED READING

Isaac Asimov, *The Chemicals of Life* (New American Library, New York, 1954). Paperback. Very simply written, but quite interesting.

G. M. Edelman, "The Structure and Function of Antibodies," *Scientific American* **223** (2), 34–53 (Aug. 1970).

H. D. Embree and H. J. DeBey, *Introduction to the Chemistry of Life* (Addison-Wesley, Reading, Mass., 1968).

Lester Grunspoon, "Marihuana," *Scientific American* **221** (6), 17–24 (Dec. 1969).

R. H. Haynes and P. C. Hanawalt, eds., *The Molecular Basis of Life—An Introduction to Molecular Biology* (W. H. Freeman, San Francisco, 1968). An excellent collection of readings from *Scientific American* containing many of the articles referred to in the footnotes.

Donald Kennedy, ed., *The Living Cell* (W. H. Freeman, San Francisco, 1965). Another excellent collection of readings from *Scientific American.* Paperback.

D. D. Koob and W. E. Boggs, *The Nature of Life* (Addison-Wesley, Reading, Mass., 1972).

A. L. Neal, *Chemistry and Biochemistry: A Comprehensive Introduction* (McGraw-Hill, New York, 1971).

Linus Pauling, *Vitamin C and the Common Cold* (W. H. Freeman, San Francisco, 1971). Paperback. Even if you disagree with Pauling's theories on vitamin C, this is delightfully written and you will learn a great deal about many related subjects.

J. E. Pike, "Prostaglandins," *Scientific American* **225** (5), 84–93 (July 1971).

J. I. Routh, D. P. Eyman, and D. J. Burton, *A Brief Introduction to General, Organic and Biochemistry* (W. B. Saunders, Philadelphia, 1971).

Symposium on Genetic Engineering, *Science and Public Affairs* **28**, No. 11 (1972).

C. Wills, "Genetic Load," *Scientific American* **222** (3), 98–107 (March 1970).

QUESTIONS AND PROBLEMS

1. Briefly describe the following classes of biochemical compounds indicating their functions and, where appropriate, drawing their general chemical structures or giving empirical formulas.

   a) carbohydrates
   b) amino acids
   c) vitamins
   d) enzymes
   e) hormones
   f) proteins
   g) polypeptides
   h) polysaccharides
   i) triglycerines
   j) antibodies
   k) lipids
   l) steroids

2. Give the apparent evolutionary argument for the following observations.

   a) Darker-skinned people originated from equatorial areas, whereas persons of races from regions nearer the poles have very light skin.

   b) Human bodies are unable to synthesize certain amino acids and vitamins.

3. Explain the analogy that fat is like a savings account for the body's energy, whereas glycogen is similar to a checking account. One could extend the analogy further by equating a glucose molecule to a $1000 check and an ATP molecule to a ten-dollar bill. Explain.

4. a) Why is the body's weight much less when excess energy is stored as fats than it would be if it were all stored as glycogen?

   b) Why is it that cows can extract useful energy from cellulose but humans cannot?

5. a) Why are the body's most complex functions usually performed by protein molecules rather than triglycerides or polysaccharides?

   b) Explain how enzymes control much of the body's chemistry.

   c) Why do most enzymes cease to function above 45°C?

6. Write general equations for the hydrolysis of polysaccharides, triglycerides, and polypeptides.

7. Indicate the nature and causes of the following diseases.

   a) rickets
   b) sickle-cell anemia
   c) pernicious anemia
   d) scurvy
   e) diabetes
   f) pellagra
   g) hemophilia
   h) hypoglycemia

8. a) Explain the role of hydrogen bonding in holding the DNA molecule together.

   b) Why is it necessary that the bonding discussed in (a) be rather weak?

   c) Explain how the genetic code is preserved when cells divide.

   d) Briefly outline the method by which the genetic code of DNA controls the construction of protein molecules.

9. Briefly explain the following terms.

   a) base pairs
   b) corpus luteum
   c) LH
   d) codons
   e) m-RNA
   f) nucleotides

10. What are the functions of the following biochemicals?

   a) insulin
   b) pepsin
   c) vitamin D
   d) salivary amylase
   e) epinephrine

11. a) What is an alkaloid?

    b) List several alkaloids and indicate the plants that contain each of them.

    c) Draw the structures of some of the amphetamines and note the portions of the structures that are similar to that of epinephrine.

    d) In what way is the chemical structure of THC (the active ingredient of marijuana) different from most other mind-altering drugs?

16 WHERE DO WE STAND?

Throughout the book we have discussed many environmental problems that involve chemistry. After reading this litany of chemical horrors that *may* occur, one may become rather depressed about prospects for the future. However, there are many reasons for hope that things will get better. Particularly in the wake of Rachel Carson's *Silent Spring*, a broad spectrum of people started to think about the dangers of man's influence on nature. Concern about the environment grew throughout the 1960s and, in April 1970, we celebrated the first Earth Day. Many conservation and environmental organizations grew in numbers and strength. Citizens began to question developments that had previously been meekly accepted as "progress" and demanded that actions be taken to clean up the environment. No longer were industries, freeways, faster aircraft, or even Olympic Games universally assumed to be beneficial.

Government at all levels has responded. Since 1969 both the Council on Environmental Quality (CEQ) and the Environmental Protection Agency (EPA) have been established. CEQ is an advisory organization to the federal government and EPA investigates environmental problems, monitors indicators of environmental quality, and establishes guidelines for control of pollutants. Similar agencies have been set up by many state and local governments.

The U.S. Congress has had an active role. They have investigated and exposed a variety of problems by holding hearings on such topics as phosphates and eutrophication, toxic metals, drugs and birth control, electric power generation in the southwestern United States, etc. They have enacted strict laws to prevent some types of environmental degradation. Among other things they

- wrote the law requiring the filing of "environmental impact" statements before large federal projects can be started
- specified the emissions standards for 1972–1974, 1975, and 1976 model automobiles
- cut off funds for government-sponsored development of the supersonic transport (SST) until the problem can be studied further.

Many states and local communities have passed laws for the protection and enhancement of environmental quality. The state of Maryland, for example, has established a Power Plant Siting Fund for a tax on electric energy sold in the state. The fund supports research on the impact of power plants, conducts studies to identify the most ideal locations for power plants, and makes acquisition of sites for future power-plant construction. Communities throughout the country have passed laws banning the sale of detergents containing phosphates, nonreturnable beverage containers, etc. Volunteer groups and private citizens have gotten into the act, too, by setting up recycling centers, avoiding excessive use of energy, water, and other resources, growing their own foods, and cutting down on the use of products such as persistent pesticides that would damage the environment.

Despite these hopeful developments, many problems remain to be solved. Below is our evaluation of the status of some of the more important problems:

1   Population

The population of the United States and many other highly developed nations seems to be stabilizing. Growth will continue for a few years (because of disproportionate numbers of young people in the population), but growth rates will approach zero. Unfortunately, in the developing nations that can least afford it, growth rates remain high. As yet there is no satisfactory chemical contraceptive for mass-scale use. Much research is in progress on various classes of compounds such as the prostaglandins that may prove to be effective. Improved contraceptive methods will probably be developed, but many years of testing will be needed before they can be used. Then will come the more difficult tasks of distributing the knowledge and materials needed and obtaining culture acceptance of the need for birth control and the use of the method. In the meantime, population growth will be straining or completely overwhelming our ability to provide even minimal resources for a decent standard of living for all people.

2   Nuclear Weapons

For many years the *Bulletin of the Atomic Scientists* (now called *Science and Public Affairs*) has shown a clock face on its cover that indicates the danger of nuclear war—the closer to midnight, the greater the threat. During the Cuban missile crisis of Oct. 1962, the clock was set at one minute 'til midnight, but today it has receded to 11:48, reflecting some optimism in the light of developments since then—the Limited Test-Ban Treaty of 1963, the 1972 U.S.–Soviet agreements on limitations of strategic weapons, and the growing detentes between the U.S. and China and the Soviet Union. However, one must not be complacent as long as weapons exist. France and China have not signed the test-ban treaty and, during 1973, both held atmospheric weapons tests. Several other nations have the capability of producing weapons in a rather short time if they decide to. The growth of nuclear energy industry means that enormous quantities of fissile material are in circulation. It's next to impossible to keep inventories of the material carefully enough to prevent or detect theft of the amount of material necessary to make nuclear devices. The problem will grow with the introduction of breeder reactors that will provide vast quantities of ^{239}Pu.

3 World Food Supply

The "green revolution" has done a great deal to increase the supply of cereal grains needed by millions of hungry people. But supplies of proteins are woefully inadequate as demonstrated in 1973 by the international disruption caused by failure of the Peruvian anchovy catch and the embargo on export of U.S. soybeans. What must be done? We Americans could help by eating less meat and more soybean products, but that won't solve the problem. Research must be done to increase yields of soybean crops. We probably cannot greatly increase the supplies of proteins from the world fish catch. However, we could probably greatly increase the productivity of the seas by learning how to do "fish farming," or "aquaculture," i.e., by growing aquatic species of our own choosing in appropriate areas of oceans and lakes.

4 Energy

There is plenty of oil and coal to supply the world's energy needs for the next few decades. (The immediate "energy crisis" in the U.S. arose from problems of distribution, the 1973 war in the Middle East, and other economic factors.) The supply of petroleum will be depleted at some time in the future, probably within a century. Furthermore, petroleum is too valuable as a starting material for the synthesis of a wide range of products (plastics, rubber, etc.) to be "wasted" by combustion. U.S. domestic supplies of oil are running so low that we will have to import large quantities in the future if alternate energy supplies are not developed. We have plenty of coal reserves, but the production and use of coal is generally very damaging to the environment—strip mining, acid-mine drainage, SO_2, and particulates released to the air. If we are to use much coal in the future, better ways of recovering and using it must be devised, e.g., coal gasification, sulfur removal from coal or stack gases, and

more efficient electrostatic precipitators. Since it would be a mistake at this time to depend only on one or two future energy sources, the U.S. is pursuing research on many forms of energy. A reorganization of federal agencies during 1973 will bring most research and planning for energy supplies into a single agency that will include fission and fusion and coal gasification. A branch of the National Science Foundation supports research on geothermal and solar energy.

5 Air Pollution

Much was being done about air pollution, primarily by EPA and the Clean Air Act of 1970, prior to the energy crisis of 1973–1974. Restrictions on the sulfur content of fuels and on emissions by automobiles, power plants and industries were beginning to bring about modest improvements in air quality of some cities. However, in the wake of the petroleum shortage of 1973–74, great pressures have developed for the relaxation of atmospheric emission limits and automobile emission standards. Many power plants that had switched to cleaner fuels were asked to switch back to coal. If these trends continue, the modest gains that had been made will quickly be lost and the atmosphere may become worse than it was in many areas.

6 Water Pollution

Although many laws have been passed and large sums are being spent for improved sewage treatment, the clean-up of America's waters will probably take many years. We haven't really found a good solution to the problem of phosphates in detergent or of eutrophication generally. There are so many chemicals in the sediments of lakes and rivers now that it would probably take years for the water to be cleaned up if we stopped all dumping tomorrow. A lot of new ideas and experiments are needed in the use of nutrients in sewage.

7 Pesticides

The environmental movement really began in response to the problems caused by persistent pesticides. Much progress is being made. People are at least sufficiently aware of the dangers to be more careful in the use of pesticides. Most uses of DDT in the U.S. have been halted, but this does not solve the problem since DDT is but one example of the many persistent pesticides. A lot of research has been done on alternative pest-control methods. One of the most promising is sex attractants. It will be most interesting to see if they will be effective against the gypsy moth that has devastated forests in New England and mid-Atlantic states and is now heading farther south.

8 Toxic Substances

Lead, mercury, and asbestos are clearly recognized as toxic substances. Many uses of lead have been curtailed, but lead poisoning will continue as long as lead-based paints remain on interior surfaces and the dust and air near highways retain high concentrations of lead. Lead additives to gasoline will apparently be slowly phased out, but many officials feel that the EPA timetable for lead removal is much too slow. Some obvious sources of mercury release have been halted, but that does not solve the problem. Vast quantities of mercury remain in the sediments of lakes and rivers as a potential hazard for many years in the future. Many less obvious sources continue to release mercury, e.g. the combustion of coal and municipal trash, which release most of the amounts of this volatile element present in the coal or consumer products burned. Now that the hazards posed by asbestos are established, uses of the mineral will probably be carefully regulated in the future. These are just three of the myriad of elements or substances in our environment that are toxic—what about all of the others? We know that many are toxic in large amounts, but what are their effects at very low levels? For example, we know that cadmium in large amounts causes hypertension, ouch-ouch disease, and other afflictions. But does it have serious effects at the low levels present in the atmosphere and water? If so, what can we do to control its release and entry into food chains and the atmosphere? Much research on these questions is needed. A new branch of the National Science Foundation, the RANN program (Research Applied to National Needs) is sponsoring extensive study of many of these questions regarding many types of toxic substances.

9 Food Additives

In discussing nuclear energy, we noted that, despite years of study of radiation effects, we still have very poor knowledge of the effects of extremely low radiation doses. In view of the great attention surrounding the radiation controversy, one might suppose that we have excellent knowledge of the effects of most other things that we come in contact with on a daily basis. That is not the case—we argue so much about radiation effects because we know relatively quite a bit about them. By contrast we know almost nothing about the long-term effects of the hundreds of chemicals we ingest everyday as food additives. However, there are indications that additives are going to be considered much more carefully in the future. The artificial sweeteners, cyclamates, were removed from the market in 1970 and the animal growth stimulant DES was banned from use in 1973. Recently, several food coloring agents were banned (including the purple ink used by meat inspectors!). It appears that the Food and Drug Administration is taking a much harder look at additives than they have in the past and it's likely that others will be prohibited from use unless they can be justified by virtue of their necessity and effectiveness, and proven to be harmless.

10 Recycling and Resources

Recycling has not progressed nearly as fast as one would have predicted in 1970. Many volunteer recycling centers have closed. Local governments have been slow to finance the effort. This is very discouraging since waste materials are a resource-out-of-place—they are both a nuisance and a pollutant as they normally exist, but they contain much-needed resources. The considerable technology needed to extract the resources—metals, energy, nutrients—already exists, but we are very slow to put the methods into practice. Perhaps we will eventually be forced to do so by shortages of virgin sources of the energy and materials. Until we do so, let's at least bury our trash in well-marked piles that can be "mined" by future generations.

11 Transportation

Despite years of strong opposition by many lobbies, it appears that the federal "Highway Trust Fund" has finally been opened up for use by other modes of transportation such as urban mass-transit systems. That is a welcome development, as automobiles have had very adverse effects upon air quality and land use in cities. But much remains to be done, particularly the improvement of rail service for passengers between nearby cities. One most encouraging development was the halt of development of the SST until its environmental effects could be assessed. This is the first example of a major item of "progress" being held up until it can be shown with some confidence whether or not it will cause serious damage to the environment. Let's hope that we will see many more such examples in the future!

In summary, we have come a long way just by becoming more aware of the severe problems that man's activities can cause for our environment. In spite of this awareness and the actions that have been taken to protect the environment, however, many problems remain to be solved. A great deal of scientific and engineering research is needed to find solutions to these problems. In view of past excesses and complete faith in "progress," let us hope that the fruits of research will be used with more thought about their ultimate effects in the future.

APPENDIX A
REVIEW OF MATHEMATICAL METHODS

A–1 Exponents

In science and engineering one often uses very large and very small numbers. To avoid writing long strings of zeroes, we can use scientific expressions involving exponents. You are no doubt familiar with squares and cubes:

Square: $\qquad x^2 = x \times x,$

Cube: $\qquad x^3 = x \times x \times x.$ $\qquad\qquad$ (A–1)

The exponent (or "power") to which the number is raised tells us how many times the number is to be multiplied by itself. A negative exponent indicates the inverse of the quantity with the positive exponent, e.g.,

$$a^{-3} = 1/a^3 = 1/(a \times a \times a). \qquad\qquad (A–2)$$

Fractional exponents indicate that one is to take "roots," e.g.,

Square root: $\qquad x^{1/2} = \sqrt{x}, \qquad 25^{1/2} = \sqrt{25} = 5,$ $\qquad\qquad$ (A–3)

Cube root: $\qquad x^{1/3} = \sqrt[3]{x}, \qquad 27^{1/3} = \sqrt[3]{27} = 3.$

If an algebraic expression contains a given quantity raised to several different powers, those terms can be combined rather simply. For example,

$$\overbrace{x^3 \times x^1 \times x^2}^{} = ?$$
$$\underbrace{x \times x \times x}_{} \times x \times x \times \underbrace{x \times x}_{} = x^6. \qquad (A–4)$$

Note that to carry out the indicated multiplication, one needs only to add the exponents. (Of course, this could not be done if the terms were mixed; e.g., we cannot simplify $x^n a^m$.) In division involving a quantity raised to various powers, we may simply subtract the exponents:

$$\frac{x^3}{x^2} = \frac{x \times x \times x}{x \times x} = x^1 = x. \qquad\qquad (A–5)$$

If an algebraic expression contains a given quantity raised to several different powers in the numerator and denominator, we can simplify it by subtracting the sum of the exponents in the denominator from the sum of those in the numerator:

$$\frac{x^2 \times x^3 \times x^6}{x^1 \times x^4} = \frac{x^{2+3+6}}{x^{1+4}} = \frac{x^{11}}{x^5} = x^{11-5} = x^6. \qquad (A–6)$$

If we raise an exponential quantity to a power, the operation is equivalent to multiplying the exponents:

$$(x^2)^3 = x^2 \times x^2 \times x^2 = x \times x \times x \times x \times x \times x = x^6, (A–7)$$

that is,

$$(x^2)^3 = x^{2 \times 3}.$$

A–2 SCIENTIFIC NOTATION

In chemistry and other sciences we frequently encounter very large and very small numbers. For example, in a rather small water droplet, there are about 2,000,000,000,000,000,000,000 individual water molecules. We can obtain some idea of the enormous size of this number if we consider dividing the molecules of the drop equally among all of the people on Earth. Each person would receive about 60,000,000,000 molecules—still a large number, but smaller than the national debt, the largest number most people are familiar with! The radii of atoms are typically about 0.00000001 cm, and the small center of an atom, the nucleus, is about 10,000 times smaller.

Clearly, it would be awkward to write down all the zeroes each time we perform calculations involving such numbers; therefore, we normally express them as a number between 1 and 10 multiplied by ten raised to an integer (whole number) power. Note that

$10^0 = 1$ (any positive number raised to the zeroth power = 1),

$10^1 = 10,$

$10^2 = 100,$

$10^6 = 1,000,000;$

that is, 10 raised to the nth power, where n is a positive integer, equals 1 followed by n zeroes.

Also note that

$10^{-1} = 1/10^1 = 1/10 = 0.1,$

$10^{-2} = 1/10^2 = 0.01$

$10^{-3} = 0.001.$

That is, $10^{-n} = $ a decimal point followed by $n - 1$ zeroes and a one. We can thus express large and small numbers as follows:

$6000 = 6 \times 1000 = 6 \times 10^3,$

$53,400 = 5.34 \times 10,000 = 5.34 \times 10^4,$

$0.008 = 8 \times 0.001 = 8 \times 10^{-3},$

$0.00056 = 5.6 \times 0.0001 = 5.6 \times 10^{-4}.$

Now suppose we are asked to evaluate an expression containing several large and small numbers:

$$\frac{4000 \times 0.005 \times 300}{0.02 \times 50} = ?$$

We can rewrite it using powers of ten:

$$\frac{4 \times 10^3 \times 5 \times 10^{-3} \times 3 \times 10^2}{2 \times 10^{-2} \times 5 \times 10^1} = ?$$

It may seem easier to group the powers of ten and, separately, the other factors:

$$\frac{4 \times 5 \times 3}{2 \times 5} \times \frac{10^3 \times 10^{-3} \times 10^2}{10^{-2} \times 10^1} = 6 \times \frac{10^2}{10^{-1}}$$
$$= 6 \times 10^3 = 6000$$

In this particular example, we probably did not save much effort by using powers of ten; however, if the problem included numbers of cosmic or atomic size there would be a considerable reduction of effort. Also, once one gets used to this form of calculation one generally makes fewer decimal errors.

A-3 LOGARITHMS

Instead of expressing a quantity as some number times ten raised to some power, we can simply express the quantity as ten raised to a power, for example,

$3,160,000 = 3.16 \times 10^6.$

But

$3.16 = \sqrt{10} = 10^{1/2} = 10^{0.5}.$

Thus we can write the number as

$10^{0.5} \times 10^6.$

Recalling that we can add exponents to simplify the expression we have

$10^{0.5} \times 10^6 = 10^{6.5}.$

Therefore

$3,160,000 = 10^{6.5}$

By the definition of logarithms, we can say that 6.5 is the *logarithm of 3,160,000 to the base 10*, which we can write in equation form:

$\log_{10} 3,160,000 = 6.5.$

In a more general form, if

$x = 10^y$ (A–8)

then

$\log_{10} x = y.$

Normally we drop the subscript 10; unless otherwise noted, log usually means $\log_{10}$.

We can raise ten to any power, not just integers (whole numbers) or simple fractions as we have done above. Any number between 1 and 10 can be expressed as ten raised to some number between 0 and 1, since $10^0 = 1$ and $10^1 = 10$. For example:

$$2 = 10^{0.301} \quad 3 = 10^{0.477} \quad 5 = 10^{0.699} \quad 8 = 10^{0.903}$$

or, in logarithmic form:

$$\log 2 = 0.301 \quad \log 3 = 0.477 \quad \log 5 = 0.699$$
$$\log 8 = 0.903.$$

Logarithms for all numbers between 1 and 10 can be obtained from logarithm tables (e.g., mathematical tables from the *Handbook of Chemistry and Physics*) or read from the L scale of a slide rule (see Fig. A–1).

Logarithms can be used very effectively to carry out complicated multiplications, divisions, and various operations with exponents.

Multiplication

$$x = 10^a, \qquad y = 10^b, \qquad x \times y = 10^{a+b},$$
$$\text{(A–9)}$$
$$\log x = a, \qquad \log y = b, \qquad \log (x \times y) = a + b.$$

Example: $z = 0.00036 \times 674,000$
i.e.,

$x = 0.00036$ and $y = 674,000$.

First, we must compute $\log x$ and $\log y$. It's probably easiest to do if we first convert x and y to scientific notation:

$$x = 3.6 \times 10^{-4} \quad \text{and} \quad y = 6.74 \times 10^5.$$

Now, express 3.6 and 6.74 as ten raised to a power. You can determine the powers by setting the hairline of a slide rule over the number on the D scale and reading the power off the L scale (see below). In this particular example we find that:

$$3.6 = 10^{0.556} \quad \text{and} \quad 6.74 = 10^{0.829}.$$

Now we can write

$$x = 10^{0.556} \times 10^{-4} \quad \text{and} \quad y = 10^{0.829} \times 10^5.$$

We can simplify the expressions by adding the exponents on the tens:

$$x = 10^{-3.444} \quad \text{and} \quad y = 10^{5.829}.$$

These expressions are of the form $x = 10^a$; therefore, $\log x = a$, i.e.:

$$\log x = -3.444 \quad \text{and} \quad \log y = 5.829.$$

Now carry out the multiplication by adding the logarithms:

$$\log z = \log x + \log y$$
$$= -3.444 + 5.829 = 2.385,$$
that is, $z = 10^{2.385}$.

It's probably easiest to determine the numerical value of z by separating $10^{2.385}$ into a product of ten to an integral power times ten to a number between 0 and 1, i.e.:

$$z = 10^{0.385} \times 10^2.$$

Now all we need do is convert $10^{0.385}$ to a number between 1 and 10. This can be done on a slide rule by placing the hairline over 0.385 on the L scale and reading 2.43 off the D scale.

$$z = 2.43 \times 10^2 = 243.$$

When you first do a calculation like this one, it's undoubtedly much more work than simply multiplying the two numbers together. But as you gain experience, you can make some kinds of manipulations faster with logarithms. There are some cases where logarithms have to be used, e.g., calculations involving the pH of solutions (see Chapter 8).

Operations with exponents:

$$z = x^a, \qquad \log z = a \times \log x \qquad \text{(A–10)}$$

Example: $z = \sqrt[3]{565} = 565^{1/3}$,

$$\log 565 = 2.752,$$
$$\log z = 2.752 \times 1/3 = 0.917,$$
$$z = 8.26.$$

Example: $z = 0.041^5$, $\quad \log 0.041 = -1.387$,

$$\log z = 5 \times \log 0.041 = 5 \times -1.387$$
$$\log z = -6.935 = 0.065 - 7,$$
$$z = 1.16 \times 10^{-7} = 0.000000116.$$

Logarithms have several other important uses besides those illustrated in the above computations. For example, the slide rule is based on logarithms—the only scale on most slide rules that is marked off in equal-sized divisions is the L scale. It represents the logarithms (to base 10) of the numbers on the

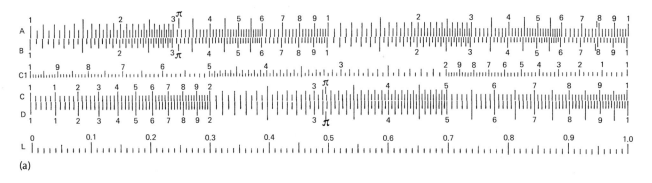

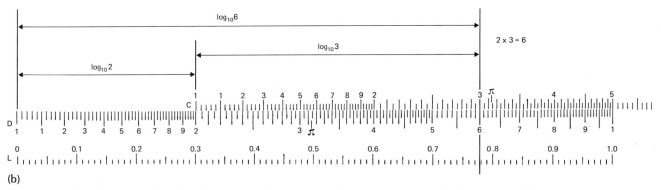

Fig. A–1 Scales of a simple slide rule. (a) Complete slide rule: C and D scales used for multiplication and division; L scale shows $\log_{10}$ of numbers on D scale; A and B have squares of numbers on D and C scales, respectively; C1 scale is inverse of C scale. (b) Illustration of multiplication of 2×3, equivalent to addition of logarithms.

D scale. As shown in Fig. A–1, the multiplication of a number on the D scale by a number on the C scale, which gives the answer on the D scale, is equivalent to addition of the logarithms of the two numbers. Similarly, division on a slide rule amounts to a subtraction of the logarithms of the denominator from that of the numerator. It will not be necessary to know how to use a slide rule to work the problems in this book, but it may be convenient.

Logarithms are sometimes useful in the construction of graphs. Some quantities vary over such wide ranges of magnitude that it is difficult to show both very large and very small values on an ordinary linear graph. In such cases, one can plot the logarithm of the value instead of the value itself. Such a graph is called a semilogarithmic plot. An example of this is shown in Fig. 6–8, where the logarithm of the cosmic abundance of each element (i.e., the estimated abundance of each element in the solar system) is plotted versus the atomic number of the element on a scale in which the abundance of

silicon is defined as 10^6. On a linear plot of these data with the hydrogen point on scale, all the other points except that for He would be so close to zero that we could not read them. We see that the plot covers an enormous range of values, from an abundance of about 3×10^{10} for hydrogen to a value of about 10^{-2} for tantalum. Occasionally one may want both axes of the graph to be logarithmic (called log-log plots). Graph paper for both kinds of graphs is available commercially.

PROBLEMS

1. Simplify the following expressions as far as possible.

a) $x^2 \times \sqrt{y \times x}$

b) $3x^{5/2} \times \sqrt{x} + \dfrac{6x^{7/2}}{\sqrt{x}}$ c) $\dfrac{(x^{3/2})^4}{x^2}$

2. Express the following in exponential notation.

   a) 400   b) 573,000

   c) 0.0003   d) 0.00258

3. Calculate the logarithms of the numbers given in Problem 2. *Note:*

   $$4 = 10^{0.602} \qquad\qquad 5.73 = 10^{0.759}$$

   $$3 = 10^{0.478} \qquad\qquad 2.58 = 10^{0.412}$$

4. a) Rearrange the equation

   $$t°F = \frac{9}{5}t°C + 32$$

   to give $t°C$ in terms of $t°F$.

   b) If $t = 25°C$, what is the Fahrenheit temperature?

   c) If $t = 113°F$, what is the Centigrade temperature?

B–1 INTERNATIONAL SYSTEM OF UNITS (SI UNITS)

In our daily lives we are most familiar with the British units of measurement—feet, pounds, gallons, bushels, etc. Although we are generally able to cope with this system because of long acquaintance with it, most of us would admit that it is a cumbersome system because it requires us to carry so many conversion factors around in our heads—12 inches per foot, 5280 feet per mile, and so on. Furthermore, the units sometimes depend on who is doing the measuring—a sailor's mile is different from a landlubber's, a jeweller's pound is different from a grocer's, and an imperial gallon is larger than the U.S. gallon. This system, annoying for the supermarket shopper, is impossible for a scientist, especially when he wishes to compare his results with those of scientists in other countries, most of whom do not use the British system. Most scientific work is therefore reported in terms of the *International System of Units* (abbreviated SI), a modern version of the metric system.

The metric system originated in France in 1791. The fundamental unit of *length*, the *meter*, was defined as one ten-millionth (that is, 10^{-7}) of the distance from the North Pole to the equator along the meridian passing through Dunkirk, Paris, and Barcelona. For comparison with laboratory measurements, the French made a standard meter bar (roughly the length of the British yard) by measuring a portion of the meridian from northern to southern France and computing the distance from the equator to the pole. From present-day measurements of the meridian we know that the original

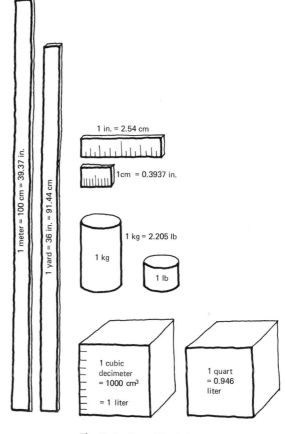

Fig. B–1 SI and English units.

Table B–1. SI and English Units

Common prefixes attached to fundamental units

Prefix	Symbol	Multiples and Submultiples	Common example of length
giga	G	10^9 = 1,000,000,000	
mega	M	10^6 = 1,000,000	
kilo	k	10^3 = 1,000	kilometer = 1000 meters
deci	d	10^{-1} = 0.1	decimeter = 0.1 meter
centi	c	10^{-2} = 0.01	centimeter = 0.01 meter
milli	m	10^{-3} = 0.001	millimeter = 0.001 meter
micro	μ	10^{-6} = 0.000001	micrometer = 0.000001 meter (or "micron")
nano	n	10^{-9} = 0.000000001	
pico	p	10^{-12} = 0.000000000001	

SI and English Equivalents
1 meter (m = 39.37 inches (in.)
1 in. = 2.54 centimeters (cm)*
1 in. = 25.4 millimeters (mm)*
1 kilogram (kg) = 2.205 pounds (lb)
1 lb = 454 grams (g)
1 ounce = 28.35 g
1 liter (l) = 1 dm^3 = 1.057 quart (qt)

*   Indicates exact relationship; i.e., all following digits of conversion factor would be zeroes.

estimate was incorrect by only about 2000 meters—an error of only 2 parts of 10,000!

The French created larger and smaller units by use of the decimal system—multiplication and division by factors of ten as in the American monetary system. Thus, one-tenth of a meter is a *decimeter* (dm), one-hundredth is a *centimeter* (cm) and so on. Greek prefixes are used for multiples of the meter and Latin for the subdivisions. Some of the common units are shown in Fig. B–1.

How do the SI units (Table B–1) relate to our familiar British units? As noted above, the meter is a little larger than the yard—39.37 inches. The unit that we shall probably use most is the centimeter, one-hundredth of a meter. One inch is equal to 2.54 cm. One foot is approximately 30 cm (= 300 mm = 3 dm = 0.3 m).

For many years, the length of the meter was defined by markings on a platinum-iridium bar kept at the International Bureau of Weights and Measures in Paris, France. This was not a satisfactory standard. A scientist wishing to make a very accurate length measurement in a distant laboratory would have to go to France, make a secondary standard meter bar by comparison with the primary standard, and take his

secondary standard back to his or her laboratory. The environment of the secondary standard would have to be carefully controlled during the trip so that it would not change in length because of changes in temperature, pressure, etc.

A standard based on natural physical properties of some material that a scientist anywhere could reproduce in his or her laboratory was adopted in 1960 at the 11th General (International) Conference on Weights and Measures. That conference defined a new standard: 1 meter = 1,650,763.73 wavelengths of the orange-red light emitted by atoms of [86]Kr (see discussion of light in Ch. 7). This standard can be used very accurately in any laboratory on Earth—or, for that matter, on the moon or farther out in space.

The official SI unit of *volume* is the *cubic decimeter*, i.e., 1000 cubic centimeters. Although no longer a standard, the *liter* will probably continue to be used by scientists. The liter is defined as equal to the cubic decimeter. The U.S. quart is slightly smaller than the liter (1 qt = 0.946 liter).

The SI unit of *mass* is the *kilogram*, which is still defined by duplicate platinum-iridium standards kept in Paris and at the U.S. National Bureau of Standards in Gaithersburg, Maryland. The kilogram is 2.2 times as large as the pound, but in the laboratory, we normally use amounts of material that have a mass of only a few grams.

Although the terms are often used interchangeably, *mass* and *weight* are *not* the same. However, they do have about the same numerical values at sea level on Earth. "Weight" measures the force of local gravity on an object. A man would weigh slightly less on a high mountain than at sea level and about one-sixth as much on the moon's surface. On the other hand, the mass of an object is an inherent property, not dependent on the place at which it is measured.

B–2 TEMPERATURE SCALES

Scientists generally use the *Centigrade* or *Celsius* temperature scales, which are the same for all practical purposes. Both are indicated by °C. The Centigrade scale is defined by two fixed points: 0°C is the freezing point of water and 100°C is boiling point of water at 1 atmosphere pressure (i.e., the normal pressure of the atmosphere at sea level, about 14.7 lb per square inch). Although we commonly use a mercury-in-glass thermometer to determine temperatures between and beyond the fixed points, they are actually defined by the volume occupied by a gas at constant pressure (see Chapter 2).

In Fig. B–2 we compare the *Fahrenheit* scale (probably the most familiar) and the Centigrade scale. The interval between the freezing and boiling points of water is divided into 180 Fahrenheit degrees and 100 Centigrade degrees; thus, a Centigrade degree is 180/100, or 9/5, as large as a Fahrenheit degree. Since 32°F = 0°C, we can convert from one to the other with the equations

$$t°F = (t°C \times \tfrac{9}{5}) + 32$$

or

$$t°C = (t°F - 32) \times \tfrac{5}{9}.$$

Note that −40° is the same on both scales. According to the above equations, normal room temperature, about 72°F, is about 22°C.

There is no upper limit on temperature but there is a lower limit, below which we cannot reduce the temperature. At this "absolute zero" of temperature, the motion of molecules and atoms is at a minimum. Absolute zero occurs at −273.15°C. For many purposes it is convenient to use a temperature scale that has its zero at absolute zero. The degrees of the *Kelvin* scale (designated K, without the degree mark) are

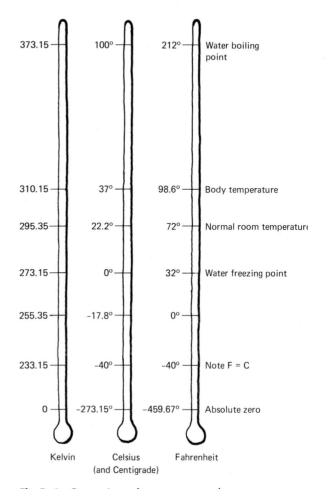

Fig. B–2 Comparison of temperature scales.

the same size as for Centigrade, but the zero of the Kelvin scale is −273.15°C. One can convert from Centigrade to Kelvin temperatures by simply adding 273.15:

$$t \text{ K} = t°C + 273.15.$$

On this scale, the freezing and boiling points of water occur at 273.15 K and 373.15 K, respectively. Many chemical measurements are performed and reported in both scales. For example, 20°C might also be reported as 293 K.

B–3 ERRORS AND SIGNIFICANT DIGITS

If someone measured an object with a ruler on which the smallest divisions were millimeters and reported that its length was 37.415 mm, we hope you would feel he was stretching the point a bit. With such a ruler, he could be certain only that the length was greater than 37 mm and less than 38 mm. By looking very carefully at the fraction of the distance covered between the 37 and 38 mm marks, he could estimate the number of tenths of a millimeter beyond 37 mm and perhaps report it as 37.4 mm, keeping in mind that his result could be off as much as about 0.1 mm; that is, the actual length would be somewhere between 37.3 and 37.5 mm. Thus, in 37.415 mm, only the first three digits have meaning and the last two are nonsense, since we cannot determine them with the instrument used. We would say that the number has three significant digits, namely, 3, 7, and 4. In reporting the result, we should leave off any insignificant digits.

It is very important for scientists, engineers, machinists, and so forth to know the uncertainties of the numbers they use. If someone reports that the length of an object is 2 cm, it may be important to know how accurately the 2 cm has been measured. If the person reporting the value has used the proper number of significant digits, one in this case, then we know only that the true length is between 1.5 and 2.5 cm; that is, it is closer to 2 cm than to 1 or 3 cm. If the value were known more accurately, it should be reported at 2.0 or 2.00 to indicate the accuracy of the measurement. Unless we know otherwise, we should assume that the uncertainty of a number is about one unit of the last digit reported. For example, if the number is 2.36, the possible error is ± 0.01 and the true result lies somewhere between 2.35 and 2.37.

If we add or subtract a series of numbers, we should round off the answer to retain only the number of decimal places in the component that has the fewest number of decimal places.

Examples:
```
   3.02            13.548
  76.1           − 12.1
   0.037           1.348
  79.157
```

Round to: 79.2 (three significant digits)   1.3
                                          (two sig. digits)

Nearly all scientific observations contain some error. We know the true results only within some limits of error, unless we are making a very simple observation such as counting discrete objects. Therefore, in most scientific papers, it is not enough simply to report a value in terms of significant digits. Most final results are reported along with the researcher's estimate of his uncertainty.

There are various ways in which limits of uncertainty are given, but the most common is the *standard deviation*. When the standard deviation is used, there is a two-thirds chance that the true value lies within the limits and a one-third chance that it lies outside the range. For example, if the value is given as 3.3 ± 0.2, there is a 67% probability that the true result lies between 3.1 and 3.5 and a 33% probability that it is outside that range.

There is a great range of accuracy with which numbers can be or need to be measured. Fundamental constants that we use frequently should be known (and, when necessary, expressed) to a high degree of precision. The speed of light in vacuum, for example, is

$$(2.997925 \pm 0.000003) \times 10^{10} \text{ cm/s,}$$

with an uncertainty of only three parts in three million. Atomic weights (see Chapter 2) are generally known to about eight significant digits. On the other hand, it is not nessary to measure a quantity such as the sulfur dioxide concentration in the atmosphere to better than about 10% accuracy, as it varies greatly with time and location.

SUGGESTED READING

National Bureau of Standards, "Brief History and Use of the English and Metric Systems of Measurement with a Chart of the Modernized Metric System," N.B.S. Special Publication 304A, 1968.

QUESTIONS AND PROBLEMS

1. How many yards are there in a mile? How many meters in a kilometer? Which calculation is easier? What does this comparison suggest to you about our present system of units in the United States?

2. Earth is, on the average, about 93 million miles from the sun. What is that distance in kilometers? How long does it take light to get from the sun to Earth?

3. How many meters are there in one mile? How many kilometers? When the speedometer of a European car reads 90 km/hr, how many mi/hr is it traveling?

4. Round off the answer to each problem to the proper number of significant digits.

```
a)   23.5      b)   501.8     c)   63.7
     0.624        − 18.            29.86
  + 13.71                       +  1.004
```

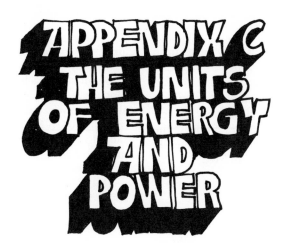

APPENDIX C THE UNITS OF ENERGY AND POWER

The basic SI unit of energy is the *joule*. The equation for the kinetic energy, E, of a moving object of mass m and velocity v is

$$E = \tfrac{1}{2}mv^2. \qquad (C-1)$$

If mass is expressed in kg and velocity in m/s, the energy is given in joules:

$$\text{joules} = \text{kg} \times (\text{m/s})^2. \qquad (C-2)$$

The units on the right-hand side of Eq. (C-2) are also called the *mks* system, i.e., the system in which the units meter-kilogram-second are used for distance-mass-time. Since chemists usually work with a few grams of material occupying only a few cubic centimeters, it is more convenient for them to use the *cgs* system, in which the basic units are centimeter-gram-second. The basic energy unit in the cgs system is the *erg*:

$$\text{ergs} = \text{g} \times (\text{cm/s})^2. \qquad (C-3)$$

By comparing Eqs. (C-2) and (C-3), we find that

$$1 \text{ joule} = 10^7 \text{ ergs}.$$

Energy, or *work*, can also be defined as *force* times the *distance* over which it operates. In *cgs* units, the unit of force is the *dyne* and energy is given by

$$\text{energy (ergs)} = \text{force (dynes)} \times \text{distance (cm)}. \qquad (C-4)$$

The force of gravity upon an object of mass m (measured in grams) varies from place to place, but it is about $980\,m$ dynes at the surface of the earth. Thus when an object drops through a vertical distance h (cm), it acquires an amount of energy given by

$$\text{energy (ergs)} = m \text{ (g)} \times 980 \text{ dynes/g} \times h \text{ (cm)}. \qquad (C-5)$$

If we work entirely with cgs units, the erg is the only energy unit that we need to know about. However, when we read about energy or environmental problems in newspaper or magazine articles, we encounter a variety of other energy units, such as the following.

- Electrical energy is measured in kilowatt-hours, kWh (A kilowatt is a unit of *power*, energy per unit time; see below.)

- The heat content of fuels is given in British thermal units, BTU's (The capacity of air conditioners is given in terms of BTU's of heat removed per hour.)

- The energy content of food is expressed in Calories (kilocalories)

- Energies involved in processes of individual atoms or molecules are often given in electron volts, eV.

- Nuclear processes of individual atoms typically involve energies that are one million times greater than those of atomic and molecular processes, so that they are generally expressed in kiloelectron volts, keV, or million electron volts, MeV.

- Since, according to Einstein's equation $E = mc^2$ (see Chapter 4), mass can be converted to energy, mass is also an energy unit. For example, 1 g = 9 × 10²⁰ ergs. In an individual atom, conversion of 1 amu of mass releases 931 MeV of energy.

- The energy released by the explosion of a nuclear weapon is measured in terms of the weight of chemical explosive, TNT (trinitrotoluene), that would release the same amount of energy. Small weapons are measured in *kilotons*, kT, and large ones, in *megatons*, MT.

These various units of energy can be converted to ergs as shown in Table C–1.

To make it easy for you to convert energies given in one unit to any other common energy unit, Table C–2 contains a grid with all the conversion factors listed to two significant digits. To convert from one unit to another, say from calories to ergs, follow the calorie row across to the column for ergs. The number 4.2×10^7 indicates that 1 cal = 4.2×10^7 ergs.

Power is amount of energy released or used per unit time; thus power has units of ergs/s or other energy units divided by time. You are probably most familiar with the power unit *watt*, which is one joule (or 10^7 ergs) per second. Most light bulbs in your house use from 60 to 150 watts of electric power. Toasters and irons draw much more power, typically 900 to 1200 watts. The power used by an appliance can also be determined from the voltage of the electricity and the current drawn:

Power (watts) = voltage (volts) × current (amperes). (C − 6)

Most line voltage in the U.S. is 110 or 115 volts. An appliance that draws 10 amps of current at 115 V thus uses power at the rate of 10 amps × 115 V = 1150 watts, or 1.15 kilowatts (kW). When the appliance is operated for one hour, the total energy used is 1.15 kilowatt hours (kWh).

Another familiar unit of power, used especially with motors or automobile engines, is *horsepower* (hp). One hp is equivalent to 746 watts or 0.746 kW. A typical modern steam-electric plant produces about 500,000 kW of electric power.

Table C–2 Interconversion Factors for Common Energy Units

	erg	eV	BTU	cal	kWh	amu
1 erg	1	6.2×10^{11}	9.5×10^{-11}	2.4×10^{-8}	2.8×10^{-14}	670
1 eV	1.6×10^{-12}	1	1.5×10^{-22}	3.8×10^{-20}	4.4×10^{-26}	1.1×10^{-9}
1 BTU	1.1×10^{10}	6.6×10^{21}	1	252	2.9×10^{-4}	7.1×10^{12}
1 calorie	4.2×10^7	2.6×10^{19}	4.0×10^{-3}	1	1.2×10^{-6}	2.8×10^{10}
1 kWh	3.6×10^{13}	2.2×10^{25}	3.4×10^3	8.6×10^5	1	2.4×10^{16}
1 amu mass	1.5×10^{-3}	9.3×10^8	1.4×10^{-13}	3.6×10^{-11}	4.1×10^{-17}	1
1 g mass	9.0×10^{20}	5.6×10^{32}	8.5×10^{10}	2.2×10^{13}	2.5×10^{-7}	6.0×10^{23}
1 g ²³⁵U fission	1.3×10^{18}	8.1×10^{29}	1.3×10^7	3.2×10^{10}	3.8×10^4	9.1×10^{20}
1 ton TNT	4.3×10^{16}	2.7×10^{28}	4.2×10^6	1.0×10^9	1.2×10^3	2.9×10^{19}
1 ton coal	2.7×10^{17}	1.7×10^{29}	2.6×10^7	6.7×10^9	7.7×10^3	1.8×10^{20}
1 barrel crude oil	6.1×10^{16}	3.8×10^{28}	5.9×10^6	1.5×10^9	1.8×10^3	4.2×10^{19}
1 ft³ natural gas	1.1×10^{13}	6.9×10^{24}	1.1×10^3	2.8×10^5	0.32	7.7×10^{15}
1 gal gasoline	1.2×10^{15}	7.7×10^{26}	1.2×10^5	3.0×10^7	35	8.3×10^{17}

Table C–1 Conversions of Energy Units to Ergs

1 calorie = heat required to raise temperature of 1 g of water by
$1°C = 4.18 \times 10^7$ ergs (1 Calorie = 1000 calories)
1 BTU = heat required to raise temperature of 1 lb of water by 1°F =
1.052×10^{10} ergs
1 kWh = 3.6×10^{13} ergs
1 g mass = 9.00×10^{20} ergs
1 amu mass = 1.66×10^{-24} g = 1.50×10^{-3} erg
1 eV = 1.60×10^{-12} erg
1 MeV = 10^6 eV = 1.6×10^{-6} erg
1 ton TNT = 4.3×10^{16} ergs
1 kg ^{235}U fission = 1.3×10^{21} ergs

The following are approximate values for energy released in combustion of fuels, varying somewhat from one sample to another.

1 ton of coal (2000 lb) = 910 kg coal = 2.7×10^{17} ergs
1 barrel of crude oil (42 gal) = 6.1×10^{16} ergs
1 ft^3 natural gas = $1.09 = 10^{13}$ ergs
1 gal gasoline = 1.2×10^{15} ergs

1 g mass	1 g ^{235}U fission	tons TNT	tons coal	barrels oil	ft^3 natural gas	gal gasoline
1.1×10^{-21}	7.7×10^{-19}	2.3×10^{-17}	3.7×10^{-18}	1.6×10^{-17}	9.1×10^{-14}	8.1×10^{-16}
1.8×10^{-33}	1.2×10^{-30}	3.7×10^{-29}	5.8×10^{-30}	2.6×10^{-29}	1.4×10^{-25}	1.3×10^{-27}
1.2×10^{-11}	7.9×10^{-9}	2.4×10^{-7}	3.8×10^{-8}	1.7×10^{-7}	9.2×10^{-4}	8.9×10^{-6}
4.7×10^{-14}	3.1×10^{-11}	9.6×10^{-10}	1.5×10^{-10}	6.8×10^{-10}	3.6×10^{-6}	3.4×10^{-8}
4.0×10^{-8}	2.6×10^{-5}	8.1×10^{-4}	1.3×10^{-4}	5.7×10^{-4}	3.1	2.9×10^{-2}
1.7×10^{-24}	1.1×10^{-21}	3.4×10^{-20}	5.4×10^{-21}	2.4×10^{-20}	1.3×10^{-16}	1.2×10^{-18}
1	672	2.1×10^4	3.2×10^3	1.5×10^4	7.8×10^7	7.3×10^5
1.5×10^{-3}	1	30	4.7	21	1.1×10^5	1.0×10^3
4.8×10^{-5}	0.033	1	0.16	0.70	3.8×10^3	35
3.1×10^{-4}	0.21	6.2	1	4.4	2.4×10^4	220
6.7×10^{-5}	0.048	1.4	0.23	1	5.3×10^3	49
1.3×10^{-8}	9.1×10^{-6}	2.6×10^{-4}	4.2×10^{-5}	1.9×10^{-4}	1	8.9×10^{-3}
1.4×10^{-6}	9.6×10^4	2.9×10^{-2}	4.6×10^{-3}	2.0×10^{-2}	120	1

APPENDIX D CONCENTRATION UNITS

The concentrations of trace elements or compounds are often given in *parts per million*, ppm. If the sample in question is a solid or liquid, the value means parts per million by *weight*. In order to obtain a feeling for the meaning of one ppm, Dale Jenkins, director of the ecology program in the Smithsonian Institution's Office of Environmental Sciences, has defined it as "the world's driest martini: one ppm of vermouth would be equivalent to one ounce of vermouth in 7800 gallons of gin." Also, it should be noted that a concentration of 10,000 ppm is the same as 1% by weight.

In the gas phase, ppm generally refers to numbers of molecules (or atoms). The concentration of CO_2 is about 320 ppm, meaning that 320 out of every one million molecules of air are CO_2 molecules. This is equivalent to saying that 0.032% of the molecules in the air are CO_2. Air quality criteria for gaseous pollutants are often given in ppm. For example, a photochemical oxidant concentration of 0.10 ppm or greater is considered to be a smog condition.

On the other hand, concentrations of particulate matter in the atmosphere are usually quoted in terms of weight per unit volume, for example, $\mu g/m^3$. Often it is of interest to compute concentrations of gases in the same terms. Suppose we want to calculate the weight concentration of a gas having molecular weight M. Let's assume that the air has a temperature of 25°C and a pressure of 1 atm. If air behaves like a perfect gas, 1 m³ (or 10^3 l) contains $1000/24.4 = 41.0$ moles of air. If the pollutant gas has a concentration of 1 ppm, then 1 m³ of air contains 41×10^{-6} mole of pollutant weighing $M \times 41 \times 10^{-6}$ g, or $41 M$ μg. Thus we can relate the two concentration units by the equation

$$\text{concentration } (\mu g/cm^3) = \text{concentration (ppm)} \times 41\,M$$

For example, an SO_2 ($M = 64$) concentration of 1 ppm is converted to

$$\text{concentration } SO_2 = 1 \times 41 \times 64 = 2620\ \mu g/m^3.$$

Note that this conversion must be altered slightly (with use of the perfect gas law) if pressure and temperature are different from the values assumed above.

In the air pollution episode that blanketed the northeastern United States during Thanksgiving week, 1966, the SO_2 concentration actually exceeded 1 ppm (or 2620 $\mu g/m^3$) for brief periods in New York City and possibly other major urban areas of the Northeast.

GLOSSARY

Aerosol. A suspension of tiny liquid droplets or fine solid particles in the air.

Albedo. Fraction of sunlight reflected after striking a planet or a particular surface; e.g., Earth has an albedo of about 34%.

Alkaloid. Nitrogen-containing compounds present in many plants although they have no known functions in the plants. Many drugs and hallucinogens are alkaloids. Other common examples are nicotine in tobacco and caffeine in coffee.

Allotropes. Different forms of a pure element that can exist in the same physical state because of different crystal or molecular structures. Examples: gray and white tin (both solids); red, white, and black phosphorus (all solids); O_2 and O_3 (ozone).

Alpha decay (α decay). Emission of an α particle (the nucleus of a ^{4}He atom) from the nucleus. Commonly occurs only in elements above $_{83}$Bi.

Alveolar sacs. Air cells in the lungs.

Amalgam. Alloy of mercury with another metal.

Amino acid. Organic acid containing nitrogen; general formula,

$$\begin{array}{ccc} & NH_2 & O \\ & | & \diagup \diagup \\ R— & CH— & C— OH. \end{array}$$

Proteins consist of many amino acids linked together.

Amphetamines. "Psychic energizers" or "pep pills" with chemical structures and physiological effects similar to epinephrine (adrenalin); sometimes prescribed by doctors (e.g., for weight loss), but heavily abused as street drugs, e.g., "speed," "splash."

Anion. Negatively charged ion; e.g., Cl^-, NO_3^-, SO_4^{2-}.

Aromatic. Containing benzene rings or similar resonant ring structures.

Aqueous. Watery.

Asbestosis. Fibrosis disease of lungs caused by inhalation of large amounts of asbestos fibers.

Atomic number (Z). Number of protons in the nucleus; also, number of positive charges on nucleus, number of electrons on a neutral atom of element, or number of the element in the periodic table.

Atomic weight (mass). *Chemical:* average relative mass of atoms of an element (with its normal distribution of isotopes) on a scale defined by $^{12}C \equiv 12.00 \ldots$ amu. *Isotopic:* relative mass of a particular isotope of an element, also on the scale defined by $^{12}C \equiv 12.000 \ldots$ amu.

Avogadro's number. The number of atoms in a gram-atom of any element or the number of molecules in a gram-mole of any compound. Has the value 6.02×10^{23}.

Avitamintosis. Disease caused by deficiency of one or more vitamins.

Barbiturates. Chemical derivatives of barbituric acid; used medically as sedatives or illicitly as "downers."

Berylliosis. A degenerative, sometimes fatal, lung disease caused by inhalation of Be compounds.

Beta (β) decay. Radioactive decay in which an electron is emitted from the nucleus, which thus increases its atomic number (Z) by one unit.

Bilharzia. Severe parasitic disease prevalent in Africa, Asia, and South America; also called *schistosomiasis.*

Biodegradable. Able to be decomposed by natural chemicals and organisms of the environment.

Biological oxygen demand (B.O.D.). Amount of oxygen required for biological decomposition of organic material in water. Water with a high B.O.D. value has a large potential for depletion of oxygen from water bodies.

Biosphere. The total mass of all living material on Earth.

Black hole. Hypothesized very dense remains of stars having such enormous gravitational fields that no light can escape from them.

Breeder reactor. Nuclear fission reactor of special type designed to convert fertile material (e.g., ^{238}U) to fissile material (^{239}Pu) while producing useful power.

Buffer. Solution or compound that can neutralize large amounts of acid or base with little resulting change of pH.

Calefaction. The warming of water, as in thermal pollution.

Carbohydrates. Class of compounds having the general formula $C_x(H_2O)_y$; includes sugars, starch, and cellulose.

Carcinogenic. Cancer producing, generally as observed in animal studies.

Catalyst. Substance that speeds up reactions without itself being used up.

Catalytic cracking. Breaking down large hydrocarbon molecules (in petroleum) in the presence of a catalyst.

Catalytic reforming. Rearranging the structures of hydrocarbon molecules (in petroleum) in the presence of a catalyst.

Cation. Positively charged ion; e.g., Na^+, Ca^{2+}, NH_4^+.

Chain reaction. Reaction that, once initiated, perpetuates itself; e.g., nuclear fission, which is caused by neutrons emitted in previous fission events.

Chelate. Complex formed between a metal ion and a large ligand such as EDTA that literally wraps itself around the central ion.

Chromophore. A chemical structure that, if present in a molecule, usually causes the molecule to be colored.

Chromosome. Material, composed largely of DNA, present in cell nuclei; carries genetic information through division of cells.

Cilia. Hair-like structures in the lungs that wave back and forth, transporting mucus and trapped particles out of the lung.

Codon. A three-unit series of bases attached to one end of a transfer-RNA that is a "code" indicating the amino acid group being transported on the other end.

Coenzyme. A substance that must be present for an enzyme to perform its function. Some vitamins are coenzymes.

Complex ion. A metal ion surrounded by several attached ligands that may be neutral or negatively charged. Example: $Zn(NH_3)_4^{2+}$.

Compound. A substance made up of identical molecules, each of which has specific numbers of atoms of various elements.

Condensation nucleus. An atmospheric particle that serves as a site for condensation of water vapor as the start of water droplet formation.

Consumers. *Ecological:* animal species which obtain nutrients and energy by eating plants ("producers") or other animals.

Critical mass. The amount of fissile material (e.g., ^{235}U) present in the minimum size of reactor or nuclear fission weapon that is large enough to sustain a chain reaction.

Denaturation. The partial breakdown of protein by heat, acids, etc.

Density. Physical property of a substance, the mass per unit volume.

Deuterium (^2D or ^2H). The rare stable isotope of hydrogen of mass number $A = 2$ in which the nuclei contain one proton and one neutron; makes up 0.015% of natural terrestrial hydrogen.

Dew point. The temperature to which air must be cooled to the point that it is saturated and, upon further cooling, causes condensation of liquid water (dew).

DNA. Deoxyribose nucleic acid, the large molecules that occur in chromosomes and contain the genetic code.

Ductile. Able to be drawn into wires; e.g., copper is a ductile metal.

Electron capture. Radioactive decay in which nucleus captures an electron from one of its orbitals, thus reducing its atomic number Z by one unit.

Element. All atoms having a particular atomic number, Z.

Endemic disease. A disease affecting people of a particular region.

Endothermic reaction. A reaction that requires the input of heat to make it occur.

Enzyme. Catalyst for a biochemical reaction.

Epilimnion. Upper layer of a thermally stratified lake.

Essential amino acids. Amino acids that cannot be synthesized in the human body and thus must be obtained in food.

Essential element. Element that is required for growth and maintenance of a biological species.

Estrogen. Female sex hormone.

Eutrophication. The aging process of a lake in which it receives so many nutrients that it becomes filled with algae, whose decomposition often seriously depletes oxygen in bottom layers.

Evaporation. Transformation of a substance from the liquid to the gaseous state.

Exothermic reaction. Reaction that releases heat.

Fallout. Material from the atmosphere deposited on ground or sea surface; e.g., radioactive particles deposited following explosion of a nuclear weapon.

Fertile material. Material such as ^{238}U that is not itself fissile with low-energy neutrons but which can be converted into fissile material (e.g., ^{239}Pu) by irradiation with neutrons.

Fissile material. Material that undergoes fission when struck by low-energy neutrons (^{233}U, ^{235}U, and ^{239}Pu).

Fission. *Nuclear:* split of a nucleus into two nuclei of about half the original mass and a few neutrons, accompanied by release of enormous energy, about 200 MeV.

Fission product. Isotopes, most of which are radioactive, produced in nuclear fission of heavy elements.

Freezing nucleus. Solid particles that readily initiate freezing of supercooled droplets of water.

Fusion reactor. Reactors in which light nuclei combine ("fuse") to form heavier ones, with the release of energy; not yet successfully operated.

Gamma (γ) ray. High-energy (usually > 0.1 MeV) photon of electromagnetic radiation emitted by radioactive isotopes.

Gene. Portions of chromosomes carrying certain types of genetic information.

Genetic code. Information carried in the DNA of chromosomes that determines all of the characteristics of the individual plant or animal, in part by governing the synthesis of enzymes and other proteins.

Genetic effect (of radiation). Change of the genetic material (i.e., mutations) of the body that determines the charactersitics of offspring.

Gram-atomic-weight (gram-atom). Mass of an element in grams equal to the atomic weight of the element. A gram-atom of any element contains the same number of atoms (Avogadro's number) as a gram-atom of any other element.

Gram-molecular-weight (gram-mole or mole). Mass of a compound in grams equal to the molecular weight of the compound. A mole of any compound contains the same number of molecules as a mole of any other compound.

Greenhouse effect. the warming of Earth's atmosphere by the trapping of IR radiation by excess carbon dioxide released by the burning of fossil fuels.

Half-life. The time over which half of the atoms of a particular radioactive species decay.

Heat island. Term referring to the fact that, because of man's activities, average temperatures are usually a few degrees higher in cities than in the surrounding countryside.

Heat of vaporization. The amount of heat needed to convert a given mass of substance from the liquid to the gas phase. In the case of water, the value is 540 cal/g at 100°C and is also called the "latent heat of steam."

Heterocyclic. Molecular structure in which there is a ring made up of more than one kind of atom—most commonly, rings made up of carbon atoms along with N, S, or O.

Hormone. Chemical secreted by a gland for chemical regulation of various body functions; some hormones are proteins and others are steroids.

Hydrogen bond. Weak bond between a hydrogen atom of one molecule and an electronegative atom of another molecule.

Hydrolysis. Reaction with water, especially reaction of the anion of a weak acid with H_2O to form undissociated acid and hydroxide ions or reaction of the cation of a weak base with H_2O to form undissociated base and hydrogen ions.

Hydrophyllic. Having a strong attraction for water.

Hydrophobic. Having no attraction for water.

Hydrosphere. The oceans, lakes, rivers, ground water systems, clouds, and water vapor in the atmosphere that include all of the water at or near Earth's surface.

Hygroscopic. Having a strong tendency to absorb water vapor from the atmosphere.

Hyperglycemia. Diabetic condition in which insufficient production of insulin causes excessive glucose concentration in the bloodstream.

Hypertension. High blood pressure.

Hypervitamintosis. Condition resulting from excessive intake of one or more types of vitamins.

Hypoglycemia. Condition in which excessive production of insulin causes abnormally low glucose concentration in the bloodstream.

Hypolimnion. Low layer of a thermally stratified lake.

Implosion. Inward directed explosion, e.g., as used to "squeeze" ^{239}Pu in a nuclear weapon to make it a supercritical mass.

Infrared (IR). Radiations with wavelengths greater than visible light. IR photons have lower energy and frequency than photons of visible light.

Inversion. *Temperature*: Temperature profile in the lower troposphere in which temperature increases with altitude, just the opposite of the usual cooling at higher altitude.

Ion. Electrically charged atom or group of atoms bonded together.

Ionization potential. The amount of energy required to remove an electron from an atom.

Isomers. Molecules having the same number and types of atoms, but with different arrangements of the atoms.

Isomerization. Rearrangement of a molecular structure without a change of its empirical formula.

Isotopes. Atoms of the same element (i.e., same atomic number, Z), but with different mass numbers A, because of different numbers of neutrons in the nucleus; for example, $^{107}_{47}Ag$ and $^{109}_{47}Ag$ are two isotopes of silver.

Isotopic abundance. The fraction of atoms in a normal terrestrial sample of an element that are a particular isotope.

Lachrymator. A chemical that causes irritation of nasal passages, tearing of eyes, etc.

Latent heat. See *Heat of vaporization.*

Ligand. Neutral or negatively charged atom or group of atoms attached to a central metal ion in a complex ion.

Lipid. Class of biochemical compounds that are soluble in non-polar solvents; includes fats and steroids.

Magma chamber. Chamber beneath Earth's surface containing molten rock minerals, gases, steam, etc.

Malleable. Able to be rolled or beaten into thin sheets; e.g., gold is a malleable metal.

Malnutrition. Diet lacking sufficient amounts of some essential nutrients such as protein.

Mass number, A. Total number of nucleons (neutrons and protons) in the nucleus.

Metabolism. The breakdown of substances in food in the body with the release of energy contained in the substances.

Meteorite. Rocky or metallic objects that fall to Earth from somewhere in space. (Meteors are smaller objects that completely burn up in passing through Earth's atmosphere.)

Methemoglobinemia. Disease in which hemoglobin is unable to transport the amount of oxygen required by cells of the body; caused particularly in infants by water with unusually high concentrations of nitrates.

Microwave. Electromagnetic radiation with wavelengths greater than those of infrared radiations.

Miscible. Able to be mixed in any proportions to form true solutions; e.g., ethyl alcohol and water are miscible, whereas mixtures of water and ether form separate phases.

Mitochondria. Sites in cells at which glucose, fatty acids, proteins, etc. are broken down and the energy released is used to manufacture ATP.

Moderator. *Nuclear*: material in a reactor (usually also the coolant) that slows down the neutrons released in fission.

Mole. See *Gram-molecular weight.*

Monomer.   The basic chemical unit from which long polymers are constructed.

Mutation.   A change in the genetic code caused by radiation, heat, chemicals, etc.

Narcotic analgesic.   Pain-killing drug that dulls the senses and often induces sleep; in large doses causes stupor, coma or convulsions; frequently addictive.

Neutron star.   Small dense remains of a star; thought to be made up largely of neutrons.

Nuclear fission.   See *Fission*.

Nuclear fusion.   Nuclear reaction in which very light nuclei are joined together to form larger nuclei accompanied by the release of energy.

Nucleon.   A neutron or a proton.

Nucleus. *Atomic:*   The very dense, small, positively charged center of an atom making up most of the mass of the atom. *Cell:* Portion of most types of cells containing the genetic material (chromosomes) of the cell.

Octane number (ON).   Measure of the anti-knock quality of gasoline; equal to the percentage of isooctane in a mixture (with *n*-heptane) that has the same anti-knock quality as the gasoline in question.

Osmotic pressure.   Pressure developed across a semipermeable membrane when pure solvent on one side of the membrane attempts to dilute a solution on the other side.

Oxidation.   Increase of the oxidation number (or *positive valence*) of a chemical species; older and more limited definition: reaction with oxygen; e.g., combustion, rusting.

Oxidizing agent.   A substance capable of oxidizing other substances.

Peptide.   Polymer made up of amino acid units; protein.

Pheromone.   Chemical emitted by members of a species to influence behavior of other members of same species; e.g., sex attractants.

Photon.   Individual "packet" or "quantum" of electromagnetic radiation such as visible light.

Photosynthesis.   Process in which plants absorb sunlight and use the energy to synthesize sugars and other compounds from starting materials such as carbon dioxide and water; source of all energy in food chains.

Polymer.   Very long-chain molecules made by linking together smaller molecular units called monomers; examples: starch, cellulose, protein, polyethylene.

Polymerization.   The process of joining together small molecules to form long chains.

Positron $(\beta+)$.   Anti-particle of the electron, having same mass and amount of charge, but with positive charge instead of negative; emitted from nuclei in positron decay, by which atomic number of the nucleus decreases by one unit.

Precipitate.   (*noun*) Solid, insoluble material deposited out of solution; (*verb*) to form a precipitate.

Producers. *Ecological*:   Plant species that produce high energy compounds and other nutrients by photosynthesis.

Progesterone.   Hormone involved in female reproductive cycle; artificial progesterone-like compounds are an ingredient in some types of birth-control pills.

Protein.   Very important class of biochemical compounds made up of long chains of amino acids.

Pulsar.   Body in outer space that emits pulses of radiation at regular intervals; may be the same as a neutron star.

Quantization.   Principle according to which certain variables (e.g., energy, angular momentum) are able to assume only certain values and none in between.

r-process.   Neutron capture on a fast ("rapid") time scale, a process involved in nucleosynthesis in stars.

Rad.   A measure of exposure to radiation; deposition of 100 ergs per gram of tissue.

Reducing agent.   Chemical substance capable of reducing other substances.

Reduction.   Decrease of the oxidation number (or positive valence) of an element.

Relative humidity.   The percentage of saturation of water vapor in the atmosphere; the actual pressure of water vapor in the air divided by the vapor pressure of water vapor in equilibrium with liquid water at the temperature of the air.

Rem.   A measure of the expected damage to tissue resulting from exposure to radiation; rems = rad × relative biological equivalent of the radiation.

Residual oil.   High-molecular weight hydrocarbon fraction of crude oil remaining after distillation of more volatile fractions, used in power plants and large central heating units; also called " #6 oil" or "Bunker C."

Residence time.   Average length of time that a given species remains in a certain portion of the environment; e.g., particles in the stratosphere, Ca^{2+} ions in the oceans.

Resonance.   The averaging of various chemical structures that differ only in the placement of electrons in chemical bonds; generally gives added stability to the chemical structure.

Ribosome.   Site in a cell at which enzymes or other proteins are synthesized.

Rickets.   Disease involving insufficient bone formation as a result of vitamin D deficiency.

RNA. Ribonucleic acid; similar in structure to DNA, its role in cells is the translation of the genetic code of DNA into the manufacture of enzymes and other proteins.

s-process. Neutron capture on a *slow* time scale, a major type of reaction of nucleosynthesis.

Saccharide. Sugar; a *mono*saccharide is a simple 5-, 6-, or 7-carbon sugar; a *di*saccharide is two such sugars linked together; *poly*-saccharides are long chains of sugars and include glycogen, starch, and cellulose.

Salinity. Mass of dissolved salts per unit volume of water; often quoted in terms of *parts per thousand* (‰), the number of grams of salts per kg of water. Ocean water has a salinity of about 35‰.

Saponification. Reaction of fats with strong bases such as NaOH to form alkali metal salts of fatty acids which constitute soap.

Saturated solution. Solution containing as high a concentration of some dissolved species as will dissolve at the given temperature.

Schistosomiasis. An endemic, parasitic disease usually obtained from contact with infested waters; common to Africa, Asia, and South America.

Sex attractants. Chemicals emitted by one sex of a species that attracts members of the opposite sex; a type of pheromone.

Solubility. The maximum concentration of a compound that can dissolve in a solvent at a given temperature.

Somatic effect. An effect upon the body (e.g., as caused by radiation exposure) to be distinguished from genetic effects upon offspring of the exposed individual.

Sonic boom. High pressure shock wave in the air caused by object moving with a velocity equal to or greater than the speed of sound.

Spectrum. A distribution of intensities of radiations (e.g., light) as a function of their energies or wavelengths.

Stratosphere. Portion of the atmosphere from an altitude of about 9–12 km to about 50 km.

Sublimation. Evaporation of a solid to form gas-phase; for example, solid carbon dioxide ("dry ice") sublimes to form gaseous CO_2 without going through a liquid phase at pressures below 5 atmospheres.

Supercooled liquid. Liquid cooled to a temperature below its normal freezing point; unstable phase that frequently freezes upon shock, addition of a seed crystal, etc.

Supernova. Explosion of a collapsing star.

Supersaturated. *Solution*: containing a concentration of substance exceeding the solubility. *Atmospheric*: humidity exceeding the water vapor pressure at the given temperature.

Synergistic effect. Effect caused by two or more agents acting together that is more intense than the sum of their effects when acting separately.

Teratogenic. Causing the formation of great deformities; e.g., the flipper-like extremities of children born of women who took Thalidomide during pregnancy.

Thermocline. Boundary between the epilimnion and hypolimnion in a thermally stratified lake.

Thermonuclear weapon. Nuclear weapon set off by a fission trigger, whose high temperature initiates fusion reactions between nuclei of light elements, which release most of the energy of the weapon.

Tritium (^{3}T or ^{3}H). Radioactive isotope of hydrogen with *mass number $A = 3$*; decays (with a 12-yr half life) by β^- emission to form ^{3}He.

Tropopause. Boundary between the stratosphere and the troposphere.

Troposphere. The lower portion of the atmosphere, up to an altitude of about 9 to 12 km.

Ultraviolet (UV). Radiation with wavelengths shorter than visible light.

Valence shell. Outermost shell of the electronic structure of an atom that contains electrons.

Vapor pressure. The pressure exerted by vapor in equilibrium with its liquid phase.

Vaporization. Change of state from the liquid to the gas phase.

Vitamin. Substance required by the body in very small quantities; except for vitamin D, must be obtained in food as they cannot be synthesized in the body.

White dwarf. Final stage of evolution of stars of about the size of the sun or smaller.

X-ray. Penetrating radiation of wavelengths shorter than ultraviolet, with energy typically of a few keV.

ANSWERS TO SELECTED PROBLEMS AND QUESTIONS

(**Note:**   We have not given answers to "definition" questions. Most of the answers to them can be found in the glossary or within the chapters. Also, we have commented only on certain of the discussion questions.)

Chapter 2

2. a) 82 protons and 126 neutrons

   b) To transform $^{208}_{82}Pb_{126}$ to $^{197}_{79}Au_{118}$, remove 3 protons and 8 neutrons.

3. a) 20; 20   b) 18; 18

4. a) $H_2 + Cl_2 \longrightarrow 2HCl$

   b) $Br_2 + Zn \longrightarrow ZnBr_2$

   c) $4V + 5O_2 \longrightarrow 2V_2O_5$ (or $2V + 5/2\ O_2 \longrightarrow V_2O_5$)

   d) $C + 2S \longrightarrow CS_2$

   e) $2Y + 3F_2 \longrightarrow 2YF_3$

5. a) 0.025   b) 1.5×10^{22} c) 3.3×10^{-22} g

6. a) 44.1   b) 64.06   c) 119.0   d) 17.0   e) 159.7

7. a) 0.70 or 70%   b) 7.0 g   c) 0.7%

8. 263

9. 69.722

10. a) $2H_2 + O_2 \longrightarrow 2H_2O$ b) hydrogen   c) 0.375 mole; 6.75 g

11. $NaNO_3$

Chapter 3

3. The water level will remain at the brim since the volume of water displaced by the floating ice is equal to the volume of water produced when the ice melts. (This assumes that the ice is floating freely in the glass.)

5. Pure elements: c) graphite (carbon)   h) carbon

    Pure compounds: b) sodium tartrate

                     f)   sodium bicarbonate, $NaHCO_3$

                     g) ethylene glycol

6. a) 0.26 cm³   b) 1.7×10^{-23} cm³ or 17 Å³

7. a) 3.3 l   b) 15.5 l   c) 42.7 l   d) 95.6 l

8. 9.41

9. a) 3.9 g   b) 53.5 g   c) In mercury, but not in water.

   d) 3.0 g water; 41.0 g Hg; Ta would not float in either liquid.

10. a) 0.86 M b) 0.018 M

11. a) 13 mm Hg   b) 12.3 g   c) 15°C

12. a) $S + O_2 \longrightarrow SO_2$ b) one-half mole of each   c) 16 g   d) 11.2 l

13. Because water boils at a lower temperature at the reduced atmospheric pressure at high

altitude. The rates of chemical reactions occurring in the egg are much slower at the reduced temperature; thus, it takes much longer to obtain the usual soft- or hard-boiled egg.

Chapter 4

1. a) Steam turbine connected to a generator produces electricity which is used to electrolyze water to make H_2 and O_2 gas.

   b) Sunlight is used to heat water to make steam which is used in a turbine to generate electricity.

   c) The rotation of the rotor of a generator produces electricity which is used in a cyclotron to accelerate protons to such high velocity (close to the speed of light) that the protons acquire additional mass.

   d) A nuclear reactor heats water to form steam that drives a turbine connected to a generator that produces electricity which is used to run an electric motor.

2. Eat food mostly from the lowest levels of food chains, i.e. plants.

5. a) 6×10^{18} ergs   b) 20 min   c) About 920,000

6. 3.9×10^{20} ergs; about 10,000 tons of TNT.

7. Assuming that the energy flux of the noontime sun is about 1 kw/m^2, $21,600, if all of the electrical energy is converted to radiant (light) energy, but fluorescent lights are only 20% efficient and incandescent, only 4%. Taking these efficiencies into account it would cost $108,000 with the former or $540,000 with the latter.

8. a) 120 watts   b) About 40,000 BTU/hr

9. 0.019 g

10. a) 62%   b) 152 tons/hr   c) 1.8×10^6 l/min   d) 17,000 l/min

Chapter 5

2. a) $^{60}_{27}Co_{33} \longrightarrow \beta^- + ^{60}_{28}Ni_{32}$ b) $^{22}_{11}Na_{11} \longrightarrow \beta^+ + ^{22}_{10}Ne_{12}$

   c) $^{41}_{20}Ca_{21} \longrightarrow ^{41}_{19}K_{22}$ d) $^{239}_{94}Pu_{145} \longrightarrow ^{4}_{2}He_2 + ^{235}_{92}U_{143}$

   e) $^{236}_{92}U_{144} \longrightarrow ^{99}_{42}Mo_{57} + ^{134}_{50}Sn_{84} + 3\,^{1}_{0}n_1$

3. 15.1 MeV

4. 38.4 MeV

5. Emit an electron.

6. An excess of neutrons over protons is needed in the heavy elements to hold the nuclei together in spite of electrostatic repulsion of the protons from each other.

7. 1-MeV γ ray > 1-MeV β^- > 5-MeV α.

8. The α emitter would be the more dangerous inside the body and the γ-ray emitter at 1-m distance.

9. Because of electrostatic repulsion among all of the protons in the nucleus.

10. a) 5 g   b) Thyroid gland

11. a) β^- decay   b) $^{137}_{54}\text{Xe} \longrightarrow \beta^- + \,^{137}_{55}\text{Cs}$

12. It is produced by α decay of uranium and other heavy elements.

14. All would be present in local fallout; (b), (c), (e), (g), and (h) in tropospheric fallout; only (c) and (h) and, to a limited extent, (b) and (e) in stratospheric fallout.

15. By inserting or withdrawing control rods containing Cd or B, both of which absorb neutrons strongly.

Chapter 6

1. To give the nuclei enough kinetic energy to overcome the electrostatic repulsion between the positively charged nuclei that keeps them apart.

2. a) 3.3×10^{-5} g   b) 3.7×10^{29} ergs   c) 50,000 yr

4. a) 5.5 g/cm^3 b) 1.4 g/cm^3 c) 4.6 km

5. 9.7

6. Elements above Bi cannot be made by neutron capture on a slow time scale (s-process).

7. a) There was about twice as much ^{238}U when the elements were first formed, whereas the ^{235}U has decayed by about seven half lives (a factor of $2^7 = 128$); therefore, the original ratio was about $0.00725 \times (128/2) = 0.46$.

   b) Easier

Chapter 7

1. a) Increasing energy: radio waves < microwaves < IR < red light < blue light < UV < X rays < γ rays.

   b) Increasing wavelength—just the opposite of (a)

   c) Increasing frequency—same order as for energy.

4. 4 in $n = 4$, 5 in $n = 5$; 32 in $n = 4$, 50 in $n = 5$.

6. Na: $1s^2 2s^2 2p^6 3s^1$; Sc: $1s^2 2s^2 2p^6 3s^2 3p^6 4s^2 3d^1$;

   Ne: $1s^2 2s^2 2p^6$; K$^+$: $1s^2 2s^2 2p^6 3s^2 3p^6$;

   Cl$^-$: $1s^2 2s^2 2p^6 3s^2 3p^6$; Li$^+$: $1s^2$.

8. a) 1.85 eV   b) 1.24×10^{-10} cm or 0.0124 Å

9. a) 0.087 Å   b) 0.00064 Å   c) 20-keV protons

10. a) Closer because of the greater attraction of the 2$^+$ nucleus of He   b) He$^+$
    c) $v = 9.86 \times 10^{15}$; $\lambda = 304$ Å

11. $s, p, d, f, g,$ and h.

12. 2; 6, 10, and 14.

13. a) $1s^2 2s^2 2p^6 3s^2 3p^6 4s^2 3d^4$; No, same as predicted except $4s^1 3d^5$ instead of $4s^2 3d^4$.

    b) $1s^2 2s^2 2p^6 3s^2 3p^6 4s^2 3d^{10} 4p^6 5s^2 4d^{10} 5p^6 6s^2 4f^{14} 5d^{10} 6p^6 7s^2 5f^{14} 6d^{10} 7p^4$

14. The weight ratio La/O in La$_2$O$_3$ is $277.8/48 = 5.79$. If its formula were assumed to be LaO, one would calculate the atomic weight of La as $5.79 \times 16 = 92.6$ which, within the limits of error of measurements made at the time, is 92.

15. c) At $_4$Be and $_{12}$Mg, an s subshell becomes filled, so the electron added in going to $_5$B or $_{13}$Al must be placed in the previously empty p subshell.

20. a), c), and e) ionic; (b) and (d) covalent

21. a) Sc_2O_3 b) Ga_2Te_3 c) SbH_3 d) GeO_2 e) CCl_4
    f) RbBr   g) CaO   h) Y_2S_3 i) H_2Se j) MgI_2

22. a) K^+, because it has the highest nuclear charge and, thus, the greatest attraction for the 18 electrons.
    b) Cl^-, for the same reason.

Chapter 8

1. 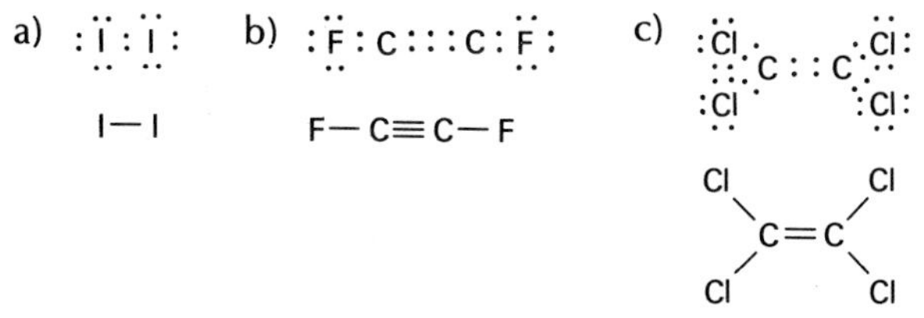 a) :I̤:I̤:   b) :F̤:C:::C:F̤:   c)

    I—I   F—C≡C—F

d)   e)

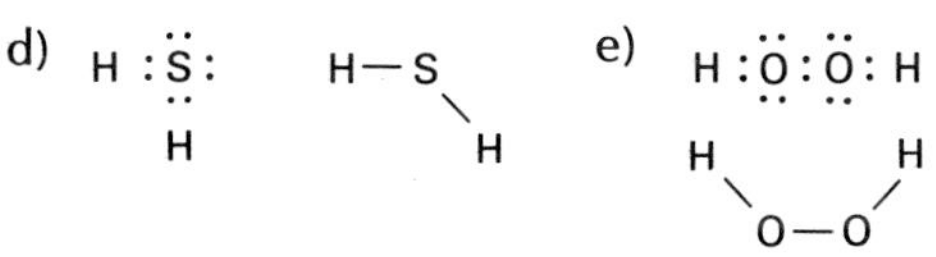

3. a) K$^+$(aq) + OH$^-$ (aq) + H$^+$ (aq) + Br$^-$(aq) ⇌ K$^+$(aq) + Br$^-$(aq) + H$_2$O (l)
    Net: OH$^-$ (aq) + H$^+$ (aq) ⇌ H$_2$O (l)

   b) 2H$^+$ (aq) + 2Cl$^-$(aq) + MgCO$_3$(s) ⟶CO$_2$(g) + Mg^{2+} (aq) + 2Cl$^-$(aq) + H$_2$O (l)
    Net: 2H$^+$ (aq) + MgCO$_3$(s) ⟶CO$_2$(g) + Mg^{2+} (aq) + H$_2$O (l)

   c) Na$^+$(aq) + OH$^-$ (aq) + HC$_2$H$_3$O$_2$(aq) ⟶Na$^+$(aq) + C$_2$H$_3$O$_2{}^-$ (aq) + H$_2$O (l)
    Net: OH$^-$ (aq) + HC$_2$H$_3$O$_2$ (aq) ⟶ C$_2$H$_3$O$_2^-$ (aq) + H$_2$O (l)

   d) K$_2$S(s) $\xrightarrow{\text{H}_2\text{O}}$ 2K$^+$ (aq) + S^{2-} (aq)
    followed by hydrolysis: S^{2-} (aq) + H$_2$O (l) ⇌ HS$^-$ (aq) + OH$^-$ (aq)

   e) NH$_4$Cl(s) $\xrightarrow{\text{H}_2\text{O}}$ NH$_4^+$(aq) + Cl$^-$ (aq)
    followed by hydrolysis: NH$_4^+$ (aq) + H$_2$O (l) ⇌ NH$_3$ (aq) + H$_3$O$^+$ (aq)

4. 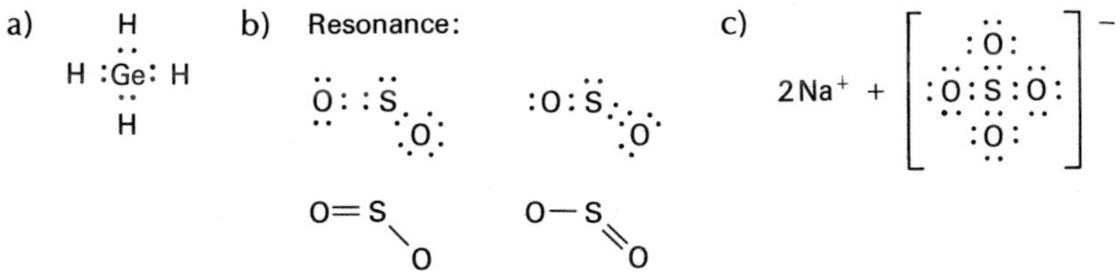 a)   b) Resonance: c)

d) $H^+ + NO_3^-$

e) H H   f) $K^+ + :\ddot{O}:H^-$

:N:N:

H H

Resonance of NO_3^-:

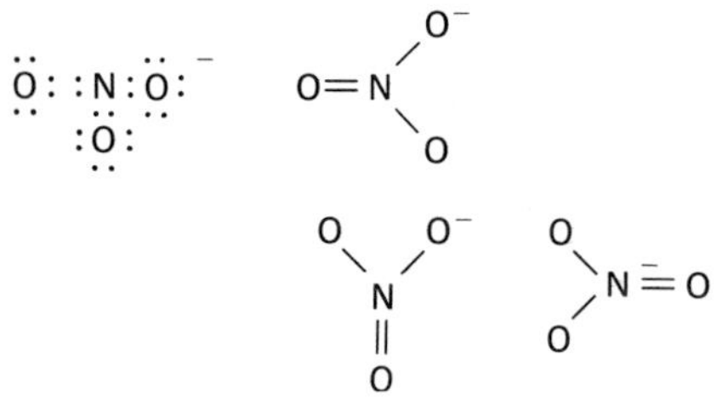

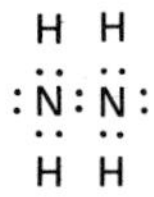

$\ddot{O}::N:\ddot{O}:^-$ $O=N\begin{smallmatrix}O^-\\\\O\end{smallmatrix}$

:Ö:

$\begin{smallmatrix}O\\\\N\\\|\\O\end{smallmatrix}O^-$ $\begin{smallmatrix}O\\\\N\end{smallmatrix}N=O$

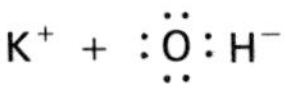

g)  $:\ddot{C}l:\ddot{O}:^-$ h) 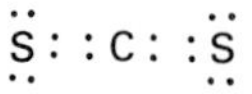 $\ddot{S}::C::\ddot{S}$

5. a) $v = 3.95 \times 10^{14} \text{ sec}^{-1}$; $E = 1.64$ eV

   b) i) 0.0015 cm   ii) $v = 2 \times 10^{13} \text{ sec}^{-1}$ iii) 0.083 eV

   c) 12.2 cm

6. 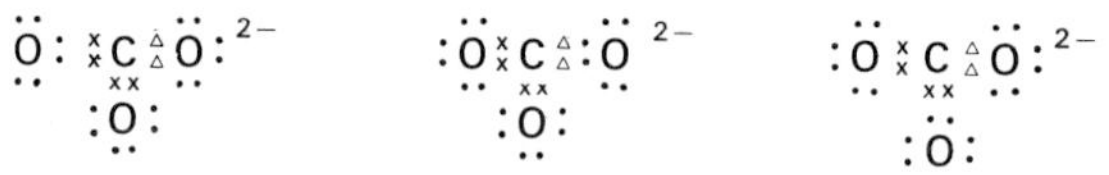 $\ddot{O}: \times C \triangle \ddot{O}:^{2-}$ $:\ddot{O}\times C\triangle:\ddot{O}\ ^{2-}$ $:\ddot{O}\times C\ddot{O}:^{2-}$

    :Ö: :Ö: :Ö:

9. a) $\text{pH} = -\log M_{H^+}$ b) pH = 0   c) pH = 15

10. a) increase   b) decrease   c) increase   d) increase

11. a) pH = 2   b) 0.71 M   c) 13.85

12. a) 0.0042 M   b) 11.6

13. $[H_2S] = 0.01 M; [H^+] = [HS^-] = 3.3 \times 10^{-5} M; [S^{2-}] = 10^{-14} M; [OH^-] = 3 \times 10^{-10} M.$

Chapter 9

4. b) 0.083 eV

10. a) 33%   b) 2.6 km

11. a) 2.3 mm   b) 11%

12. a) 1.6×10^{14} g; 7%   b) 5.4×10^{12} g; 1.8×10^{11} moles; 0.3%; more than the amount present, about 290%.

13. a) 18 yr   b) 767 yr

Chapter 10

6. Element 117 would probably be similar to the known halogens, gaining an electron to form 1− ions. Element 118 would be similar to the noble gases and thus would probably form few compounds. Element 119 would be a highly reactive alkali metal, losing a single electron to form 1+ ions.

7. a) $2Li(s) + 2H_2O\ (l) \rightarrow 2Li^+(aq) + 2OH^-(aq) + H_2(g)$

b) $H^+(aq) + \underline{Br^-(aq)} + \underline{K^+(aq)} + OH^-(aq) \rightarrow \underline{K^+(aq)} + \underline{Br^-(aq)} + H_2O(l)$
Net reaction: $H^+(aq) + OH^-(aq) \rightarrow H_2O(l)$

c) $Cl_2(g) + 2Na(s) \rightarrow 2NaCl(s)$

d) $BaO(s) + H_2O(l) \rightarrow Ba^{2+}(aq) + 2OH^-(aq)$

e) No reaction

f) $Ca^{2+}(aq) + CO_3^{2-}(aq) \rightarrow CaCO_3(s)$

g) $Na_2CO_3(s) + 2H^+(aq) + 2Cl^-(aq) \rightarrow 2Na^+(aq) + 2Cl^-(aq) + CO_2(g) + H_2O(l)$

h) $2SO_2(g) + O_2(g) \rightarrow 2SO_3$ (g, l, or s, depending on temperature)

i) $4Al(s) + 3O_2(g) \rightarrow 2Al_2O_3(s)$

8. a)

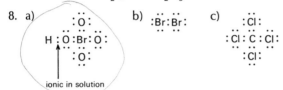

ionic in solution

b) :Br:Br:

c)

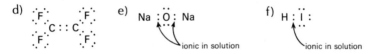

d)

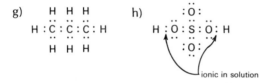

e) Na :Ö: Na
ionic in solution

f) H : Ï :
ionic in solution

g)   H H H
   H : C : C : C : H
      H H H

h)   :Ö:
   H : Ö : S : Ö : H
           :Ö:
ionic in solution

9. a) 0   b) 1+   c) 6+   d) 3+   e) 4+   f) 1−   g) 3−

11. a) and b) are reducing agents; c), d), e), and f) are oxidizing agents

15. a) $880 \mu g/m^3$ b) Steady-state model, $R_{in} = 1.76 \times 10^7 g/hr = R_{out}$; $R_{out} = V_{out} \times C_{SO_2}$;
$C_{SO_2} = R_{out}/V_{out} = 293 \mu g/m^3$

Chapter 11

1. a) 1+   b) 3−   c) 4+   d) 3+   e) 5+   f) 5+   g) 5+   h) 3−
i) formally 5− if C considered 4+

2. N has been oxidized from 3− to 0. Cr is reduced from 6+ to 3+.

7. Arsenic is the same chemical group (VA) as phosphorus; thus, one would expect As to be rather similar chemically and, therefore, to be an impurity in phosphates.

11. 35%

12. 8.8%

Chapter 12

2. a) 70%   b) 0.23 ton

5. a) $Cr_2O_3(s) + 6H^+(aq) \rightarrow 2Cr^{3+} + 3H_2O(l)$

   b) $Cr_2O_3(s) + 2OH^-(aq) + 3H_2O(l) \rightarrow 2Cr(OH)_4^-(aq)$

6. a) $2Al(s) + Cr_2O_3(s) \rightarrow 2Cr(s) + Al_2O_3(s)$

   b) 1.9 kg   c) 6×10^{-35} ppm

13. a) $3 \times 10^{-40}M$ b) $6 \times 10^{-32}\mu g/l$ c) 6×10^{-35} ppm

    d) Increase. At lower pH there would be a higher concentration of H^+, some of which would combine with S^{2-} to form HS^-, thus reducing the S^{2-} concentration, allowing more Hg^{2+} to come into solution.

15. Assuming that tires have diameters of 60 cm, widths of 20 cm and that about 1 cm of tread is worn off over 20,000 miles, and that the car goes about 15 miles per gallon, we estimate the Cd concentration to be about $0.0005\ \mu g/m^3$, much smaller than the observed value.

Chapter 13

2. (structures not shown below can be found in text)

a) $CH_3CH_2CH_2OH$

   n-propyl alcohol

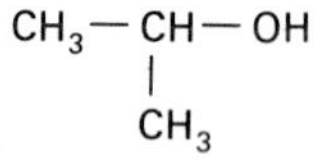

   $CH_3 - CH - OH$
   $\quad\quad\quad |$
   $\quad\quad\quad CH_3$

   isopropyl alcohol

c) 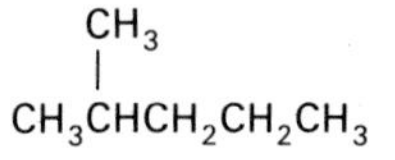 CH_3
   $\quad |$
   $CH_3CHCH_2CH_2CH_3$

d) 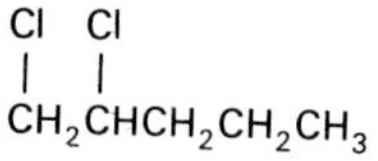 Cl Cl
   $\;\;|\;\;\;|$
   $CH_2CHCH_2CH_2CH_3$

e) 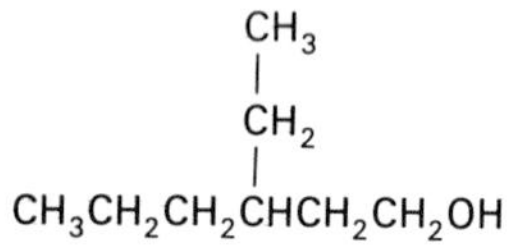 CH_3
   $\quad\quad |$
   $\quad\quad CH_2$
   $\quad\quad |$
   $CH_3CH_2CH_2CHCH_2CH_2OH$

g) 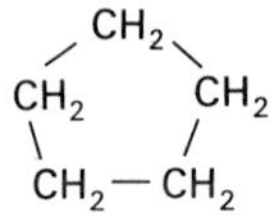 CH_2
   $CH_2\quad CH_2$
   $CH_2 - CH_2$

l) $CH_3CH_2 - O - CH_2CH_2CH_3$

m) 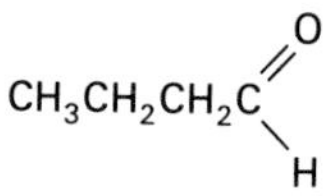 $CH_3CH_2CH_2C\overset{\displaystyle O}{\underset{\displaystyle H}{}}$

n) 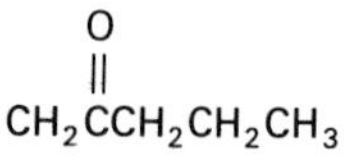 $\quad\quad O$
   $\quad\quad ||$
   $CH_2CCH_2CH_2CH_3$

q) 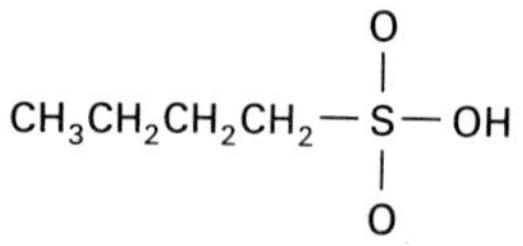 $CH_3CH_2CH_2CH_2 - \overset{\displaystyle O}{\underset{\displaystyle O}{S}} - OH$

r) $CH_3CH_2CHCH_2CH_3$
   $\quad\quad\quad |$
   $\quad\quad\quad NH_2$

s) $\quad\quad Br$
   $\quad\quad |$
   $CH_3 - C - CH_3$
   $\quad\quad |$
   $\quad\quad Br$

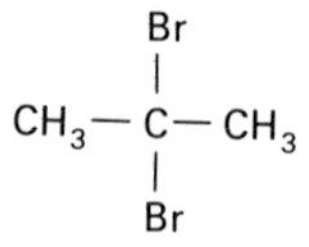

3. $CH_3CH_2CH_2CH_2CH_2CH_3$ $CH_3CHCH_2CH_2CH_3$ $CH_3CH_2CHCH_2CH_3$
   $\quad\quad\quad\quad\quad\quad\quad\quad\quad\quad CH_3$ $\quad\quad\quad\quad CH_3$

   $CH_3CH - CHCH_3$
   $\quad\quad |\quad\quad |$
   $\quad\quad CH_3\ CH_3$

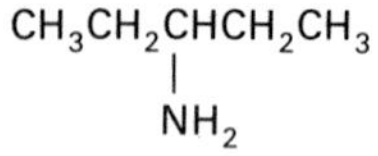

   $\quad\quad CH_3$
   $\quad\quad |$
   $CH_3CCH_2CH_3$
   $\quad\quad |$
   $\quad\quad CH_3$

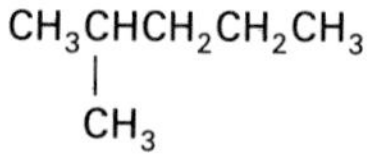

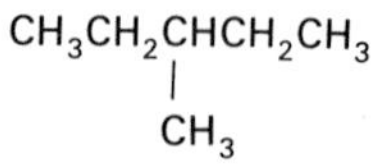

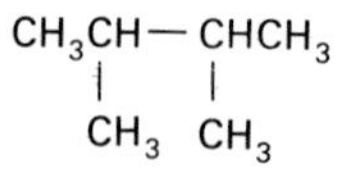

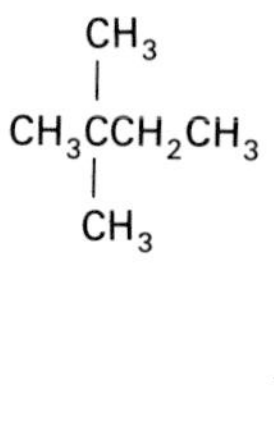

5. 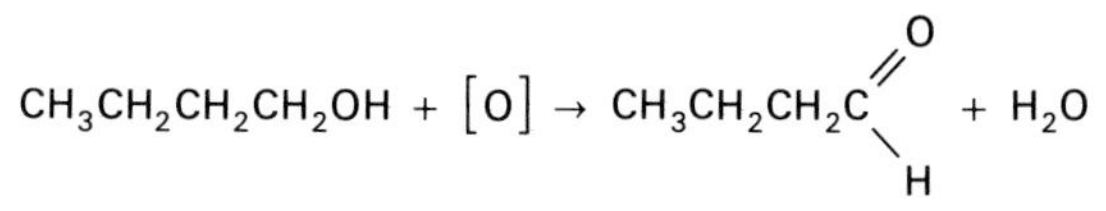 $CH_3CH_2CH_2CH_2OH + [O] \rightarrow CH_3CH_2CH_2C{\overset{O}{\underset{H}{\Vert}}} + H_2O$

1-butanal

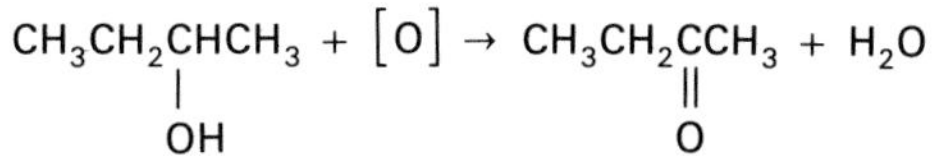

$CH_3CH_2\underset{\underset{OH}{|}}{C}HCH_3 + [O] \rightarrow CH_3CH_2\underset{\underset{O}{\Vert}}{C}CH_3 + H_2O$

2-butanone

9. 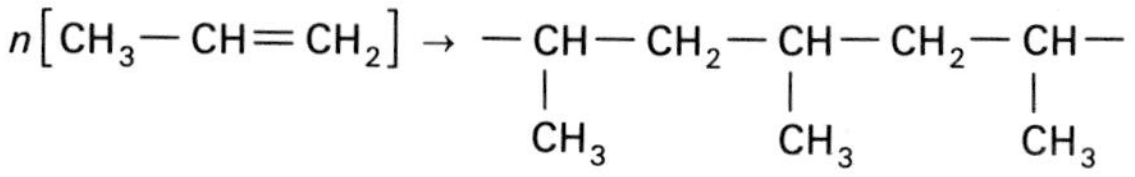 $n[CH_3-CH{=}CH_2] \rightarrow -\underset{\underset{CH_3}{|}}{C}H-CH_2-\underset{\underset{CH_3}{|}}{C}H-CH_2-\underset{\underset{CH_3}{|}}{C}H-$

Chapter 14

4. 64%

5. a) $2C_7H_{18}(l) + 23O_2(g) \rightarrow 14CO_2(g) + 18H_2O(g)$

   b) 11.5   c) 3.6   d) 15.6

7. a) 88   b) About 2 g/gal

12. a) 6.2 $\mu g/m^3$ b) 2.1 $\mu g/m^3$ c) Typically 1 to 2 $\mu g/m^3$, so our calculation is quite reasonable.

INDEX

Italic page numbers denote major definitions or discussions or tabulations.